Lecture Notes in Physics

380

I. Tuominen D. Moss
G. Rüdiger (Eds.)

The Sun and Cool Stars: activity, magnetism, dynamos

Proceedings of Colloquium No. 130
of the International Astronomical Union
Held in Helsinki, Finland, 17–20 July 1990

Springer-Verlag
Berlin Heidelberg GmbH

Editors

I. Tuominen
Observatory and Astrophysics Lab., University of Helsinki
Tähtitorninmäki, SF-00130 Helsinki, Finland

D. Moss
Mathematics Department, The University
Manchester M13 9PL, United Kingdom

G. Rüdiger
Sternwarte Babelsberg und
Astrophysikalisches Observatorium Potsdam
Rosa-Luxemburg-Str. 17a, O-1591 Potsdam, FRG

This book was processed by the authors using the $\TeX$ macro package from Springer-Verlag Berlin Heidelberg GmbH.

ISBN 978-3-662-13811-3 ISBN 978-3-540-46483-9 (eBook)
DOI 10.1007/978-3-540-46483-9

Originally published by Springer-Verlag Berlin Heidelberg New York in 1991
Softcover reprint of the hardcover 1st edition 1991

2153/3140-543210 – Printed on acid-free paper

Foreword

The solar cycle has been studied for over 100 years in its more obvious manifestations. However, it is only in the last decade or so that it has been possible to verify the plausible assumption that the Sun is a typical, albeit rather unspectacular, lower main sequence star, in the sense that analogues of the solar cycle are common if not universal amongst stars of similar age and mass. These developments have generated a new area of astronomy, loosely termed the "solar-stellar connection", that has shown almost exponential growth during this period. IAU Symposium 102 in Zürich in 1982 was perhaps the first IAU meeting to be substantially devoted to stellar activity and it may be of some interest to compare the Proceedings of that meeting with these to assess the progress (or otherwise) made since then! During the intervening period a great quantity of data has been amassed by a variety of observational techniques, both traditional and innovative, using the entire accessible spectrum. New measures and indices of stellar activity have been developed (spectroscopic, photometric, polarimetric) and, perhaps most remarkably, the union of state-of-the-art observational technology with the power of supercomputers is now making possible the direct modelling of temperature and magnetic features on stellar surfaces.

At the same time our knowledge of processes occurring within the Sun that are related to the solar cycle has increased substantially. At one time our sole knowledge of the solar velocity fields came from studying the different rotation laws of various surface features. Now helioseismological studies are revealing more directly the angular velocity law in the solar interior, and they have also helped to establish the position of the base of the convection zone – essential information for theoretical modelling. Our ability to measure surface flows has increased, although this area is not without controversy.

On the theoretical front, there has been an increasing consensus (with one or two dissenters) that the engine of solar and stellar activity is a "turbulent dynamo". The modern concept of such a dynamo goes back to the work of Parker, Babcock and Leighton, but in the last decade numerical and analytical investigations have burgeoned. Two basic approaches can be identified, the highly parameterized "mean field" dynamos introduced by Steenbeck, Krause and Rädler and the computationally much more expensive simulations pioneered by Gilman and Miller. A major problem with studies of the solar cycle and solar dynamo has been that the Sun is just one realization for one set of parameters of an active star. Any dynamo/turbulence theory has enough parameters that achieving a fit of a model to one object cannot provide a conclusive test of the theory. (Nevertheless, even this limited objective has not yet been achieved!) Now that enough active stars have been observed to begin to deduce some meaningful statistical relations between the various parameters – cycle periods, rotation periods, spectral type, etc. – there is a real possibility of subjecting candidate dynamo theories to a proper examination.

Bearing all these factors in mind, one of our main intentions when planning this meeting was to bring observers and theoreticians face to face. This would, at the very least, give everyone an opportunity to appreciate each other's problems, and with this insight to reflect on the direction of their own work. The ultimate goal must be, of course, to confront theory and observation and to examine whether theoretical models are well founded. We thus deliberately gave a substantial emphasis to theoretical studies, including those on the basic theories of magnetohydrodynamic turbulence that underpin any model of a dynamo operating in a stellar convection zone. It should give those who use mean field models some pause for thought to realize how little is known about the turbulent fluid properties that they casually subsume into turbulent transport coefficients, α-effect and so on! These contributions are found in Section I. Section II deals with stellar dynamos in general and specializations to the solar case are found in Section III, together with relevant observational contributions. The lengthy Section IV is given over to the various studies of stellar activity and its relation to the solar cycle. Generally, the ratio of theory to observation is a decreasing function of page number.

Readers of this volume must surely be impressed by the progress, both theoretical and observational, that has been achieved in the last 10 years. Nevertheless a heretic might be excused for asking whether observers and theoreticians are always talking about the same objects! In the solar case there have been substantial efforts to model the "real Sun", but even then the most sophisticated models represent a physical system that is by orders of magnitude less complex than reality, and fail to reproduce successfully all the basic features of the solar cycle. For other stars, in the continuing game of leapfrog between observations and theory, the observations are a clear jump ahead.

After all the discussions about how the solar cycle is similar to those of other stars, and the current inability of theory to produce convincing models of these cycles, it was refreshing to end the meeting by observing one phenomenon that plausibly is unique to the Sun and is also highly predictable – the total solar eclipse of the morning of Sunday July 22.

Some of the contributions were received as TeX files that had been prepared by the authors using the Springer TeX macro. Others were retyped or reformatted during the editing process; the Editors accept responsibility for any errors that were introduced at this stage. It was not possible to include all the submitted material in these Proceedings. The Editors apologize to those authors whose contributions do not appear.

The assistance of Mr. Reino Anttila in retyping a substantial part of the text of this volume is gratefully acknowledged. Many people contributed in one way or another to the success of the meeting. The Editors' thanks are due to them; their help made possible the eventual production of this volume.

The organizers of the Colloquium acknowledge financial support from the Ministry of Education of Finland, and additional smaller contributions from the International Astronomical Union, University of Helsinki, Amer Group Ltd, Outokumpu Electronics, Kemira Group, and others, which made possible participation from 26 countries worldwide.

Helsinki, January 1991 The Editors

Contents

I

Convection and turbulence as the basis of magnetic activity

- Nature of the turbulence in solar and stellar convection zone
- Rotational influence on turbulence
- Simulations vs. mean-field approach
- Mixing length theory, global convection, turbulence models

The Confrontation of Mean-Field Theories with Numerical Simulations

Fritz Krause

Astrophysikalisches Observatorium Potsdam, O-1560 Potsdam, Fed. Rep. of Germany

Abstract: The investigation of magnetic phenomena on the Sun is confronted with the problem of turbulently moving electrically conducting media in rotating objects. One way of attacking this complicated problem, which was at first successful, was the development of mean-field magnetohydrodynamics, where the cooperative action of the small-scale turbulence was taken into account by certain average effects. The development of more and more powerful computers now offers possibilities of calculating the small-scale phenomena in a direct way. This paper is an attempt to compare these approaches, at least as far as there are comparable results.

1. Introduction

Dynamos operating in real cosmical objects are closely connected with turbulent motions, which are of highly complicated structure both temporally and spatially. Thus it can clearly be seen that a detailed description of the dynamo process demands an unmanageable set of data. The way out of this dilemma was discovered long ago at the beginning of the theory of turbulence: equations for the mean-fields are derived and the fluctuating fields enter via averaging processes into the equations. Reynolds stresses and eddy viscosity are examples. Analogously mean-field magnetohydrodynamics has been developed. The mean cross-product of the fluctuating velocity field with the fluctuating magnetic field, the turbulent electromotive force $\overline{\mathbf{u}' \times \mathbf{B}'}$, appears as the analogue of the Reynolds stresses. The turbulent magnetic diffusivity and the α-effect, which proved to be the key for the solution of the dynamo problem, are the best known effects from the concerted action of the small scale fields. With physically reasonable assumptions dynamo models for the solar magnetic field and its cycle have been constructed, which describe surprisingly well the basic features of this phenomenon.

Parallel to this line of investigation there have been attempts undertaken to attack the dynamo problem of the Sun, i.e. the problem of magneto-convection in rotating objects, by direct numerical simulation. However, the high degree of complexity of the problem makes it necessary to include parametrization of the collective phenomena here also, but at a lower level than in mean-field theories. It should be noted that a sufficiently good modelling of the basic features of the solar magnetic field has not yet been achieved.

The continually growing power of computers makes parametrizations more and more superfluous. However, an appraisal of the results, represented by a large set of data, needs eventually the calculation of certain averages, i.e. parameters like eddy viscosity, α-effect, etc. must be calculated by averaging operations over the data sets. There is now the possibility of comparing this new information concerning the mean-field parameters derived by mean-field methods (mainly analytical) with the values derived from numerical simulations. Thus we have the possibility of confirming such basic relations as that between α and the helicity. However, a general comparison will meet with difficulties, since basic assumptions for mean-field theories do not necessarily coincide with those of the numerical simulations.

2. Magneto-convection in spherical shells

That the magnetic phenomena on the Sun are closely related to the turbulent motions in the convection zone is, for example, clearly to be seen from the enhanced decay of sunspots: if the decay time of the magnetic field is calculated according to $t_{decay} = \mu\sigma d^2$, where d is the diameter of the sunspot, $\mu = \mu_o$ the vacuum permeability, and σ is the electrical conductivity, we find a time scale of some thousand years, whereas in reality we see a decay time of some weeks.

This destructive action of turbulence is immediately accepted; however it has to be suspected that turbulence may also have some constructive features: Cowling's theorem states that axisymmetric (or, more general, simple-structured) motions cannot provide dynamo excitation. For example, the differential rotation cannot generate and maintain the solar magnetic field. Consequently, if the solar magnetic field is dynamo generated, the turbulent convective motions must play a key role.

First hints at the constructive action of turbulence were found in the fifties: in a paper from 1951 L. Biermann showed that an anisotropic turbulence in a rotating body must generate a differential rotation. Thus a large scale motion is generated from small-scale turbulence. A few years later E.N. Parker (1955) claimed that the turbulence of an electrically conducting media in a rotating body excites a dynamo: a large scale magnetic field is generated by this "cyclonic turbulence".

The investigation of the magnetic phenomena on the Sun is thus confronted with the problem of turbulently moving electrically conducting media in a rotating object. The basic equations are given by the conservation laws of mass, momentum, and energy

$$\frac{\partial \ln \varrho}{\partial t} + (\mathbf{u} \cdot \nabla) \ln \varrho + \text{div } \mathbf{u} = 0, \tag{1}$$

$$\frac{\partial \mathbf{u}}{\partial t} + (\mathbf{u} \cdot \nabla)\mathbf{u} = -\frac{p}{\varrho}\nabla \ln p + \mathbf{g} - 2\Omega \times \mathbf{u} + \frac{1}{\varrho}\mathbf{j} \times \mathbf{B} + \frac{1}{\varrho} \text{Div } \tau, \tag{2}$$

$$\frac{\partial e}{\partial t} + (\mathbf{u} \cdot \nabla)e = -\frac{p}{\varrho} \operatorname{div} \mathbf{u} + \frac{\kappa}{\varrho} \Delta e + Q_{visc} + Q_{joul}, \tag{3}$$

together with the induction equation

$$\frac{\partial \mathbf{B}}{\partial t} = \operatorname{curl}(\mathbf{u} \times \mathbf{B}) + \frac{1}{\mu\sigma} \Delta \mathbf{B}. \tag{4}$$

In addition, an equation of state and boundary conditions have to be taken into account. Here denotes ϱ the mass density, $\mathbf{u}$ the velocity, p the pressure, $\mathbf{g}$ the gravity, Ω the angular velocity of the overall rotation, $\mathbf{B}$ the magnetic field, $\mathbf{j}$ the current density, e the specific internal energy, κ the heat conduction and τ the viscosity tensor defined by

$$\tau_{ij} = \nu\varrho(u_{i,j} + u_{j,i} - \tfrac{2}{3}\delta_{ij} \operatorname{div} \mathbf{u}), \tag{5}$$

with the kinematic viscosity ν. Finally define Q_{visc} and Q_{joule} the heat production by viscosity and electrical resistivity. These quantities are quadratic expressions in the velocity gradient and the current density, respectively.

In particular, the boundary conditions specify that heat energy flows into the spherical shell under consideration at the bottom and leaves it at the surface, for example by the emission of radiation. The equations (1)–(4) have the simple solution

$$\mathbf{u} = \mathbf{B} = \mathbf{j} = 0, \frac{\partial \varrho}{\partial t} = 0, \tag{6}$$

with e determined by

$$\frac{\partial e}{\partial t} = \frac{\kappa}{\varrho} \Delta e, \tag{7}$$

together with the appropriate boundary conditions. In this case the transport of energy is simply carried on by molecular conduction.

However, this solution generally proves to be unstable. First the velocity field $\mathbf{u} = 0$ becomes unstable, and energy transport occurs by convective motions – hot material rises towards the surface, cold material sinks down to the bottom. In addition, since the medium in the solar convection zone is electrically conducting, the magnetic field $\mathbf{B} = 0$ also becomes unstable and dynamo excitation of the solar magnetic field sets in.

The scheme represented in Fig. 1 illustrates the physical situation. Transport processes of this kind can show two basic bifurcations: firstly the thermal instability at the critical Rayleigh number and secondly the dynamo instability at the critical value of the dynamo number. The system evolves from a stove via a heat engine into a power station.

It is now clearly to be seen that the considered problem requires the solution of the full set (1)–(4) of nonlinear differential equations, i.e. a problem of unmanageably high complexity.

In order to attack this problem methods already developed in the theory of turbulence have been applied, namely the derivation of equations controlling the

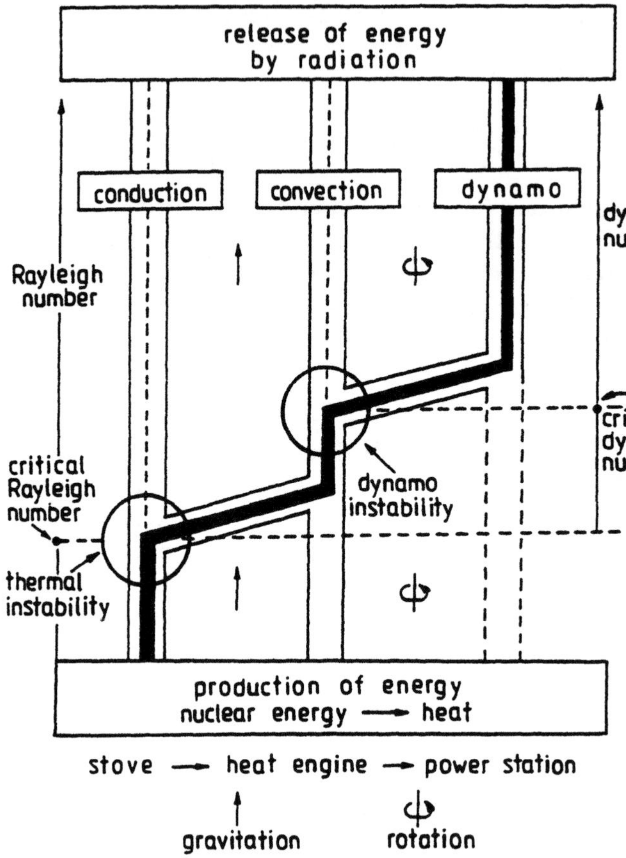

Fig. 1. The transport of energy in a star like the Sun is characterized by two basic bifurcations: the thermal instability marks the onset of convection and the dynamo instability that of the excitation of a magnetic field.

mean quantities. The nonlinear interactions of the fields involved in this energy transport provide new cooperative processes which enter the mean-field equations. The best known examples are characterized by the eddy viscosity and the α-effect.

The goal here is to determine the space-time behaviour by determining stable solutions of the mean field equations. The other way - possible in the future because of the growing power of computers - is the direct determination of the (irregular, unstable) solutions of the original equations. Basically both methods must lead to the same results and so there must be the possibility of checking one against the other.

3. The mean-field concept

The physical quantities will be represented by the sum of a mean part, denoted by a bar, and a fluctuating part, denoted by a dash:

$$\mathbf{u} = \overline{\mathbf{u}} + \mathbf{u}', \quad \mathbf{B} = \overline{\mathbf{B}} + \mathbf{B}', \quad p = \overline{p} + p', \dots \tag{8}$$

The averaging procedure need not to be specified, however some general rules have to be fulfilled, the so-called Reynolds rules:

$$\overline{F'} = 0, \quad \overline{\overline{F}} = \overline{F}, \tag{9}$$

$$\overline{FG} = \overline{F} \cdot \overline{G} + \overline{F'G'}, \ \overline{\overline{F} \cdot \overline{G}} = \overline{F} \cdot \overline{G}, \ \overline{\overline{F} \cdot G'} = 0, \tag{10}$$

where F and G denote arbitrary random fields.

These rules are fulfilled for statistical averages. For real measured quantities time or space averages have to be used. In the latter cases the validity of (9) and (10) will at best be guaranteed approximately; in particular the condition $\overline{F'} = 0$ is questionable. This is also the case if unstable solutions of the system (1)–(4) have been determined by numerical integration.

Because of the first rule in (10) the nonlinear terms in the equations (1)–(4) give rise to additional terms in the mean-field equations. We will illuminate this procedure with the Navier-Stokes equation for incompressible turbulence and the induction equation, assuming a weak magnetic field, i.e. that

$$\frac{\overline{B^2}}{2\mu} < \frac{\varrho}{2}\overline{u^2}; \tag{11}$$

the Lorentz force is neglected in (2).

According to (10) we have

$$\overline{(\mathbf{u} \cdot \nabla)\mathbf{u}} = (\overline{\mathbf{u}} \cdot \nabla)\overline{\mathbf{u}} + \overline{(\mathbf{u}' \cdot \nabla)\mathbf{u}'}, \tag{12}$$

$$\overline{\mathbf{u} \times \mathbf{B}} = \overline{\mathbf{u}} \times \overline{\mathbf{B}} + \overline{\mathbf{u}' \times \mathbf{B}'}, \tag{13}$$

and so find from the original equations

$$\frac{\partial \mathbf{u}}{\partial t} + (\mathbf{u} \cdot \nabla)\mathbf{u} = -\nabla \frac{p}{\varrho} + \mathbf{F} + \nu \Delta \mathbf{u}, \tag{14}$$

$$\frac{\partial \mathbf{B}}{\partial t} = \mathrm{curl}(\mathbf{u} \times \mathbf{B}) + \frac{1}{\mu\sigma} \Delta \mathbf{B}, \tag{15}$$

the following mean-field equations

$$\frac{\partial \overline{\mathbf{u}}}{\partial t} + (\overline{\mathbf{u}} \cdot \nabla)\overline{\mathbf{u}} = -\nabla \frac{\overline{p}}{\varrho} + \overline{\mathbf{F}} - \overline{(\mathbf{u}' \cdot \nabla)\mathbf{u}'} + \nu \Delta \overline{\mathbf{u}}, \tag{16}$$

$$\frac{\partial \overline{\mathbf{B}}}{\partial t} = \mathrm{curl}(\overline{\mathbf{u}} \times \overline{\mathbf{B}}) + \mathrm{curl}(\overline{\mathbf{u}' \times \mathbf{B}'}) + \frac{1}{\mu\sigma}\Delta\overline{\mathbf{B}}. \tag{17}$$

Two additional terms are present in the equations (16) and (17):

$$-\overline{(\mathbf{u}' \cdot \nabla)\mathbf{u}'} = \frac{\partial Q_{ij}}{\partial x_j}, \tag{18}$$

with the correlation tensor

$$Q_{ij}(\mathbf{x}, t) = \overline{\mathbf{u}'_i(\mathbf{x}, t)\mathbf{u}'_j(\mathbf{x}, t)} \tag{19}$$

and the turbulent electromotive force

$$\mathcal{E} = \overline{\mathbf{u}' \times \mathbf{B}'}. \tag{20}$$

The appearance of these two new quantities mathematically means that the system is not closed, that is additional relations have to be derived which relate these second order statistical moments back to the mean fields. This problem leads deep into the theory of turbulence with all its solved and unsolved problems.

However, by making a few very natural assumptions, it is possible to derive in a rather simple but rigorous way to quite general explicit expressions for these quantities. For the case under consideration the turbulent motion is given. Subtracting (17) from (15) we find

$$\frac{\partial \mathbf{B}'}{\partial t} - \mathrm{curl}(\overline{\mathbf{u}} \times \mathbf{B}') - \frac{1}{\mu\sigma}\Delta\mathbf{B}' = \mathrm{curl}(\overline{\mathbf{u}} \times \mathbf{B}') + \mathrm{curl}(\mathbf{u}' \times \mathbf{B}' - \overline{\mathbf{u}' \times \mathbf{B}'}) \tag{21}$$

Without explicit integration we can see that $\mathbf{B}'$ is a linear functional of $\overline{\mathbf{B}}$, and the same is true of $\mathcal{E}$ (c.f. Krause and Rädler, 1980). In addition, we assume local dependence, i.e. the characteristic space and time scales of the fluctuations, say correlation length λ_{cor} and correlation time τ_{cor}, are small compared with the corresponding scales $\overline{\lambda}$, $\overline{\tau}$ of the mean fields:

$$\lambda_{cor} \ll \overline{\lambda}, \quad \tau_{cor} \ll \overline{\tau}. \tag{22}$$

Hence $\mathcal{E}$ can be represented by the expression

$$\mathcal{E}_i = a_{ij} \cdot \overline{B}_j + b_{ijk}\frac{\partial \overline{B}_j}{\partial x_k}, \tag{23}$$

where the pseudo-tensors a_{ij}, b_{ijk} depend on the properties of the turbulent motion, i.e. on the physical quantities influencing the turbulence.

The simplest, therefore basic, model is that of an absolutely structureless turbulence, i.e. a turbulence that is homogeneous, isotropic and steady. In this case only the two tensors characterizing the Euclidean space, namely the Kronecker tensor δ_{ij} and the Levi-Civita tensor ϵ_{ijk}, are available; hence

$$a_{ij} = \alpha\delta_{ij}; \quad b_{ijk} = \beta\epsilon_{ijk}, \tag{24}$$

and so from (23)

$$\mathcal{E} = \alpha\overline{\mathbf{B}} - \beta\,\mathrm{curl}\,\overline{\mathbf{B}}. \tag{25}$$

The properties of the turbulence are now contained in the pseudo-scalar α and the scalar β: the best known effects, the α-effect and the turbulent magnetic diffusion, characterized by the corresponding diffusivity β, are thus derived in a simple way. It should be noted that this derivation is a rigorous one, but based on the assumption (11) of weak magnetic field, and (22) of locality. A pseudo-scalar α is not simply available in real systems. It needs at least rotation and stratification. The latter may be due to gravity and then we can construct a pseudo-scalar

$$\alpha = \alpha_o(\mathbf{g} \cdot \Omega). \tag{26}$$

Eventually this does mean that the α-effect is quite a natural effect in a rotating systems. However, it has to be taken into account that the system is now anisotropic and the complete expression for a_{ij} reads

$$a_{ij} = \alpha_o(\mathbf{g} \cdot \Omega)\delta_{ij} + \alpha_1 g_i \Omega_j + \alpha_2 g_j \Omega_i \tag{27}$$

with scalars $\alpha_o, \alpha_1, \alpha_2$ determined by the turbulence. The original form (25) has to be replaced by a much more complex expression with a certain number of parameters (Rädler, 1980). This is also true if the restriction (11) is removed when nonlinear effects allow for further tensorial constructions.

The method illuminated here for the turbulent emf $\mathcal{E}$ has also been developed for the Reynolds stresses $-\varrho Q_{ij}$ which appear in the Navier-Stokes equation for the mean velocity field. In particular, a relation

$$Q_{ij} = \Lambda_{ijk}\Omega_k, \tag{28}$$

has similarly been considered in the context of differential rotation (Rüdiger, 1989). With the required symmetry of Λ_{ijk} one finds as the simplest form in a stratified medium (Λ-effect)

$$\Lambda_{ijk} = \Lambda(\epsilon_{ipk}g_j + \epsilon_{jpk}g_i)g_p \tag{29}$$

4. Success of mean-field models

Especially in connection with the question of the origin of the magnetic fields of cosmic object the mean-field concept has proved itself to be very effective. This need not be discussed here in detail. As a striking example Fig. 2 shows the observed butterfly diagram of the sunspot phenomena and one derived from an $\alpha\omega$-dynamo model based on mean-field magnetohydrodynamics (Steenbeck and Krause, 1969).

The, at least qualitative, agreement showing the latitudinal migration of the magnetic activity towards the equator is obvious. During the last 20 years many

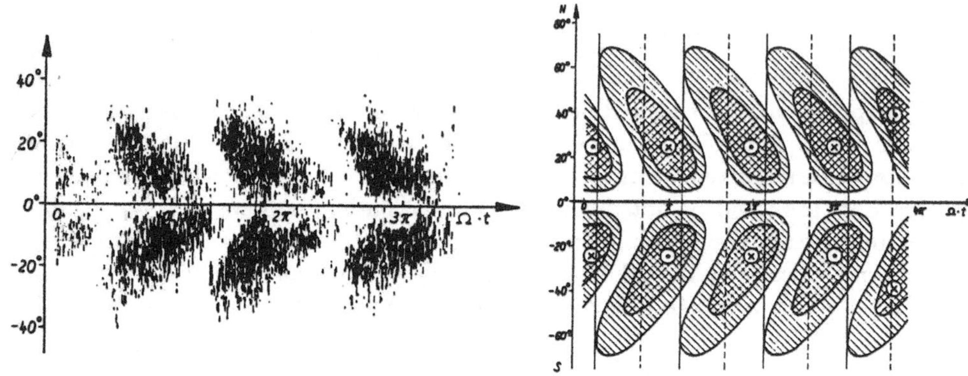

Fig. 2. Representation of the butterfly diagram of the Sun and of a kinematic $\alpha\omega$-dynamo model. A modelling of basic features of the solar cycle is possible, however, α-effect and differential rotation are chosen independently here.

models of this kind have been elaborated with the intention of optimizing the fit to observation. Here a selection of papers may be quoted: Roberts (1972), Roberts and Stix (1972), Yoshimura (1975a, b), Rädler (1980, 1986); and in addition there are monographs discussing the issue: Moffatt (1978), Krause and Rädler (1980), Zeldovich *et al.* (1983). Recently, observations have provided more data and in this way more conditions on the free parameters can be imposed (Brandenburg *et al.*, 1990). For example, one important new challenge to dynamo theory is given by the statements concerning the dependence of the differential rotation with depth, which are derived by means of helioseismology.

The accuracy of the relation (25) becomes questionable if nonlinear effects are taken into account, i.e. if the condition (11) is dropped and the magnetic field significantly influences the motion. A simple modelling is possible by reducing α with the growing magnetic field strength (α-quenching). These investigations show that the stability of a solution in the nonlinear regime is decisive in deciding which magnetic field will be excited and maintained by the dynamo. The growth rates determined from the linear problem are of less significance (Krause and Meinel, 1988; Brandenburg *et al.*, 1989).

A scenario is thus possible that predicts stable regular magnetic fields with certain symmetries (axisymmetry, equatorial symmetry) just beyond the marginal dynamo number, but for values of C that substantially exceed the marginal number, non-symmetric or even irregular solutions have to be expected.

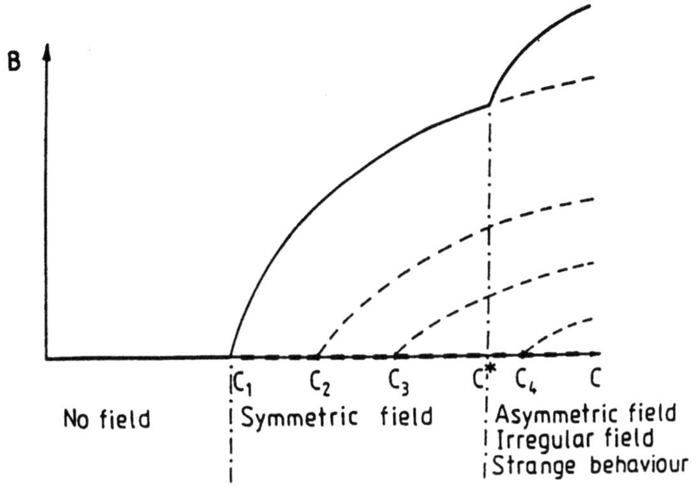

Fig. 3. Schematic dependence of dynamo excited cosmic magnetic fields on the dynamo number C. The solution bifurcating at the marginal dynamo number proves to be stable and so will be excited. Solutions bifurcating at higher values C are generally unstable and therefore never realized. The first solution reflects to some extend the symmetries of the cosmical object operating as a dynamo. For higher dynamo numbers symmetry breaking and irregularity have to be expected.

5. Numerical simulations of the solar cycle

The first attempts to derive solutions of the hydrodynamic equations (1)–(3) by direct numerical integration were carried out by P. Gilman in the early seventies (Gilman, 1972). The model consisted of a rotating spherical shell heated from below and the Boussinesq approximation was adopted. As the first step it was possible to find differential rotation, caused by the convection, that basically corresponded to that observed at the solar surface. In the next step the calculations of the hydromagnetic equations (1)–(4) were brought to the point that dynamo excitation of a magnetic field occurs (Gilman and Miller, 1981; Gilman, 1983).

The magnetic fields excited in these models deviate significantly from those observed on the Sun. The models of Gilman (1983) do show cyclic behaviour with time, but, in contrast to the solar field, the toroidal belts migrate toward the poles and not to the equator. A number of reasons can be listed, that might be responsible for these discrepancies:

(i) First of all, Boussinesq approximation must be mentioned, which does not allow the steep density gradient in the solar convection zone to be modelled.

(ii) The number of grid points is restricted by the capacity of the available computer. In the calculations quoted here the cell size lies above that of the super granules. The physics of the subgrid scales is taken into account by a parametrization according to the mean-field concept. In this case there is simply an eddy viscosity and turbulent magnetic viscosity.

(iii) If the boundary conditions are taken into account in a rigorous way they have to guarantee a fit to a solution of the problem outside the sphere. Because magnetic fields are long range physical quantities, the description of their influence generally requires non-local boundary conditions. For example, mean-field models such as those mentioned in section 3 consider the sphere to be embedded in an insulating space. If a grid point method is used for integration non-local boundary conditions require an additional mathematical effort. Therefore the local boundary condition $\mathbf{B}_{tan} = 0$ is used in all cases. There is no physical justification.

(iv) The most important reason may arise from the fact that differential rotation represented by, say, the Λ-effect and the α-effect are caused by the same convective motion. In the simulations both effects are simultaneously generated and the excited magnetic field is produced by their common action. In the mean-field model quoted in section 3 that shows a nice agreement with the real sun, the differential rotation and α-effect are independently chosen: there is a rotational shear at the bottom of the convection zone and an α-effect in the upper regions. The differential rotation that is generated in the models of Gilman does not show this structure.

Further efforts in the field of numerical simulations have been directed towards removing of the Boussinesq conditions. G. Glatzmaier attacked the problem using the anelastic approximation (Glatzmaier, 1984, 1985a, b), which allows more compressibility effects to be taken into account. However, it is still restricted to the low Mach number regime. The final results of these investigations are very similar to those of Gilman. In particular, a poleward migrating field was still found.

6. Numerical experiments

A new development has started in recent years with the elaboration of computer codes for fully compressible convection. Firstly two-dimensional models were considered (e.g. Chan *et al.*, 1982; Hurlburt *et al.*, 1984), and these have recently been extended to three dimensions (Chan and Sofia, 1986; Stein and Nordlund, 1989).

These investigations are carried out in rectangular boxes. They do not really pretend to simulate the solar cycle, but rather are numerical experiments carried out in order to study convection, including the effects of magnetic fields (Hurlburt and Toomre, 1988; Nordlund and Stein, 1989). With the aim of comparing results with those of mean-field magnetohydrodynamics Brandenburg *et al.* (1990) simulated conditions where the α-effect and turbulent magnetic diffusivity may be determined.

The most interesting question clearly is whether the well-known relation

$$\alpha = -\frac{\tau_{cor}}{3}\overline{\mathbf{u}' \cdot \mathrm{curl}\,\mathbf{u}'} \tag{30}$$

can be confirmed. This relation has been derived on the basis of the second order correlation approximation in the high-conductivity limit (c.f. e.g. Krause and

Rädler, 1980). It relates the parameter α to the helicity $\overline{\mathbf{u}' \cdot \operatorname{curl} \mathbf{u}'}$ of the turbulent motions. It should be noted that the proportionality of α to $(-\overline{\mathbf{u}' \cdot \operatorname{curl} \mathbf{u}'})$ has been confirmed in an experiment with liquid sodium (Steenbeck *et al.*, 1968).

At first glance the results derived by Brandenburg *et al.* (1990) contradict relation (30) in so far as the calculated α is directly proportional to the helicity. What can be the reason? The discrepancy is surely due to the fact that the calculation of α in (30) is based on isotropic conditions where the boundaries are distant. Both conditions are clearly violated in the numerical experiment. The direction of gravity is preferred in the model and downdrafts extend from the top to the bottom of the layer (cf. Stein and Nordlund, 1989).

However, homogeneity is guaranteed in the horizontal planes. If the anisotropy is taken into account, relation (25) must be replaced by

$$\mathcal{E} = \alpha_V \mathbf{B}_V + \alpha_H \mathbf{B}_H. \tag{31}$$

A similar derivation to that leading to (30) then gives

$$\alpha_H = -\frac{1}{3} \int_0^\infty \overline{u'_z(\mathbf{x},t) \cdot (\operatorname{curl} \mathbf{u}'(\mathbf{x},t+\tau))_z}\, d\tau \tag{32}$$

which may be evaluated as

$$\alpha_H = -\frac{\tau_{cor}}{3} \overline{\mathbf{u}'_z \cdot (\operatorname{curl} \mathbf{u}')_z}. \tag{33}$$

The horizontal α is related to the vertical helicity. The latter relation is, indeed, confirmed by the above mentioned numerical experiment.

The authors continued their investigations by introducing a non-vanishing mean-current and a gradient of the mean velocity. As a result a comparison with the standard expressions of the turbulent diffusivities becomes possible, which gives a fairly good agreement.

7. Conclusion

The comparison of mean-field results with those of numerical experiments is not in any case possible in a simple way.

Mean-field theories are generally based on assumptions that provide a practicable, or even comfortable, process of analytical deduction. The latter include the use of (i) statistical averages, (ii) two-scale property, i.e. weak variations of the mean quantities in space and time over scales of the fluctuations, and (iii) distant boundaries. In addition it has to be noted that a certain closure is always used, i.e. a decision concerning the neglect of higher order statistical moments.

Numerical simulations are generally restricted by the efficiency of the computer. Even the best of today's computers are not capable of solving a sufficiently well posed problem of magnetoconvection in a spherical shell. Restriction of the model, e.g. to a rectangular box, to artificial boundary conditions, etc., have to be taken into account. In addition, devices have to be introduced to guarantee numerical stability.

For these reasons the discrepancies revealed by a comparison are not surprising. In future modelling of situations corresponding more closely to reality by numerical simulations will become increasingly possible. Then it will become possible to compensate for the weak aspects of mean-field theories, especially the badly known parameters and the difficulties with nonlinearities.

References

Biermann, L.: 1951, *Z. Astrophys.* **28**, 304

Brandenburg, A., Krause, F., Meinel, R., Moss, D., Tuominen, I.: 1989, *Astron. Astrophys.* **213**, 411

Brandenburg, A., Nordlund, Å., Pulkkinen, P., Stein, R.T., Tuominen, I.: 1990, *Astron. Astrophys.* **232**, 277

Chan, K.L., Sofia, S.: 1986, *Astrophys. J.* **307**, 222

Chan, K.L., Sofia, S., Wolff, C.L.: 1982, *Astrophys. J.* **263**, 935

Gilman, P.A.: 1972, *Solar Phys.* **27**, 3

Gilman, P.A.: 1983, *Astrophys. J. Suppl. Ser.* **53**, 243

Gilman, P.A., Miller, J.: 1981, *Astrophys. J. Suppl. Ser.* **46**, 211

Glatzmaier, G.A.: 1984, *J. Comput. Phys.* **55**, 461

Glatzmaier, G.A.: 1985a, *Astrophys. J.* **291**, 300

Glatzmaier, G.A.: 1985b, *Geophys. Astrophys. Fluid Dyn.* **31**, 137

Hurlburt, N.E., Toomre, J., Massaguer, J.M.: 1984, *Astrophys. J.* **282**, 557

Krause, F., Meinel, R.: 1988, *Geophys. Astrophys. Fluid Dynamics* **43**, 95

Krause, F., Rädler, K.-H.: 1980, *Mean-Field Magnetohydrodynamics and Dynamo Theory*, Akademie-Verlag Berlin, Pergamon Press, Oxford

Moffatt, H.K.: 1978, *Magnetic Field Generation in Electrically Conducting Fluids*, Cambridge University Press

Nordlund, Å., Stein, R.F.: 1989, in *Solar and Stellar Granulation*, eds. R. Rutten and G. Severino, Kluwer Acad. Publ.

Parker, E.N.: 1955, *Astrophys. J.* **322**, 293

Rädler, K.-H.: 1980, *Astron. Nachr.* **301**, 101

Rädler, K.-H.: 1986, *Astron. Nachr.* **307**, 89

Roberts, P.H.: 1972, *Phil. Trans. Roy. Soc.* **A 272**, 663

Roberts, P.H., Stix, M.: 1972, *Astron. Astrophys.* **18**, 453

Rüdiger, G.: 1989, *Differential Rotation and Stellar Convection*, Akademie Verlag Berlin, Gordon and Breach

Steenbeck, M., Krause F.: 1969, *Astron. Nachr.* **291**, 49

Steenbeck, M., Kirko, I.M., Gailitis, A., Klawina, A.P., Krause, F., Laumanis, I.J., Lielausis, O.A.: 1968, *Dokl. Akad. Nauk* **180**, 326

Stein, R.F., Nordlund, Å.: 1989, *Astrophys. J. Letters* **342**, L95

Yoshimura, H.: 1975a, *Astrophys. J.* **201**, 740

Yoshimura, H.: 1975b, *Astrophys. J. Suppl. Series* **294**, 467

Zeldovich, Ya.B., Ruzmaikin, A.A., Sokoloff, D.D.: 1983, *Magnetic Fields in Astrophysics*, Gordon and Breach

On Numerical Studies of Solar/Stellar Convection

Kwing L. Chan[1] and Katsuyo Serizawa[2]

[1]Applied Research Corporation Landover, Maryland, USA
[2]Iowa State University Ames, Iowa, USA

1. Introduction

It has become a cliché to say that convection lies at the foundation of the activities, magnetism, and dynamos of the Sun and cool stars – the title of this colloquium. However, it has not been possible, until recently, to study theoretically these phenomena and their connections with quantitative rigor. The new development owes much to the rapid growth in the power of computers and the spread of sophisticated numerical techniques.

Since the last decade numerical studies of solar and stellar convection problems thrive. The trend towards a greater use of computers in these studies will likely become stronger. Now is a good time to reflect on the numerical approach itself, so that we may understand it better and make better use of it in the future. Here we look at two aspects of this methodology. First (Section 2), we examine its relationships to the analytical approach and observation, for which we make a differentiation between numerical experimentation and simulation. It is obvious that the ultimate test of any theory, numerical or not, is to confront observations, but not generally realized is that the connection to analytical studies is an important requisite for the healthy development of numerical studies. This point is to be elaborated. Second (Section 3), we discuss the prevalence of pitfalls in the numerical approach; two examples of current interest in convection computations are analysed in detail. In Section 4, we go on to suggest that one of these pitfalls might have caused the discrepancy between the results of numerical modelling of the solar differential rotation and helioseismology.

For economy of space and to avoid repetition, a review of the literature is not provided here. The readers are referred to Zahn (1987) for a recent review on the subject of solar/stellar convection. An updated survey of the literature can be found in Chan *et al.* (1990).

2. Connections to the analytical approach and observations

2.1 Numerical experimentation vs. simulation

In many reports of numerical studies, the words 'numerical experiment' and 'simulation' are used interchangeably, indicating that the differences between these two modes of numerical undertaking are not generally appreciated. When such confusion occurs, it is often accompanied by a failure to acknowledge the physical and numerical limitations of the computation and to qualify the applicability of the results. This is so because numerical experiment and simulation serve very different purposes, invoke different techniques, and generate different kinds of products. Numerical experiments have close link to the development of analytical concepts and theories while simulations are directed towards confronting or interpreting observations. A contrasting comparison of these two modes of study is summarized in Table 1.

Table 1. Numerical experimentation versus simulation

	experimentation	simulation
objective	study principles	reproduce observational results
application	extrapolation to inaccessible ranges	confront observations
model	idealized	realistic
prerequisite	know limitation of computation	also need realistic parameterizations
code design	for 10–20 runs	for 1–2 runs
it failed	prerequisite violated	difficult to identify cause

There is much similarity between numerical and laboratory experiments. They share the procedure of varying the experimental setup or external parameters with the intention to find patterns of scaling relationships for a well-defined physical system or process. The aim is to study the underlying principles. An important application of these principles is to provide estimates for situations which are not yet accessible by the experiments. To quite the contrary, the aim of a simulation is to reproduce or predict the occurrence of an observational event; it has no freedom to transgress the boundaries of the constraints. A good example is weather prediction; based on known principles and conditions, its objective is to model the detailed distributions and evolutions of the weather quantities faithfully. For this reason, the inputs to a simulation must be as realistic as possible and the recipes for the parameterized processes must be reliable.

A computer code designed according to these requirements imposes very heavy demand on the computational resources. Only a very limited number of cases can be computed; sometimes only a fraction of a case can be completed. The design of numerical experiments is quite the opposite. To make easy the extraction of the principles underlying the process of interest, the numerical model should be as clean from secondary effects as possible. It is not constrained to match the conditions and parameters to those of a realistic situation (quite often that is not feasible). Most importantly, the parameters should be chosen within the range where the numerical code performs efficiently and accurately, so that the model can produce reliable results. It is also important to realize that this range of validity depends on the objective and scope of the study, so extreme care is needed to avoid overlooking the limitations of the model.

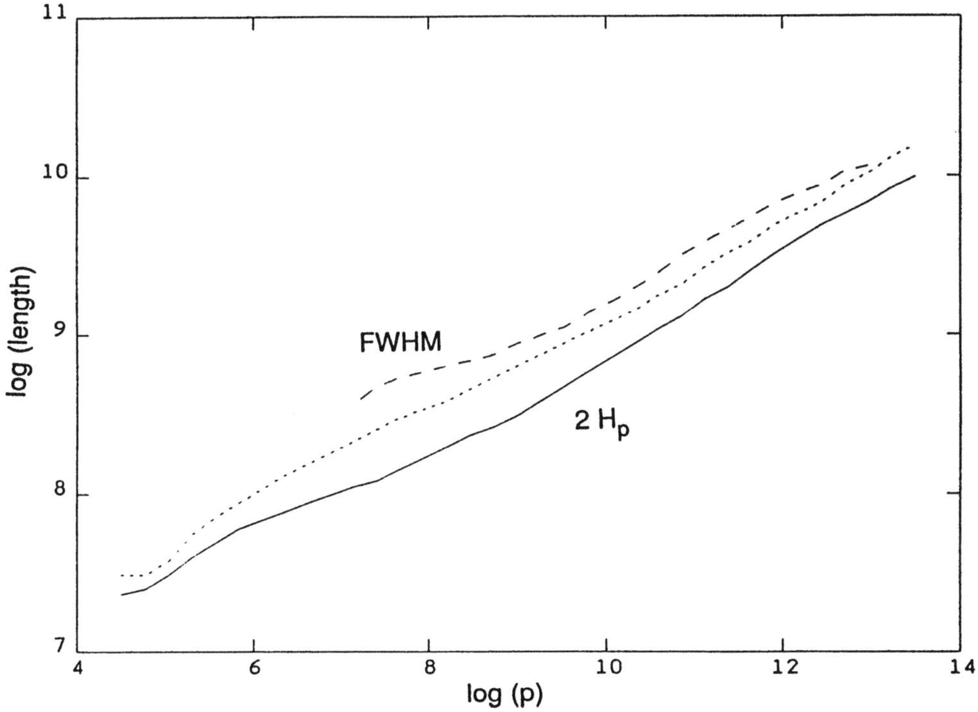

Fig. 1. Comparison of the FWNM of the convective modes with other length scales

When a simulation fails to reproduce observational results correctly, claim is often made that a certain theory is ruled out by the observations. Quite often, such claims are perfunctory. The real culprit may lie in the numerical model, in the input, or in the parameterizations. While the detection of a failure is straightforward in a simulation, its cause or causes are generally difficult to identify. On the other hand, the failure of a numerical experiment is not so clear-cut. As long as the procedure and setup of the experiment is properly documented and followed, the results of the study can be considered to be valid data points. 'Failure' may occur in the interpretation of the results, as inferences are made beyond the numerical and physical limitations of the model.

Therefore, if a numerical study happens to get lost between experimentation and simulation, it is highly susceptible to generating confusing results and misleading claims.

2.2 Connecting numerical experiments to analytical results – an example

Since a theoretical framework cannot be based only on pictures and graphs, it is useful and important to relate numerical works to established concepts and analytical results. This includes several aspects: (i) guide the design of the numerical model; (ii) help to avoid pitfalls; (iii) add insights; and (iv) provide inputs for further analytical study.

Here we discuss a case based on the numerical experiments performed by Chan and Sofia (1986, 1989). Motivated by the controversy over the validity of the mixing length theory (MLT), they studied compressible convection in three-dimensional convective layers with depths around 5 pressure scale heights. Their experiments varied the input flux to the layer, the thermal properties of the gas, and the size of the numerical grids. Quite invariant to these variations, the vertical autocorrelation lengths of the vertical velocity, the temperature fluctuation, and the entropy fluctuation were found to be scaled by the pressure scale height. This provides support to the mixing length picture.

More interestingly, the width and shape of the autocorrelation curves of Chan and Sofia (see Figure 1 of their 1986 paper) have striking similarity to those of the convective modes obtained by Narasimha and Antia (see Figure 1 of their 1982 paper). These authors obtained the convective modes from a solar convection zone model calculated by a mixing length theory which approaches the standard MLT in the deeper layer of the convection zone. They showed that the convective flux can be consistently reconstructed from a paralinear superposition of the statistically independent unstable modes, provided the effects of turbulent conductivity and viscosity are taken into account. The sizes of these modes are more or less scaled by the local pressure scale heights as shown in Figure 1 (kindly provided by Dr. Narasimha). In this figure, the solid curve shows the length of 2 pressure scale heights, the dashed curve shows the FWHM (full-width-half-maximum) of the modes, and the dotted curve shows the mixing length used for the background solar model.

The close resemblance of the auto-correlation curves and the convective modes prompted us to interpret the vertical coherence of the fluid motion showing up in the numerical calculations as a trace of the localized linear modes which drive the flow. The turbulence generated by the nonlinear effects may be important in controlling the amplitudes of these modes by providing a dissipation mechanism, but the cause and the general pattern of the flow may be linked back to these modes. Such a scenario is quite consistent with the analytical model of turbulent convection developed by Canuto and his co-workers (see his article in these Proceedings and references therein). In fact, even the generation of the strong downflow filaments (Massaguer and Zahn, 1980; Hurlburt *et al.*, 1984) and their coalescence in successively deeper layers (Stein and Nordlund, 1989) may be the result of a not-so-random superposition of the modes arranged by nonlinear locking.

Besides providing bases for developing useful insights, analytical results can also help to guide us through the treacheries of computation, a subject to be discussed next.

Table 2. Examples of pitfalls in numerical study of convection

• inappropriate approximations
numerical – how much physics is caused to be lost?
physical – viscosity modelling suitable?
• ill-posed conditions
background structure – consistent with dynamics?
boundaries – open vs. closed top and bottom
• insufficient relaxation
integration time > thermal relaxation time?
• over-confidence and over-interpretation

3. Pitfalls in the numerical approach

More than any other approach, numerical studies are extremely susceptible to pitfalls. The large amount of numbers generated by a calculation makes it very difficult to analyse the meanings and to detect errors. Errors can occur in any of the different stages of a study. A summary of the common sources is listed in Table 2. Some of pitfalls are quite well-known, but some are subtler. Here, we discuss two important, possibly controversial topics in the computation of solar/stellar convection.

3.1 Open versus closed boundaries at top and bottom

Under most circumstances, the domain of a calculation can only be part of the region where interactions take place. This makes the choice of boundary conditions a very difficult problem. The appropriate choice of course depends on the situation. In some studies, closed upper and lower boundaries which keep the fluid to stay inside a layer are used because the system is well-defined, and the conservation of various quantities can more readily be checked. In many other studies, it is argued that open boundaries which allow fluid to go in and out of the domain would represent a situation more realistically. In these cases, large oscillations of the layers generally occur. Oscillations do occur in closed-boundary cases but with much smaller amplitude. This makes us wonder if some of the large amplitude oscillations are caused by the open boundaries.

An open horizontal boundary is usually created by making $\partial \boldsymbol{v}/\partial z = 0$, and p, T = constant where $\boldsymbol{v}$ is the velocity vector, z is the vertical coordinate, p is the pressure, and T is the temperature. The values of the pressure at the top and bottom are chosen such that the initial state of the gas is in hydrostatic equilibrium. Trouble arises as the convection is turned on, then the mean distribution of the gas

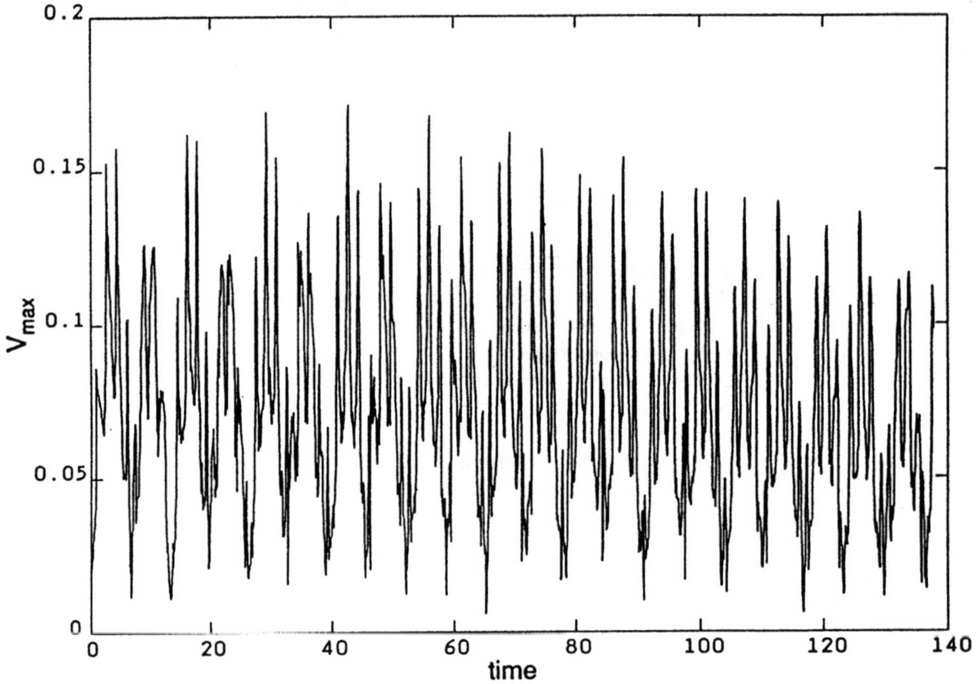

Fig. 2. Absolute value of the maximum vertical velocity (unit velocity = sound speed at the top) vs. time = depth of the layer/unit velocity)

will be changed by the development of convection. The structure of the whole layer will evolve towards a new configuration to be determined self-consistently with the convection (see, for example, Figure 1b of Sofia and Chan, 1984). The original values of the pressure used for the upper and lower boundaries most likely will not be consistent with the new configuration. Will this generate spurious oscillations? To explore this possibility, we performed some simple numerical experiments.

Oscillations can be excited by the convective turbulence; to avoid contamination, we study a convectively stable layer with open boundaries which are created by the conditions as described above ($\gamma = 5/3$, $T/T = 2$, polytropic index = 2, Prandtl number Pr = 1; the subscripts 'bot' and 'top' denote values at bottom and top respectively). When everything is in balance, no significant motion is generated, as should be. However, what if the boundary pressure is perturbed, as may be caused by an adjustment of the overall mean structure? This is imitated by a 2% shift of the boundary pressure at the bottom of the layer. The first thing that happens is the appearance of a lot of vertical oscillations (see Figure 2), and they persist for a very long time.

When the viscosity of the gas is increased by a factor of 100, the damping of the motion can be traced (solid curve in Figure 3), but a constant vertical flow through the layer remains, carrying with it an artificial energy flow. We can see the appeal and advantages of allowing flows to go through the boundaries of a domain, but such extraneous behaviour is quite undesirable. Is there some way to improve the situation? One thing to try is to seal off an end so that no constant net flow can go through. We therefore close the top boundary in the above model (set $v_z = 0$) and re-do the experiment. The outcome is illustrated by the dashed curve in Figure 3. The amplitude of the boundary induced oscillations becomes much smaller, and indeed no net flow is sustained. This way, the total amount of material inside the layer can automatically adjust for a new mean structure with less side-effects.

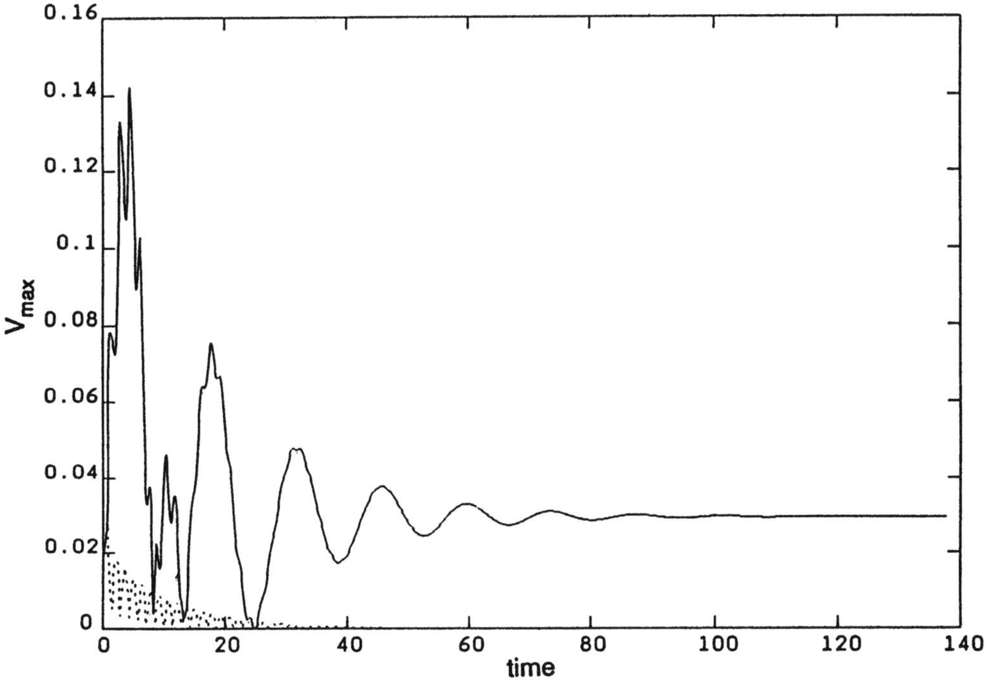

Fig. 3. Absolute values of maximum vertical velocities vs. time for the fully open (solid curve) and half open (dashed curve) boundary cases.

3.2 Significance of thermal relaxation

Now we turn to a computational requirement that is generally acknowledged but not always observed. The convective velocity, temperature fluctuation etc. depend on the structure of the convection zone (e.g. the entropy gradient), and vice versa. When the structure is evolving, the convection is evolving, and there is no guaranteed way to choose an initial structure that is sufficiently close to the developed state. To obtain consistency between the convective process and the structure of the convection zone, it is necessary to integrate the model beyond a thermal relaxation (Kelvin-Helmholtz) time scale. This time scale is given by

$$\tau_{\rm th} = \int e dz/F, \tag{1}$$

where e is the internal energy density and F is the energy flux.

Here, we illustrate the significance of this time scale by looking at the relaxation of a two-dimensional calculation (adequate for our purpose). The model starts with a slightly superadiabatic, polytropic distribution ($\gamma = 5/3$, $T_{\rm bot}/T_{\rm top} = 8$) whose energy content can be estimated as

$$\int e dz \sim p_{\rm bot}\, d/(2\gamma - 1), \tag{2}$$

where d is the total depth of the layer. (Note that the estimate provided by Eq. 34 of Chan and Sofia, 1986, contains an extraneous factor Z at the denominator). A constant flux is applied at the lower boundary, and the entropy is fixed at the upper boundary. Some small velocities are used to generate the initial perturbation.

The evolution of the height distribution of the horizontally averaged entropy is shown in Figure 4. The long dot-dashed, short dot-dashed, dashed, solid, and dotted curves show the distributions at the instances 1/4, 1/2, 3/4, 1, and 5/4 respectively; a trace of the initial entropy distribution can be detected near the bottom of the figure. All of the distributions are quite flat in the deeper region of the convection zone, but the steady adiabat is not obtained until the integration goes beyond 3/4.

The evolution of the energy flux flowing out the top of the region is shown by the solid curve in Figure 5; the dashed line shows the fixed input flux. It is clear that the two fluxes do not match until about. Therefore the layer is not in statistical energy balance during the transient phase which lasts about one thermal relaxation time.

The effect of thermal imbalance on the reliability of numerical results again depends on the problem of interest. For properties that are sensitive to the structure of the convection zone, the results of calculations that do not satisfy the relaxation requirement should be viewed with caution.

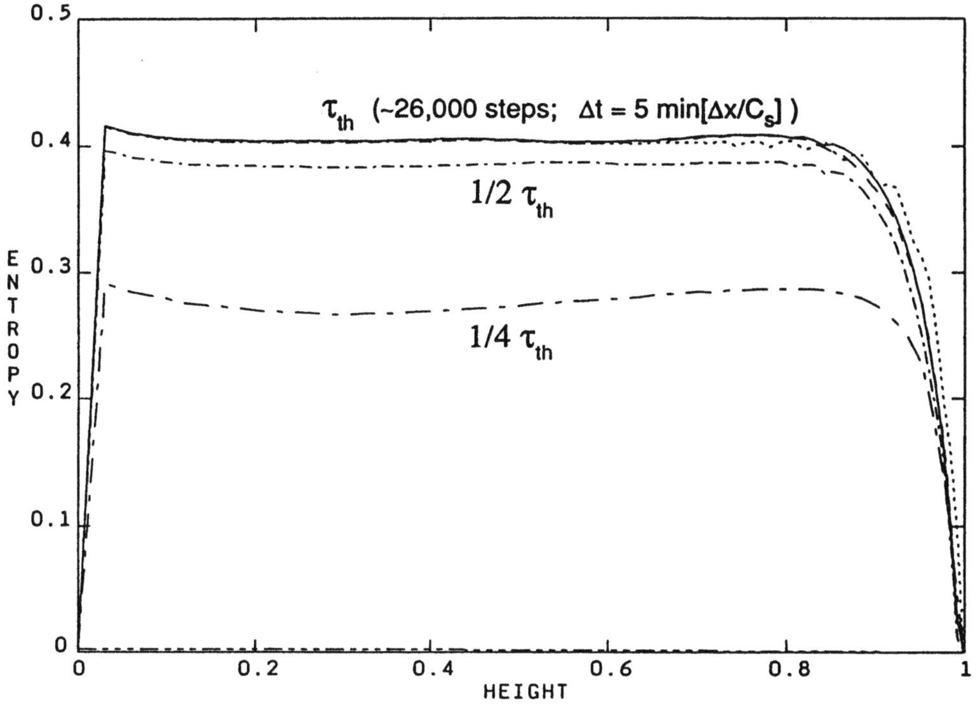

Fig. 4. Thermal relaxation of the entropy distribution

4. Taylor columns in numerical models of solar differential rotation

The lack of thermal relaxation may be the problem that plagues existing numerical models of the solar differential rotation (Glatzmaier, 1985; Gilman and Miller, 1986). These models predicted angular velocity contours mostly parallel to the rotation axis (Taylor columns), but recent results from helioseismology indicate that the contours are more like radial in the solar convection zone (Brown *et al.*, 1989; Libbrecht, 1989; also see the article by Goode in these Proceedings). This discrepancy is a puzzle. For this, let us examine the conditions under which Taylor columns are generated. In quasi-stationary situation the momentum balance in a rotating shell is governed by:

$$(1/\rho)[\nabla p + \rho \boldsymbol{g} + \nabla \cdot (\rho \boldsymbol{v}\boldsymbol{v} - \boldsymbol{\sigma})] + 2\boldsymbol{\Omega} \times \boldsymbol{v} = 0, \tag{3}$$

where $\boldsymbol{v}$ is the velocity, $\boldsymbol{g}$ is the gravitational acceleration, $\boldsymbol{\sigma}$ is the viscous stress tensor, and $\boldsymbol{\Omega}$ is the angular velocity of the shell. If the Coriolis term *really* dominates,

$$\nabla \times (\boldsymbol{\Omega} \times \boldsymbol{v}) = \boldsymbol{\Omega}(\nabla \cdot \boldsymbol{v}) - \boldsymbol{\Omega} \cdot (\nabla \boldsymbol{v}) + \boldsymbol{v} \cdot (\nabla \boldsymbol{\Omega}) - \boldsymbol{v}(\nabla \cdot \boldsymbol{\Omega})$$
$$= -\Omega \partial_z \boldsymbol{v} = 0, \qquad (4)$$

and the Taylor columns follow. This happens if: (i), i.e., where L is the characteristic length, is the kinematic viscosity, and Ta is the Taylor number, so that the viscous terms are unimportant; (ii) , i.e. heliostrophic, where V is the characteristic velocity associated with the scale L, so that the nonlinear advection terms also are unimportant; (iii), i.e. close to isentropic, since density and pressure fluctuations where S is the entropy. In the global scale, the first two conditions are generally believed to be satisfied, but the last condition is highly questionable (Durney, 1987). As indicated by the non-existence of Taylor columns in the solar convection zone, this may just be the condition that has been violated in the numerical models.

In the light of recent results from helioseismology the entropy gradient in the solar convection zone therefore appears to be a crucial factor of the mechanism that generates the global differential rotation. It is important to determine the distribution of the gradient correctly, consistently with the convection. However, this is extremely difficult because according to our previous discussion, this takes a Kelvin-Helmholtz time scale. While the hydrostatic adjustment requires only a few years (Spiegel, 1987), the thermal relaxation takes about 10 years. Existing numerical models can only compute through a hydrostatic adjustment time scale. As a result, the entropy distribution obtained by a model represents only an initial state of a very long transient.

There are two ways that allow global scale motions to develop in the numerical models despite the relative meagerness of the buoyance term compared to the Coriolis terms. First, the buoyance driving is performed in smaller scales where the hindering effects of the Coriolis terms are smaller, and then energy is transferred to the global scale through nonlinear interactions (Gilman and Miller, 1986). Second, kinetic energy is generated at the expense of the entropy stratification which becomes even less superadiabatic (Glatzmaier, 1987). The kinetic energy residing in the global scale can then be organized by the Coriolis force into Taylor columns.

At this point, it is also important to point out that while a mixing length type theory may provide a good estimate for the structure of the upper convective region where the Coriolis effects are unimportant, it is not applicable to regimes where the Coriolis effects are large. Most likely, when the Coriolis terms become significant, the superadiabatic gradient needs to be larger than that given by MLT, so that the hindering effects of rotation can be countered.

5. Summary

In this article, we reflect upon the numerical methodology of studying solar/stellar convection and stress the importance and difficulty of avoiding pitfalls. While these seemingly technical "fine points" are usually ignored in preference to the final numerical products, their appreciation is crucial to the proper comprehension of the numerical results. In developing our discussion, we touch upon a variety of subjects; the major points are recapitulated here so that the readers can more easily see their interconnections:

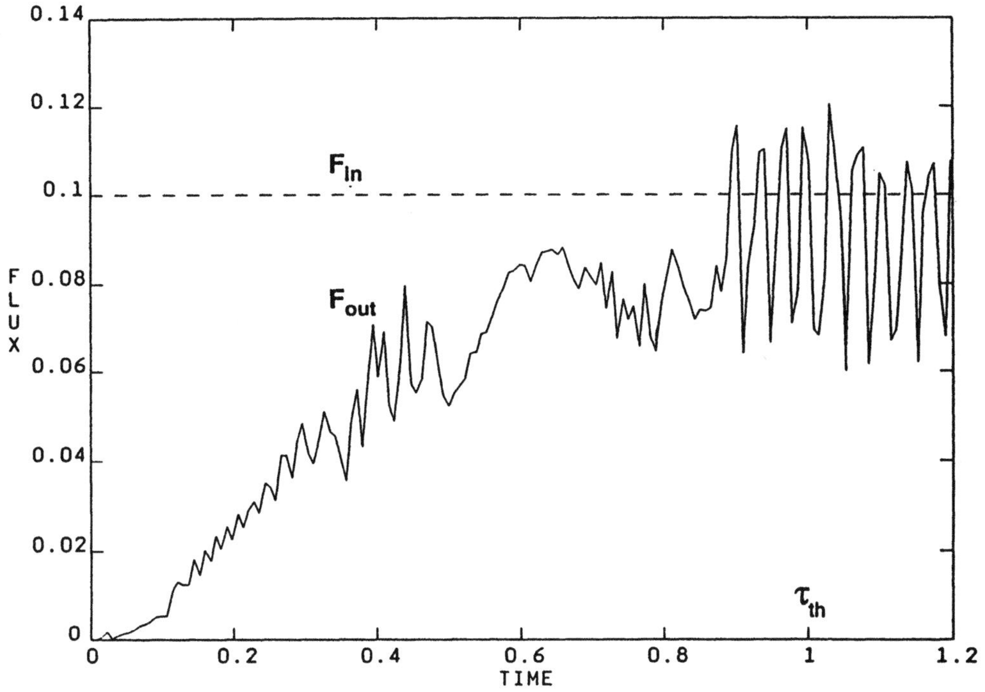

Fig. 5. Thermal relaxation of the energy flux

(i) Numerical studies can be divided into two classes: experimentation and simulation, each serving very different purposes. Conscientious effort to specify and taylor a numerical study to one of these classes can greatly clarify the contribution and limitation of the work.

(ii) The numerical approach is prone to pitfalls. The cardinal one is probably overconfidence; it leads to overlooking the numerical traps and to overinterpreting the results.

(iii) Numerical studies, at least so far, have to base on the concepts and understandings developed by analytical or semi-analytical studies. It is important to acknowledge and utilize this wealth of resources for consolidating the numerical results.

(iv) As an example, we point out that the existence of strong autocorrelations in the turbulent quantities within a pressure scale height is compatible with the dominance of localized convective modes obtained by a semi-analytical theory. The numerical and analytical results reinforce each other.

(v) Among the many possible trouble spots in the numerical study of solar/stellar convection, we discuss two in detail: first, the hazard of open upper and lower boundaries; second, the peril of not achieving thermal relaxation. We test for these problems in idealised situations, so that the effects can be cleanly demonstrated. It is not clear to what extent such problems might have (or have not) affected existing numerical studies, as the situations may be very different. Our discussion is intended only to be a caution.

(vi) The disagreement between the results of numerical modelling of the solar differential rotation and helioseismology is presently an important puzzle. We speculate that this may be caused by the lack of thermal relaxation in the numerical models, since we believe a correct entropy gradient can break the dominance of the Coriolis effects.

(vii) Due to the difficulty of obtaining thermal relaxation in a numerical model of the solar convection zone, it is probably more useful, at least at the moment, to explore further analytical and semi-analytical models for the solar differential rotation.

Acknowledgements. KLC thanks Dr. Ilkka Tuominen for the enjoyable meeting and NSF for support (AST-8815457). KS thanks NASA for a summer student research fellowship at Goddard Space Flight Center. Furthermore, we thank Drs. D. Narasimha, Hans Mayer, and Detlef Gigas for helpful discussions. Part of the computing was performed at the Pittsburgh Supercomputing Center.

References

Brown, T.M., Christensen-Dalsgaard, J., Dziembowski, W.A., Goode, P., Gough, D.O., Morrow, C.A.: 1989, *Astrophys. J.* **343**, 526

Chan, K.L., Sofia, S.: 1986, *Astrophys. J.* **307**, 222

Chan, K.L., Sofia, S.: 1989, *Astrophys. J.* **336**, 1022

Chan, K.L., Nordlund, A., Steffen, M., Stein, R.F.: 1990, in *Solar interior and atmosphere*, eds. A.N. Cox and W.C. Livingston, U. Arizona Press, Tucson, in press

Durney, B.R.: 1987, in *The internal solar angular velocity*, eds. B.R. Durney and S. Sofia, Reidel, Dordrecht, p. 235

Gilman, P.A., Miller, J.: 1986, *Astrophys. J. Suppl.* **61**, 585

Glatzmeier, G. A.: 1985, *Astrophys. J.* **291**, 300

Glatzmeier, G. A.: 1987, in *The internal solar angular velocity*, eds. B.R. Durney and S. Sofia, Reidel, Dordrecht, p. 263

Hurlburt, N.E., Toomre, J., Massaguer, J.M.: 1984, *Astrophys. J.* **282**, 557

Libbrecht, K.G.: 1989, *Astrophys. J.* **336**, 1092

Massaguer, J.M., Zahn, J.-P.: 1980, *Astron. Astrophys.* **87**, 315

Narasimha, D., Antia, H.M.: 1982, *Astrophys. J.* **262**, 358

Sofia, S., Chan, K.L.: 1984, *Astrophys. J.* **282**, 550

Spiegel, E.A.: 1987, in *The internal solar angular velocity*, eds. B.R. Durney and S. Sofia, Reidel, Dordrecht, p. 321

Stein, R.F., Nordlund, A.: 1989, *Astrophys. J. (Letters)* **342**, L95

Zahn, J.-P.: 1987, in *Solar and stellar physics*, eds. E.H. Schröter and M.S. Schüssler, Springer-Verlag, New York, p. 55

Turbulent Convection: A New Model

V. M. Canuto

NASA Goddard Space Flight Center
Institute for Space Studies
2880 Broadway
New York, NY 10025

Abstract: We use the latest models of turbulence to compute a new expression for the turbulent convective flux, F_c. The new values of F_c are up to ten times larger than those given by the mixing length theory, MLT. Astrophysical considerations indicate that the new model fares better with observational data than the MLT.

1. Introduction

In stellar interiors, a great deal of heat is transported by convection rather than by radiation. Consequently, a reliable model for the convective heat flux F_c is needed. The model adopted thus far in astrophysics, the mixing length theory, MLT (Böhm-Vitense, 1958), can yield results in agreement with the data provided the expression for F_c(MLT) is increased by an arbitrary amount $\alpha > 1$ which is not unique for all stars. The situation is clearly unsatisfactory.

To construct a more reliable model for convective turbulence one must: 1) forsake the intuitive physical picture of the MLT and couch the problem in terms of modern turbulence theory, 2) specify the exact nature of the approximations underlying the MLT, and, finally, 3) begin a systematic removal of these approximations. We have initiated this program and have found that MLT model is based on the following assumptions:

(a) Whereas the turbulent kinetic energy is known to be distributed among eddies of all sizes, the MLT takes into account only one eddy that contains all of this energy. That is, the MLT assumes an eddy energy spectrum $E(k)$ of the form

$$E(k) \propto \delta(k - k_0), \tag{1}$$

centered around the smallest wavenumber k_0 corresponding to the largest eddy compatible with the geometry of the system. (The integral of $E(k)$ over all k yields the turbulent kinetic energy per unit mass.)

(b) The basic difficulty of turbulence is the non-linear term (NLT) that distributes the energy over a wide range of eddies. Although the net action of the NLT on a given eddy is in general a complex set of processes, the two main effects are *cascade* and *backscatter*: given an eddy of size $\sim 1/k$, the effect of all the smaller eddies is to remove energy from it (cascade), as well as to return some of that energy (backscatter), i.e., schematically we have

$$\text{NLT} = \text{Cascade} + \text{Backscatter}. \tag{2}$$

The MLT is equivalent to taking

$$\text{NLT} = \text{Cascade}, \tag{3}$$

an approximation that clearly underestimates the convective flux.

(c) The nonlinear interactions operate on time scales that depend on the size of the eddies, i.e., the eddy correlation time scale $1/n_c(k)$ is a strong function of k: eddies with dimensions comparable with the size of the system are dominated primarily by the stirring process, while medium size eddies resulting from repeated break-up processes have lost memory of the nature of the source and have a correlation time scale independent of the source. We have found that the MLT assumes that

$$n_c(k) \sim n_s(k), \tag{4}$$

where $1/n_s(k)$ is the time scale characterizing the instability of the source. It is clear that Eq. (4) is true for a limited group of large eddies only.

We shall remove approximations (a)-(c) by trading the MLT formalism in favor of a turbulence theory that has become available thanks to recent progress in turbulence modeling (DIA, Kraichnan, 1964; EDQNM, Lesieur, 1987; RNG, Yakhot and Orszag, 1986; GISS model, Canuto *et al.*, 1985, 1987).

2. A new derivation of the MLT flux

Consider the Fourier transformed equations for the velocity and temperature fluctuations $u_i(\mathbf{k})$ and $\theta(\mathbf{k})$, where $i = 1,2,3$ (Ledoux *et al.*, 1961; Yamaguchi, 1963; Nakano *et al.*, 1979). (The time dependence of u_i and is omitted for brevity and the summation convention is adopted.)

$$(\frac{\partial}{\partial t} + \nu k^2)u_i(\boldsymbol{k}) + \text{NLT}_1(\boldsymbol{k})_i = g\alpha\lambda_j \Pi_{ij}(\boldsymbol{k})\theta(\boldsymbol{k}), \tag{5a}$$

$$(\frac{\partial}{\partial t} + \chi k^2)\theta(\boldsymbol{k}) + \text{NLT}_2(\boldsymbol{k}) = \beta\lambda_i u_i(\boldsymbol{k}), \tag{5b}$$

where $\alpha = 1/T$, ν is the kinematic viscosity, χ is the thermometric conductivity (thermal conductivity $K = c_p\rho\chi$), $\beta = T(\nabla - \nabla_{\rm ad})/H_p$ is the superadiabatic temperature gradient and $\lambda = (0,0,1)$. The non-linear terms $\text{NLT}(\boldsymbol{k})$ are defined as

$$\text{NLT}_1(\boldsymbol{k}) = \Lambda_{inm}(\boldsymbol{k})\sum_{\boldsymbol{q}} u_n(\boldsymbol{k}-\boldsymbol{q})u_m(\boldsymbol{q}), \tag{5c}$$

$$\text{NLT}_2(\boldsymbol{k}) = ik_n \sum_{\boldsymbol{q}} u(\boldsymbol{k}-\boldsymbol{q})\theta(\boldsymbol{q}), \tag{5d}$$

where $\Pi_{ij} = \delta_{ij} - k_i k_j/k^2$ and $2\Lambda_{inm} = i(k_n\Pi_{im} + k_m\Pi_{in})$.

The most important terms are the non-linear terms which couple velocity fields $u_i(\boldsymbol{k})$ with different wavenumbers, i.e., eddies of different sizes, and thus determine the resulting energy spectrum of the turbulent eddies. The reliability of any turbulence model depends crucially on how successfully the non-linear terms are treated. Before discussing that, however, there is a basic physical property of the NLT that must be stressed. Suppose one assumes that NLT($\boldsymbol{k}$) = 0. In this case, there cannot be a stationary solution of Eqs. (5). Rather, the velocity and temperature fields grow as $(u_i, \theta) \sim \exp(n_s t)$. Substitution into (5a-b) yields the rate $n_s(k)$

$$2n_s(k) = -(\nu + \chi)k^2 + [(\nu - \chi)^2 k^4 + 4g\alpha\beta x(1+x)^{-1}]^{1/2}, \tag{6}$$

where $x = (k_x^2 + k_y^2)/k_z^2$. Since $n_s(k) > 0$ for a wide interval of wavenumbers, it follows that when NLT(k) = 0, the system is unstable under perturbations, i.e., the kinetic energy grows exponentially in time, which is clearly unphysical. Let us now switch on the non-linear interactions. Their most distinguishing feature is that their integral over all wavevectors is zero (Batchelor, 1953),

$$\int \mathrm{NLT}(\boldsymbol{k}) d^3\boldsymbol{k} = 0, \tag{7}$$

which explains why they are called transfer terms: *the NLT do not generate or dissipate energy, they merely spread that which is available (from the external sources) among eddies of different sizes*, mostly but not exclusively, from the energy-rich large eddies to the energy-poor small ones. It follows that a stationary state can be achieved only in the presence of non-zero NLT. In summary, we shall write

$$NLT = 0, \quad \partial/\partial t \neq 0, \tag{8a}$$

$$NLT \neq 0, \quad \partial/\partial t = 0. \tag{8b}$$

After a model for the NLT has been chosen, one can solve Eqs. (5a-b) and determine the stationary energy spectrum. How well that final state compares with experimental data depends of course on how correctly the NLT have been modeled. Let us consider the following choice

$$\partial/\partial t = 0, \quad \mathrm{NLT}_1(\boldsymbol{k})_i = \nu_t(k)k^2 u_i(\boldsymbol{k}), \tag{9}$$

where ν_t is a turbulent viscosity. Then, Eqs. (5a-b), with $i = 3$, $u_3(k) = w(k)$, yield

$$[\nu + \nu_t(k)][\chi + \chi_t(k)]k^4 = g\alpha\beta\frac{x}{x+1}, \tag{10}$$

where, in analogy with the case of turbulent viscosity, $\mathrm{NLT}_2(\boldsymbol{k}) = \chi_t(k)k^2\theta(\boldsymbol{k})$. Equation (10) is a statement about time scales: the right hand side represents the square of the natural frequency of the system, the Brunt-Väisälä frequency, which is required to be equal to the one provided by the non-linear interactions. To solve (10), let us first note that available evidence from different types of turbulence indicates that the so-called turbulent Prandtl number $\sigma_t = \nu_t/\chi_t$ is of order unity or less. Considering further that for large eddies $\nu_t \gg \nu$, and introducing σ_t, Eq. (10) yields

$$\frac{\chi_t}{\chi} = \tfrac{1}{2}[(1 + \lambda\Sigma)^{1/2} - 1], \tag{11}$$

with $\Sigma = 4A^2(\nabla_r - \nabla_{rad})$, where the variable A is given by Cox and Giuli (1968, see Eq. 14.70), while the dimensionless constant λ (of order unity) is defined as

$$\lambda = 162\,\sigma_t^{-1}\frac{x+1}{x}(k_0 d)^{-4}. \tag{12}$$

We note that χ_t/χ is just the convective efficiency, Γ (see Cox and Giuli, Eq. 10.106). Multiplying (5b) by w and averaging gives the convective flux $F_c/c_p\rho = \langle\theta w\rangle$,

$$\langle w\theta\rangle = \beta k_0^{-2}\chi^{-1}(1+\chi_t/\chi)^{-1}\langle w^2\rangle, \tag{13}$$

where the kinetic energy K is given by $2K = (1+1/x)w^2$ (Canuto and Hartke, 1986). Next, we need to express K in terms of the other variables. This problem has been repeatedly addressed in the turbulence literature and several expressions have been proposed. Among them, the following two are particularly relevant (Canuto *et al.*, 1988)

$$\nu_t = \xi_3 K^2\epsilon^{-1}, \quad \nu_t = \xi_1\epsilon^{1/3}k_0^{-4/3}, \tag{14}$$

where ϵ is the rate of energy input per unit mass into the turbulent flow and $\xi_1 = 0.053$, $\xi_3 = 0.09$. If we write, $F_c/c_p\rho = \langle w\theta\rangle = \beta\chi_t = \beta\chi\Phi$, we have $\epsilon = g\alpha\langle w\theta\rangle = g\alpha\beta\chi\Phi$.

Expressing K in terms of ν_t and ϵ, ϵ in terms of $\langle w\theta\rangle$, and the final result in terms of Φ, we obtain, after some algebra

$$\Phi = 2\frac{x}{x+1}\sigma_t^2(\lambda\Sigma)^{-1}[(1+\lambda\Sigma)^{1/2}-1]^3. \tag{15}$$

Within factors of order unity, (15) coincides with the MLT expression derived by Gough and Weiss (1976),

$$\Phi = (9/8\Sigma)[(1+\Sigma)^{1/2}-1]^3. \tag{16}$$

Table 1.

S	Φ_{MLT}	Φ_1	Φ_1/Φ_{MLT}
10^5	53	115	2.17
10^6	173	513	2.96
10^7	560	2.0×10^3	3.57
10^8	1.8×10^3	7.1×10^3	3.94
10^9	5.6×10^3	2.4×10^4	4.28
10^{10}	1.8×10^4	8.1×10^4	4.50
10^{12}	1.8×10^5	8.55×10^5	4.75
10^{14}	1.8×10^6	8.76×10^6	4.87
10^{16}	1.8×10^7	8.84×10^7	4.91

Table 2.

S	K	Φ_2	Φ_2/Φ_{MLT}
10^4	5.3×10^3	40	2.74
10^6	8.8×10^5	1070	6.35
10^8	9.8×10^7	1.5×10^4	8.33
10^{10}	1.0×10^{10}	1.6×10^5	9.09
10^{12}	1.0×10^{12}	1.7×10^6	9.66
10^{14}	1.0×10^{14}	1.7×10^7	9.66
10^{16}	1.0×10^{16}	1.7×10^8	9.66

3. New models of turbulence

We shall begin by adopting the EDQNM result for the energy spectrum $E(\boldsymbol{k})$, (Lesieur, 1987)

$$[\partial/\partial t - 2n_s(k)]E(k,t) = T(k,t), \tag{17}$$

with the transfer $T(k,t)$ given by

$$T(k) = \int\int dpdqE(q)[k^2E(p) - p^2E(k)]a(p,q,k)\tau(k,p,q). \tag{18}$$

Here, $a(p,q,k) = (xy + z^3)/q$, and x, y, and z are the cosines of the angles opposite $\boldsymbol{k}$, $\boldsymbol{p}$, and $\boldsymbol{q}$, respectively. The integrations over p and q are such that $\boldsymbol{k} = \boldsymbol{p} + \boldsymbol{q}$. The function τ is the correlation time scale governing the non-linear interactions. Once τ is specified, Eq. (17) may be solved for $E(k)$. The convective flux is then computed via

$$F_c = \frac{c_p\rho}{g\alpha}\int_0^\infty 2E(k)[n_s(k) + \nu k^2]dk. \tag{19}$$

3.1 First model

It is well-known from the EDQNM model that the transfer $T(k)$ represents a variety of physical processes among eddies. The most basic process is the energy cascade from large to small eddies which is usually represented by a turbulent viscosity $\nu_t(k)$ which, via a closure, must be related to $\theta(k)$. If one approximates $T(k)$ by ($y(k)$ is the mean square vorticity)

$$T(k) \to -2\frac{\partial}{\partial k}y(k)\nu_t(k) \tag{20}$$

one has, after integrating (17) from 0 to k, the following equation for $E(k)$

$$\epsilon(k) = [\nu + \nu_t(k)]\int_0^k 2p^2E(p)dp, \tag{21a}$$

$$\epsilon(k) = 2\int_0^k [n_s(p) + \nu p^2]dp. \tag{21b}$$

The physical interpretation of (21a) is clear: the energy $\epsilon(k)$ injected into the wavenumber interval $[0,k]$ is dissipated partly by molecular viscosity and partly by turbulent viscosity ν_t. Canuto *et al.* (1987) have solved the above equations for $E(k)$ and then calculated the convective flux. The results are presented in Table 1 in which Φ_{mlt} is given by Eq. (16) and Φ_1 is computed using the solution of (21). The variable S is related to Σ introduced earlier by $S = (81/2)\Sigma$.

3.2 Second model

In the second model, we obtain the spectrum $E(k)$ by solving Eq. (17) without any approximation to $T(k)$. Once $E(k)$ is obtained, the flux is again computed using (19). The results are shown in Table 2.

The turbulent kinetic energy per unit mass K is given in units of $(\chi/\Lambda)^2$, where Λ is the mixing length. The values of Φ can be represented by the following expression

$$\Phi = a\Sigma^m[(1+b\Sigma)^n - 1]^p, \tag{22}$$

with $a = 24.868$, $b = 9.7666 \times 10^{-2}$, $m = 0.14972$, $n = 0.18931$, and $p = 1.8503$. As one can see, in the limit of high convective efficiency, $\Sigma \gg 1$,

$$\Phi(\text{this model})/\Phi(\text{MLT}) \approx 10. \tag{23}$$

In other words, the new convective flux is up to ten times larger than the MLT value. This result has been recently confirmed by a detailed numerical simulation of turbulent convection (Cabot *et al.*, 1990).

4. Astrophysical implications

Since the new model naturally yields a convective flux larger than the MLT, the corresponding value of α ($\Lambda = \alpha H_p$) will be correspondingly smaller and thus more in accord with the basic tenets of incompressibility adopted in all the previous models. Since $F_c \propto \Lambda^2 \propto \alpha^2$, we estimate that

$$\alpha_{\text{new}} \sim \alpha_{\text{MLT}}/(5-8)^{1/2} \tag{24}$$

Since $\alpha_{\text{MLT}} = 1.4$, we predict $\alpha_{\text{new}} = 0.5 - 0.6$. Indeed, detailed evolutionary studies indicate that $\alpha_{\text{new}} = 0.7$ (Canuto and Mazzitelli, 1991). It may also be noted that the new larger convective flux may have important consequences in helioseismology (L. Paternò, private communication) and in the determination of the age of globular clusters.

References

Batchelor, G.K.: 1953, *The Theory of Homogeneous Turbulence*, Cambridge Univ. Press
Böhm-Vitense, E.: 1958, *Zeitschr. Astrophys.* **46**, 108
Cabot, W., Hubickyj, O., Pollack, J.B., Cassen, P., Canuto, V.M.: 1990, *Geophys. and Astrophys. Fluid Dyn.* **53**, 1
Canuto, V.M., Goldman, I.:1985, *Phys. Rev. Letters* **54**, 430
Canuto, V.M., Goldman, I., Chasnov, J.: 1987, *Phys. Fluids* **30**, 3391
Canuto, V.M., Hartke, G.J.: 1986, *Astron. Astrophys.* **168**, 89
Canuto, V.M., Goldman, I., Chasnov, J.: 1988, *Astron. and Astrophys.* **200**, 291
Canuto, V.M., Mazzitelli, I.: 1991, *Astrophys. J.*, (in press).
Cox, J. P., Giuli, R. T.: 1968, *Principles of Stellar Structure*, Gordon and Breach
Gough, D.O., Weiss, N.O.: 1976, *Mon. Not. Roy. Astron. Soc.* **176**, 589
Kraichnan, R.H.: 1964, *Phys. Fluids* **7**, 1048
Lesieur, M.: 1987, *Turbulence in Fluids*, M. Nijhoff Publ.
Ledoux, P., Schwarzschild, M., Spiegel, E.A.: 1961, *Astrophys. J.*, **133**, 184
Nakano, T., Fukushima, T., Unno, W., and Kondo, M.: 1979, *Publ. Astron. Soc. Japan* **31**, 713
Yakhot, V. and Orszag, S.A.: 1986, *Jour. of Sci. Comp.* **1**, 3
Yamaguchi, S.: 1963, *Publ. Astron. Soc. Japan* **15**, 412

Large Scale Convection in Stars : Towards a Model for the Action of Coherent Structures

Michel Rieutord [1,2] and Jean-Paul Zahn [1]

[1]Observatoire Midi Pyrenees, 14 av. E. Belin, 31400 Toulouse, France

[2]CERFACS, 42 av. Coriolis, 31057 Toulouse, France

Abstract: We show that, representing the descending fluid in a convection zone by a porous medium, the differential rotation of the (rising) fluid is very close to that in an axisymmetric model of the convection zone with anisotropic viscosity.

1. Introduction

The computation of large scale convective flows in astrophysical objects is certainly one of the great challenge of contemporary astrophysics. The computation of such flows is a basic requirement for understanding the dynamical behaviour of stars or planets. The huge Reynolds number of the flow, due to the large size of the objects, makes the flow highly turbulent and so very difficult to compute. The main problem is that we do not have, for the moment, a complete theory of turbulence.

Until now two approaches have been considered: firstly the mean field approach and secondly direct numerical simulations. These two approaches possesses some major drawbacks which need to be recapitulated.

The mean field approach (see Rüdiger, 1989 and references therein) is concerned, in the case of stars, with the determination of the long term evolution of the mean axisymmetric fields. Indeed, we wish to know the nature of the mechanism that maintains the differential rotation or governs the magnetic activity. The mean is thus defined as to smooth out longitudinal and short-time dependence of the phenomena. For instance, if we are interested in the velocity field, we set :

$$\boldsymbol{u} = \boldsymbol{U} + \boldsymbol{u}', \tag{1}$$

where $\boldsymbol{u}'$ stands for the fluctuation around the mean $\boldsymbol{U} = \langle \boldsymbol{u} \rangle$. The evolution of the mean $\boldsymbol{U}$ can be derived, and, in the case of an incompressible fluid, gives the Reynolds equation

$$\partial_t U_i + U_j \nabla_j U_i = -\nabla_i P + \nu \Delta U_i - \partial_j R_{ij}, \tag{2}$$

where $R_{ij} = \langle u'_i u'_j \rangle$ is the Reynolds stress tensor. This last quantity cannot be computed unless we have a theory of turbulence that tells us the relation between this tensor and $\boldsymbol{U}$. In the case where separation of scales applies (i.e. $|\boldsymbol{u}'| \ll |\boldsymbol{U}|$), then the equation for $\boldsymbol{u}'$ can be made linear and solved for some given forcing. An expression for R_{ij} is then obtained. The main question is then about the validity of such an approximation. Indeed, it is questionable if this approximation can be applied to large scale non-axisymmetric modes. This is the main drawback of the method.

The second approach, the numerical one, is still concerned with solving (2), but now the mean is taken on a much smaller scale: essentially it is a smoothing of all the small scales not resolved by the numerical scheme. Here, as above, the relation between the Reynolds stress tensor and the mean velocity field is not known. However, such relation may be easier to obtain, since we are now dealing with small scales which have certainly been derived and nothing proves to be satisfactory in all cases. Let us recall a few existing (subgrid) models.

First, Smagorinski's model (1963) where :

$$R_{ij} = -C_S \Delta^2 \sqrt{S_{ij} S^{ij}} S_{ij}, \tag{3}$$

with $S_{ij} = (\partial_i U_j + \partial_j U_i)$. This model has been tested against direct numerical simulations and turned out to be able to represent only 16% of the true R_{ij} (Fertziger, 1985). We can also mention the $K\varepsilon$ model (see Peyret and Taylor, 1983), the renormalization group approach (Yakhot and Orszag, 1986) and a new version of Smagorinski's approach by Leith (1989). The main drawback with these models is their lack of predictability: either because, like the $K\varepsilon$ model, they have constants that need adjusting or, like Smagorinski's, they lack universality, leaving their predictions questionable.

2. The role of coherent structures?

Coherent structures (we refer here mainly to vortex tubes) play an important role which remains to be understood. It is well known that their presence reflects the nongaussianity of the turbulence statistics which creates all the difficulties of the statistical approaches. One of the difficulties of modelling the action of such objects is due to their "two-scaledness", that is to say that such an object as a vortex tube has a length of the order of the integral scale whilst it has a width of a few dissipation scales (see for instance Vincent and Meneguzzi, 1990). In large eddy simulations, such objects would appear as singularities of the flow since their thickness is not resolved while their length is. We may also note, from the spectral point of view, that such structures constitute a direct link between small and large scales and thus allow a direct exchange of energy.

We shall model these structures in a very crude way. We assume that they are purely passive. First we make them solid and we fix them in space (no advection by the flow). They thus make a porous medium through which the fluid passes.

To see the first consequences of such an assumption we computed the differential rotation of the solar convection zone (SCZ) when it is modeled by such a medium. We note that a porous medium was also used by Bretherton and Spiegel (1968) to model the SCZ for studying the solar spin-down. titlea3.The solar convection zone as a porous medium? Let us make the following assumptions : the SCZ is a porous medium where the solid part stands for the descending fluid (the plumes) and the fluid saturating this porous medium stands for the rising fluid; it is injected at the base of the SCZ and extracted at the top (which is our boundary conditions on the flow).

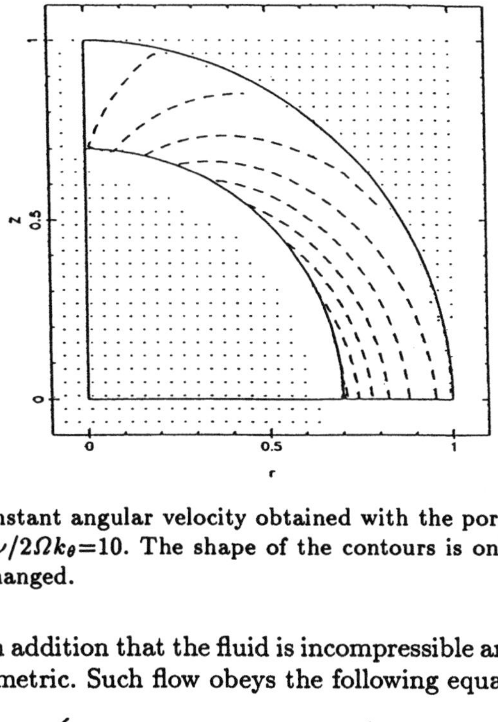

Fig. 1. Contours of constant angular velocity obtained with the porous medium model. The scaling factor $E=\nu/2\Omega k_\theta=10$. The shape of the contours is only slightly modified when k_r/k_θ or E are changed.

We shall assume in addition that the fluid is incompressible and that the motion is steady and axisymmetric. Such flow obeys the following equations:

$$\begin{cases} 2\boldsymbol{\Omega} \times \boldsymbol{v} = -\nabla P + \nu[k]^{-1}\boldsymbol{v}, \\ \operatorname{div} \boldsymbol{v} = 0, \end{cases} \tag{4}$$

where $\nu[k]^{-1}\boldsymbol{v}$ is the Darcy force characteristic of porous media (see Cushman 1990). $[k]$ is the permeability tensor which we take in the form

$$\begin{bmatrix} k_r & 0 & 0 \\ 0 & k_\theta & 0 \\ 0 & 0 & k_\phi \end{bmatrix} . \tag{5}$$

The anisotropy of the medium is then characterized by the ratio k_r/k_θ ($k_\theta = k_\phi$) which we take to 10: the medium is more permeable in the vertical direction than in the horizontal one. The resulting differential rotation is plotted on Fig. 1. The shape of the contours only slightly depends on the anisotropy. In Fig. 2 we

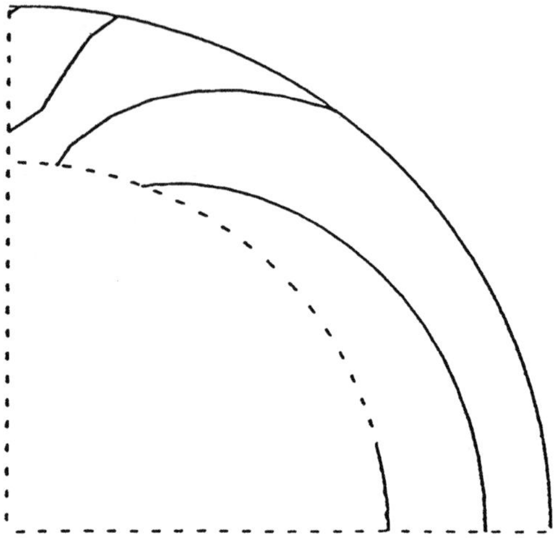

Fig. 2. Contours of constant angular velocity obtained with a numerical integration of anelastic axisymmetric equations of motion, using an anisotropic viscosity (20% of anisotropy).

The anisotropy of the medium is then characterized by the ratio k_r/k_θ ($k_\theta = k_\phi$) which we take to 10: the medium is more permeable in the vertical direction than in the horizontal one. The resulting differential rotation is plotted on Fig. 1. The shape of the contours only slightly depends on the anisotropy. In Fig. 2 we plotted the differential rotation obtained by Brandenburg and Rieutord (1990) using an axisymmetric, anelastic model with an anisotropic viscous tensor (the horizontal viscosity being larger than the vertical one). The two calculations give a similar distribution of angular velocity with a positive gradient ($\partial\Omega/\partial r > 0$). In both models the Coriolis force is the main contributor to the distribution, but the remarkable fact is that the model presented here is extremely simple. The next steps in this direction of investigation will include relaxing the restriction to a solid porous medium (coherent structures should be advected by the flow), computing their statistics, and allowing the structures to interact as a source of field.

References

Brandenburg, A., Rieutord, M.: 1990, in preparation.

Bretherton, F., Spiegel, E.A.: 1968, *Astrophys. J. Lett.* **153**, L77

Cushman, A.: 1990, *Dynamics of fluids in hierarchical porous media*, Academic Press

Fertziger, J.: 1985, in *Theoretical Approaches to Turbulence*, eds. D.L. Dwoyer, M.Y. Hussaini, R.G. Voigt, Springer

Leith, C.: 1989, *Phys. Fluids A* **2**, 297

Peyret, R., Taylor, T.D.: 1983, *Computational Methods for Fluid Flow*, Springer

Rüdiger, G.: 1989, *Differential Rotation and Stellar Convection*, Gordon and Breach, New York

Smagorinski, J.: 1963, *Mon. Weather Rev.* **91**, 94

Vincent, A., Meneguzzi, M.: 1990, to appear in *J. Fluid Mech.*

Yakhot, V., Orszag, S.: 1986, *J. Scientific Computing* **1**, 3

Magnetoconvection Patterns in Rotating Convection Zones

Paul H. Roberts

Center for Earth and Planetary Interiors
Institute of Geophysics and Planetary Physics
University of California, Los Angeles, California, 90024

Abstract: In addition to the well-known granulation and supergranulation of the solar convection zone (the "SCZ"), the presence of so-called "giant cells" has been postulated. These are supposed span the entire thickness of the SCZ and to stretch from pole to pole in a sequence of elongated cells like a "cartridge belt" or a bunch of "bananas" strung uniformly round the Sun. Conclusive evidence for the existence of such giant cells is still lacking, despite strenuous observational efforts to find them. After analyses of sunspot motion, Ribes and others believe that convective motions near the solar surface occurs in a pattern that is the antithesis of the cartridge belt: a system of "toroidal" or "doughnut" cells, girdling the Sun in a sequence that extends from one pole to the other. Galloway, Jones and Roberts have recently tried to meet the resulting theoretical challenge, with the mixed success reported in this paper.

A significant parameter of rotating magnetoconvection is the Elsasser number, $\Lambda = \sigma B^2/2\Omega\rho$, where $\mathbf{B}$ is the magnetic field (predominantly zonal in the Sun), Ω is the angular velocity, ρ is the density and σ is the (turbulent) electrical conductivity. Intuitively one expects that, when $\Lambda \ll 1$, Coriolis forces will dominate Lorentz forces and, because of the Proudman–Taylor theorem, cells will be elongated in the direction of $\boldsymbol{\Omega}$ and will form the cartridge belt pattern; when $\Lambda \gg 1$, one anticipates that the magnetic field will be decisive, and that the convection cells will align themselves with $\mathbf{B}$ in the toroidal pattern.

Jones, Roberts and Galloway (1990) have sought the preferred pattern of convection in an apparently simple system intended to mimic regions near the solar equator: a plane horizontal layer of compressible conducting fluid rotates about a horizontal axis perpendicular to an ambient uniform horizontal magnetic field. They determined the onset of convection, and demonstrated that the convection is necessarily time-dependent, a result that holds for all Λ, including even the non-magnetic case $\Lambda = 0$. Apart from this interesting difference, the model behaves in much the same way as the corresponding incompressible (Boussinesq) model, for which convection at onset is direct (time independent). In particular, even for large variations in the local value of Λ across the layer, the convection pattern is the same at all depths, and shows no preference for the toroidal pattern. To simulate the large superadiabatic gradient in the upper convection zone, they also examined a two-zone model in which the Rayleigh number, R, in the upper zone is ten times that in the lower zone. Two kinds of convective modes were discovered, one highly concentrated in the upper zone, one predominant in the lower zone. The latter resemble banana cells; the former the doughnuts.

Jones and Roberts (1990) studied the one zone model (modified by the addition of vertical side walls parallel to the applied field) at finite convective amplitudes, and showed that the mean Lorentz force introduced by the convection creates a large "geostrophic" flow parallel to $\mathbf{B}$, this flow being a function of height only. In certain circumstances (stress–free side walls) an increase in R leads to symmetry-breaking in the Boussinesq model. One mode is concentrated near the surface and moves as a wave in one direction; the other is concentrated near the base of the convection zone and moves in the opposite direction. Jones and Roberts studied the unfolding of this symmetry-breaking through compressibility. The relationship of this research to the Ribes hypothesis is discussed in this article.

1. Background

The question of what kind of large-scale convection occurs in the Sun is of considerable interest not only in understanding how the Sun creates its magnetic field but also in providing an explanation of the internal rotation of the Sun determined by solar seismology. Although there have been advances recently on both observational and theoretical fronts, complete elucidation is not imminent. The present contribution, the result of joint investigations by Galloway, Jones and Roberts, will (it is felt) add a little insight to the theory. In what follows, the paper of Jones, Roberts and Galloway (1990) will be called "Paper 1"; its sequel, Jones and Roberts (1990), will be "Paper 2". We will not repeat the detailed analysis and results of these papers. We shall merely focus on a few salient points that, it is hoped, will be of interest.

The existence of two scales of motion on the Sun has long been known: the granulation and the supergranulation. A third scale of motion was postulated in 1968 by Simon and Weiss: "giant cells" that span the whole thickness of the solar convection zone ("SCZ"). Their hypothesis has generally been supported by the results of numerical simulations of convection in compressible atmospheres, even ones that extend over several scale-heights; it has usually been found that convection is dominated by cells that extend across the entire depth of the convecting layer.

When a convective layer rotates rapidly, slow motions within it try to accommodate themselves to the Proudman–Taylor theorem, which states that the steady infinitesimal motion of an inviscid fluid is two–dimensional with respect to the direction of the angular velocity of rotation, $\mathbf{\Omega}$. The pattern of convection therefore tends to be in rolls parallel to the axis of rotation. This led to the speculation that large-scale convection occurs in the SCZ in a sequence of rolls, each of which extends from pole to pole, and which together form a cartridge belt pattern round the solar equator [Figure 1(a)]. While the conditions of the Proudman–Taylor theorem are not precisely satisfied, there have been a number of theoretical calculations (e.g. Roberts, 1968; Busse, 1970) and laboratory experiments (e.g. Carrigan and Busse, 1983; Hart *et al.*, 1986) that have confirmed the dominant role of the Proudman-Taylor constraint. Furthermore, numerical experiments, explicitly designed to simulate the SCZ by a spherical fluid shell, have exhibited [for both incompressible (Boussinesq) and compressible shells] clear indications of a cartridge belt pattern at equatorial latitudes. Associated with this pattern there is, at finite convective amplitudes, a zonal motion that is approximately "geostrophic", i.e. approximately constant on cylinders parallel to the rotation axis. The existence of this motion is in contradiction with recent results from solar seismology on the internal rotation of the Sun.

The strong theoretical and experimental expectation that cartridge belt convection exists in the Sun led observers to make a determined attempt to establish, by analysing the Sun's surface motion, that such giant cells actually are present. This search proved to be largely fruitless, and the situation became even more mysterious when Ribes *et al.* (1985) claimed that, superimposed on the granulation and supergranulation seen on the Sun's surface, there exists a pattern of toroidal cells, each of which encircles the Sun, and which together form a sequence that extends from one pole to the other [Figure 1(b)].

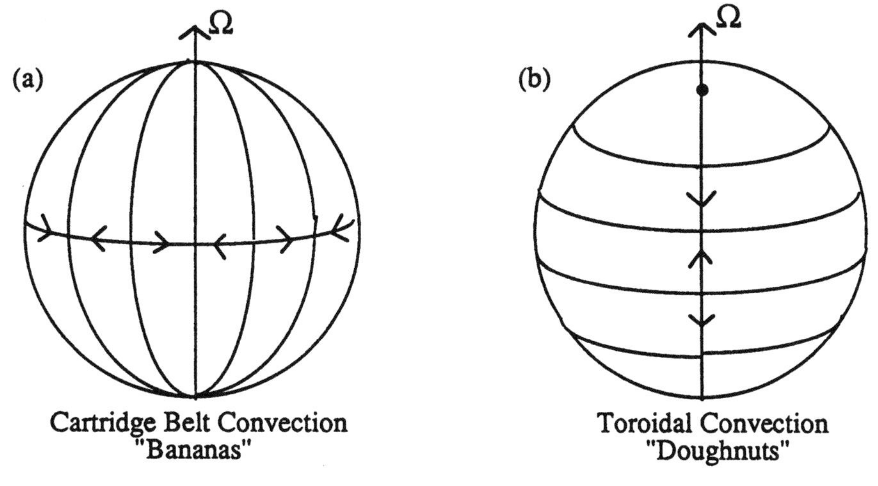

Fig. 1. Types of convection in a sphere.

The idea that initiated the following program of research is very simple. Magnetohydrodynamics ("MHD", for short) boasts its own Proudman–Taylor theorem, which states that steady infinitesimal motions in a perfectly conducting fluid are two-dimensional with respect to the direction of the magnetic field, **B**. When a layer of electrically conducting fluid lies in a magnetic field, slow motions within it try to accommodate themselves to the MHD theorem by forming a pattern of convection in rolls parallel to the magnetic field. Since **B** is predominatly toroidal in the Sun, there would, in the absence of rotation, be a strong tendency for large–scale convection in the SCZ to occur in a sequence of tori of the type envisaged by Ribes *et al.* In short, there are two extremes: if B is "large" and Ω is "small", we expect doughnuts threaded onto the solar axis but, if B is "small" and Ω is "large", we expect bananas girdling the equator. And remarkably a compressible layer like the SCZ can, at one and the same time, possess both tendencies. This may be understood in the following way: The correct dimensionless parameter that quantifies the adjectives "small" and "large" above is the Elsasser number,

$$\Lambda = \frac{\sigma B^2}{2\Omega\rho}, \tag{1.1}$$

where ρ is density, and σ is the magnetic conductivity, which in the Sun is effectively the turbulent conductivity. [In what follows, we shall often use in place of σ the "magnetic diffusivity", $\eta = 1/\mu\sigma$, where μ is the permeability of free space.] When one forms a dimensionless number like Λ, he usually assigns it a single value by inserting "representative" magnitudes of B, σ and ρ. In astrophysical reality, the effective σ depends on position, being large in stably stratified regions where molecular transport coefficients are appropriate, and small in turbulent regions such as the SCZ where the magnetic field is, like the fluid motions, on a very small length scale and therefore ohmically dissipates electromagnetic energy at a greatly enhanced rate. Even more pronounced than the variation of η within the SCZ, is the variation through seven orders of magnitude of ρ. This is so great that, if we temporarily regard Λ as a function of position that depends on the local values of B, σ and ρ, we may well believe that Λ is small at the bottom of the SCZ but is large at its top. In such a case one must expect that the convection "wants" to obey the Proudman-Taylor theorem at the bottom of the SCZ, but wants to encourage toroidal motions at the top! The main objective of the two papers was to explore this idea.

Initially it was unclear how a linear stability analysis could either support, or cast doubt on, our idea. For, in such a theory, all perturbations away from the basic conduction solution are proportional to

$$\exp[im\phi], \tag{1.2}$$

where ϕ is longitude; bananas require $m \neq 0$ while doughnuts are axisymmetric: $m = 0$. But these m are global. One cannot envisage $m \neq 0$ at the bottom of the SCZ and $m = 0$ at the top. So an analysis of the linear problem cannot validate or invalidate our idea. Nevertheless, it is hardly a good theoretical practice to go straight to nonlinear convection before one has understood the marginal situation!

2. On making life bearable

Nearly everyone approaches MHD with that sinking feeling, "Too many parameters!" Every force and every diffusivity seems to have its own dimensionless measure, and soon they create an environmental miasma. Adding rotation makes matters worse, and including compressibility makes them much worse. With this in mind, we resolved to investigate as simple a situation as possible, consistent with retaining the key physical effects identified above. In this section we shall describe briefly the four main simplifications that we shall at various times employ.

First and foremost (simplification 1), we decided to eliminate all geometrical effects by investigating convection in a plane polytropic layer, $z_0 d \leq z \leq (1+z_0)d$, in which the conduction solution (in which the fluid velocity, $\mathbf{V}$, is zero) is

$$T_0 = \beta_0 z, \qquad p_0 = P\left(\frac{z}{d}\right)^{m+1} = \mathcal{R}\rho T, \qquad \rho_0 = \frac{P(m+1)}{gd}\left(\frac{z}{d}\right)^m, \tag{2.1}$$

where p is gas pressure, T is temperature, $\mathbf{g} = g\mathbf{1}_z$ is the effective gravitational acceleration, assumed constant ($\mathbf{1}_q$ denotes the unit vector in the direction of q increasing), $\mathcal{R}$ is the gas constant, and m is the polytropic index; β_0 is the temperature gradient, which is not to be confused with the more significant "superadiabatic" temperature gradient β:

$$\beta_0 = \frac{g}{\mathcal{R}(m+1)}, \qquad \beta = \beta_0 - \frac{g}{c_p} = \beta_0\left(\frac{m+1}{\gamma} - m\right). \tag{2.2}$$

Here γ is the ratio of specific heats (assumed constant), and $c_p = \gamma\mathcal{R}/(\gamma-1)$ is the specific heat at constant pressure. By varying z_0 one can vary the effects of compressibility from the case of the complete polytrope ($z_0 = 0$) to the incompressible – or "Boussinesq" – limit ($z_0 \to \infty$). We concentrated most on the cases $z_0 = 0.1$ and $z_0 \to \infty$. [By supposing that the effective gravitational acceleration, g, is constant, rather than the actual acceleration due to gravity, we have excluded complications arising from the centrifugal force.]

The extra suffix $_0$ on T, p, ρ, ... serves to distinguish values in the conduction state from the corresponding values

$$T = T_0 + T', \qquad p = p_0 + p', \qquad \rho = \rho_0 + \rho', \qquad \ldots, \tag{2.3}$$

of these variables when convection occurs. In the linear convective stability problem, $\mathbf{V}$, T', p', ρ', ... are infinitesimal quantities whose squares and products may be neglected, and which may therefore be divided into non-interacting Fourier modes proportional to [in place of (1.2)]

$$\exp\left[ik\frac{x}{d} + il\frac{y}{d} + (\sigma + i\omega)\frac{\eta t}{d^2}\right], \tag{2.4}$$

where k and l are the horizontal wavenumbers, and $\sigma + i\omega$ is the growth rate; all these constants are real and dimensionless. The use of σ both for growth rate and conductivity should not cause confusion since we shall usually replace the conductivity by the equivalent diffusivity η.

We decided to simulate the region near the equator in the spherical problem, and took therefore both $\mathbf{B}$ and $\mathbf{\Omega}$ horizontal and perpendicular to each other, so that we have

$$\mathbf{\Omega} = \Omega \mathbf{1}_y, \qquad \mathbf{B}_0 = B_0 \mathbf{1}_x, \qquad \mathbf{V}_0 = 0. \tag{2.5}$$

The doughnut cells then correspond to $k = 0$ and the banana cells to $l = 0$. See Figure 2 on which the correspondence between coordinates is also shown.

Simplification 2 is the "anelastic approximation", i.e. the linearization of the thermodynamic relations about a reference state, a procedure that can be defended if $\beta \ll \beta_0$. The technical burden is eased, e.g. the Coriolis force, $2\rho\mathbf{\Omega}\times\mathbf{V}$ is replaced by the simpler $2\rho_0\mathbf{\Omega}\times\mathbf{V}$. Most important is the fact that the mass conservation equation is reduced to

$$\nabla\cdot(\rho_0 \mathbf{V}) = 0. \tag{2.6}$$

As a result, sound waves are "filtered out" of the system; we need only worry about longer time scales.

Simplification 3 is also to do with timescales. If $B = 0$, inertial-gravity waves can propagate in the layer, and we shall suppose not only that $\beta/\beta_0 \ll 1$ (see above) but also that the buoyancy frequency is much less than rotation frequency:

$$N \equiv \left(\frac{g\beta}{d\beta_0}\right)^{1/2} \ll 2\Omega. \tag{2.7}$$

When $\Omega = 0$, the layer can transmit Alfvén waves of frequency V_A/d, where $V_A = B_0/\sqrt{(\mu\rho)}$ is the Alfvén velocity. We shall suppose that

$$V_A/d \ll 2\Omega. \tag{2.8}$$

This leaves the inertial wave at much the same frequency $[O(\Omega)]$ as in the case $B = 0$, but slows the Alfvén wave down to frequencies of order $V_A^2/2\Omega d^2$. These are the timescales of interest in the thermal convection problem, and the fast, inertial waves are filtered out by omitting the inertial acceleration, $D\mathbf{V}/Dt$, in the equation of motion in comparison with the Coriolis acceleration, $2\mathbf{\Omega}\times\mathbf{V}$; here D/Dt is the motional derivative.

Simplification 4 concerns the diffusivities. We shall suppose that

$$E \equiv \frac{\nu_0}{2\Omega d^2} \ll 1, \qquad \frac{\kappa_0}{2\Omega d^2} \ll 1, \qquad \frac{\eta}{2\Omega d^2} \ll 1. \tag{2.9}$$

Here $\nu_0 = \lambda/\rho_m$ and $\kappa_0 = K/\rho_m c_p$ are the kinematic viscosity and thermal diffusivity at the center of the layer; since the dynamic viscosity λ and thermal conductivity K are constants, ν and κ are inversely proportional to ρ_0. Because of (2.7) and (2.8), inequalities (2.9) do not imply that Λ and the relevant Rayleigh number, R, are large, where

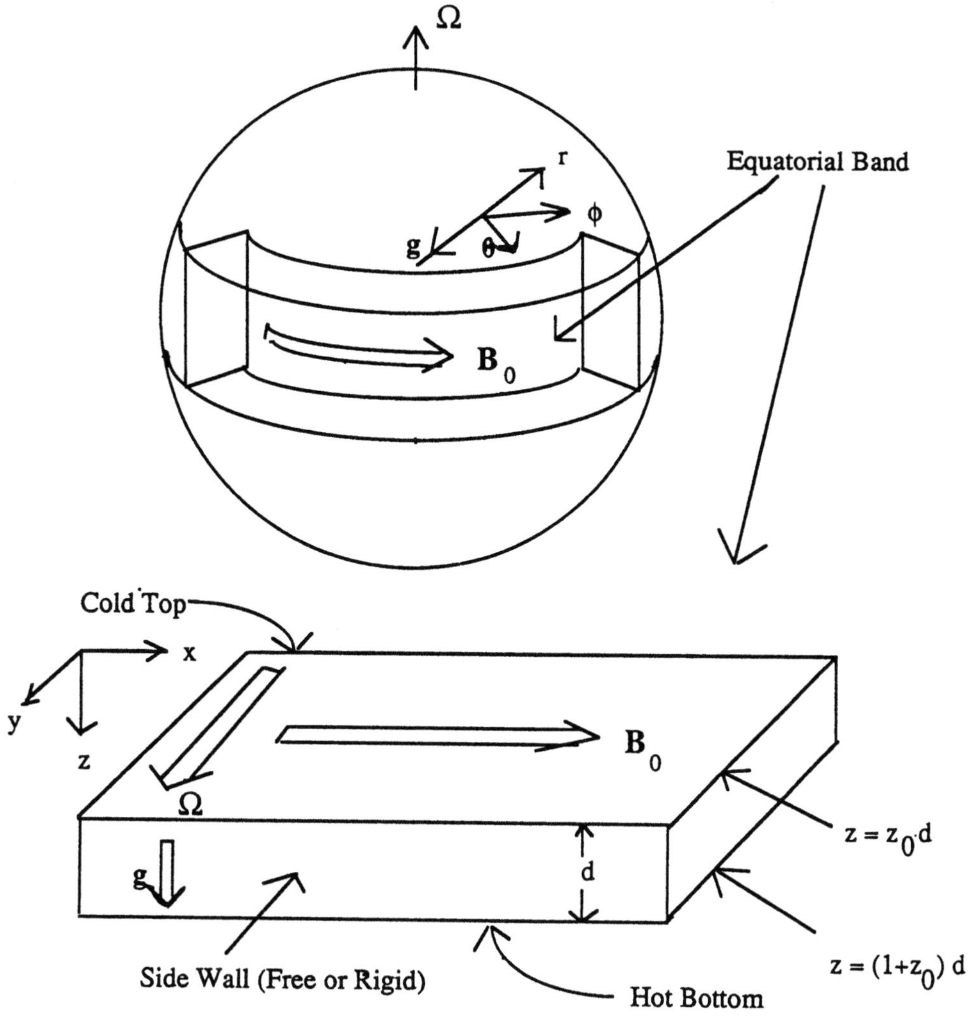

CORRESPONDENCE:

Cartesian Coordinates:	x	y	z
Spherical Coordinates:	ϕ	θ	$-r$
Fields	B_0	Ω	g

Fig. 2. Relationship between spherical and plane geometries

$$\Lambda = \frac{B^2}{2\Omega\mu\rho\eta}, \qquad R = \frac{g\beta d}{2\Omega\beta_0\kappa_0}. \tag{2.10}$$

For we may write Λ and R as O(1) ratios of small quantities:

$$\Lambda = \frac{(V_A/2\Omega d)^2}{\eta/2\Omega d^2}, \qquad R = \frac{(N/2\Omega)^2}{\kappa_0/2\Omega d^2}. \tag{2.11}$$

The main effect of (2.9) is to eliminate viscous forces, except in boundary layers. Not all unpleasantness is removed by this simplification. The polytropic state is a conduction solution only if the thermal conductivity, K, is constant, but the thermal diffusivity, $\kappa = K/\rho c_p$, enters more naturally into the convection theory. By simpification 2, this may be replaced by $\kappa_0 = K/\rho_0 c_p$, a known function of z. This kind of nuisance is unavoidable, but is largely irrelevant to the main topic.

We conclude this section with a summary of the basic equations that result from our simplifications; we write them in dimensionless form, and replace z by

$$\zeta = z/(z_0 + \tfrac{1}{2})d. \tag{2.12}$$

We then have

$$\nabla\cdot(\zeta^m \mathbf{V}) = 0, \tag{2.13}$$

$$\zeta^m \mathbf{1}_y \times \mathbf{V} = -\frac{(z_0+\frac{1}{2})}{m+1}\nabla p + \rho \mathbf{1}_z + \Lambda(\nabla\times\mathbf{B})\times(\mathbf{1}_z + \mathbf{B}) + E[\nabla^2\mathbf{V} + \tfrac{1}{3}\nabla(\nabla\cdot\mathbf{V})], \tag{2.14}$$

$$\frac{\partial \mathbf{B}}{\partial t} = \nabla\times[\mathbf{V}\times(\mathbf{1}_z + \mathbf{B}) - \nabla\times\mathbf{B}], \tag{2.15}$$

$$\frac{\zeta^m}{q\gamma}\frac{DT}{Dt} = \frac{(\gamma-1)}{q\gamma}\zeta\frac{D\rho}{Dt} - [R\zeta^m + \frac{1}{\gamma q(z_0+\frac{1}{2})}\{\rho - m(\gamma-1)\zeta^{m-1}T\}]V_z + \nabla^2 T, \tag{2.16}$$

$$p = \zeta^m T + \zeta\rho, \tag{2.17}$$

$$\nabla\cdot\mathbf{B} = 0, \tag{2.18}$$

where $q = \kappa_0/\eta$ is the ratio of diffusivities. The primes have been omitted from T', p', ρ', etc; the motional derivatives are

$$\frac{DT}{Dt} = \frac{\partial T}{\partial t} + \mathbf{V}\cdot\nabla(T_0 + T), \qquad \frac{D\rho}{Dt} = \frac{\partial \rho}{\partial t} + \mathbf{V}\cdot\nabla(\rho_0 + \rho).$$

Clearly (2.13) is the mass conservation equation (2.6), while (2.14) expresses momentum conservation; (2.15) is the magnetic induction equation, (2.16) is the heat conduction equation, (2.17) is the gas law and (2.18) expresses conservation of magnetic flux.

In the Boussinesq limit (2.13)–(2.18) reduce to the much simpler form:

$$\nabla\cdot\mathbf{V} = 0, \tag{2.19}$$

$$\mathbf{1}_y\times\mathbf{V} = -\nabla p + T\mathbf{1}_z + \Lambda(\nabla\times\mathbf{B})\times(\mathbf{1}_z + \mathbf{B}) + E\nabla^2\mathbf{V}, \tag{2.20}$$

$$\frac{\partial \mathbf{B}}{\partial t} = \nabla\times[\mathbf{V}\times(\mathbf{1}_z + \mathbf{B})] + \nabla^2\mathbf{B}, \tag{2.21}$$

$$\frac{1}{q}\frac{DT}{Dt} = \nabla^2 T - RV_z, \tag{2.22}$$

$$\nabla\cdot\mathbf{B} = 0. \tag{2.23}$$

3. The impossibility of steady modes

The elimination of steady modes is by *reductio ad absurdum*, i.e. we shall suppose that ω in (2.4) is zero. More precisely, we may suppose without loss of generality that solutions exist in which V_z is proportional to $\cos(kx + ly) \exp \sigma t$, where k, l and σ are real.

The linearized form of the governing equations (2.13)–(2.18) are, with E set zero [see (2.9)],

$$\nabla\cdot(\zeta^m \mathbf{V}) = 0, \tag{3.1}$$

$$\zeta^m \mathbf{1}_y \times \mathbf{V} = -\frac{(z_0 + \frac{1}{2})}{m+1} \nabla p + \rho \mathbf{1}_z + \Lambda(\nabla\times\mathbf{B})\times\mathbf{1}_x, \tag{3.2}$$

$$\frac{\partial \mathbf{B}}{\partial t} = \frac{\partial \mathbf{V}}{\partial x} - (\nabla\cdot\mathbf{V})\mathbf{1}_x + \nabla^2 \mathbf{B}, \tag{3.3}$$

$$\frac{\zeta^m}{q\gamma}\frac{\partial T}{\partial t} = \frac{(\gamma-1)}{q\gamma}\zeta\frac{\partial \rho}{\partial t} - R_c \zeta^m V_z + \nabla^2 T, \tag{3.4}$$

$$p = \zeta^m T + \zeta\rho, \tag{3.5}$$

$$\nabla\cdot\mathbf{B} = 0, \tag{3.6}$$

R_c being the critical Rayleigh number for marginal convection.

In steady convection, T and $-V_z$ are in phase, as (3.4) shows: since V_z is proportional to $\cos(kx+ly)\exp\sigma t$, so is $-T$. In the Boussinesq approximation, ρ depends only on T, so that ρ and V_z are in phase. The maxima of V_z in any horizontal plane mark the "centers" of the cold, heavy, descending streams; the minima correspond to the centers of the hot, light, ascending plumes. The buoyancy force $\rho\mathbf{1}_z$ has the right phase to maintain that situation.

In a rapidly rotating fluid, p and V_z are out of phase as the x-component of (3.2), namely

$$\frac{\partial p}{\partial x} = -\frac{m+1}{z_0 + \frac{1}{2}} \zeta^m V_z, \tag{3.7}$$

shows: p is proportional to $\sin(kx + ly)\exp\sigma t$ rather than to $\cos(kx + ly)\exp\sigma t$ as V_z is; this is an instance of what is sometimes called "Buys Ballot's law". The fact that p is out of phase with T is irrelevant in a Boussinesq fluid, since ρ depends only on T, and steady convection is still possible with ρ and V_z in phase, as described above. In a compressible fluid however, ρ depends both on T *and* on p; see (3.5). Thus ρ and V_z can no longer be in phase. The centers of maximum negative buoyancy no longer coincide with the maximum downward velocities. The convection pattern therefore moves, as the downward motion tries to "find" the density maxima, but creates new, translated density maxima as it does so. This result, the first surprise we encountered, does not depend on simplification 3; it is equally true for nonmagnetic compressible convection. Nevertheless, we found that the general features of the compressible modes were often not qualitatively very different from those of the corresponding Boussinesq modes, except that the former drift while the latter do not.

Our numerical results did not encourage us to believe that our original idea, that a transition from doughnut to banana convection arises through the variation with depth of ρ_0, and hence of the effective Λ, is correct. The situation might have been different had the most unstable modes been *multicellular*, rather than "unicellular", i.e. if there had been many rather than one layer of cells between the top and bottom of the layer, each of which had some degree of independence from its neighbors. Although the idea that convection preserves its coherence only over one scale height is basic to mixing-length theories of turbulent stellar convection, our marginal laminar modes appeared to show no such tendency, a fact already noted in a number of previous studies of compressible convection. Eigenfunctions are shown in Figures 3 in two cases ($\Lambda = 1$, $l = 1$, $k = 2.5$ top; $\Lambda = 16$, $l = 2$, $k = 1$ bottom), in both of which the ratio of the density at the bottom of the layer to that at the top is about 36.5, but in each of which the eigenfunctions extend right across the layer.

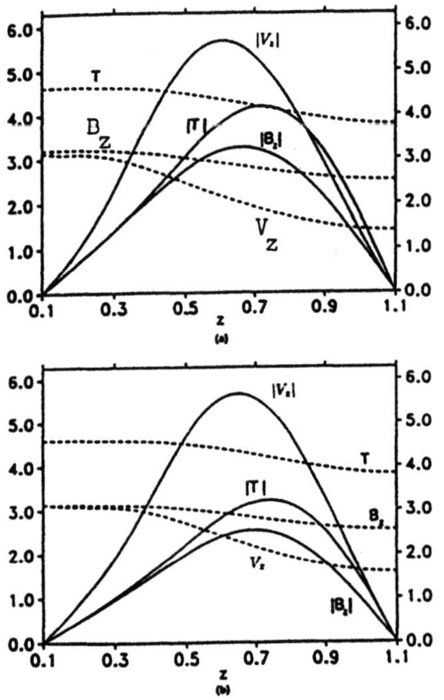

Fig. 3. One zone model

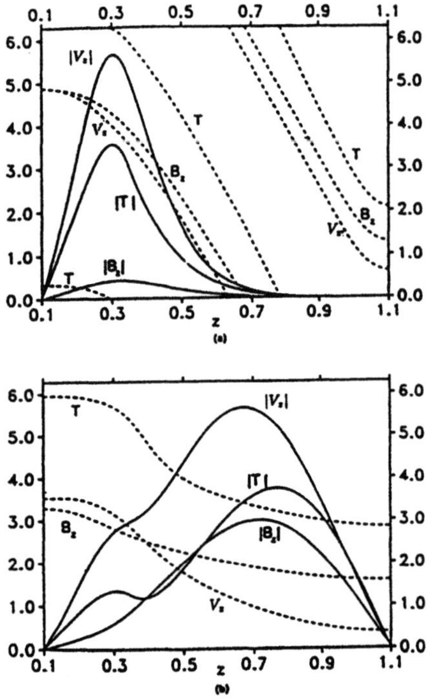

Fig. 4. Two zone model

One interesting feature emerged from the calculations however: although the phases (shown by the dashed curves) of $\boldsymbol{V}_z$ and T differ by about π at depth, as happens in the "normal" Boussinesq case (see above), they are out of phase by only $\frac{1}{2}\pi$ at the top of the layer. This suggested that, when convection is of finite amplitude convection even if that amplitude is small, the convective heat flux will be impeded near the top of the layer, so that a large mean superadiabatic temperature gradient will develop to carry heat out of the layer. With this thought in mind, we studied a two-zone model in which the effective Rayleigh number in the upper quarter of the layer ($z_0 < z < z_0 + \frac{1}{4}$) is ten times that in the lower three quarters ($z_0 + \frac{1}{4} < z < 1 + z_0$). We brought this about by a small difference in γ between the two zones; see (2.2) and (2.10). Two kinds of eigenmodes were found, one concentrated mostly in the upper 25% of the layer, the other in the lower 75%; they are sketched in Figures 4. The ratio, k/l, at criticality is less for the upper mode than for the lower, as one would expect from its greater effective Elsasser number. The effect is not as great as one might have anticipated however: $k/l \approx 1.5$ for the upper mode and ≈ 1 for the lower, but the effective Elsasser number at the center of the upper layer is almost four times its overall value $\Lambda = 4$. For each l, the lower modes are more easily for small k, while the minimum $R(= R_m(l)$, say) for that l occurs for the upper mode at higher k; $R_m(l)$ increases with l. Smallest are $R_m(3)$ and $R_m(4)$, both of which are close to 82.

The results of the two zone model encouraged us more than those of the one zone model. The hope expressed above, that the upper layer might act independently of the lower layer, may be a vain one [see the discussion at (1.2)], at least in infinitesimal convection. But, by introducing heuristically into the linear stability problem an effect that is plausibly present at finite amplitude, namely an increase in the superadiabatic temperature gradient with height, we obtained some indications that the upper layer "prefers" doughnut structures and the lower layer banana modes. There seemed to be little doubt that, to investigate this properly, we had to investigate finite amplitude convection. Further surprises were in store, as we shall now describe.

4. Finite amplitude Boussinesq convection

Our second surprise arose from what is often called the "Taylor constraint". Consider the Boussinesq case, in the weakly nonlinear range, in which $|R/R_c - 1| \ll 0$. In the limit $E \to 0$, (2.20) becomes

$$\mathbf{1}_y \times \mathbf{V} = -\nabla p + T\mathbf{1}_z + \Lambda(\nabla \times \mathbf{B}) \times (\mathbf{1}_x + \mathbf{B}), \tag{4.1}$$

from which Taylor's constraint follows:

$$\mathbf{F}(z,t) = 0, \tag{4.2}$$

where

$$\mathbf{F}(z,t) \equiv \Lambda\overline{[(\nabla \times \mathbf{B}) \times \mathbf{B}]_H}; \tag{4.3}$$

the overbar denotes the horizontal average (i.e. the average over x and y) and the $_H$ denotes horizontal components. To leading order for small $|R/R_c - 1|$, we may replace $\mathbf{B}$ in (4:3) by the eigenfunction determined from the linear eigenvalue system derived from (2.19)–(2.23) for $R = R_c$. And it is found immediately that

$$\mathbf{F}(z,t) \neq 0, \tag{4.4}$$

in contradiction with (4.2).

This kind of difficulty can hardly be described as the next surprise. It was encountered first by Taylor (1963) and is now a commonplace in the the theory of highly rotating fluids. Inequality (4.4) states that the Lorentz force contains a component that cannot be balanced by Coriolis forces, buoyancy forces or pressure gradients. It must therefore accelerate the fluid to create a growing horizontal motion that is independent of x and y, a so-called "geostrophic flow":

$$\mathbf{U} = \mathbf{U}_H(z,t). \tag{4.5}$$

When this velocity becomes sufficiently large, it will be incorrect to treat it as part of a perturbation series; instead, it will have to be included in the zero order state, i.e. the last of (2.5) will have to be amended to $\mathbf{V}_0 = \mathbf{U}$, with the result that many new terms will have to be incorporated into (2.20)–(2.22). In particular $\mathbf{U}$ will alter the magnetic field and the Lorentz force. Taylor (1963) argued that the geostrophic flow would quickly adjust itself until the new magnetic field satisfied (4.2), and he gave a prescription that thereafter uniquely determined a $\mathbf{U}$ such that the new $\mathbf{B}$, as it evolved, would continuously satisfy (4.2). This defines the "Taylor state", in which the geostrophic flow is "O(1)" (by which we mean it is independent of ν, see below) but in which (4.2) is obeyed.

Taylor's suggestion has been welcomed widely, as being both a mathematically and a physically appealing resolution of the conflict between (4.2) and (4.4). We too expected it to apply to the present problem but (and here comes the second surprise!) we found that Taylor's idea failed: Taylor states do not exist when $\Lambda > 4.080\ldots$. This is illustrated in Figure 5. The lower part of the diagram shows the results of the original linear stability problem, with $\mathbf{U}$ absent. For small Λ, banana cells are excited first at R_c but, as Λ increases, the preferred cells become "oblique", i.e. neither k nor l are zero. These two branches of the stability curve are continued as dashed lines when they are no longer preferred. Also shown is the curve, labeled R_T, defined by the Taylor states. Evidently none such exist when Λ exceeds 4 (approximately).

It was first pointed out by Roberts and Stewartson (1975), and independently by Braginsky (1975), that the conflict between (4.2) and (4.4) might be resolved by a mechanism different from Taylor's, through the term omitted in writing down (4.1): the viscosity ν.[1] It is at once clear that, since E is small, the viscous stress can be made potent only if the velocity shear created by the unbalanced $\mathbf{F}$ is large. This shear can be predominantly in a boundary layer at a wall (see below) or distributed throughout the fluid, the possibility we examine first. But we quickly find that, in either case, only the x-component of the geostrophic flow, that aligned with $\mathbf{B}_0$, arises:

$$\mathbf{U} = U(z,t)\mathbf{1}_x. \tag{4.6}$$

The horizontally averaged (2.20) gives

[1] Braginsky also pointed out that, if the walls were electrically conducting, an electromagnetic coupling with the walls would also be present, which could resolve the difficulty. Here, for simplicity, we shall confine attention to the simpler mechanism that relies on viscosity.

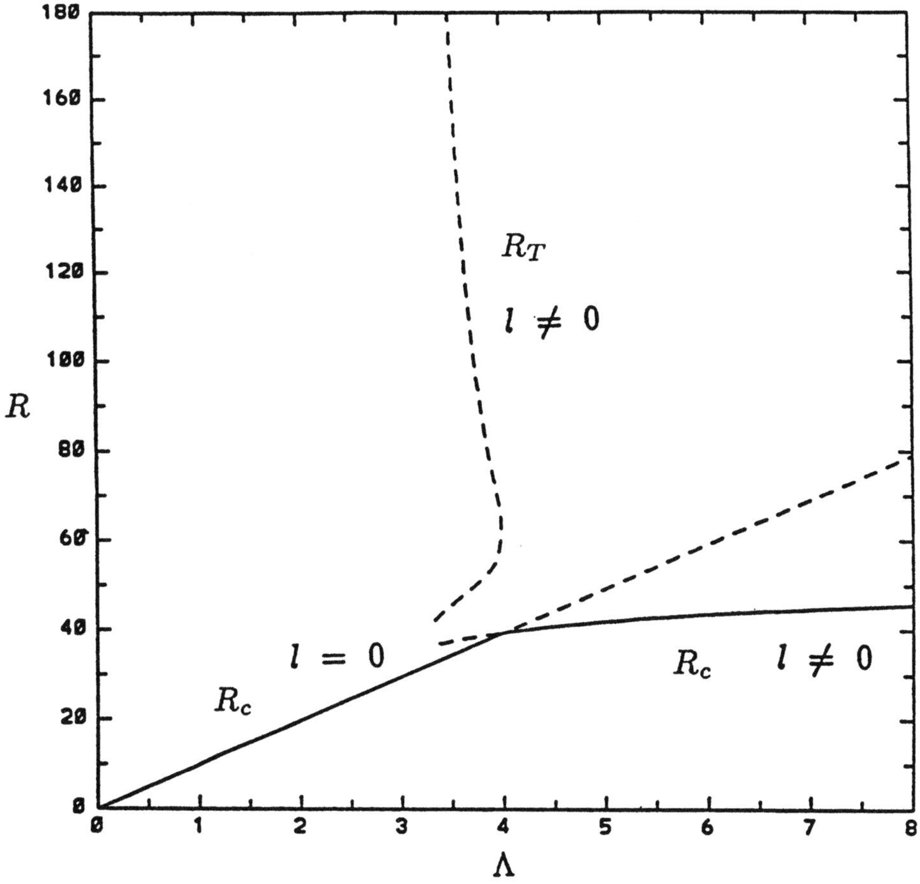

Fig. 5. Marginal and Taylor states.

$$0 = \Lambda\overline{[(\nabla\times\mathbf{B})\times\mathbf{B}]}_H + E\nabla^2\mathbf{U}, \tag{4.7}$$

and this offers one way of balancing **F**: (2.3) can be scaled as

$$T \to T_0 + E^{1/2}T, \quad p \to p_0 + E^{1/2}p, \quad \mathbf{V} \to \mathbf{U} + E^{1/2}\mathbf{V}, \quad \mathbf{B} \to \mathbf{B}_0 + E^{1/2}\mathbf{B}. \tag{4.8}$$

We then obtain what has sometimes been described as a "nonlinear eigenvalue problem", that of solving the linear equations

$$\nabla\cdot\mathbf{V} = 0, \tag{4.9}$$

$$\mathbf{1}_y\times\mathbf{V} = -\nabla p + T\mathbf{1}_z + \Lambda(\nabla\times\mathbf{B})\times\mathbf{1}_x, \tag{4.10}$$

$$\frac{\partial\mathbf{B}}{\partial t} + \mathbf{U}\cdot\nabla\mathbf{B} = \mathbf{B}\cdot\nabla\mathbf{U} + \nabla\times(\mathbf{V}\times\mathbf{1}_x) + \nabla^2\mathbf{B}, \tag{4.11}$$

$$\frac{1}{q}\left(\frac{\partial T}{\partial t} + \mathbf{U}\cdot\nabla T\right) = \nabla^2 T - RV_z, \tag{4.12}$$

$$\nabla\cdot\mathbf{B} = 0, \tag{4.13}$$

in such a way that the nonlinear restriction,

$$0 = \Lambda\overline{[(\nabla\times\mathbf{B})\times\mathbf{B}]_x} + \nabla^2 U, \tag{4.14}$$

is met. The solutions to (4.9)–(4.13) must satisfy the appropriate conditions on the upper and lower boundaries, as must the solution of (4.14) for U. As for the previous linear eigenvalue problem (the one in which $\mathbf{U}$ was assumed zero), the oblique modes $k \neq 0$ come in pairs at equal angles to $\mathbf{B}_0$, differing by the sign of l.

There is a second way in which viscosity can act, namely through Ekman layers. These exist at any point on a no–slip boundary at which $\boldsymbol{\Omega}$ has a nonzero normal component; in the geometry of Figure 2, this means that they do not exist on the horizontal boundaries, but only on "side walls". We supposed that these are impermeable, thermally insulating, and perfect electrical conductors. It is then necessary to assume that the two oblique rolls have equal amplitudes, so that (2.4) is replaced by (for example)

$$B_z(x,y,z,t) = ik\hat{B}(z)\exp[i(kx+\omega t)]\cos ly + *, \tag{4.15}$$

where the * denotes the complex conjugate of the expression preceding it and now $l > 0$. (Other variables are expanded similarly, but without the prefactor ik that B_z has, which proves to be convenient.) These Fourier modes satisfy all conditions on the side walls if their separation, L, is $n\pi/l$, where n is an integer, and we make the simplest choice: $n = 1$, i.e. the side walls are at $y = 0$ and $y = \pi/l$. Our infinite plane layer has become a "duct"; in terms of the original spherical model, we are confining the fluid between two particular latitudes.

When $\mathbf{U} = \mathbf{0}$, overstable modes ($\omega \neq 0$) are preferred only when $q > 1$. We did not concern ourselves with such q; in fact we usually assumed that $q = 1$. We concentrated on the case $\Lambda = 6$ and we here adopt the values of k and l found in Paper 1 to be the most unstable: $k = 1.117$ and $l = 1.668$. [2] If these walls are slippery, the solution is governed by (4.9)–(4.14), but when they are no-slip surfaces everything changes.

In the case of no–slip side walls, all the shear required to balance $\mathbf{F}$ takes place in Ekman layers on those walls. These shears are, at level z, of order $U(z)/\delta_\nu$, where $\delta_\nu = (\nu/\Omega)^{1/2}$ is the Ekman layer thickness. They create a stress of order $2\rho_0(\Omega\nu)^{1/2}U(z)$ which is required to offset the net unbalanced force, LF (per unit x-length), on the entire fluid between the walls at that level. Thus U must be of order $(L/\rho_0)(\Omega\nu)^{-1/2}F = \mathrm{O}((\mathrm{L}/\mu\rho_0\mathrm{d})(\Omega\nu)^{-1/2}\mathrm{B}^2)$. In fact, a more precise calculation shows that

$$U = \frac{L}{2\rho_0\mu(\Omega\nu)^{1/2}}\overline{[(\nabla\times\mathbf{B})\times\mathbf{B}]_x}. \tag{4.16}$$

[2] It was realized rather late that, even though these are direct when $\mathbf{U} = \mathbf{0}$, this may not be true when $\mathbf{U} \neq \mathbf{0}$. The diagrams given below were prepared before we made the slightly different choice of paper 2, namely $k = 1.6091$, $l = \pi$, one for which direct instability is assured. The structure of the solutions is qualitatively the same.

From this $B^2/\nu^{1/2}$ dependence, we see that, in our dimensionless units, B must be $O(E^{1/4})$, so that scaling (4.8) must be replaced by

$$T \to T_0 + E^{1/4}T,\ \ p \to p_0 + E^{1/4}p,\ \ \mathbf{V} \to \mathbf{U} + E^{1/4}\mathbf{V},\ \ \mathbf{B} \to \mathbf{B}_0 + E^{1/4}\mathbf{B}. \quad (4.17)$$

Equations (4.9)–(4.13) hold as before; merely the dimensionless form of (4.16) replaces (4.14), but (the third surprise) this has an enormous, even *qualitative*, effect on the solutions!

Consider first the case of no-slip side walls, "the rigid duct", as we call it. When R is large, the solutions are multicellular, i.e. any vertical line across the duct meets several different cells, each operating in a different band of z. This may be seen with particular clarity in Figure 6(b) where the phases of B and T are shown. They are are almost coincident throughout the layer; that of V_z is too close to that of T to be included. The presence of many cells causes the geostrophic velocity to reverse its direction equally often with height, as is seen in Figure 6(a). The amplitudes of the perturbation variables oscillate similarly. The existence of multicellular solutions is encouraging for the main objective of the present line of study, that of finding convective modes that are banana–like at the bottom but doughnut–like at the top. But this is, of course, prevented here by the symmetry of the Boussinesq solutions about the mid–plane $z = \frac{1}{2}$.

The "free duct", as we call the duct with slippery side walls, shows no such behavior. Instead it exhibits a symmetry breaking that the rigid duct apparently lacks. As R is increased from zero, the duct first becomes unstable (when $\Lambda = 6$) for $R \approx 43.637$ with, as noted earlier, $k \approx 1.117$ and $l \approx 1.668$; this point is denoted by B_1 in Figure 7. Each roll extends from one z boundary to the other, and is completely symmetric, e.g. $V_z(1-z) \equiv V_z(z)$; all other variables are similarly symmetric or antisymmetric with respect to the mid–plane of the layer. As R is increased further, this structure is maintained by the finite amplitude convection until a second critical value of R is attained. At this point, two time-dependent solutions arise, neither of which is symmetric with respect to $z = \frac{1}{2}$, but which preserve an overall symmetry in the sense that one is the reflection of the other in the mid–plane, e.g. $V^{(1)}(1-z) \equiv V^{(2)*}(z)$. These are represented in Figure 7 by the two curves emerging from B_3 and continuing to the right. Convection occurs primarily in the upper half of the layer in one solution, and primarily in the lower half for the other. This is encouraging for the main objective of the present line of study, that of explaining banana–like convection at the bottom of the layer but doughnut–like convection at the top.

The two solutions are time-dependent, and travel in the $\pm x$-directions with equal but opposite phase velocities. After undergoing this symmetry–breaking bifurcation, denoted by B_3 in Figure 7, the original symmetric solution undergoes a second bifurcation of saddle-node type, corresponding to B_4 in Figure 7, where it links to the branch of finite amplitude solutions that emerges from the second critical Rayleigh number. This point, denoted by B_2 in Figure 7, corresponds to bicellular convection of infinitesimal amplitude, i.e. convection in which there are two layers of rolls between the horizontal boundaries.

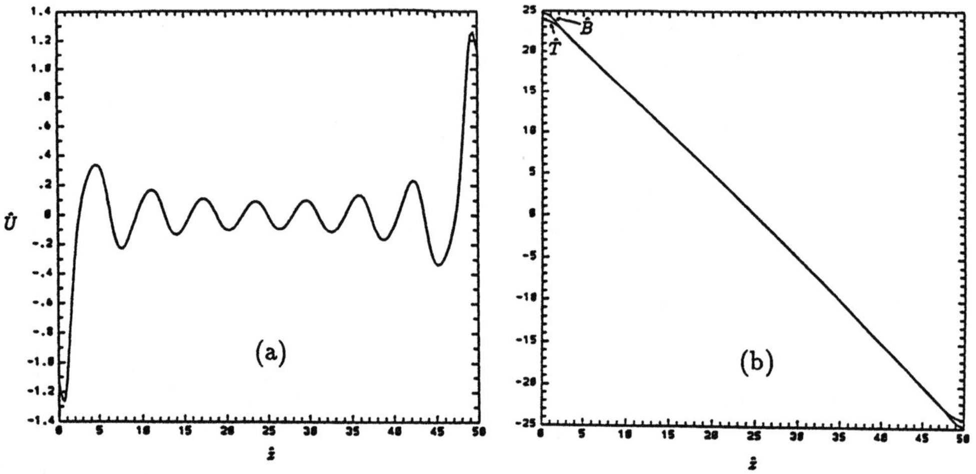

Fig. 6. Rigid Boussinesq duct (a) geostrophic flow, (b) phases of B and T. The variables carry a hat ˆ to indicate that they have been scaled; see Paper 2 for details.

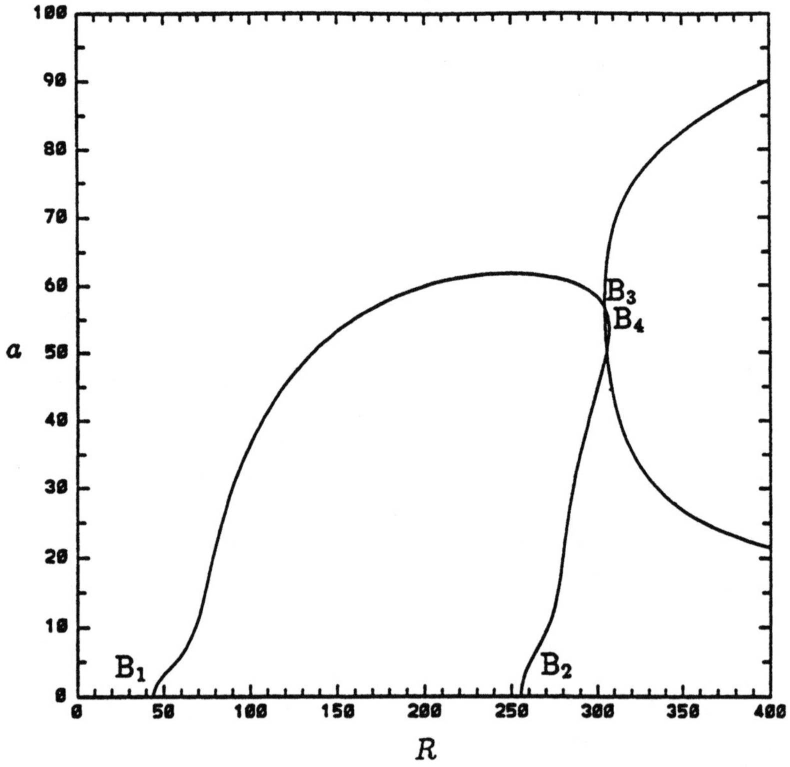

Fig. 7. Bifurcation diagram for free Boussinesq duct. The amplitude parameter a is constructed in such a way that the points representing solutions that are symmetric about $z = \frac{1}{2}$ coincide, but those which lack that symmetry are represented by different points; see Paper 2 for details.

Figures 8 give an idea of the structure of one of the two solutions when R is large. The solution hugs a boundary, in this case $z = 0$, and is of boundary layer type; ω is such that the Doppler–shifted frequency, $\omega + kU$, is nearly zero in this layer.

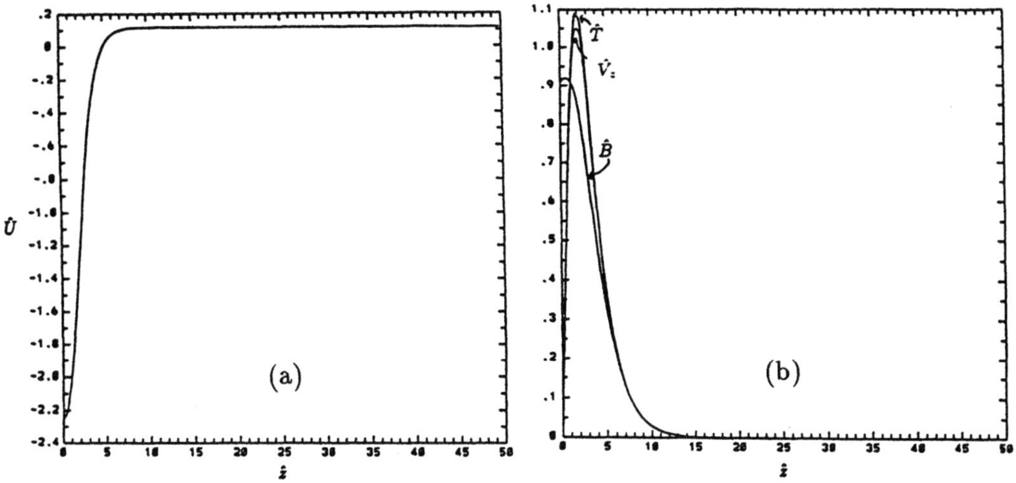

Fig. 8. Free Boussinesq duct (a) geostrophic flow, (b) amplitudes of V_z, B and T. The axes are scaled but in a different way from Figure 6; see paper 2 for details.

5. Finite amplitude compressible convection

No qualitatively new feature emerged when we studied compressible convection in a free duct. The symmetry of the Boussinesq model does not exist in a compressible fluid and, as we saw in Section 3, the convection is necessarily time-dependent. In Paper 1 we found nevertheless that the eigenfuctions of compressible convection are not totally dissimilar from those of the Boussinesq model. It is therefore not altogether surprising that, when the symmetry breaking bifurcation unfolds, one of the two branches that emerged from that pitchfork bifurcation joins smoothly to the linear unicellular solution arising at the smallest critical Rayleigh number, while the other joins to the second smallest, corresponding to bicellular convection. This is most clearly seen by the curves having alternate long and short dashes in Figure 9, which corresponds to the slightly compressible case $z_0 = 1.0$; here β_1 and β_2 correspond to the B_1 and B_2 of the Boussinesq case (full curves). It will be seen that, as R increases, one branch runs roughly parallel to each of the two branches of non-symmetric Boussinesq solutions. This is also true for the more compressible layer, $z_0 = 0.1$, denoted by the dashed curves in Figure 9, and in which b_1 and b_2 corresponds to B_1 and B_2 in the Boussinesq case. The "primary" branches, emerging from β_1 and b_1, are unicellular throughout, and correspond to convection occurring primarily in the lower part of the layer. The "secondary" branches, emerging from β_2 and b_2 are initially bicellular but, as R increases, the upper cells intensify while the lower ones evanesce, until at large R the solution is essentially unicellular, as in the Boussinesq case.

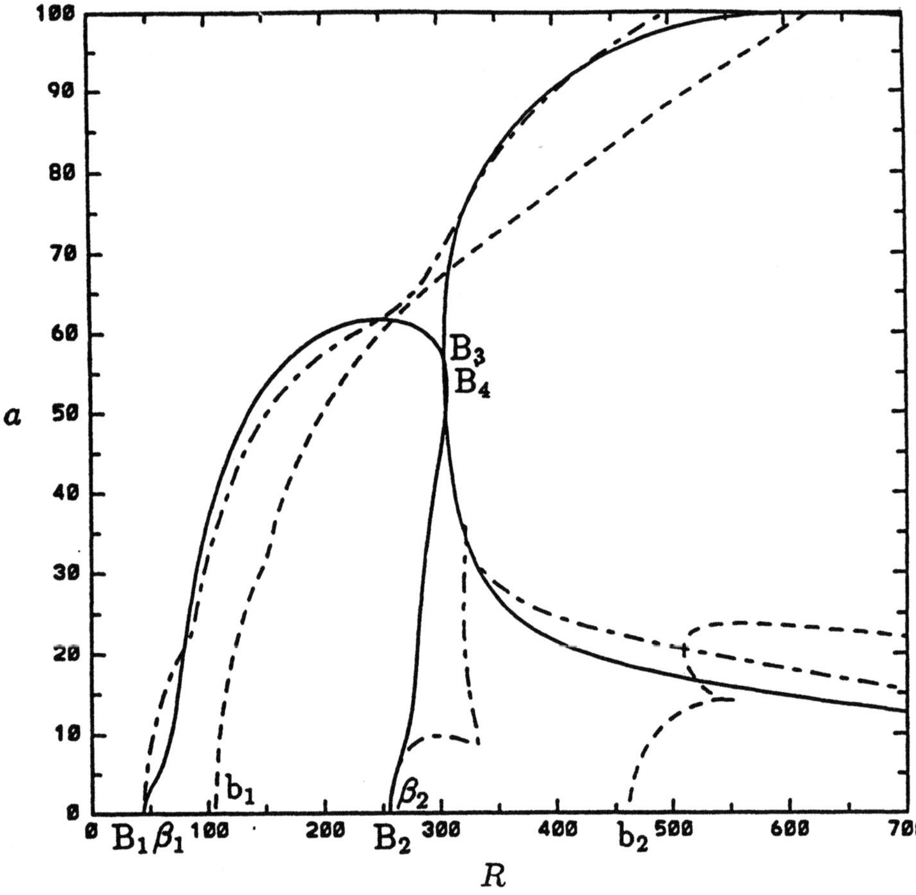

Fig. 9. Bifurcation diagram for free compressible duct.

6. Inconclusions

The present line of investigation, initially set in motion by the claims of Ribes and others (see her paper in this volume), has certainly not run along anticipated lines. There have been three major surprises:

1: Compressible convection cannot be steady in a rotating fluid;

2: Taylor's hypothesis fails;

3: Conditions at distant boundaries have profound effects on the entire flow structure.

The first surprise was pleasant in one sense, since we believe it represents a modest discovery, the first time that the impossibility of steady convection in a rotating compressible fluid has been noted and explained. In another sense it was an unpleasant surprise, since time-dependent convection is more troublesome to analyze than steady convection.

The second surprise is of interest particularly to those who have partaken in current research on the theory of the geodynamo, in which the role of Taylor's constraint is central. Currently the preferred approach to the MHD of this highly rotating system is to introduce a coupling between the Earth's core and its mantle. This guarantees a solution for the MHD state in all circumstances, one in which the geostrophic flow is in general large, e.g. in the case of viscous coupling it is proportional to $\nu^{-1/4}$, and ν is very small in the core ($E \ll 1$); see Section 4 above. This approach does not rule out Taylor states, which may occur naturally as solutions in which the geostrophic flow is not large, being independent of the assumed core-mantle coupling (e.g. independent of ν when the coupling is viscous), but it avoids the pitfall of seeking Taylor states in situations in which, perhaps, none exists. Until now, the discovery of a Taylor state has always been fortuitous; no general rules have been evolved that guarantee its existence or non-existence. Paper 2 represents the first exception to that situation: "anti-Taylor inequalities" were developed that ruled out Taylor states when Λ is too small.

The third surprise was something of a body blow to any like ourselves who hoped to use very simple models, ones that exclude all unnecessary mechanisms, to explain complex MHD behaviors. Amongst the idealizations has been our simplification 1: the elimination of geometrical effects, by supposing that the working fluid occupied a plane layer (later a duct), and by assuming that gravity and the prevailing magnetic field are uniform. In defence of this step, now clearly seen to have been too optimistic, we should point to the studies of the linear stability by Eltayeb and Roberts (1970) and Eltayeb (1972, 1975), which gave results qualitatively insensitive to the relative orientations of $\mathbf{g}$, $\boldsymbol{\Omega}$ and $\mathbf{B}_0$, and which initially encouraged us to believe that geometrical considerations would not be crucial. The discovery of the anti-Taylor inequalities for the present situation, was already a foretaste of an impending disappointment, for it demonstrates that our model behaves at finite amplitude quite differently from the plane layer Boussinesq model in which rotation is parallel to gravity, a configuration extensively studied by Roberts and Stewartson (1974, 1975), Soward (1980, 1986) and Skinner and Soward (1989), and in which Taylor states occur.

Still worse was in store however. Even within the framework of our model, the nature of the distant boundaries has repercussions on the entire MHD state. When the side walls of our duct are no-slip surfaces, the convection becomes increasingly multicellular as the Rayleigh number is increased; if they are stress-free surfaces, the symmetry of the solution is destroyed once R becomes large enough. These are drastic differences, greater than one might have anticipated even with prior experience behind one of the behavior of highly rotating systems, including their universal tendency towards two-dimensionality with respect to the direction of $\boldsymbol{\Omega}$, a tendency epitomised by the celebrated Proudman-Taylor theorem.

By calling the third surprise a "body blow", and describing it in such pessimistic terms, we have in mind the original aims of Papers 1 and 2, that of providing firm theoretical backing for the claims of Ribes and others. As a study of rotating MHD, the investigation has proved to be very exciting, the behaviors of the rigid and free ducts being especially fascinating. So much has emerged from the study that, speaking for myself (and I feel sure for my fellow authors too), the required time and effort have been well spent. It is only when one begins to question how far it has achieved its stated aim that some disappointment creeps in. The behavior of the two zone model described in Section 3 does contain more than a hint that our basic idea is correct, namely that, for the reasons expounded in Section 1, strong convection in a compressible MHD layer will, in the right range of Elsasser number, be of doughnut form near the top of the layer while being banana–like at depth. The existence of two modes in the free duct described in Section 4 is at first encouraging also, but one's excitement diminishes somewhat when (Section 5) he finds that it is the bottom mode that connects to the primary branch of solutions, the one emerging from β_1 and B_1 in Figure 9.

With hindsight it seems that simplification 1 was too drastic, and has taken us too far away from the real Sun. What should the next step be? Perhaps to investigate strong convection in a spherical model, in which the imposed $\mathbf{B}_0$ is zonal? Hopefully, the behavior of such a model will, at moderate R, resemble that of our free duct. (The rigid duct would represent better a latitude band of the spherical system where the flow is brought to rest at the confining edges of the band, a situation remote from the solar application.) We look to the future with keen anticipation!

Acknowledgements. As explained in Section 1, this paper is largely a report on work undertaken by the author jointly in collaboration with Dr. C.A. Jones. That collaboration would have been impossible without the support of the National Science Foundation of the United States, who brought us together by awarding grant DMS88–22287, for which we offer our thanks. I am also grateful to the U.S. Office of Naval Research for support under grant N00014–86–K–0641. Dr. Jones kindly criticised this paper while it was still at an early stage.

References

Braginsky, S.I.: 1975, *Geomag. Aeron.* **15**, 122
Busse, F.H.: 1970, *Astrophys. J.* **159**, 629
Carrigan, C.R., Busse, F.H.: 1983, *J. Fluid Mech.* **126**, 287
Eltayeb, I.A.: 1972, *Proc. R. Soc. Lond. A* **326**, 229
Eltayeb, I.A.: 1975, *J. Fluid Mech.* **71**, 161
Eltayeb, I.A., Roberts, P.H.: 1970, *Astrophys. J.* **162**, 699
Jones, C.A., Roberts, P.H., Galloway, D.J.: 1990, *Geophys. Astrophys. Fluid Dynam.* **53**, 145. This paper was referred to as 'Paper 1' above.
Jones, C.A., Roberts, P.H.: 1990, *Geophys. Astrophys. Fluid Dynam.*, to appear (1990). This paper was referred to as 'Paper 2' above.
Hart, J.E., Toomre, J. Deane, A. E., Hurlburt, N. E., Glatzmaier, G. A., Fichtel, G. H., Leslie, F., Fowlis, W. W., Gilman, P. A.: 1986, *Science* **234**, 61
Ribes, E., Mein, P., Mangeney, A.: 1985, *Nature* **318**, 170
Roberts, P.H.: 1968, *Phil Trans. R. Soc. Lond. A***263**, 93
Roberts, P.H., Stewartson, K.: 1974, *Phil. Trans. R. Soc. Lond. A***277**, 287
Roberts, P.H., Stewartson, K.: 1975, *J. Fluid Mech.* **68**, 447
Skinner, P.H., Soward, A.M.: 1989, *Geophys. Astrophys. Fluid Dynam.* **44**, 91
Soward, A.M.: 1980, *J. Fluid Mech.* **98**, 449
Soward, A.M.: 1986, *Geophys. Astrophys. Fluid Dynam.* **35**, 329
Taylor, J. B.: 1964, *Proc. R. Soc. Lond. A***274**, 274

Stellar Convection as a Low Prandtl Number Flow

Josep M. Massaguer

Departament de Física Aplicada Universitat Politècnica de Catalunya,
08034 Barcelona, Spain

Abstract: Thermal convection in the Sun and cool stars is often modeled with the assumption of an effective Prandtl number $\sigma \simeq 1$. Such a parameterization results in masking of the presence of internal shear layers which, for small σ, might control the large scale dynamics. In this paper we discuss the relevance of such layers in turbulent convection. Implications for heat transport – i.e. for the Nusselt number power law – are also discussed.

1. Introduction

Astrophysical plasmas are almost inviscid fluids. In stellar convection zones these fluids are optically thick almost everywhere, and radiative heat transport is there an efficient diffusion process. Therefore, thermal convection in stars is a low Prandtl number flow, with the Prandtl number $\sigma = \nu/\kappa$ defined as the ratio of the viscous, ν, and thermal, κ, molecular diffusion coefficients. In the solar case, for instance, the molecular viscosity is so small compared to the thermal diffusion that σ can reach values of $\sigma = 10^{-9}$, meaning that viscous dissipation is only relevant for layers of several centimeters thick. Nobody can expect such small Prandtl numbers to be of any relevance for describing the large scale dynamics of the solar convection zone, but viscous dissipation cannot simply be neglected as it will break the energy balance. Therefore, the real question is about modeling the viscous diffusion term or, stated otherwise, a parameterization for σ.

The relevance of the Prandtl number in fully developed turbulent convection has been widely questioned in the past – see, for instance, the footnote on page 424 of Monin and Yaglom (1971). At large scales, molecular diffusion is negligible as compared to turbulent diffusion, so it might be plausible to describe the large scale dynamics with no reference to the molecular diffusion coefficients, not even to their ratio, the Prandtl number. However the implication is not that obvious. Molecular coefficients can play a role in building up small scale structures, mostly in internal boundary layers, delimiting convective eddies, thermals or plumes.

2. On the effective Prandtl number

The most popular parameterization technique is to take instead of the molecular value its turbulent analogue, the turbulent Prandtl number $\sigma_t = \nu_t/\kappa_t$, where ν_t and κ_t are eddy transport coefficients for momentum and heat. A proper definition of the scalar quantities ν_t and κ_t can only be made for isotropic turbulence or for some prescribed geometries – free shear flows, near wall turbulence, axisymmetric jets, etc. – where the eddy coefficients have a very precise meaning; otherwise ν_t and κ_t become tensorial quantities. From a large amount of measurements for different flows and geometries the conclusion might be that σ_t takes a value close to one, with the most frequent values in the range $\sigma_t = 0.6 - 0.9$. See Launder (1978) for an authoritative review.

It is thought that taking $\sigma \simeq 1$ instead of the true σ in the Navier-Stokes equation is a safe procedure for describing the large scale dynamics, but it is not. Two flows computed with different σ values may show the same σ_t value but different large scale dynamics. The relevant question to ask is about the existence of an asymptotic value σ_{as} such that computing the flows for any $\sigma < \sigma_{as}$, everything else being unchanged, produces indistinguishable large scale dynamics. So far the question is still open, though existing numerical work puts an upper bound $\sigma_{as} < 0.1$ to this asymptotic value. In Massaguer and Mercader (1988) and Massaguer *et al.* (1990) this asymptotic limit is conjectured from the onset of an instability to be described below. In Toomre *et al.* (1990) it has been found, for highly stratified fluids, that changing the Prandtl number in the range $\sigma = 0.1 - 1$ results in a significant change in the size of the observed structures. And the latter authors also report, from an unpublished work by J. Herring that, even in the Boussinesq convection, changing σ in that range produces significant changes in the large scale structure. Neither of the previously quoted works is conclusive, and indeed the whole subject is still open. But it is remarkable that Kraichnan (1962), from a discussion based on a mixing-length treatment, conjectured for the Prandtl number of the transition between high and low Prandtl number regimes a value that he estimates "with low confidence as about 0.1". His model is based on a detailed study of the thermal and viscous boundary layers, and the estimate depends on the value of the ratio Re_T/Pe_T between the Reynolds and Pèclet numbers at the edge of the thermal layer. But he warns the reader that "if the value turns out to be greater than the nominal value 10 which we have used, the effect of the theory would be to lower the value of σ which separates the low and high Prandtl number ranges... values of Re_T/Pe_T greater than 10 would not be surprising".

Such a low separating value of σ will originate in the turbulent shear boundary layers existing at the edge of the big eddies, where large scale isotropy and homogeneity will break down. In order to quantify such an effect, we can borrow some ideas from subgrid scale modeling. As discussed, for instance, by Lessieur (1987) in a chapter on Large Eddy Simulation, the contribution to the large scale dynamics by modes with wavenumbers $k > k_c$, with k_c in the inertial range, can be described by the eddy coefficients

$$\nu_t(k_c) = 0.28\sqrt{\frac{E(k_c)}{k_c}}$$

$$\kappa_t(k_c) = 0.42\sqrt{\frac{E(k_c)}{k_c}}$$

thus giving $\sigma_t \simeq 0.67$. But the Prandtl number entering the equations is $\sigma = (\nu + \nu_t)/(\kappa + \kappa_t)$. In the limit of small σ this expression can be approximated as

$$\sigma(k_c) \simeq \frac{0.28Pe(k_c)}{1 + 0.42Pe(k_c)}$$

where $Pe = v\ell/\kappa$ is a Pèclet number defined by $v = \sqrt{k_c E(k_c)}$ and $\ell = k_c^{-1}$ and k_c is a wavenumber in the inertial range. Shear layers put a strong constraint on the value of k_c. Following a suggestion from Kraichnan's paper, we can define the edge of one such layer as a point where diffusion and advection of heat take the same value, implying $Pe(k_c) = 0.5$ and $\sigma(k_c) \simeq 0.1$ in agreement with papers quoted above. The asymptotic value σ_{as} corresponds to the maximum value of k_c.

3. Shear dynamics

Shear layers can modify scalings substantially, but including them in the balances is a hard task. In order to explain where difficulties lie, let us consider thermal convection in a Boussinesq fluid. The velocity $\boldsymbol{v}$ is a solenoidal field, and can be split into a poloidal $\boldsymbol{v}_P = \nabla \times \nabla \times (\phi \boldsymbol{e}_z)$ and a toroidal $\boldsymbol{v}_T = \nabla \times (\psi \boldsymbol{e}_z)$ component, $\boldsymbol{v} = \boldsymbol{v}_P + \boldsymbol{v}_T$, where $\boldsymbol{e}_z$ is a unit vector antiparallel to gravity and the scalars ϕ and ψ are associated to the vertical components of the velocity and vorticity fields. As a physical picture for the poloidal velocity field we can think of a vertical disk rotating around its horizontal axis. With a string of such a disks we can reproduce the flow inside rolls and hexagons, the best known convective structures. Toroidal flows can be represented by horizontal disks rotating around their vertical axis. In most contexts left-handed and right-handed toroidal flows are required so as to cancel the vertical component of the angular momentum and, depending on the geometry of the container, vertically rotating disks will be paired off either horizontally or vertically. A final patchwork with such pairs will always result in important shear stresses A crucial step in understanding shear processes in thermal convection is to realize that, besides boundary effects, *thermal convection can always be modeled as a poloidal flow.* The poloidal flow is always an exact solution of the Navier-Stokes equation, though it may be unstable. Most known scalings, with Kraichnan's exception, reflect global balances for the poloidal component. A recent attempt by Castaing *et al.* (1989) to include shear stresses is also to be mentioned.

The poloidal velocity field scales, roughly speaking, as $v_P \simeq \kappa/\ell$ and the Reynolds number takes a value $Re = v_P\ell/\kappa \simeq \sigma^{-1}$. Decreasing σ increases Re without bound, and at low σ any poloidal field is expected to be shear unstable.

The onset of this shear instability has been examined in Massaguer and Mercader (1988) and an asymptotic regime for small σ has been found. It is also worth mentioning that near the onset of instability the flow turns out to be chaotic with components $\boldsymbol{v}_P$ and $\boldsymbol{v}_T$ uncorrelated (Massaguer *et al.*, 1990). Therefore, the flow is chaotic in time and intermittent in space. The toroidal velocity field arises as a secondary flow, and its description requires a previous knowledge of the non-linear dynamics of the poloidal flow, mostly of its geometry. So it is not surprising that most early attempts at modeling turbulent convection simply neglected the contribution of the toroidal component. In Malkus' theory of turbulence, for instance, – see Spiegel's (1962) reformulation – the vertical vorticity component is treated as a damped mode, implying $\boldsymbol{v}_T = 0$. A similar difficulty arises in Canuto *et al.* (1987) extension of Ledoux, Schwarzchild & Spiegel's spectral model. A preliminary attempt to include secondary instabilities can be found in Canuto *et al.* (1990).

3.1 Dimensional analysis

Introducing a scale of inhomogeneity $\ell = k_c^{-1}$ associated with the internal boundary layers, as we have done before, results in an additional degree of freedom in the well known dimensional analysis – see Massaguer (1990) and references therein. It is widely agreed that at low σ values the large scale dynamics should be independent of the molecular viscosity ν. And the Nusselt number N – i.e.: the non-dimensional heat flux – becomes $N = N(Ra\sigma, \ell/d)$, where Ra is the Rayleigh number and d is the layer thickness. For an homogeneous and isotropic flow $\ell/d \simeq 1$, the molecular diffusion of heat should be negligible, $\kappa << \kappa_t(k_c)$, and the the heat flux cannot depend on κ. The only possible power-law for N is the classical mixing-length scaling $N \propto (Ra\sigma)^{1/2}$.

If thin internal boundary layers are present in the flow, $\ell/d << 1$, the heat flux will be partially controlled by them. Let us assume the extreme situation where the flux is completely independent of the top and bottom wall boundaries. If the temperature gradient is $\partial_z T$, the gravity g and the expansion coefficient α,

$$Ra\sigma = \frac{\alpha g d^4 \partial_z T}{\kappa^2}$$

then the only power-law for N independent of d – i.e. controlled by ℓ – is $N \propto \ell/d(Ra\sigma)^{1/4}$. At this point we must recall that for large σ values the heat flux behaves as $N \propto Ra^{1/3}$, which corresponds to a flow controlled by wall boundary layers; and also, that recent experiments with gaseous helium ($\sigma \simeq 1$) by Threlfall (1975), and Castaing *et al.* (1989) give $N \propto Ra^{2/7}$, which corresponds, according to the scaling done by the latter authors, to a shared control of both internal and wall boundary layers. The increasing value of the exponents $1/4 < 2/7 < 1/3$ corresponds to an increasing dominance of the wall boundary layers with respect to internal boundary layers – please, notice that $2/7$ is the simplest integer ratio between $1/4$ and $1/3$.

The lack of well defined criteria for choosing the exponent n in the expression $N \propto (Ra\sigma)^n$ is a real shortcoming for stellar convection. The above discussion shows n to be in the range $n = 0.25 - 0.50$, with some preference for the smaller values, but the physics depends on the evaluation of very subtle balances, and detailed modeling is required.

3.2 A final remark

Shear layers are induced by the presence of a toroidal component in the turbulent flow, and this may also result in the enhancement of the horizontal velocity components of the Reynolds stresses. Clear indications of non-equipartition of energy between the poloidal and toroidal velocity components have been reported by Massaguer *et al.* (1990). Such a large scale anisotropic transport has been invoked a number of times as a mechanism for redistributing angular momentum on the solar surface, mostly in the context of the differential rotation but, to the best of our knowledge, low σ modeling has not even been attempted.

Acknowledgements. The present work was supported by the Dirección General de Investigación Científica y Técnica (DGICYT) under grant PS87-0107. A travel grant from CIRIT is also acknowledged.

References

Canuto, V.M., Goldman, I., Chasnov, J.: 1987, *Phys. Fluids* **30**, 3391

Canuto, V.M., Cheng, Y., Hartke, G.J., Schilling, O., Chasnov, J.: 1990, *J. Fluid Mech.* (to be published)

Castaing, B., Gunaratne, G., Heslot, F.L., Kadanoff, L., Libchaber, A., Thomae, S., Wu, X.Z., Zaleski, S., Zanetti, G.: 1989, *J. Fluid Mech.* **204**, 1

Kraichnan, R.H.: 1962, *Phys. Fluids* **5**, 1374

Launder, B.E.: 1978, in *"Turbulence"*, ed. P. Bradshaw, p. 231, *Topics in Applied Physics* **12**, Springer

Lessieur, M.: 1987, *Turbulence in Fluids*, M. Nijhoff Publ., Dordrecht

Massaguer, J.M.: 1990, in *"Inside the Sun"*, eds. G. Berthomieu and M. Cribier p. 101, Kluwer.

Massaguer, J.M., Mercader, I.: 1988, *J. Fluid Mech.* **189**, 367

Massaguer, J.M., Mercader, I., Net, M.: 1990, *J. Fluid Mech.* **214**, 579

Monin, A.S., Yaglom, A.M.: 1971, *Statistical Fluid Mechanics*, vol 1, The MIT Press.

Spiegel, E.A.: 1962, in *"Mécanique de la Turbulence"*, Colloques Internationaux du CNRS à Marseille (Editions CNRS Paris). p. 181

Threlfall, D.C.: 1975, *J. Fluid Mech.* **67**, 17

Toomre, J., Brummell, N., Cattaneo, F., Hurlburt, N.E.: 1990, *Computer Phys. Commun.* **59**, 105

Convective Overshoot as a Source of Helicity

R. L. Jennings

D.A.M.T.P., Silver St., Cambridge, CB3 9EW, UK

Abstract: In order to understand what happens in the Sun when convection overshoots into the radiation zone an idealized model of penetrative convection with rotation is studied. Here we highlight two properties of the model which occur with parameters relevant to the Sun. Firstly rotation allows motions to persist far beneath the convection zone, and secondly the profile of helicity with depth is dominated by a local maxima just beneath the convection zone. This second result has consequences in dynamo theory.

1. Introduction

Overshoot at the base of the solar convection zone motivates this study of astrophysical penetrative convection, even though the simplest examples of rotating penetrative convection remain poorly understood. Stability analysis is used to determine preferred horizontal length scales, eigenfunctions and critical Rayleigh numbers, and we are particularly interested in isolating features arising in the two layer configuration that do not occur in a single layer of fluid. No magnetic fields are included here, but a helicity distribution is obtained from which an α-effect can be calculated, this effect being the cornerstone of mean-field dynamo theory (see, for example Krause and Rädler, 1980). A full account of the mathematical details appears in Jennings (1990).

2. The model

Many simplifications are adopted here, such as the Boussinesq approximation, a simple equilibrium, and a plane parallel geometry. Although formally there is initially no motion, we loosely think of there being a mean flow in the convection zone that is turbulent, while the lower layer's fluid is laminar. Thus we assign turbulent values to the thermal diffusivity and kinematic viscosity in the upper layer, and laminar values to these quantities in the lower fluid, and further consider the convective region to be only slightly superadiabatic in contrast to the lower fluid which is extremely subadiabatic. Ratios of these respective quantities yield dimensionless parameters:

$$K = \frac{\kappa_{laminar}}{\kappa_{turbulent}}, \qquad V = \frac{\nu_{laminar}}{\nu_{turbulent}}, \qquad B = \frac{\beta_{subcritical}}{\beta_{supercritical}} . \tag{1}$$

Both K and V are small and positive, while B is large and negative. Since the transition from a convectively unstable stratification to one that is stable to convection is probably very sharp in the Sun it is convenient to neglect the details of the transition and model using a step function (see Fig. 1).

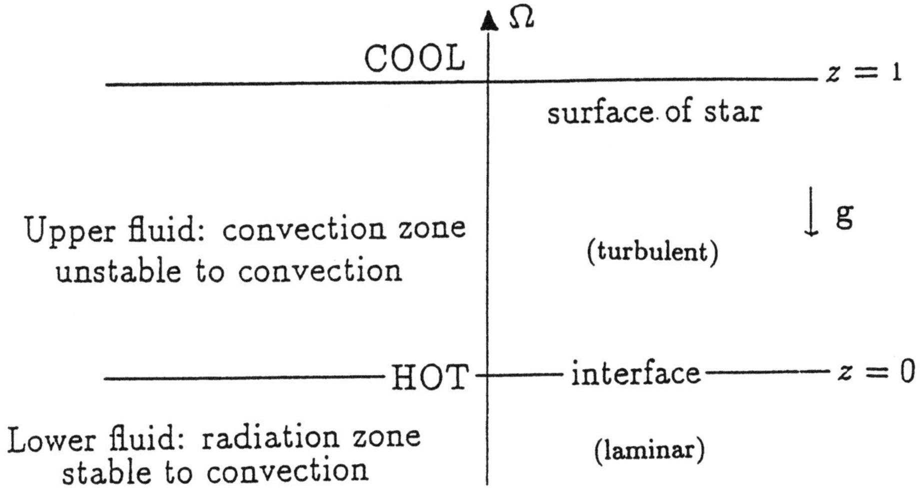

Fig. 1. The geometry; the stable layer is semi-infinite and $-\infty < x, y < \infty$.

2.1 Eigenvalues

Following the procedure in chap. 3 of Chandrasekhar (1961) we obtain equations for the vertical velocity perturbation $W(z)$ in each layer:

$$\left[\left(D^2 - a^2\right)^3 + TD^2 + Ra^2\right] W_{upper} = 0 \tag{2}$$

$$\left[\left(D^2 - a^2\right)^3 + \left(T/V^2\right) D^2 + \left(BRa^2/KV\right)\right] W_{lower} = 0 , \tag{3}$$

where $D = d/dz$, T is the Taylor number, R is the Rayleigh number, and $a^2 = a_x^2 + a_y^2$ is the horizontal wavenumber. Note that there is no difference here between x and y, and that overstable marginal solutions are not considered. Eqns. (2) and (3) together with the boundary conditions and conditions at the interface listed below define an eigenvalue problem, with non-trivial solutions for $W(z)$ only if R has a special value.

At $z = 1$ the boundary is impenetrable, stress free and isothermal, and as $z \to -\infty$ all perturbations vanish. At the interface ($z = 0$) we require the following quantities to be continuous:

- Vertical velocity;
- Horizontal velocity, (both x and y);
- Horizontal stress (both x and y);
- Normal stress;
- Temperature;
- Heat flux.

Analytic solutions for $W(z)$ in each layer which satisfy the boundary conditions at $z = 1$ and $z = -\infty$ are easily obtained. However, to satisfy the above conditions at the interface it is necessary to use numerical methods. Having fixed parameters T, V, K, B and a we seek the eigenvalue R. Varying only a, this is repeated until we find the critical value of a (a_c) at which R is minimized. This minimum value is the critical Rayleigh number R_c, and a_c gives the preferred horizontal length scale at the onset of convection.

3. Results

3.1 $V \ll 1$, an important limit

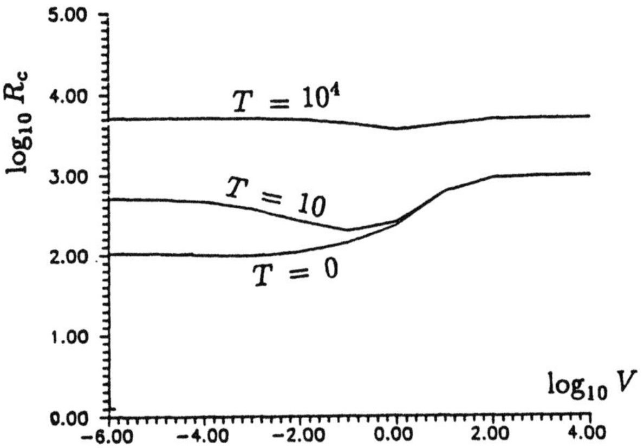

Fig. 2. Curves of $\log_{10} R_c$ versus $\log_{10} V$ at different values of T, with $K = -B = 1$ in each case.

From Fig. 2 it is seen that if $V \ll 1$ the results for no rotation ($T = 0$) are very different to those for small rotation ($T = 10$). In other words $T = 0$ is singular when V is small. To understand how this singularity arises for $V \ll 1$, consider Eq. (3) with $K = -B = 1$, which has solutions $W_{lower} \propto \exp(lz)$, where l is a root of the polynomial:

$$V^2 \left(l^2 - a^2\right)^3 + Tl^2 - VRa^2 = 0 \, . \tag{4}$$

In the case of no rotation ($T = 0$) the balance in Eq. (4) is:

$$Vl^6 - Ra^2 \approx 0\,, \tag{5}$$

from which we can deduce that there is no motion in the lower fluid except in thin viscous boundary layers of $O\left(V^{1/6}\right)$ thick. With rotation ($T \neq 0$), there are now two different balances:

$$V^2 l^4 + T \approx 0 \quad \text{and} \quad Tl^2 - VRa^2 \approx 0\,. \tag{6}$$

The first balance defines thin Ekman layers of $O\left(V^{1/2}\right)$, but the second shows that one wavenumber is very small, of $O\left(V^{1/2}\right)$, showing motions persist far beneath the interface to depths of $O\left(V^{-1/2}\right)$.

3.2 Helicity Profiles

Using the eigenfunctions we can compute the horizontally averaged helicity profile $\bar{h}(z)$;

$$\bar{h} = 2\left(\overline{WZ}\right) + a^{-2}\left[\overline{DZ\,DW} - \overline{Z\,D^2W}\right]\,, \tag{7}$$

where Z is the vertical vorticity, and bars denote horizontal averages. Eq. (7) shows that rotation is essential for the generation of helicity because without rotation Z decays to zero. An α-effect, is closely related to the flow's helicity; $\alpha \propto -h$, so the profiles $\bar{h}(z)$ provide qualitative information regarding the distribution of α in the Sun. Note however, that a formal derivation of α is via a two-scale analysis involving a large scale flow and small scale turbulence (Krause and Rädler, 1980), and therefore the relation here between $\bar{h}$ and α is not rigorous.

Estimating the parameter values which apply to the Sun in this model is not easy, but we took $K = 10^{-6}$, $B = -10^4$ and $T = 10^5$ after considering standard solar models discussed in Chaps. 1 and 2 of Priest (1982). There is considerable uncertainty with the value of V, but it is generally accepted that V is smaller than K, and we took $V = 10^{-8}$. For these parameters we find $R_c = 2 \times 10^4$ and $a_c = 8.5$ (tall thin cells), and the resulting helicity profile is shown in Fig. 3. Since helicity is a quadratic quantity its sign is unaffected by the normalization used for the eigenfunctions. The curve in Fig. 3 is normalized such that the maximum value is ± 1, and as can be seen this maximum is -1 and occurs just beneath the interface.

4. Conclusions

Two interesting results have come from this simple model which now need to be tested using better models. The first is that in the Sun motions can penetrate far below the convection zone as a consequence of rotation and V being small. Secondly, somewhat surprisingly overshoot leads to a large source of helicity just beneath the convection zone suggesting that a dynamo could work well there even

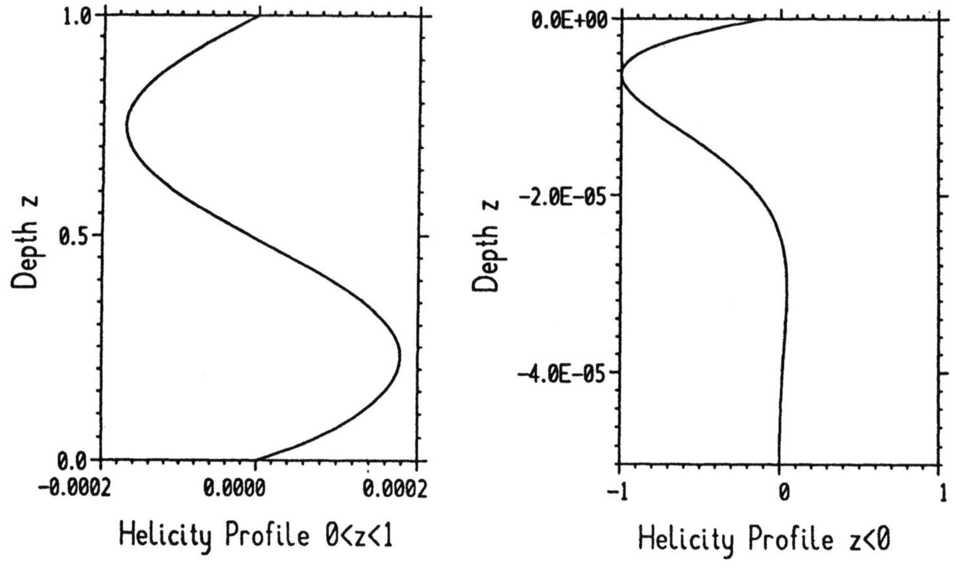

Fig. 3. Helicity profile for solar parameters. The scales of the two graphs are very different, and almost all the helicity occurs in a thin boundary layer beneath the interface. This helicity is negative so the local α-effect is positive. The shape of the profile in the upper layer is typical of the helicity that results in a single layer of convecting fluid.

though it is not the site of the most vigorous convection. This gives an α-effect which is positive in the northern hemisphere, which if correct implies that ω is negative because only when the product $\alpha\omega < 0$ do dynamo models yield the observed activity migrations from the poles to the equator.

References

Chandrasekhar, S.: 1961, *Hydrodynamic and Hydromagnetic Stability*, Clarendon Press, Oxford

Jennings, R. L.: 1990, to appear in *Geophys. Astrophys. Fluid Dyn.*

Krause, F. and Rädler, K.-H.: 1980, *Mean-Field Magnetohydrodynamics and Dynamo Theory*, Pergamon Press, Oxford

Priest. E. R.: 1982, *Solar Magnetohydrodynamics*, Reidel, Dordrecht

Topological Pumping in the Lower Overshoot Layer

Kristóf Petrovay

Eötvös University, Department of Astronomy[1]
Budapest, Kun Béla tér 2, H-1083 Hungary

Abstract: Problems associated with topological pumping in the lower overshoot layer suggest a strongly turbulent and strongly differentially rotating upper radiative zone as the seat of the dynamo and as flux reservoir.

1. Introduction

Nonlocal phenomenological models of the lower solar convective zone predict that the strongly turbulent regime is sharply bounded from below: in a very thin ($\leq$ 100 km) transitional layer the turbulent velocity falls to zero and the subadiabaticity becomes quite large (e. g. Van Ballegooijen, 1982; Pidatella and Stix, 1986). In this "sharply bounded" case the radiative zone is obviously not an appropriate place for the storage of the toroidal flux tubes responsible for the formation of active regions. On the other hand, it is well known that buoyant instabilities in the bulk of the convective zone proper should lead to the loss of flux tubes with magnetic fluxes of 10^{22} Mx (the typical active region value) and field strengths of the equipartition value ($B_{eq}^2/8\pi = \rho v^2/2$) or higher on timescales short compared to the cycle time. The only region where these problem may be avoided is the lower half of the lower overshoot layer (Van Ballegooijen, 1982), so on the basis of this exclusional reasoning it is generally supposed that the flux is stored in the lower overshoot layer.

In Section 2, however, it will be shown that if the field strength in the flux tubes does not greatly exceed 10^5 G there exists a mechanism of flux transport that will lift the flux tubes very effectively out of the lower overshoot layer into the higher parts where buoyancy may take over. This mechanism is known as *topological pumping*. So, accepting the sharply bounded case, we are faced with the severe problem of being unable to find any layer in the Sun that would be suitable for dynamo operation and flux storage. Possible ways out of this impasse are discussed in Section 3.

---

[1] Present address: Dept. of Theor. Physics, University of Oxford, Oxford, UK

2. Topological pumping

In a horizontal cross section of a turbulent flow regimes of up- and downflow are in general topologically not equivalent: one of them is connected, while the other forms isolated "parcels" (or "columns", in 3D). E. g. in the lower overshoot layer it can be shown that the downflows are isolated and the upflows are connected (Petrovay, 1990). A one-dimensional object like a horizontal flux tube can evidently not be transported in its whole extent in the direction of the isolated flows, but it can be transported as a whole by the connected flow; in this way, a net transport of magnetic flux in the direction of the connected flow is expected. This topological pumping has been extensively studied for the case of laminar flows advecting passive magnetic fields (Drobyshevski and Yuferev, 1974; Arter, 1983; Moffatt, 1983).

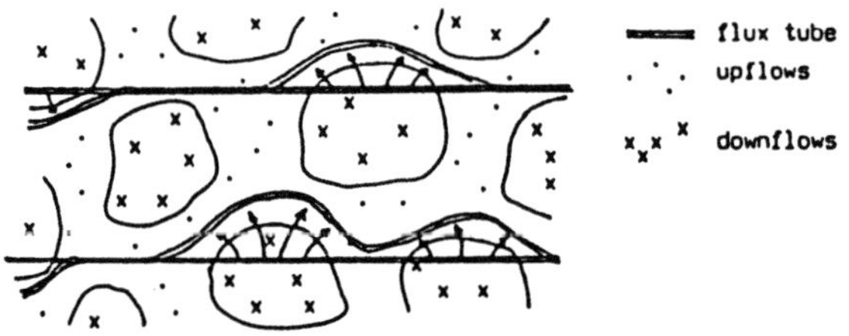

Fig. 1. Topological pumping: in a horizontal cross section of the lower overshoot layer the diverging isolated downflows sweep the flux tubes into the connected upflows. Then the upflows lift the tubes upward in their whole extent.

In order to judge if topological pumping is more effective than other transport mechanisms in the solar case, we take the typical values of different physical quantities in the lower overshoot layer and use them for reasonable estimates of the relevant forces and timescales. Our basic model is that of Van Ballegooijen (1982) fitted to the Unno *et al.* (1985) model, with crude anisotropy corrections on the basis of the results of Canuto (1989). Typical values of the physical parameters are as follows. Thickness of the layer available for flux storage: $h \sim 5 \cdot 10^8$ cm. Density: $\rho \sim 0.1\,\mathrm{g/cm^3}$. Gravity: $g \sim 5 \cdot 10^4\,\mathrm{cm/s^2}$. Scale height: $H \sim 5 \cdot 10^9$ cm. Horizontal correlation length: $5 \cdot 10^9\,\mathrm{cm} < l_h < 5 \cdot 10^{10}$ cm. R. m. s. vertical velocity: $\sigma_r \sim 3 \cdot 10^3$ cm/s. R. m. s. horizontal velocity: $\sigma_h \sim 2 \cdot 10^4$ cm/s. Typical horizontal velocities "felt" by the tube during this process are even higher than σ_h: $v_h \sim 4 \cdot 10^4$ cm/s. These high horizontal velocities arise as a consequence of the sharp boundary of the turbulent regime. The tubes are supposed to have a magnetic flux of $\Phi \sim 10^{22}$ G (the typical value for an active region).

It is found that the horizontal drag can overwhelm the curvature force acting on the tube unless the field strength is considerably stronger than 10^5 G:

$$F_{d1} \sim \rho v_h^2 B^{1/2} \Phi^{-1/2} \cong 0.5\,\mathrm{dyn/cm^3} > 0.2\,\mathrm{dyn/cm^3} \geq \frac{B^2}{4\pi l_h} \sim F_m .$$

The timescale of horizontal sweepout is

$$\tau_1 \sim \left(\frac{2\rho l_h}{F_{d1}}\right)^{1/2} \sim \text{a few days} \ll 1 \text{ month} \sim \tau_{turb},$$

so the typical flux tubes will lie in the connected flow for practically their whole extent. Now the vertical flow can sweep them upwards, out of the overshoot layer. Buoyancy cannot hinder this, except in the lowest $\sim$ 1000 km thick layer:

$$F_{d2} \sim \rho v_r^2 B^{1/2} \Phi^{-1/2} \sim 10^{-3}\text{dyn/cm}^3 > 5{\cdot}10^{-4}\text{dyn/cm}^3 \sim \rho g \Delta \nabla h/H \simeq g \Delta\rho \simeq F_b$$

The timescale of vertical sweepout is

$$\tau_2 \sim \left(\frac{2\rho h}{F_{d2}}\right)^{1/2} \sim \text{a few days} \ll 1 \text{ month} \sim \tau_{turb}.$$

In conclusion: all the conditions for the dominance of the topological pumping over other mechanisms of flux transport are fulfilled. *Strong horizontal motions in the sharply bounded case will sweep the flux tubes out of the downflows and then up from the overshoot layer on a timescale much shorter than that of turbulence.*

3. Location of the dynamo and flux reservoir

There are three possible solutions of the flux storage puzzle outlined above.

First, the flux tubes may actually be considerably stronger than 10^5 G. This however does not seem very plausible for observational and theoretical considerations summarized elsewhere (Petrovay, 1991).

Second, the buoyancy may not be so effective in transporting the tubes upward as generally thought, so flux confinement might still be possible above the lower overshoot regime. Parker's (1987) estimates however show that the thermal shadows can confine tubes with field strength of the equipartition value or stronger only if their flux is much higher than the observed value of 10^{22} Mx. Similarly, the turbulent pumping effect (Moffatt, 1983) seems to be unable to overwhelm the buoyancy for tubes with realistic properties. The simulations by Jennings *et al.* (1991) and Brandenburg *et al.* (1991) show the contrary: flux becomes concentrated in a relatively thin "cloud" a little above the lower turbulent boundary, between the regions dominated by topological and turbulent pumping, respectively; but the dependence of this result on flux tube properties is not clear. Besides, the boundary conditions used in these calculations amount to setting the buoyant flux removal timescale to infinity, and the competition of buoyancy and turbulent pumping is actually decided by their timescales – so these calculations can hardly be regarded as decisive. This problem deserves further investigation.

The most plausible solution to the flux storage problem is to accept Durney's (1989) proposal, that the layer where the solar rotational law changes so dramatically, from strong latitudinal differential rotation pervading the convective zone to nearly rigid rotation characterizing the deep interior, is situated at the top of the radiative zone, just below the overshoot layer. Kelvin-Helmholtz instabilities may be expected to generate strong turbulence here, so ***the lower boundary of the turbulent regime is not sharp:*** horizontal motions are less strong and topological pumping is ineffective.

We may conclude that the problems associated with topological pumping in the lower overshoot layer can serve as an argument in favour of a strongly turbulent and strongly differentially rotating upper radiative zone as the seat of the dynamo and as flux reservoir.

References

Arter, W.: 1983, *J. Fluid Mech.* **132**, 25

Brandenburg, A., Jennings, R., Nordlund, Å., Stein, R.F., Tuominen, I.: 1991, these Proceedings

Canuto, V. M.: 1989, *Astron. Astrophys.* **217**, 333

Drobyshevsky, E.M., Yuferev, V.S.: 1974, *J. Fluid Mech.* **65**, 33

Durney, B.R.: 1989, *Solar Phys.* **123**, 197

Jennings, R., Brandenburg, A., Nordlund, Å., Stein, R.F., Tuominen, I.: 1991, these Proceedings

Moffatt, H.K.: 1983, *Rep. Prog. Phys.* **46**, 621

Parker, E.N.: 1987, *Astrophys. J.* **321**, 984

Petrovay, K.: 1990, *Astrophys. J.* **362**, 722

Petrovay, K.: 1991, *Solar Phys.*, submitted

Pidatella, R.M., Stix, M.: 1986, *Astron. Astrophys.* **157**, 338

Unno, W., Kondo, M., Xiong, D.-R.: 1985, *P. A. S. Japan* **37**, 235

Van Ballegooijen, A. A.: 1982, *Astron. Astrophys.* **113**, 99

Diffusion of Particles, Heat and Magnetic Fields in Compressible Turbulent Media

A.Z. Dolginov and N.A. Silant'ev

A.I.Ioffe Phys.-Tech. Institute
Academy of Sciences of USSR
194021 Leningrad, USSR

Abstract: A new method for the calculation of kinetic coefficients is presented. This method allows us to obtain the distribution of scalar and vector fields (such as the temperature, the admixture particle number density and the magnetic field) in turbulent cosmic media with any value of $S = u_0\tau_0/R_0$. The explicit expression for the "turbulent" diffusivity D_T is obtained. In some cases D_T becomes negative, implying the clustering of the admixture particles in patches (a local increase of the temperature and magnetic fields). The magnetic α-effect is considered for the case $S \sim 1$.

Processes of scalar and vector field transport in turbulent medium are of great importance in most astrophysical objects. The general form of the equations which describe these processes is

$$\hat{K}f(\boldsymbol{r},t) = \hat{L}f(\boldsymbol{r},t), \qquad \hat{K} = \frac{\partial}{\partial t} - D_0\nabla^2. \tag{1}$$

(a) For the case of particle propagation $f(\boldsymbol{r},t) = f_p(\boldsymbol{r},t)$ is the number density of the particles with momentum p at the space-time point $\boldsymbol{r}, t$, and $\hat{L} = \hat{L}_p = -(\boldsymbol{u})\nabla - div\,\boldsymbol{u}(p/3)(\partial/\partial p)$. $\boldsymbol{u}(\boldsymbol{r},t)$ is the turbulent velocity of the medium. $\langle \boldsymbol{u}(\boldsymbol{r},t)\rangle = 0$. The brackets $\langle\ \ \rangle$ denote an averaging over the velocity ensemble.

(b) If one considers the particle distribution, disregarding their momenta, then $f(\boldsymbol{r},t) = N(\boldsymbol{r},t) = \int_0^\infty f_p(\boldsymbol{r},t)p^2\,dp$ and $\hat{L} = \hat{L}_N = -(\boldsymbol{u}\nabla) - div\,\boldsymbol{u}$.

(c) For the case of heat propagation $f(\boldsymbol{r},t) = T(\boldsymbol{r},t)$, $\hat{L} = \hat{L}_T = -(\boldsymbol{u}\nabla) - \mathcal{R}/\mu C_V div\,\boldsymbol{u}$,where T is the temperature, μ is the molecular weight, C_V is the specific heat capacity and $\mathcal{R}$ is the gas constant.

(d) The magnetic field distribution is determined by (1) with $f(\boldsymbol{r},t) = \boldsymbol{B}(\boldsymbol{r},t)$, the magnetic field, and $\hat{L}f = \hat{\mathcal{L}}_B\boldsymbol{B} = -(\boldsymbol{u}\nabla)\boldsymbol{B} + (\boldsymbol{B}\nabla)\boldsymbol{u} - \boldsymbol{B}div\,\boldsymbol{u}$. D_0 in (1) is the diffusion coefficient for the field considered.

Most papers devoted to the solution of (1) are restricted to the case $S^2 = (u_0\tau_0/R_0)^2 \ll 1$, where $u_0^2 = \langle u^2(\boldsymbol{r},t)\rangle/3$ and R_0 and τ_0 are the space and time scales of the turbulence. By definition the motion within the scale values of order R_0,τ_0 is more or less regular and becomes random only for scale sizes $L \gg R_0$ and $t \gg \tau_0$. The condition $S^2 \ll 1$ allows us to take into account only the correlation tensor of second rank, $\langle u_i(\boldsymbol{r},t)u_j(\boldsymbol{r},t)\rangle$ where $\boldsymbol{R} = \boldsymbol{r} - \boldsymbol{r}', \tau = t - t'$. The turbulence is assumed to be stationary and homogeneous. If $S \gtrsim 1$ the correlation tensors of higher rank become important.

In this paper we present a new method for the solution of the transport problem in a compressible turbulent media for the case $S \gtrless 1$. The Greens function of (1) is determined by

$$(\hat{K} - \hat{L})G(1,2) = \delta(1-2) \equiv \delta(\boldsymbol{r}_1 - \boldsymbol{r}_2)\delta(t_1 - t_2), \tag{2}$$

where the symbols 1, 2,... denote $\boldsymbol{r}_1$, t_1 ; $\boldsymbol{r}_2$, t_2 ;... and d1, d2,... denote $d\boldsymbol{r}_1 dt_1$; $d\boldsymbol{r}_2 dt_2$;... The integral form of (2) is

$$G(1,2) = G_0(1,2) + \int \mathrm{d}3\, G_0(1,3)\hat{L}(3)G(3,2), \quad \hat{K}G_0(1,2) = \delta(1,2). \tag{3}$$

If one uses a direct iteration of (3), beginning with $G_0(1,2)$, then a divergent series for the case $S \gtrsim 1$ is obtained. The series is valid and asymptotic for $S^2 \ll 1$. To avoid this difficulty let us take instead of (3) the two coupled equations (4) and (5), which are identical to equation (3) for any value of the operator $\hat{O}$. The particular choice of $\hat{O}$ is equivalent to the particular choice of the zero approximation for $G(1,2)$.

$$G(1,2) = M(1,2) + \int \mathrm{d}3\, G(1,3)\left[\hat{L}(3)M(3,2) - \int \mathrm{d}4\, \hat{O}(3,4)M(4,2)\right], \tag{4}$$

$$M(1,2) = G_0 + \int \mathrm{d}3 \int \mathrm{d}4\, G_0(1,3)\hat{O}(3,4)M(4,2). \tag{5}$$

The advantage of this form is that it allows the possibility of choosing the zero approximation in such a way to provide a better convergence of the iteration and it takes into account the higher rank correlation tensors. Let us choose $\hat{O}(3,4) = \langle \hat{L}(3)G_0(3,4)\hat{L}(4)\rangle$, leading to an asymptotic iteration series for $G(1,2)$ adequate for $S^2 \sim 1$. This statement can be proved by calculations with subsequent approximations. The second iteration takes into account the correlation tensors up to the fourth rank, etc. The equation for the mean Greens function is

$$\langle G(1,2)\rangle = G_0(1,2) + \int \mathrm{d}3\, G_0(1,3)\langle \hat{L}(3)G(3,2)\rangle. \tag{6}$$

Equations (5) and (6) are approximately equivalent if one inserts (3) into (6) and takes $\langle \hat{L}(3)G_0(3,4)\hat{L}(4)\rangle\langle G(4,2)$ instead of $\langle \hat{L}(3)G_0(3,4)\hat{L}(4)G(4,2)\rangle$. This approximation for $M(1,2)$ is, in many cases, sufficiently precise ($\sim$ 10 %) for large values of S^2. $G(1,2)$ depends on the particular form of $\hat{L}$. Consider the case $\hat{L} = \hat{L}_N = -(\boldsymbol{u}\nabla) - div\, \boldsymbol{u}$, as an example. Taking into account that $\langle G(R,\tau)\rangle$ is a smooth function on scales $\sim R_0, \tau_0$, one can take $\langle G\rangle$ out of the integral. This leads to a diffusion equation for $N(\boldsymbol{r},t)$ with a "turbulent" diffusivity:

$$\partial N/\partial t - (D_0 + D_T)\nabla^2 N = 0, \tag{7}$$
$$D_T = \frac{1}{3}\int d\boldsymbol{R}\int d\tau\,[\langle u_i(1)G(1,2)u_i(2)\rangle - \boldsymbol{R}\cdot\langle \boldsymbol{u}(1)G(1,2)div\,\boldsymbol{u}(2)\rangle].$$

In the first approximation we can take in (7) $G(1,2) \approx M(R,\tau)$. The kernel of (5) depends on $\boldsymbol{R}$ and τ only and $M(R,\tau)$ can be obtained in the explicit form using the Fourier transform. Consider, as a simple example, the potential turbulence $\boldsymbol{u} = \nabla\varphi$ with the correlation function

$$\langle\varphi(1)\varphi(2)\rangle = \frac{3u_0^2}{k_0^2}\frac{\sin k_0 R}{k_0 R}\exp\left(-\frac{|\tau|}{\tau_0}\right). \tag{8}$$

In this case the Fourier transform of the $M(R,\tau)$ is

$$\tilde{M}(\boldsymbol{k},\tau) = \frac{\theta(\tau)}{a}\left[\exp\left(-\frac{\tau}{2\tau_0}\right)\right]\left[\sin\frac{a\tau}{2\tau_0} + a\cos\frac{a\tau}{2\tau_0}\right], \tag{9}$$

where $\theta(\tau) = 1$ if $\tau > 0$ and $\theta(\tau) = 0$ if $\tau < 0$, $a = (4u_0^2\tau_0^2 - 1)^{1/2}$ and D_0 is assumed to be much smaller than D_T. The first integral term in (7) is always positive, but the sign of the second one depends on the particular form of $\langle \boldsymbol{u}G div\,\boldsymbol{u}\rangle$. In a compressible medium ($div\,\boldsymbol{u} \neq 0$) $div\,\boldsymbol{u}$ may be negative. Using (8), (9) and (7) we obtain

$$D_T = 2u_0^2\tau_0(2 - 9S^2)(2 + 9S^2)^{-2}, \qquad S = u_0\tau_0 k_0/3, \tag{10}$$

and so D_T becomes negative if $S^2 > 2/9$. There are cases, for example acoustic turbulence, when $D_T + D_0$ is always positive. The negative diffusion leads to formation of patches of the admixture particles. It does not contradict the laws of thermodynamics, because the physical system is thermodynamically open.

By the method described above we can obtain the equation for the average magnetic field in the case $S \gtrsim 1$. We obtain

$$\left(\frac{\partial}{\partial t} - (D_0 + D_T)\nabla^2\right)\langle\boldsymbol{B}\rangle = \alpha_T\,rot\langle\boldsymbol{B}\rangle - \beta_T\frac{\partial}{\partial t}rot\,\langle\boldsymbol{B}\rangle. \tag{11}$$

In the case of isotropic, stationary, homogeneous non-mirror symmetric turbulence

$$\langle u_i(1)u_j(2)\rangle = \delta_{ij}B_\perp(R,\tau) + R_iR_j(B_\parallel(R,\tau) - B_\perp(R,\tau))R^{-2} + e_{ijk}R_kC(R,\tau). \tag{12}$$

In this case D_T and α_τ are given by

$$D_T = -\frac{1}{6\pi^2}\int_0^\infty dq\,q^3\int_0^\infty d\tau\left[\frac{1}{2}(\tilde{M}_+ + \tilde{M}_-)\frac{\partial}{\partial q}\tilde{B}_\parallel(q,\tau) - (\tilde{M}_+ - \tilde{M}_-)\frac{1}{q}\frac{\partial}{\partial q}\hat{C}(q,\tau)\right], \tag{13}$$
$$\alpha_T = \frac{1}{3(2\pi)^3}\int d\boldsymbol{q}\int_0^\infty d\tau\left[\frac{1}{2q}(\tilde{M}_+ + \tilde{M}_-)(\delta_{ij}q^2 + q_iq_j)\tilde{B}_{ij}(\boldsymbol{q},\tau) - \right.$$

$$q(\tilde{M}_+ - \tilde{M}_-)\frac{\partial}{\partial q}\tilde{C}(q,\tau)\Big] ,$$

$$B_{ij}(\boldsymbol{R},\tau) = \langle u_i(1)uj(2)\rangle .$$

The expression for α_T differs from that for β_T only by an additional factor τ^{-1} in the integral.

$$\tilde{M}_\pm(q,\tau) = \left[a_\pm^{-1}\sin(a_\pm\tau/2\tau_0) + \cos(a_\pm\tau/2\tau_0)\right]\exp(-\tau/2\tau_0)\,\theta(\tau) \qquad (14)$$

$$a_\pm = (4u_0^2\tau_0^2q^2 - 1 \pm 8C_0\tau_0^2 q)^{1/2},$$

$$C(0,\tau) = C_0\, exp(-|\tau|/\tau_0).$$

In an incompressible medium with $B_{||}(0,\tau) = u_0^2\, exp(-|\tau|/\tau_0)$, $c_0 = \gamma u_0^2 k_0/2$, $|\gamma| \leq 1$, $S = u_0\tau_0 k_0/3$ we have

$$D_T = \frac{2u_0^2\tau_0(2+9(1+\gamma^2)S^2)}{(2+9S^2)^2 - 81\gamma^2S^4}, \quad \alpha_T = \frac{4u_0^2k_0(1+9S^2)\gamma}{(2+9S^2) - 81\gamma^2S^4}. \qquad (15)$$

Equations (15) are valid to within an error of $\sim$ 20 % up to $S = 1$.

References

Dolginov, A.Z., Silant'ev, N.A.: 1987, *J. Exp. Theor. Phys.* JETP **93**, 159

Dolginov, A.Z., Silant'ev, N.A.: 1990, *Astron. Astrophys.*

Dolginov, A.Z., Sliant'ev, N.A.: 1991, *Geoph. Astroph. Fluid Dynamics* (submitted)

Silant'ev, N.A., Korolkov, M.D.: 1990, *Astron. Nachr.* **311**, 107

Silant'ev, N.A.: 1988, *Proc. Int. Meet. on "Physics and Evolution of Stars"*, Leningrad, Nauka

The Spatial Structure of Homogeneous Turbulence at High Reynolds Number

A. Vincent and M. Meneguzzi

CERFACS, 42 av. G. Coriolis, 31057 Toulouse, France

Abstract: A direct numerical simulation at resolution 240^3 is used to obtain a statistically stationary three-dimensional homogeneous and isotropic turbulent field at a Reynolds number around 1000 ($R_\lambda \approx 150$). The energy spectrum displays an inertial subrange extending over more than one decade. In the physical space, the strongest vorticity is organized in very elongated thin tubes. The typical length of a tube is around the integral scale of the flow and its thickness is a few dissipation scales.

In the past, direct numerical simulations of the three-dimensional Navier-Stokes equation have been restricted to modest Reynolds numbers (Orszag and Patterson (1972); Siggia and Patterson (1978); Siggia (1981)). These simulations have mostly given properties on the dissipation scale of a turbulent flow. It now becomes feasible, on computers of the latest generation, to reach Reynolds numbers at which there exists an inertial subrange in the energy spectrum (Kerr, 1985; Yamamoto and Hosokawa, 1988; She *et al.*, 1988, 1990). The object of our work is to study a more extended inertial subrange.

We solve the Navier-Stokes equation for incompressible fluids in rotational form

$$\frac{\partial \boldsymbol{v}}{\partial t} = \boldsymbol{v} \times \boldsymbol{\omega} - \nabla(p + v^2/2) + \nu \nabla^2 \boldsymbol{v} + \boldsymbol{f} \qquad (1)$$

with the continuity equation $\nabla \cdot \boldsymbol{v} = 0$. $\boldsymbol{v}$ is the velocity field, $\boldsymbol{\omega} = \nabla \times \boldsymbol{v}$ the vorticity, p the pressure, ν the kinematic viscosity and $\boldsymbol{f}$ a force field. Since we are interested in (statistically) homogeneous turbulent flows, we take periodic boundary conditions in all directions, with period 2π. A pseudospectral method is used to compute the right-hand side of this equation (see Gottlieb and Orszag, 1977). The time marching is performed using a second order finite-difference scheme. We chose to force the field in a deterministic way, as was done by Kerr (1985). All Fourier modes with wavenumber components equal to 1 or 0 are forced with a constant amplitude $\boldsymbol{f}$ independent of $\boldsymbol{k}$.

The calculation presented here was done with 240^3 Fourier components, and a viscosity of 10^{-3}. This gives $R \approx 1000$ ($R_\lambda \approx 150$) (Vincent and Meneguzzi, 1990). After integration over a few tens of turnover times we obtain a statistically stationary regime, in the sense that the mean kinetic energy density fluctuates by a few percent around a constant value. Our time unit is such that the large eddy turnover time is 0.35. The energy spectrum $E(k)$ at a given time is shown in Fig. 1 at time 14. A power law range can be seen for $k < 25$, with an exponent a little larger than $-5/3$. The dashed line shows the same spectrum multiplied by $k^{5/3}$. The small increase in the energy spectrum at the highest wavenumbers is a numerical effect due to a lack of resolution of the very small scales in the dissipative range. It is not due to aliasing and does not affect the inertial range. We have redone a part of our runs with aliasing removed by the method of shifted grids described in Patterson and Orszag (1971), and we did not find any significant difference in the results. The fact that this is an inertial range is confirmed by inspection of the energy flux spectrum, shown in Fig. 2. The energy flux $\Phi(k)$, which is the rate of energy transferred per unit mass by all modes with $k' < k$ to modes with $k' > k$, is constant in the inertial range. This is seen in Fig. 2 to occupy a little more than one decade in wavenumber space.

A 3D view of the vorticity field is given in Fig. 3. The vorticity at each grid point is represented by a vector, here so small that individual vectors can hardly be seen. Vectors are only plotted if their modulus is larger than a given threshold. One can see that the vorticity is organized in thin elongated tubes, as previously reported by Siggia (1981), Kerr (1985) and She *et al.* (1990). The length of these tubes seems to be of order 1 , i.e. of the same order as the integral scale (the cube size is 2π). Their thickness is of the order of a few dissipation scales, here a few grid points. This is confirmed by a more detailed analysis, e.g. a cut through a typical vorticity tube, which shows clearly that its characteristic thickness is a few grid points. The dissipation scale, l_D, in our simulation is of the order of the mesh size, while the Taylor microscale is approximately ten mesh sizes. Therefore, the characteristic tube thickness seems to be intermediate between these two lengths.

From these visualizations, one is led to the conclusion that these vorticity tubes, which seem to be the basic structure of three-dimensional homogeneous turbulence, involve all the scales of the flow. We have done the same kind of pictures of the vorticity field after removing all dissipation range scales. The large scales to which the forcing is applied were also removed. One can see the external regions of the tubes. A single tube along with its environment has been extracted from the vorticity field by a pattern recognition technique (Fig. 4) (Villasenor and Vincent, 1990). Some helical structure can be seen, as noted by She *et al.* (1990). This might also correspond to secondary instabilities and is closely related to the dynamic of the tube itself.

Spatial correlations between vorticity and rate of strain were reported by Kerr (1985). At each grid point, we computed the eigenvalues $\lambda_1, \lambda_2, \lambda_3$ of the rate of strain tensor $S_{ij} = \frac{1}{2}(\partial_i v_j + \partial_j v_i)$. λ_1 is always negative and λ_3 is always positive ($\nabla \cdot \boldsymbol{v} = 0$ implies $\lambda_1 + \lambda_2 + \lambda_3 = 0$). The intermediate eigenvalue λ_2 is found to be positive in approximately two thirds of the cases, in agreement with the results of the analysis of Kerr (1985) and Ashurst *et al.* (1987). Three-dimensional plots of the eigenvector fields are used to study the correlations with vorticity. Let us call $\boldsymbol{e_1}, \boldsymbol{e_2}, \boldsymbol{e_3}$ the eigenvectors associated with λ_1, λ_2, λ_3 respectively, each eigenvector being given the amplitude and sign of its associated eigenvalue. The highest rate of strain is seen to be generally in the vicinity of vorticity tubes. The eigenvector $\boldsymbol{e_3}$ associated with the positive eigenvalue, i.e. the stretching direction, is perpendicular to the vorticity. The eigenvector $\boldsymbol{e_1}$, associated with the negative eigenvalue of the strain, is also perpendicular to the vorticity. The intermediate eigenvector $\boldsymbol{e_2}$ is the one which is generally aligned with vorticity. Our conclusions are drawn from fully three-dimensional examination of the flow using rotation, zoom and clipping. We could not find an example of $\boldsymbol{e_3}$ aligned with vorticity.

The general alignment of vorticity with $\boldsymbol{e_2}$ reflects the quasi two-dimensionality of the vortex tubes. Indeed, in two dimensions, $\lambda_2 = 0, \lambda_3 = -\lambda_1$ and ω would be perpendicular to the plane $(\boldsymbol{e_1}, \boldsymbol{e_3})$. A discussion of this correlation of vorticity with the intermediate rate of strain direction is presented in Ashurst *et al.* (1987), where an explanation is proposed using a model due to Vieillefosse (1982, 1984). According to calculations using this model, a vorticity tube is created along the direction $\boldsymbol{e_3}$ of maximum stretching, but very soon becomes aligned with $\boldsymbol{e_2}$. We are presently following vorticity tubes back in time until the time of their formation. Moreover, we observe that the maxima of velocity form a sheet perpendicular to the axis of the tube. In Fig. 5 the projection of the velocity field on a plane perpendicular to a vorticity tube is shown. Our preliminary impression is that the tubes are generated by an instability due to the shear associated with the velocity sheets. If this is the case, the above picture would not be correct. The results of this investigation will be presented in a forthcoming paper. Moreover, we observe that the large velocity regions are sheets located near and parallel to the axis of the tubes but directed perpendicularly to it.

Acknowledgements. The calculations presented were done on the Cray-2 of CCVR (Centre de Calcul Vectoriel pour la Recherche), Palaiseau, France.

References

Ashurst, W.T., Kerstein, A.R., Kerr, R.M., Gibson C.H.: 1987, *Phys. Fluids* **30**, 2343

Gottlieb, D., Orszag, S.A.: 1977, *SIAM CBMS-NSF Regional Conference Series* in Applied Math. "*Numerical Analysis of Spectral Methods.*"

Kerr, R.: 1985, *J. Fluid Mech.* **153**, 31

Orszag, S.A., Patterson, G.S.: 1972, *Phys. Rev. Lett.* **28**, 76

Patterson, G.S., Orszag, S.A.: 1971, *Phys. Fluids.* **14**, 2538

She, Z.S., Jackson, E., Orszag S.A.: 1988, *J. Sci. Computing* **3**, 407

She, Z.S., Jackson, E., Orszag S.A.: 1990, *Nature* **344**, 226

Siggia, E.D., Patterson, G.S.: 1978, *J. Fluid Mech.* **86**, 567

Siggia, E.D.: 1981, *J. Fluid Mech.* **107**, 375

Yamamoto, K., Hosokawa, I.: 1988, *J. Phys. Soc. Japan* **57**, 1532

Vieillefosse, P.: 1982, *J. de Physique* **43**, 837

Vieillefosse, P.: 1984, *Physica* **125A**, 150

Vincent, A.P., Meneguzzi, M.: 1990, to appear in *J. Fluid Mech.*

Villasenor, J.D., Vincent, A.P.: 1990, submitted to *Computer Vision, Graphics, and Image Processing* (submitted).

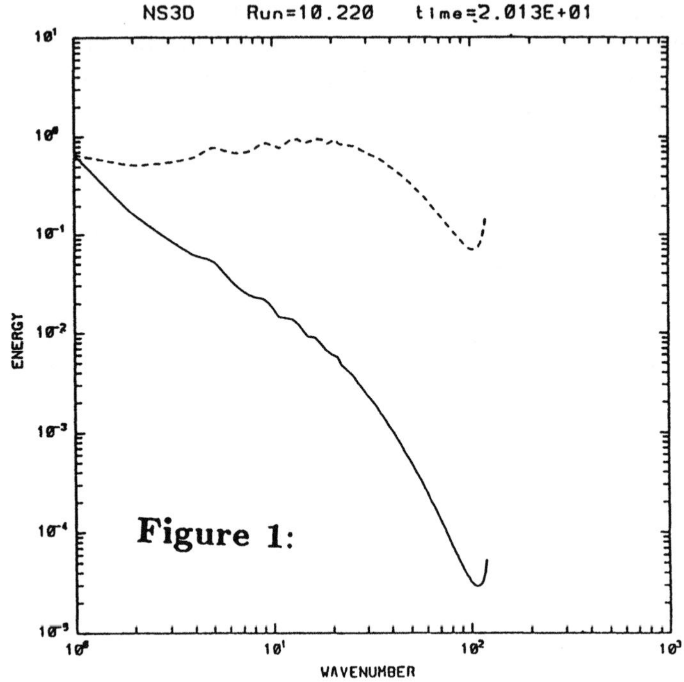

Fig. 1. Energy spectrum at a given time (continuous curve). Same energy spectrum multiplied by $k^{5/3}$ (dashed curve).

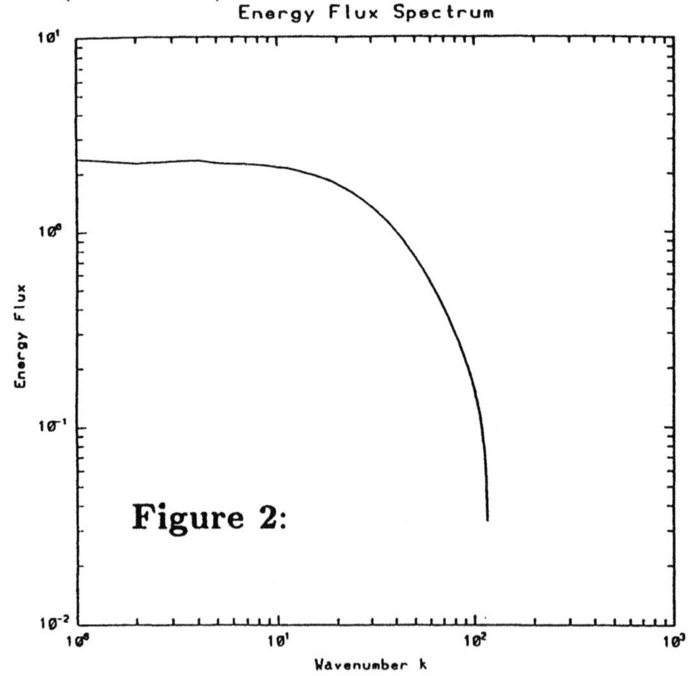

Fig. 2. Energy flux corresponding to the spectrum of Fig. 1.

Figure 3:

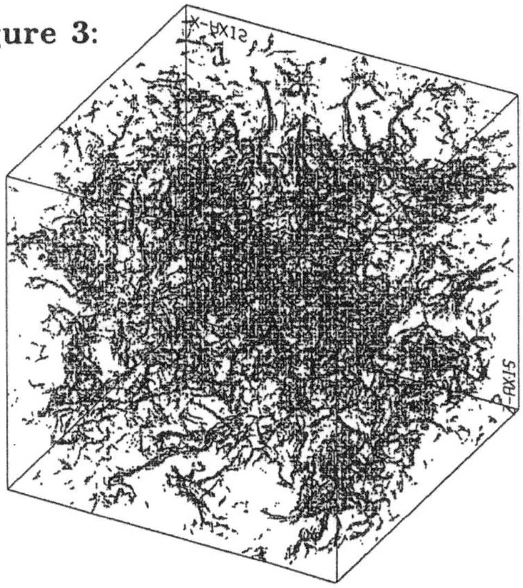

Fig. 3. View of the vorticity field, represented by a vector of length proportional to the vorticity amplitude at each grid point. Only vectors larger than a given threshold value are shown.

Fig. 4. A view of a single vorticity tube after it has been extracted by a pattern recognition technique.

Figure 4:

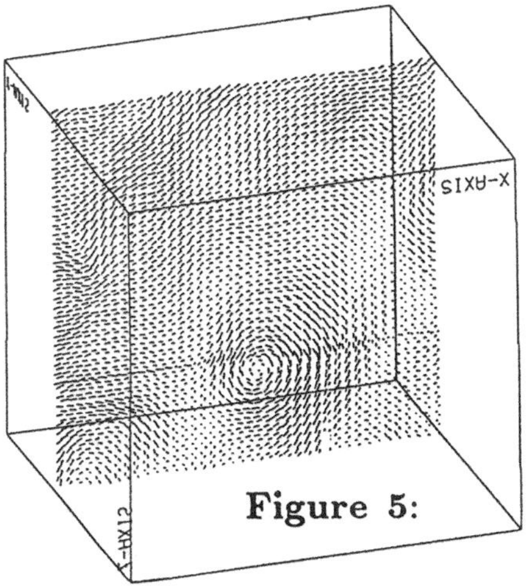

Figure 5:

Fig. 5. Projection of the velocity field on a plane perpendicular to a particular vorticity tube.

Compressible MHD in Spherical Geometry

Lorenzo Valdettaro and Maurice Meneguzzi

CERFACS, 42 Av. G. Coriolis, 31057 Toulouse CEDEX, France

Abstract: The generation of magnetic field by a conducting, compressible fluid inside a spherical shell is studied by direct numerical simulations. A pseudo-spectral method is used in order to resolve accurately all the scales present in the problem. The range of parameters considered is the following: a unit Prandtl number, Rayleigh numbers up to 100 times critical, Taylor number 625, an aspect ratio of 2, a Mach number slightly less than 1, and pressure and temperature scale heights of the order of the thickness of the shell. A dynamo effect is observed for magnetic Prandtl numbers larger than 1. We present the properties of the turbulent flow, the role of the helicity and of the differential rotation in the enhancement of the magnetic field, and the spectral properties of the flow fields.

1. The reference state

We consider a compressible fluid inside a spherical shell heated from within. We take as thermodynamic equation the perfect gas law. The equations are solved using a pseudo-spectral semi-implicit method (Valdettaro and Meneguzzi, 1989). No model is used for the small scales; the resolution used is 64 points in latitude, 128 points in longitude and 65 points in radius.

The reference state is determined as follows:

We suppose that the equilibrium quantities depend only on radius, and that the gravity is due to the inner core, i.e. we neglect the self-gravitation. Therefore the temperature, density and gravity profiles are of the form:

$$g(r) = \frac{a_0}{r^2}, \qquad T(r) = a_1 + \frac{a_2}{r}, \qquad \rho(r) = a_3 T^m,$$

with a_0, a_1, a_2 and a_3 arbitrary constants; m is the polytropic index. The constants are determined by giving the values of the following quantities:

The Rayleigh number

$$R_a \equiv - \left. \frac{dS/C_p}{dr} \frac{gh^3}{\kappa\nu} \right|_{top}$$

where S is the entropy, C_p is the specific heat at constant pressure, h is the thickness of the shell, g is the gravity and κ and ν are the thermal conductivity and the kinematic viscosity. The Rayleigh number is the parameter which governs the onset of convection and of turbulent motions. We have taken $R_a = 100R_{ac}$, R_{ac} being the critical Rayleigh number for the onset of convection.

The stratification rate $\chi \equiv \rho_{top}/\rho_{bottom}$. We take a constant density, i.e. $\chi = 1$. Consequently the polytropic index m is equal to zero.

The entropy gradient

$$\varepsilon \equiv -\left.\frac{dS/C_p}{d\log r}\right|_{top} .$$

This quantity measures the relative importance of compressibility, since it is proportional to the square of the Mach number (Gilman and Glatzmaier, 1981). We choose a value of order 1 ($\varepsilon = 3$) in order to have non negligible compressible effects. It turns out from the computations that the average Mach number is slightly less than 1.

The other quantities which define the initial state are:

The Taylor number $T_a = \Omega^2 d^4/\nu^2$; we have taken a moderate rotation rate: $T_a = 625$. The aspect ratio β, i.e. the ratio between the outer and the inner radius of the shell; we have chosen the value $\beta = 2$. The Prandtl number: we have assumed the kinematic viscosity and the thermal diffusivity constant and equal. Therefore $P_r \equiv \nu/\kappa$ is equal to 1. This assumption is in good agreement with the solar context if we assume that both ν and κ there can be represented there as eddy diffusivities. Finally we have assumed a perfect monoatomic gas, i.e. $c_p = 5/2$, $c_v = 3/2$.

The boundary conditions we impose are the following: we specify the temperature at the bottom and at the top of the shell. The ratio between the inner and outer temperature is fixed by the parameters we have defined above: $T_{bottom}/T_{top} = 1 + 2\varepsilon\beta c_p/(c_v - m) = 21$. We have thus between 2 and 3 temperature scale heights. For the velocity we consider rigid boundary conditions, i.e. $\boldsymbol{v} = 0$. Finally for the magnetic field we take a perfect conductor at the bottom of the shell (i.e. the radial components of the magnetic field and of the electric current vanish there) and we allow only for a radial component of the magnetic field at the top boundary.

2. The properties of the flow

We let the fluid evolve from the initial state defined above and without magnetic field until a statistically stationary state is achieved. We summarize here the characteristics of the flow we obtain.

The entropy profile, averaged in latitude and longitude, is shown in Fig. 1. As expected, the flow field acts to maintain a constant entropy inside the shell. The central adiabatic part is surrounded by the inner and outer thermal boundary layers, which represent roughly 10% of the total width. In Fig. 2 we plot the density profile. We find, as in Massaguer and Zahn (1980), a reversal of the density perturbation near the top of the shell, indicating a negative buoyancy force there. This effect does not exist in the incompressible case.

The efficiency of the convection is measured by the Nusselt number which is the ratio of the total superadiabatic energy flux divided by the purely conductive energy flux. The profile we obtain is strongly dependent on the radius, going from a minimum of about 3 near the boundaries to a maximum value of 15 at a middle radius.

The effects of compressibility are measured by the average Mach number, which in our case is around 0.3. The maximum Mach number fluctuates in time around a value of 0.6. The flow is therefore entirely subsonic. The Reynolds number averaged over the sphere is around 60, with a maximum value of 140. The Rossby number $R_o = R_e/\sqrt{T_a}$, which measures the importance of rotation is slightly less than 1.

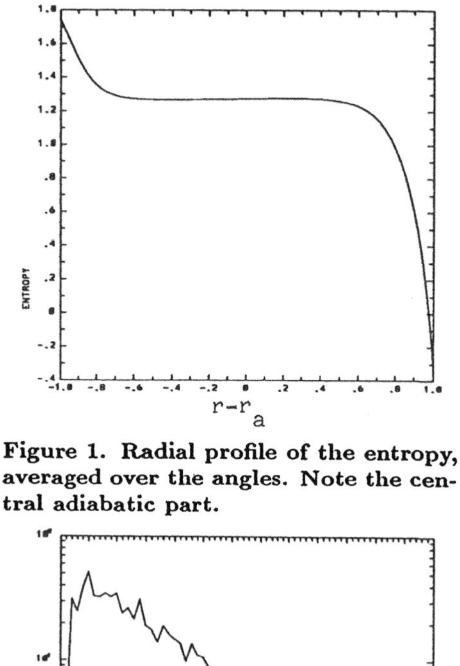

Figure 1. Radial profile of the entropy, averaged over the angles. Note the central adiabatic part.

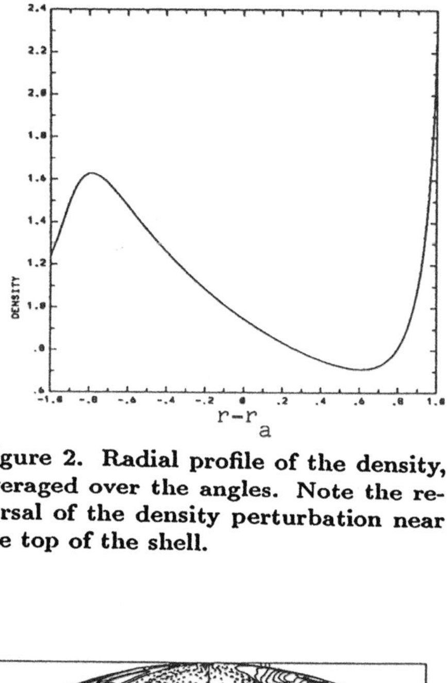

Figure 2. Radial profile of the density, averaged over the angles. Note the reversal of the density perturbation near the top of the shell.

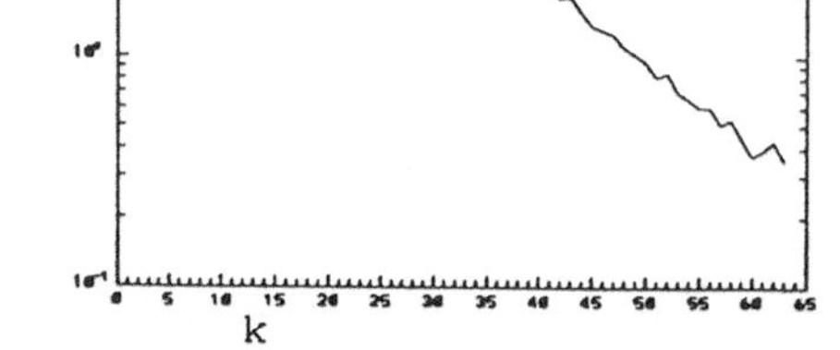

Figure 3. Horizontal spectrum of the radial component of the velocity at the middle radius.

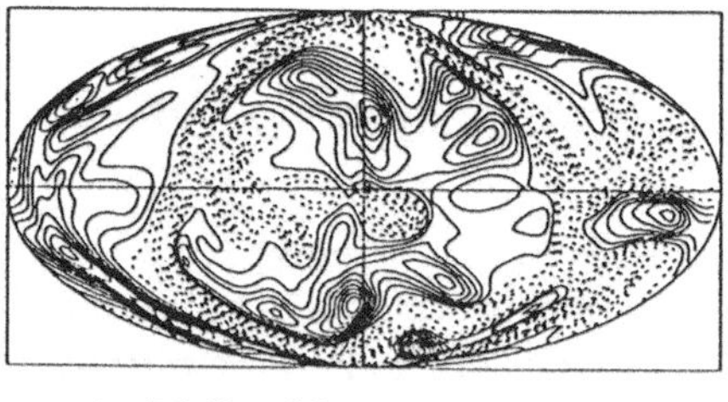

Figure 4. Mollweide projection on the sphere of the isocontours of the radial velocity at middle radius. Continuous lines: V_r pointing toward the exterior; broken lines: V_r pointing toward the interior.

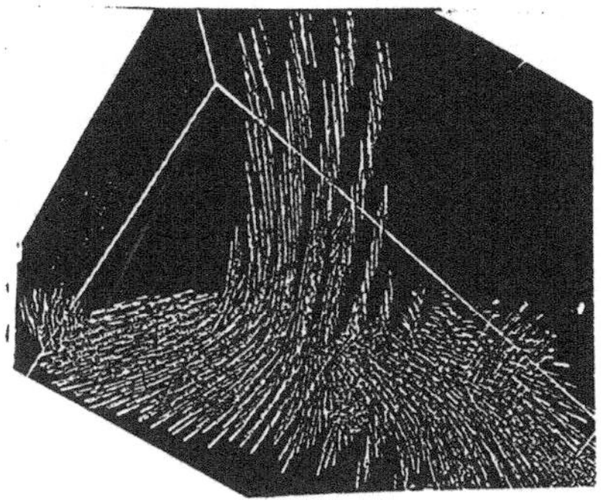

Figure 5. View of a descending plume. The velocity vectors are represented by a vector of length proportional to the velocity. Only vectors larger than a given threshold are shown. The outer part of the shell is on top.

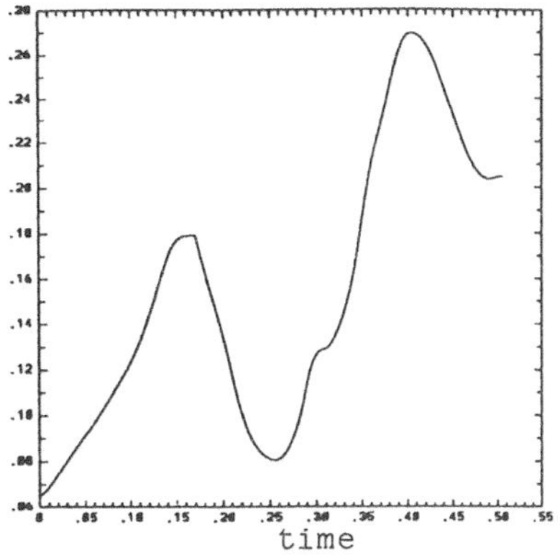

Figure 6. Evolution in time of the total magnetic energy during the nonlinear phase.

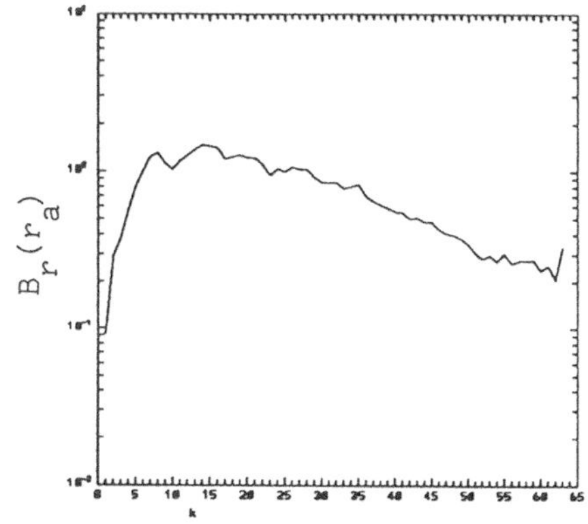

Figure 7. Horizontal spectrum of the radial magnetic field at middle radius.

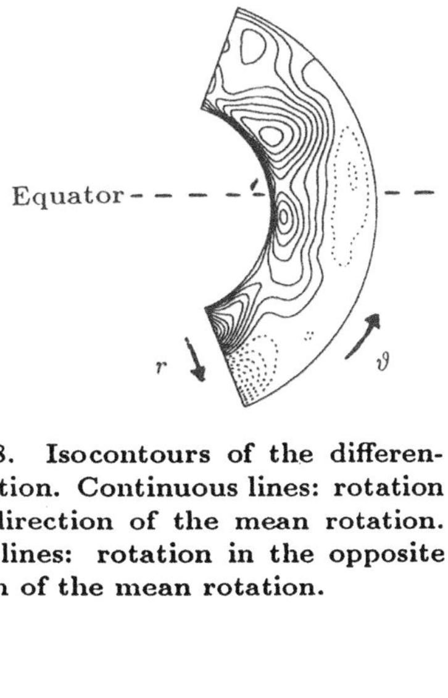

Figure 8. Isocontours of the differential rotation. Continuous lines: rotation in the direction of the mean rotation. Broken lines: rotation in the opposite direction of the mean rotation.

Figure 9. Isocontours of the toroidal component of the magnetic field. Cut in a meridional plane. Note the strong magnetic tube at the base of the convection zone.

The fluid displays motions in a wide range of scales; this is clearly seen from the spectrum of the radial velocity, which is plotted in Fig. 3. The large scale motions are organised in the form of convection cells. In Fig. 4 are shown the isocontours of radial velocity at the middle radius. We see that there are both ascending and descending cells; however, the most intense cells are descending. In Fig. 5 is shown a descending plume: it has the form of a thin tube, enlarged at the base, and extending throughout the depth of the shell.

3. The dynamo effect

After the statistically stationary state has developed we have introduced a seed magnetic field of very low intensity ($E_m/E_{kin} \simeq 10^{-5}$) and we let it evolve in time. At the beginning we have frozen the fluid and we let the magnetic field evolve alone. The equation for the magnetic field becomes linear and the evolution is exponential in time (it is the "kinematic dynamo"). We find, for a magnetic Prandtl number $P_m = 2.5$, an intensification of the magnetic field with a small growth rate, of the order of the thermal diffusion time. We have thus a slightly supercritical dynamo effect. Higher values of the magnetic Prandtl number would have lead to stronger dynamo effects, but we could not increase P_m above that value, without increasing the resolution of the computations.

We have then let the fluid evolve together with the magnetic field for 0.4 thermal times, which is much larger than the eddy turnover time (of the order of $0.03t_{th}$ in our calculations) but it is only of the order of the growth rate of the magnetic field we expect from the linear analysis. In Fig. 6 we plot the evolution of the magnetic energy in time. We clearly see an increase, which is however much lower than those observed in other direct simulations (see e.g. Meneguzzi and Pouquet, 1989; Brandenburg *et al.*, 1991). Although we have not reached the stage of the nonlinear saturation of the magnetic field, we are able to explore and interpret the growth phase.

In Fig. 7 is plotted the spectrum of the radial component of the magnetic field, at middle radius. As expected (Meneguzzi *et al.*, 1981) the magnetic field spectrum appears to be more flat than that of the velocity field (Fig. 3). This means that the magnetic field displays structures on smaller scale than the velocity field.

The two major factors which tend to increase the magnetic field are the differential rotation $\langle v_\phi \rangle / r \sin\vartheta$, which generates strong toroidal magnetic field, and helicity $\langle \boldsymbol{v} \cdot \boldsymbol{\omega} \rangle$, which generates poloidal fields from toroidal fields. We plot in Fig. 8 a typical profile of the differential rotation we obtain. We clearly see a strong generation of differential rotation near the base of the convection zone. As expected, this strong differential rotation generates a strong toroidal magnetic field. This is shown in Fig. 9, where the contour lines of B_ϕ are plotted in a meridional plane. We see the section of a strong magnetic tube at the base of the convection zone.

The relative helicity ($H \equiv \langle \boldsymbol{v} \cdot \boldsymbol{\omega} \rangle / (\langle v^2 \rangle \langle \omega^2 \rangle)^{1/2}$) in our run fluctuates between a minimum of 1% and a maximum of 6%. This value is small, although of the same order of magnitude of that found in Meneguzzi and Pouquet (1989) in

Boussinesq approximation, and could explain the small growth rate we obtain for the magnetic field.

In conclusion, our three-dimensional simulation suggests a consistent compressible dynamo, based at the bottom of the convection zone, probably due to an $\alpha - \Omega$ mechanism. This mechanism (Parker, 1955) has been the most successful one for the solar dynamo. Applications of our results to the solar context are however not possible, due to the large difference between the solar parameters and those we can afford with present computers. We think that the interest of such direct simulations is to give a better insight to the basic mechanisms and lead to a better theoretical understanding of the dynamo problem.

References

Brandenburg, A., Jennings, R., Nordlund, Å., Stein, R.F., Tuominen, I.: 1991, these Proceedings

Gilman, P.A., Glatzmaier, G.A.: 1981, *Astrophys. J. Suppl.* **45**, 335

Massaguer, J.M., Zahn, J.P.: 1980, *Astron. Astrophys.* **87**, 315

Meneguzzi, M., Frisch, U., Pouquet, A.: 1981, *Phys. Rev. Letters* **47**, 1060

Meneguzzi, M., Pouquet, A.: 1989, *J. Fluid Mech.* **205**, 297

Parker, E.N.: 1955, *Astrophys. J.* **121**, 491

Valdettaro, L., Meneguzzi, M.: 1989, in *Proceedings from the Workshop on Supercomputing Tools for Science and Engineering*, Pisa

The Role of Overshoot in Solar Activity: A Direct Simulation of the Dynamo

A. Brandenburg [1,2], R.L. Jennings [3], Å. Nordlund [4],

R.F. Stein [5], I. Tuominen [1]

[1]Observatory and Astrophysics Laboratory, University of Helsinki, Finland
[2]NORDITA, Blegdamsvej 17, DK-2100 Copenhagen, Denmark
[3]Dept. of Appl. Math. and Theor. Physics, Cambridge CB3 9EW, UK
[4]Copenhagen University Observatory, DK-1350 Copenhagen K, Denmark
[5]Dept. of Physics and Astronomy, Michigan State University, USA

Abstract: We investigate convective overshoot in a layer of electrically conducting fluid. The radiative conductivity is assumed to be larger in the lower part of the layer which makes it stable to convective motions, yet penetrative convection from the upper layer can occur. The numerical resolution is 63^3 gridpoints. We observe a dynamo effect for magnetic Reynolds numbers around one thousand when a magnetic seed field is rapidly concentrated to form flux tubes. Later the average magnetic field is expelled from the convectively unstable regions, but it accumulates in the interface between the convection zone and the radiative interior.

1. Introduction

The nature of the solar dynamo is unclear. Does the dynamo operate in the entire convection zone, or in the overshoot layer beneath? Is magnetic buoyancy a "problem" for the dynamo? Is the solar dynamo fast or slow?

We investigate these questions using a direct simulation of turbulent hydromagnetic convection. The presence of a stably stratified overshoot layer is modelled by assuming the radiative diffusivity in the lower half of the simulated domain to be larger than that in the upper unstably stratified part (cf. Hurlburt *et al.*, 1986).

2. The basic equations

We solve the induction equation and the equations for conservation of momentum, energy, and mass

$$\frac{\partial \boldsymbol{B}}{\partial t} = \operatorname{curl}(\boldsymbol{u} \times \boldsymbol{B}) + \eta \nabla^2 \boldsymbol{B},$$

$$\frac{\mathrm{D}\boldsymbol{u}}{\mathrm{D}t} = -\frac{1}{\rho}\nabla p + \boldsymbol{g} - 2\boldsymbol{\Omega} \times \boldsymbol{u} + \frac{1}{\rho}\boldsymbol{J} \times \boldsymbol{B} + \frac{1}{\rho}\mathrm{Div}\,\boldsymbol{\tau},$$

$$\frac{\mathrm{D}e}{\mathrm{D}t} = -\frac{p}{\rho}\operatorname{div}\boldsymbol{u} + \frac{1}{\rho}\nabla(\mathcal{K}\nabla e) + \tau_{ij}u_{i,j} + \frac{\eta}{\rho}\boldsymbol{J}^2,$$

$$\frac{\mathrm{D}\ln\rho}{\mathrm{D}t} = -\operatorname{div}\boldsymbol{u},$$

For details of the notation see Paper I (Brandenburg *et al.*, 1990). We assume the upper and lower boundary to be a stress-free perfect conductor. A constant radiative flux is imposed at the bottom and the surface temperature is kept constant. In the current implementation of the code one time step takes for 63^3 grid points about 3.4 sec of CPU time. With the four processors on the Cray-XMP/432 the real system time can be of the order of only one second per time step.

The governing nondimensional parameters in our simulation are a Taylor number of 10^5 (based on the thickness of the unstable layer), a Rayleigh number of 10^6 (approx. 50 times supercritical), a Prandtl number of 0.2, and a magnetic Prandtl number $\nu/\eta = 4$. The density contrast is approximately 1:10. The resulting Reynolds number is around 300 (based on rms-velocity) and the Chandrasekhar number 3×10^3. The Rossby number is approximately unity and the Elsasser number about ten.

3. Results

We find large scale coherent magnetic structures similar to vorticity tubes seen in homogeneous turbulence (She *et al.*, 1990; Vincent and Meneguzzi, 1990). In a video animation we observed that magnetic flux tubes are pulled downwards and wounded up close to the interface between the convection zone and the radiative interior. A dynamo effect is found for magnetic Prandtl numbers larger than unity. The magnetic Reynolds numbers are in our case much higher than those for dynamo action in Bousinesq convection (Meneguzzi and Pouquet, 1989). The initial growth rate of magnetic energy is comparable with the convective turnover time and we can therefore speak of a fast dynamo.

The magnetic energy has a maximum at the interface. This is also the location where both induction effects and Ohmic dissipation are largest. The fact that there is no accumulation of magnetic flux in the upper layers leads us to believe that the perfectly conducting upper boundary is not artificially suppressing magnetic buoyancy, as suggested by Petrovay (1991) during the conference. However, we often observe that tubes organize themselves vertically in the upper unstably stratified part and thus escape the buoyancy mechanism to work efficiently.

We have plotted in Fig. 1 vectors of vorticity $\boldsymbol{\omega} = \mathrm{curl}\, \boldsymbol{u}$, electric current $\boldsymbol{J}$, and magnetic field $\boldsymbol{B}$. Vectors are plotted only where their magnitude exceeds a certain threshold.

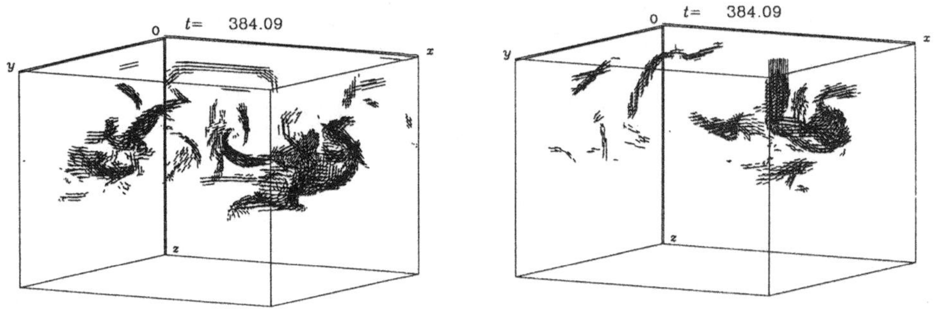

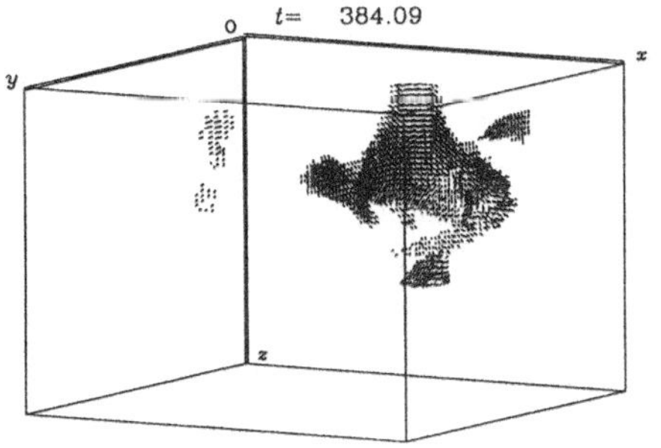

Fig. 1. Formation of tubes, sheets and current sheets: the vorticity vectors are organized into tubes whilst the electric current forms sheets around these vorticity tubes. The magnetic field forms tubes too and resembles fox tails swishing above the interface! Note that the magnetic field vectors are, in some cases, aligned with the vorticity vectors.

References

Brandenburg, A., Nordlund, Å., Pulkkinen, P., Stein, R.F., Tuominen, I.: 1990, *Astron. Astrophys.* 232, 277 (Paper I)

Hurlburt, N.E., Toomre, J., Massaguer, J.M.: 1986, *Astrophys. J.* **311**, 563

Meneguzzi, M., Pouquet, A.: 1989, *J. Fluid. Mech.* **205**, 297

Petrovay, K.: 1991, these Proceedings

She, Z.-S., Jackson, E., Orszag, S. A.: 1990, *Nature* **344**, 226

Vincent, A., Meneguzzi, M.: 1990, *J. of Fluid Mech.*, in press

Vector Potential Magnetic Null Points

Klaus Galsgaard and Åke Nordlund

Copenhagen University Observatory,
Øster Voldgade 3, 1350 Copenhagen K, Denmark

Recent observations (Tarbell *et al.*, 1990; see also Ruzmaikin *et al.* these Proceedings) show that the surface distribution of magnetic fields on the solar surface is selfsimilar, with an approximately constant fractal dimension over a large range of horizontal scales. Also, recent ideas about flare energy release assumes a "fibrous" corona, with the energy release consisting of many small scale events (Vlahoz, 1989) Prompted partly by these observations, we investigate the properties of self-similar three-dimensional magnetic fields.

1. Magnetic null points

The global connectivity of non-trivial 3-D magnetic fields is strongly influenced by the presence of critical points (null points) where all components of the magnetic field vanish. In a neighbourhood of such points, the magnetic field is characterized by the magnetic field gradient matrix $\underline{\underline{\mathbf{G}}} = B_{i,j}$;

$$\mathbf{B}(\mathbf{r}) = \sum_j B_{i,j}\delta r_j = \underline{\underline{\mathbf{G}}} \cdot \delta\mathbf{r} \,. \tag{1}$$

where $\underline{\underline{\mathbf{G}}}$, again, is completely characterized by its eigenvalues and eigenvectors. Since $div(\mathbf{B}) = 0$, the sum of the eigenvalues ($= trace(\underline{\underline{\mathbf{G}}})$) vanishes. Thus there are only two, non-trivially different, combinations of eigenvalues; those with all eigenvalues real (one with the sign differing from the other two), and those with one real and two complex conjugate eigenvalues. Corresponding to these two cases, there are two different classes of field structure near magnetic null points; non-spiralling field lines corresponding to the case with all real eigenvalues, and spiraling field lines corresponding to the case with complex eigenvalues. In both cases, the field lines in one direction asymptotically approach an axis along the eigenvector belonging to the largest (real) eigenvalue, while in the other direction the field lines asymptotically approach the plane containing the other two eigenvectors (Fukao *et al.*, 1975; Green, 1988).

2. Field representation

We construct schematic, but reasonably complicated ("non-textbook") magnetic fields, by representing the field by the real part of a complex exponential series. Specifically, here, we write $\mathbf{B}$ in terms of a vector potential $\mathbf{A}$

$$\mathbf{B} = \nabla \times \mathbf{A} \,, \tag{2}$$

where

$$\mathbf{B}(\mathbf{r}) = \mathrm{Re}\left\{ \sum_i \sum_j \mathbf{f}_i \mathbf{A}_j e^{\mathbf{g}_i \mathbf{k}_j \cdot \mathbf{r}} \right\} . \tag{3}$$

j indexes groups of components with similar wave numbers $\mathbf{k}_j$, and i indexes several sets of these groups, scaled with wavenumber scaling factors g_i and amplitude scaling factors f_i.

For a force-free field, the current is proportional to the magnetic field

$$\mathbf{I} = (\nabla \times \mathbf{B}) = \alpha \mathbf{B} \, . \tag{4}$$

For the complex exponential series, this may be shown to imply that the size of α must be equal to the length of the wavevector, $\alpha = |\mathbf{k}|$. A more detailed analysis shows that $\mathbf{A}$, in the force free case, must lie in the plane orthogonal to $\mathbf{k}$, and must consist of a certain combination of two linearly independent vectors in that plane.

3. Examples of 3-D topologies

To construct force-free fields with single (periodic) null points, we use groups of three terms with different $\mathbf{k}$ directions. Sets of such groups, with increasing wavevector and decreasing amplitude, where each group fulfills the force-free condition, produce fields that are not exactly force-free, but must be in a rather low energy state.

The panels in Figure 1. show two examples of such fields. The first column is a sequence where one null point from each group remains in the same location. The second column is a sequence where the null points for the $i >= 2$ groups are displaced by 1/8 period with respect to the null points of the $i = 1$ group. The plots are made by first finding all null points in the region, then placing a small number of starting points at a small distance from the null point, around the main axis, and trace these in both directions. As illustrated by the figure, both these selfsimilar fields exhibit a complicated field structure with many null points. Whether or not superpositions of fluctuations on several scales results in many null points depends on the relative amplitudes of the fluctuations. If the field gradients near a null point of a large scale component are smaller than the gradients of the smaller scale components, then in a neighbourhood of the null point of the large scale fluctuation the small scale fluctuations are able to create a number of null points. These form a cluster of "satellite" null-points around the null point of the large scale fluctuation (cf. Galsgaard and Nordlund, 1990ab). Presumably, for a field with a continuous spectrum of fluctuations, there is a critical spectrum slope, below which the field topology is vastly more complicated than above.

References

Fukao, S., Masayuki, U., Takao, T.: 1975, *Rep. Ionos. Jpn.* **29**, 133

Galsgaard, K., Nordlund, Å.: 1990a, in *The Dynamic Sun*, ed. L. Dezsõ, Debrecen, p. 146

Galsgaard, K., Nordlund, Å.: 1990b, in *Mechanisms of Chromospheric and Coronal Heating*, ed. P. Ulmschneider *et al.*, Springer, Berlin, in press

Green, J.M.: 1988, *J.Geophys. Res* **93**, 8583

Tarbell, T. *et al.*: 1990, *Solar Photosphere: Structure, Convection and Magnetic Fields*, J.O. Stenflo (ed.), IAU Symp. 138, Kluwer, Dordrecht, p. 147

Vlahoz, L.: 1989, *Solar Phys.* **121**, 431

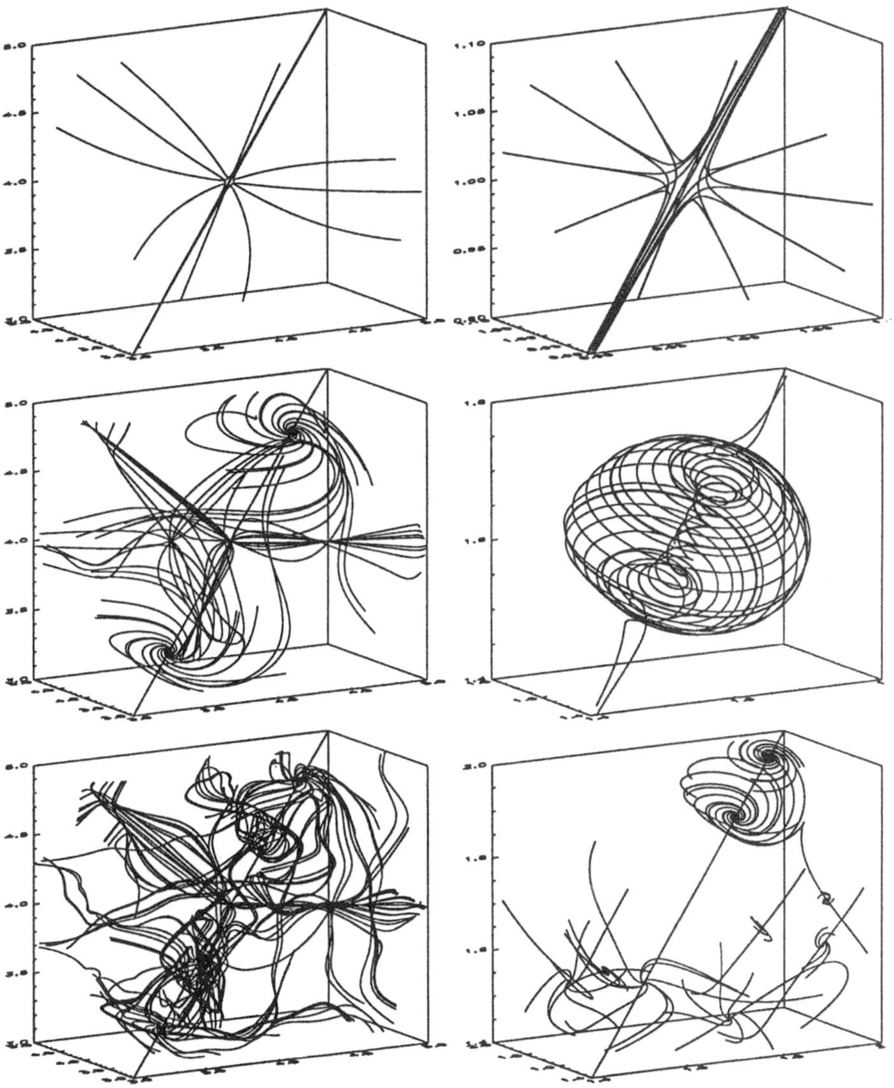

Fig. 1. The panels show two series of magnetic fields generated from groups of three vector potentials with increasing wave number and decreasing amplitude. The left column shows a sequence of magnetic field line plots where one null point from each group with $i >= 2$ is located at the same position as a null point from $i = 1$ group. The field is defined by the following wavevectors, vector potentials and scaling constants:
$\mathbf{k}_1$=[(0,1),(0,0),(0,0)], $\mathbf{A}_1 = [(0,0),(1,1),(-1,1)]$,
$\mathbf{k}_2$=[(0,0),(0,1),(0,0)], $\mathbf{A}_2 = [(-1,1),(0,0),(1,1)]$,
$\mathbf{k}_3$=[(0,0),(0,0),(0,1)], $\mathbf{A}_3 = [(1,1),(-1,1),(0,0)]$,
g_i=(1,4,16), f_i=(1,.13,.0169), i=1,3.
The plots in the right column are made from the same wavevectors $\mathbf{k}$, and scaling constants g_i, f_i but with these vector potentials:
$\mathbf{A}_1$=[(0,0),(1,0),(0,1)], $\mathbf{A}_2$=[(0,1),(0,0),(1,0)], $\mathbf{A}_3$=[(1,0),(0,1),(0,0)].
The last plot in this series has been simplified, by following the field lines for a shorter distance than for the other plots.

Magnetic Tubes in Overshooting Compressible Convection

R. L. Jennings[1], A. Brandenburg[2], Å. Nordlund[3],

R. F. Stein[4], I. Tuominen[2]

[1]D.A.M.T.P., Silver St., Cambridge, CB3 9EW, UK
[2]Observatory and Astrophysics Laboratory, University of Helsinki, Finland
[3]Copenhagen University Observatory, DK-1350 Copenhagen K, Denmark
[4]Department of Physics and Astronomy, Michigan State University, USA

Abstract: A magnetic tube is introduced into turbulent compressible penetrative convection. After being strongly advected, most of the magnetic flux is stored in the overshoot region. With rotation there are meridional travelling waves.

We investigate the evolution of a strong toroidal magnetic field initially beneath the solar convection zone by simulating highly supercritical compressible convection with overshoot, see Hurlburt *et al.* (1986). Once the convection is well developed the magnetic field is added. The geometry is Cartesian, and we only consider two-dimensional solutions ($\partial/\partial y = 0$). The xz plane is convectively unstable for $0 \le z \le 1$, but stably stratified for $1 < z \le 2$. We assume that solutions are periodic in x and discretize using 63^2 gridpoints. A similar model is described more fully by Brandenburg *et al.* (1990).

An initial magnetic field $B_y(x,z)\hat{\mathbf{y}} = B_0 \exp\left[-100\{(x-1)^2 + (z-1.25)^2\}\right]\hat{\mathbf{y}}$ represents the flux tube, which is advected and diffused according to the induction equation:

$$\rho\,[\partial/\partial t + (\mathbf{u}\cdot\nabla)]\,(B_y/\rho) = \eta\nabla^2 B_y\,. \qquad (1)$$

A magnetic pressure force is all that the Lorentz force yields, but this leads to an initial pressure discontinuity. Alternatively, the tube can be inserted such that the density or temperature is initially discontinuous, yet in each case sound waves rapidly smooth any steep gradients and there is a progression towards a final state in which most of the flux is stored in the overshoot region, see Figs. 2 and 3.

Including rotation makes the flow move poleward as a travelling wave. Inserted magnetic tubes also moves poleward, albeit rather slowly. This is suggestive of the latitudinal migration of magnetic features during the solar cycle. A detailed paper is in preparation.

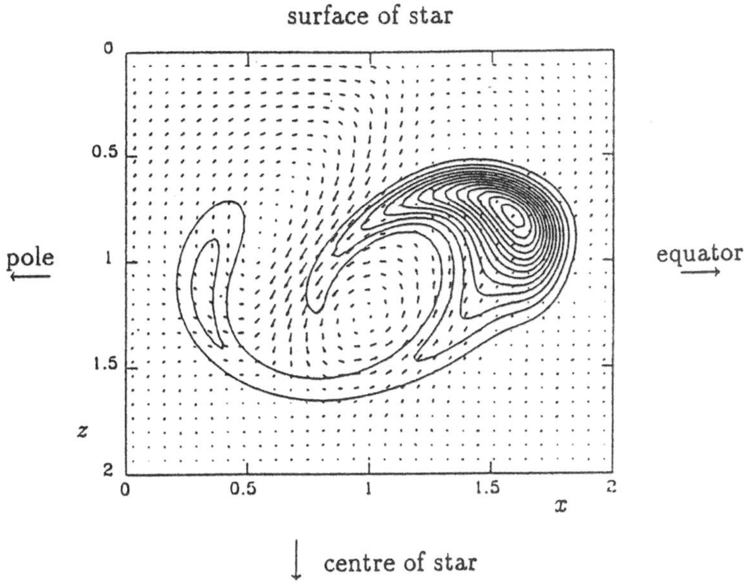

Fig. 1. Magnetic buoyancy makes the tube rise and expand. However, most of the tube's expansion is due to the decreasing density as it rises. Contours are of B_y, and arrows show the flow $\mathbf{u}$.

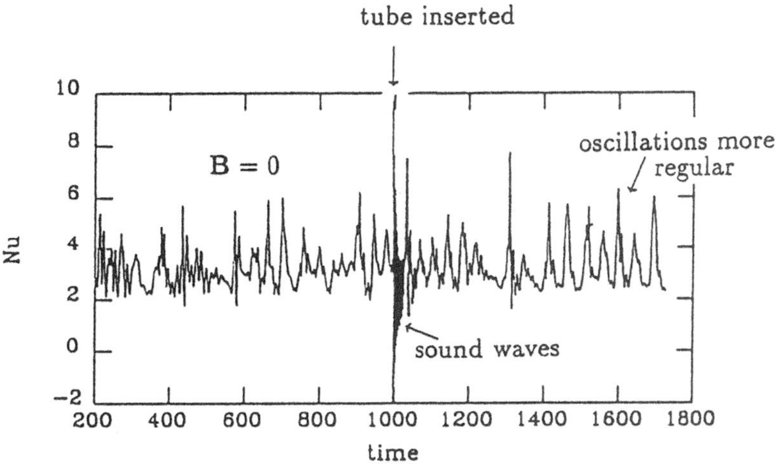

Fig. 2. A timeseries of the Nusselt number showing aperiodic oscillations with periods characteristic of internal gravity waves.

$t = 1427$ (squeeze)

$t = 1452$ (relax)

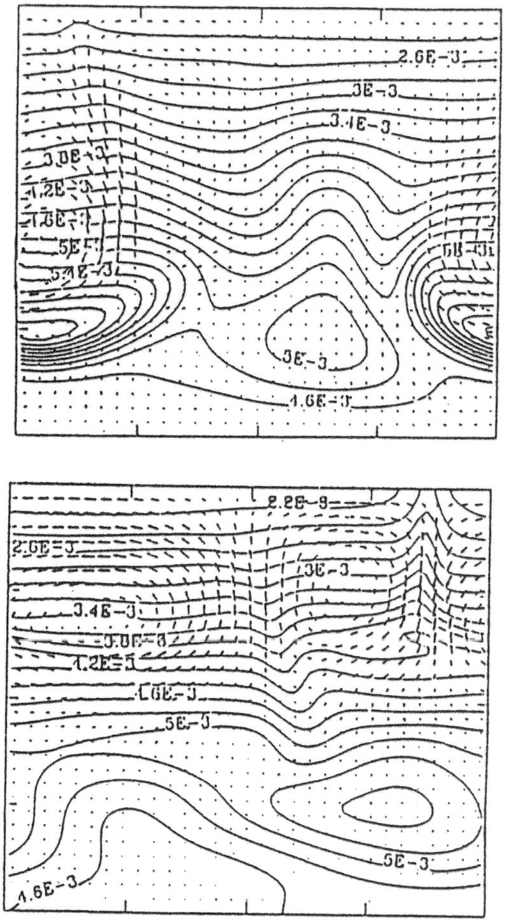

Fig. 3. Descending plumes compress the stored field, which then relaxes before the next compression. The frequency of compressions is compatible with that of gravity waves.

References

Brandenburg, A., Nordlund, Å., Pulkkinen, P., Stein, R.F., Tuominen, I.: 1990, *Astron. Astrophys.* **232**, 277

Hurlburt, N.E., Toomre, J., Massaguer, J.M.: 1986, *Astrophys. J.* **311**, 563

Convective Cores in Stellar Models

I.W. Roxburgh and M. Monteiro

Astronomy Unit, Queen Mary and Westfield College, University of London

Abstract: The condition for homogeneous radiative stellar models to be marginally stable to convection at the centre is investigated for the family of models where the opacity κ and energy generation ε are given by power laws in temperature and density $\kappa = \kappa_0 \rho^{\alpha} T^{-\beta}$, $\varepsilon = \varepsilon_0 \rho T^{\eta}$. The Naur-Osterbrock (1953) condition $6\eta > 6 + 10\beta - 15\alpha$ is a necessary but not sufficient condition. A better estimate is obtained by taking the effective polytropic index n = dlogP/dlogT - 1 to be a linear function of temperature T throughout the model. This gives the condition

$$6\eta = 10\beta - 15\alpha + \frac{12 + 4\beta}{1+\alpha}$$

The predictions of this condition agree well with results for a set of stellar models $0 \le \alpha \le 1$, $0 \le \beta \le 5$.

1 Analysis.

The equations of stellar structure for homogeneous radiative stellar models with an ideal gas equation of state and the opacity and energy generation given by power laws are:

$$\frac{dP}{dr} = -\frac{GM_r}{r^2}\rho \qquad \frac{dT}{dr} = -\frac{3\kappa\rho L_r}{16\pi ac T^3 r^2} \qquad \frac{dM_r}{dr} = 4\pi r^2 \rho \qquad \frac{dL_r}{dr} = 4\pi r^2 \varepsilon\rho$$

$$P = \frac{\Re}{\mu}\rho T \qquad \kappa = \kappa_0 \rho^{\alpha} T^{-\beta} \qquad \varepsilon = \varepsilon_0 \rho T^{\eta}$$

Dimensionless variables θ, π, ϕ, λ, ζ are defined by $T = T_c\theta$, $P = P_0\pi$, $M_r = M_0\phi$, $L = L_0\lambda$, $r = R_0\,\zeta$ where T_c is the central value of T, and P_0, M_0, L_0, R_0 are suitably chosen scaling constants, the equations of stellar structure reduce to:

$$\frac{d\pi}{d\zeta} = -\frac{\pi\,\phi}{\theta\,\zeta^2} \qquad \frac{d\theta}{d\zeta} = -\frac{\pi^{1+\alpha}\lambda}{\theta^{4+\alpha+\beta}\zeta^2} \qquad \frac{d\phi}{d\zeta} = \frac{\pi\,\zeta^2}{\theta} \qquad \frac{d\lambda}{d\zeta} = \pi^2\,\theta^{\eta-2}\,\zeta^2$$

The central boundary conditions take the form $\theta = 1$, $\pi = \pi_0$, $\phi = 0$, $\lambda = 0$, $\zeta = 0$.
With $g_0 = \pi_0^{2+\alpha}$ the solution for θ and π can be developed as a series in ζ as

$$\theta = 1 - \pi_0 g_0 \zeta^2 + \frac{\pi_0^2 g_0}{360}[\,11 + 5\alpha + g_0(3\eta - 26 - 5\alpha - 5\beta)]\,\zeta^4 + \ldots.$$

$$\pi = \pi_0 - \frac{\pi_0^2}{6}\zeta^2 + \frac{\pi_0^3(1 - g_0)}{45}\zeta^4 + \ldots$$

The polytropic index n is defined by

$$n+1 = \frac{d\log P}{d\log T} = \frac{d\log\pi}{d\log\theta} = \frac{\theta}{\pi}\frac{d\pi}{d\zeta}\frac{1}{d\theta/d\zeta} = \frac{\theta^{4+\alpha+\beta}\phi}{\pi^{1+\alpha}\lambda}$$

which therefore has the series expansion

$$n+1 = \frac{1}{g_0}\{ 1 + \frac{p_0}{60}[22 + 10\alpha + 2g_0(3\eta -26 - 5\alpha - 5\beta) - 6(1 -g_0)]\zeta^2 + ...$$

If the centre is marginally unstable to convection then n+1 = 2.5 at $\zeta = 0$, $g_0 = 2/5$ and the series expansion for (n+1) in powers of $(1-\theta)$ and the value of $d(n+1)/d\theta$ are

$$n+1 = \frac{5}{2} - \frac{1}{4}(6 + 10\beta - 15\alpha - 6\eta)(1 - \theta) + O((1-\theta)^2)$$

$$\frac{d(n+1)}{d\theta} = \frac{1}{4}(6 + 10\beta - 15\alpha - 6\eta) \qquad \text{at } \theta = 1,\ \zeta = 0.$$

The Naur-Osterbrock (1954) condition follows from requiring that $d(n+1)/d\theta < 0$ at $\zeta = 0$

$$6\eta > 6 + 10\beta - 15\alpha \qquad \text{(N-O)}$$

This is only a necessary condition. A more accurate condition is obtained by requiring that $d(n+1)/d\theta$ is such that (n+1) increases from 2.5 at the $\theta = 1$ to the surface value at $\theta = 0$ (cf Roxburgh 1985). The value of n+1 in the surface layers is readily determined since as $\theta \to 0$, $\lambda \to \lambda_0$, $\phi \to \phi_0$ where λ_0 and ϕ_0 are constants, hence

$$\frac{d\pi}{d\theta} \to \frac{\theta^{3+\alpha+\beta}\phi_0}{\pi^{\alpha}\lambda_0}, \qquad \pi \to \theta^{(4+\beta+\alpha)/(1+\alpha)}\frac{(1+\alpha)\phi_0}{(4+\beta+\alpha)\lambda_0}, \qquad \text{as } \theta \to 0$$

$$n + 1 \to \frac{4+\beta+\alpha}{1+\alpha} \qquad \text{as } \theta \to 0$$

If the centre is marginally stable to convection then (n+1) = 2.5 at $\theta = 1$; hence the average value of $d(n+1)/d\theta$ throughout the model is

$$\frac{d\overline{(n+1)}}{d\theta} = \frac{5}{2} - \frac{4+\beta+\alpha}{1+\alpha}$$

If we now take $d(n+1)/d\theta$ to be constant throughout the model, this mean value is equal to the value at the centre, hence

$$6\eta = 10\beta - 15\alpha + \frac{12 + 4\beta}{1+\alpha} \qquad \text{(R-M)}$$

This is the new criterion to estimate whether stellar models have convective cores.

2 Results.

The equations of stellar structure in the dimensionless form given above were solved to determine η such that n+1 = 2.5 at $\zeta = 0$, given α and β. The solutions are very sensitive to the assumed value of η; if η is too small (n+1) rapidly decreases, whereas if η is too large (n+1) rapidly increases, the requirement that n+1 should tend to

$(4+\alpha+\beta)/(1+\alpha)$ for large ζ (small θ) determines η without too much difficulty. The results are shown in Figure 1. In Figure 2 these numerical results are compared with the predictions of the Naur-Osterbrock condition and the new condition. The predictions of the new condition are in satisfactory agreement whereas the Naur-Osterbrock condition underestimates the value of η required for the existence of a convective core.

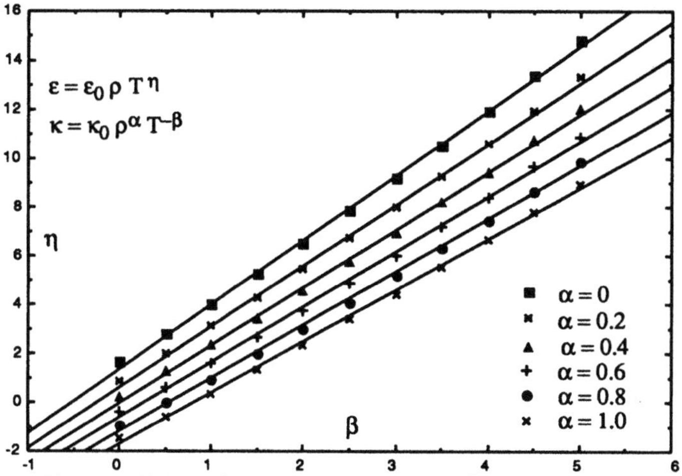

Figure 1. Values of η as a function of α and β, such that radiative stellar models are marginally unstable to convection at the centre.

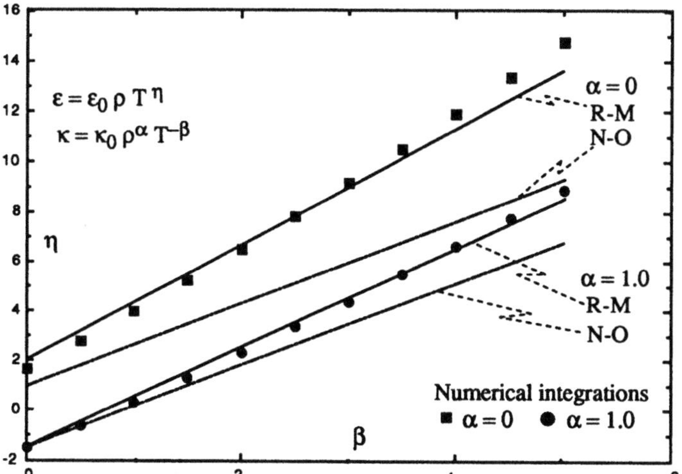

Figure 2. Values of η from numerical integrations, the new (R-M) condition and old Naur Osterbrock (N-O) condition.

References

Naur, P., and Osterbrock, D.E.: 1953, "Convective Cores in Stars", *Astrophys J.* **117**, 306.

Roxburgh, I,W,: 1985, "Present Problems of the Solar Interior", *Solar Physics*, **100**, 21.

Rotational Effects on Reynolds Stresses in the Solar Convection Zone

P. Pulkkinen[1], I. Tuominen[2], A. Brandenburg[2,3], Å. Nordlund[4], R.F. Stein[5]

[1]Dept. of Theoretical Physics, University of Helsinki, Finland
[2]Observatory and Astrophysics Laboratory, University of Helsinki, Finland
[3]NORDITA, Blegdamsvej 17, DK-2100 Copenhagen, Denmark
[4]Copenhagen University Observatory, Denmark
[5]Dept. of Physics and Astronomy, Michigan State University, USA

Abstract: Three-dimensional hydrodynamic simulations are carried out in a rectangular box. The angle between gravity and rotation axis is kept as an external parameter in order to study the latitude-dependence of convection. Special attention is given to the horizontal Reynolds stress and the Λ-effect (Rüdiger, 1989). The results of the simulations are compared with observations and theory and a good agreement is found.

1. Introduction

In traditional mean-field turbulence theory one writes the velocity as a sum of a mean-field part and a fluctuation around the mean, i.e. $\mathbf{u} = \langle \mathbf{u} \rangle + \mathbf{u}'$. This leads to the appearance of the Reynolds stress tensor $Q_{ij} = \langle u'_i u'_j \rangle$ in the equation of momentum. Q_{ij} may be expressed in terms of $\langle \mathbf{u} \rangle$. In the presence of rotation $\Omega = \langle u_\phi \rangle / r \sin\theta$ gives the dominant contribution. According to Λ-effect theory (Rüdiger 1989) the cross-correlations $Q_{r\phi}$ and $Q_{\theta\phi}$ can be written in the form

$$Q_{r\phi} = \nu_T \Omega \left(-\frac{r}{\Omega}\frac{\partial \Omega}{\partial r} + V^{(0)} + V^{(1)} \sin^2\theta + V^{(2)} \sin^4\theta + \ldots\right) \sin\theta, \tag{1}$$

$$Q_{\theta\phi} = \nu_T \Omega w_1 \left(1 + \frac{w_2}{w_1} \sin^2\theta + \ldots\right) \cos\theta \sin^2\theta. \tag{2}$$

In this work we compute the cross-correlations directly from numerical simulations and thus obtain the V- and w-coefficients. We solve the equations of mass, momentum, and energy by using a modified version of the code by Nordlund and Stein (1990):

$$\frac{\partial \rho}{\partial t} + \nabla \cdot (\rho \mathbf{u}) = 0, \tag{3}$$

$$\rho \frac{\partial \mathbf{u}}{\partial t} + \rho(\mathbf{u} \cdot \nabla)\mathbf{u} = -\nabla p + \mathbf{g} - 2\Omega \times \mathbf{u} + \nabla \cdot \boldsymbol{\tau}, \tag{4}$$

$$\rho \frac{\partial e}{\partial t} + \rho(\mathbf{u} \cdot \nabla)e = \nabla \cdot (k \nabla T) - p \nabla \cdot \mathbf{u} + \Phi_{\mathrm{visc}}. \tag{5}$$

2. Method and results

We simulate stellar convection in a three-dimensional box which is located in the convective zone. The orientation of the box is such that its z-axis is parallel to gravity, the x-axis points opposite to the θ-direction, and the y-axis points towards east.

The box is divided into 31^3 mesh points at which the differential equations are stepped forward in time. In our model the density stratification is weak (density contrast is 1.5) and therefore, only part of the convective zone can be covered by the box. An important parameter is the Rossby number which here is close to unity. This is compatible with the value at the bottom of the solar convective zone.

The top and bottom boundaries are assumed to be impenetrable and stress free. We require a constant radiative flux at the bottom and assume the temperature to be fixed at the top. The boundaries in the x- and y-directions are periodic.

Table 1. Comparison of values of the w-coefficients

	w_1	w_2	w_2/w_1
Boussinesq-ansatz	-1.1	$+0.1$	-0.1
Virtanen (1989)	$+1.0$	$+5.6$	5.6
This work	$+0.2$	$+1.0$	5.0
Tuominen and Rüdiger (1989)	$+4.5$	$+2.9$	0.6

Seven different latitudes at the southern hemisphere are considered separately (-90, -75, -60, -45, -30, -15, and 0 degrees). $Q_{r\phi}$ ($\widehat{=} -Q_{yz}$) and $Q_{\theta\phi}$ ($\widehat{=} -Q_{xy}$) are calculated at these latitudes and the free parameters in equations (1) and (2) are adjusted to give the best fit. The error bars are a measure of the fluctuations. The fit for $Q_{r\phi}$ is shown in Fig. 1. In this fit the first term on the right-hand side of equation (1) was removed. At all depths of the box we find the same qualitative behaviour of the V-coefficients: $V^{(0)}$ and $V^{(2)}$ are negative and $V^{(1)}$ is positive. A similar result was obtained by Tuominen and Rüdiger (1989) at the middle of the convective zone. The fit for $Q_{\theta\phi}$ is shown in Fig. 2. In table 1 the ratio w_2/w_1 from our simulations is compared with those from other studies. The Boussinesq-ansatz does not include any Λ-effect while the work by Virtanen (1989, see also Tuominen, 1990) is a fit to Greenwich data and shows good agreement with the present work. Tuominen and Rüdiger base their numerical model on the differential rotation profile of Stenflo (1989) and also have positive w-coefficients, although their latitudinal profile of $Q_{\theta\phi}$ is somewhat different.

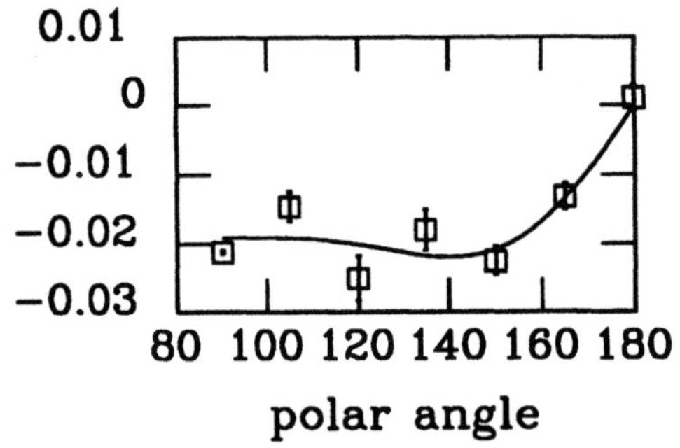

Fig. 1. The fit for $Q_{r\phi}$. The values from simulations (squares) were calculated at the mid-layer of the box.

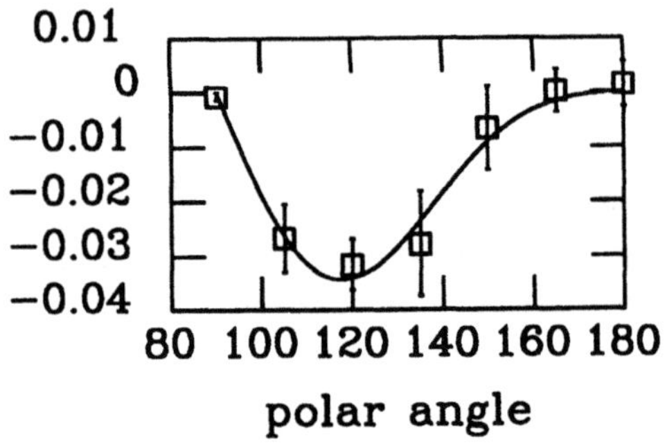

Fig. 2. The fit for $Q_{\theta\phi}$. The solid line is the fit and the squares denote the values from simulations which were calculated at the base of the box.

3. Conclusions

The main results of this work seem promising: the latitudinal dependence of the computed stresses show similar behaviour with the Λ-theory, and the horizontal Reynolds stress agrees with observations. The results for the horizontal Reynolds stress show that the Boussinesq-ansatz is insufficient as it gives the wrong sign for the horizontal cross-correlation. The Λ-theory introduces the V- and w- terms in equations (1) and (2) and gives the "correct" sign and shape for the latitude-dependence of the correlations. It should also be noticed that the production of this kind of effect does not need a global model. This suggests that the main generator of the Λ-effect is the Coriolis force. The presence of impenetrable boundary conditions in our model is, however, unrealistic and deserves further investigation in the future.

References

Nordlund, Å., Stein, R. F.: 1990, *Comp. Phys. Comm.* **59**, 119

Rüdiger, G.: 1989, *Differential Rotation and Stellar Convection: Sun and Solar-type stars*, Akademie-Verlag Berlin

Stenflo, J.: 1989, *Astron. Astrophys.* **210**, 403

Tuominen, I., Rüdiger, G.: 1989, *Astron. Astrophys.* **217**, 217

Tuominen, I.: 1990, in *The dynamic Sun*, ed. Dezsõ, Publ. Debrecen Heliophys. Obs. **7**, p. 27

Virtanen, H.: 1989, Licentiate thesis, University Helsinki

Ward, F.: 1965, *Astrophys. J.* **141**, 534

II

Current stellar dynamo theory

- The nonlinear mean-field dynamo
- Stability: regular or chaotic solutions?
- Intermittency

Stable and Unstable Solutions of the Nonlinear Dynamo Problem

Reinhard Meinel

Astrophysikalisches Observatorium Potsdam, O-1560 Potsdam, Fed. Rep. of Germany

Abstract: A local potential approach to nonlinear dynamo models which allows the use of variational techniques to investigate the problem of stability is introduced. The method applies at least to quasi-kinematic dynamo models, i.e. to models which include the back-reaction of the magnetic field on the fluid motion in a simplified way. A special application leads to a previously investigated one-dimensional dynamo model which shows a coexistence of a periodic solution (limit cycle) with two stable steady solutions of opposite polarities. The inclusion of some small-amplitude noise leads to interesting transition phenomena which may be of relevance to explain the behaviour of astrophysical dynamos. A simple dynamical system with a two-dimensional phase-space is used for illustration.

1. Introduction

The kinematic dynamo theory can be regarded as a theory which investigates the stability of hydrodynamic configurations in electrically conducting media with respect to small magnetic fields $\boldsymbol{B}$. In other words, the stability of the state $\boldsymbol{B} \equiv 0$ is analyzed. If this state free of magnetic field proves to be unstable only the full set of coupled magnetohydrodynamic equations can give an answer to the question which state is realized then. Again the problem of stability is important. Only solutions of the magnetohydrodynamic equations which are stable with respect to small (hydrodynamic and magnetic) perturbations are relevant for a description of long-living magnetic configurations in astrophysical objects.

Most of the results presented so far in the field of nonlinear dynamo theory belong to a kind of models which may be called quasi-kinematic mean-field dynamo models. In these models the back-reaction of the magnetic field on the fluid motion is described by more or less well-founded assumptions about the dependence of hydrodynamic functions on the (mean) magnetic field. As kinematic dynamo models these models lead to differential equations for the magnetic field alone, i.e. the velocity field – including its dependence on the magnetic field – is considered as given. These models are characterized thus by a disproportion in the degree of physical accuracy: they represent the electrodynamics quite accurately but are rather crude with respect to the hydrodynamics. Nevertheless a lot of interesting bifurcations to steady, periodic, and quasiperiodic solutions have been found for such models (e.g. Brandenburg *et al.*, 1989; Schmitt and Schüssler, 1989; Jennings *et al.*, 1990).

In the present paper we show that all these models can be derived from a simple variational principle. We follow a method of Glansdorff and Prigogine (1971) who developed a variational technique for dissipative systems. A specialization leads back to a one-dimensional dynamo model that was introduced by Krause and Meinel (1988). We compare the results of Meinel and Brandenburg (1990) for this model with the properties of a simple dynamical system. In this way a possible explanation of an irregular time-behaviour of astrophysical dynamos (including the geo-dynamo) becomes more plausible.

2. A variational principle

Consider the following Lagrange density depending on two vector functions $\boldsymbol{B}$ and $\boldsymbol{B}_0$:

$$\mathcal{L}(\boldsymbol{B},\boldsymbol{B}_0)=\tfrac{1}{2}\eta(\operatorname{curl}\boldsymbol{B})^2-(\boldsymbol{u}\times\boldsymbol{B}_0+\mathcal{E})\cdot\operatorname{curl}\boldsymbol{B}+\boldsymbol{B}\cdot\dot{\boldsymbol{B}}_0 \tag{1}$$

where the positive-definite scalar function η and the vector functions $\boldsymbol{u}$ and $\mathcal{E}$ may depend on $\boldsymbol{B}_0$ and the coordinates explicitly:

$$\eta=\eta(\boldsymbol{x},\boldsymbol{B}_0),\qquad \boldsymbol{u}=\boldsymbol{u}(\boldsymbol{x},\boldsymbol{B}_0),\qquad \mathcal{E}=\mathcal{E}(\boldsymbol{x},\boldsymbol{B}_0). \tag{2}$$

A dot denotes a time derivative.

Now we calculate the variational derivative of $\mathcal{L}$ with respect to $\boldsymbol{B}$, take it at $\boldsymbol{B}=\boldsymbol{B}_0$, and set it equal to zero:

$$\left.\frac{\delta\mathcal{L}}{\delta\boldsymbol{B}}\right|_{\mathbf{B}=\mathbf{B}_0}=0. \tag{3}$$

In components $\delta\mathcal{L}/\delta\boldsymbol{B}$ reads

$$\left(\frac{\delta\mathcal{L}}{\delta\boldsymbol{B}}\right)_i=\frac{\partial\mathcal{L}}{\partial B_i}-\left(\frac{\partial\mathcal{L}}{\partial B_{i,j}}\right)_{,j} \tag{4}$$

where the usual summation convention is used, and $B_{i,j}$ means $\partial B_i/\partial x_j$.

Equation (3) is equivalent to

$$\dot{\boldsymbol{B}}_0=\operatorname{curl}(\boldsymbol{u}\times\boldsymbol{B}_0+\mathcal{E})-\operatorname{curl}(\eta\operatorname{curl}\boldsymbol{B}_0) \tag{5}$$

which is just the standard form of the induction equation of mean-field magnetohydrodynamics (cf. Krause and Rädler 1980). $\boldsymbol{B}_0$ is the mean magnetic field, $\boldsymbol{u}$ the mean velocity field, $\mathcal{E}$ the mean electromotive force describing the large-scale induction effects of small-scale turbulent motions, and η the magnetic diffusivity.

Together with (2) various types of nonlinear dynamo models are included in this formalism. The so-called α-quenching, for instance, is described by

$$\mathcal{E} = \alpha \boldsymbol{B}_0 \tag{6}$$

with an α-coefficient depending on $\boldsymbol{B}_0$ such that α vanishes as $|\boldsymbol{B}_0| \to \infty$ (Rüdiger, 1974). The effects of magnetic buoyancy, as another example, can be modelled by

$$\boldsymbol{u} = \boldsymbol{u}_0 + \boldsymbol{u}_{\mathrm{B}}(\boldsymbol{B}_0); \tag{7}$$

for further details see Moss et al. (1990).

The Lagrange density $\mathcal{L}(\boldsymbol{B}, \boldsymbol{B}_0)$ defined in (1) can be interpreted as a "local potential" in the sense of Glansdorff and Prigogine (1971). To show this we consider a finite volume V and restrict ourselves, for simplicity, to perfect conductor boundary conditions, i.e. we assume that the tangential components of the electric field $\boldsymbol{E} = \eta \operatorname{curl} \boldsymbol{B}_0 - (\boldsymbol{u} \times \boldsymbol{B}_0 + \mathcal{E})$ vanish at the boundary. We consider the functional

$$\Phi(\boldsymbol{B}, \boldsymbol{B}_0) = \int_V \mathcal{L}(\boldsymbol{B}, \boldsymbol{B}_0) dV \tag{8}$$

and find for $\boldsymbol{B}_0$ satisfying (5) and $\boldsymbol{B} \neq \boldsymbol{B}_0$

$$\Delta\Phi \equiv \Phi(\boldsymbol{B}, \boldsymbol{B}_0) - \Phi(\boldsymbol{B}_0, \boldsymbol{B}_0) = \tfrac{1}{2} \int_V \eta[\operatorname{curl}(\boldsymbol{B} - \boldsymbol{B}_0)]^2 dV > 0, \tag{9}$$

i.e. $\Phi(\boldsymbol{B}, \boldsymbol{B}_0)$ has an absolute minimum at $\boldsymbol{B} = \boldsymbol{B}_0$. This minimum property can be employed to investigate (5) by means of variational techniques, e.g. the Ritz method. In this way a derivation of the Galerkin expansion is also possible which automatically guarantees its convergence (cf. Glansdorff and Prigogine, 1971).

3. Reduction to a one-dimensional model

3.1 The local potential

Now we assume that the magnetic field depends only on one spatial (cartesian) coordinate, say z, and that the component B_z vanishes:

$$\boldsymbol{B} = [B_x(z,t), \quad B_y(z,t), \quad 0]. \tag{10}$$

The components B_x and B_y can be combined to form the complex function

$$B \equiv B_x + \mathrm{i} B_y. \tag{11}$$

Inserting (10) into (1) we obtain with (6) and $\boldsymbol{u} \equiv \boldsymbol{0}$

$$\mathcal{L}(B, B_0) = \tfrac{1}{2} \left[\eta B' B^{*\prime} + \mathrm{i}\alpha (B_0 B^{*\prime} - B_0^* B') + B \dot{B}_0^* + B^* \dot{B}_0 \right], \tag{12}$$

where an asterisk denotes complex conjugation and a prime denotes differentiation with respect to z. Assuming

$$\eta = \text{const} > 0, \quad \alpha = \alpha_0 f(B_0^* B_0), \quad \alpha_0 = \text{const} > 0, \tag{13}$$

and introducing dimensionless space- and time-coordinates we find

$$\mathcal{L}(B, B_0) = B' B^{*\prime} + \mathrm{i} C f(B_0^* B_0)(B_0 B^{*\prime} - B_0^* B') + B \dot{B}_0^* + B^* \dot{B}_0. \tag{14}$$

(We have omitted the unnecessary factor $\frac{1}{2}$.) The dimensionless "dynamo number" C is defined by

$$C = \frac{\alpha_0 d}{\eta} \tag{15}$$

where d is the thickness of the electrically conducting slab under consideration.

Equation (3) reads now

$$\left. \frac{\delta \mathcal{L}}{\delta B^*} \right|_{B=B_0} = 0 \tag{16}$$

with

$$\frac{\delta \mathcal{L}}{\delta B^*} = \frac{\partial \mathcal{L}}{\partial B^*} - \left(\frac{\partial \mathcal{L}}{\partial B^{*\prime}} \right)' \tag{17}$$

and leads to

$$\dot{B}_0 = B_0'' + \mathrm{i} C [f(B_0^* B_0) B_0]'. \tag{18}$$

(Note that the variational derivative of $\mathcal{L}$ with respect to B gives the complex conjugate of (18).)

We assume that B_0 and B vanish at the boundaries $z = 0$ and $z = 1$ of the slab:

$$B_0(0, t) = B_0(1, t) = 0, \tag{19}$$

$$B(0, t) = B(1, t) = 0. \tag{20}$$

These boundary conditions correspond to a vacuum (free of magnetic field) for $|z| > 1$.

Now we consider the functional

$$\Phi(B, B_0) = \int_0^1 \mathcal{L}(B, B_0) dz \tag{21}$$

and calculate

$$\Delta\Phi \equiv \Phi(B, B_0) - \Phi(B_0, B_0) \tag{22}$$

for $B \neq B_0$. With (18), (19) and (20) we obtain by integration by parts

$$\Delta\Phi = \int_0^1 (B' - B_0')(B' - B_0')^* dz > 0. \tag{23}$$

Thus Φ is again a "local potential". The one-dimensional nonlinear dynamo model (18), (19) was investigated by Krause and Meinel (1988) and Meinel and Brandenburg (1990).

3.2 Stable and unstable solutions

Various solutions of (18), (19) have been obtained for the following assumptions about the function f:

$$f = 1 - B^*B, \tag{24}$$

$$f = \frac{1}{(1+B^*B)^2}, \tag{25}$$

$$f = \frac{1}{1+(B^*B)^\kappa}, \quad \kappa \geq 1. \tag{26}$$

At first, for $C > 2\pi$, there is always a stable steady solution B_s. It bifurcates from the solution $B \equiv 0$ at $C = C_{\text{crit}} = 2\pi$. At $C = 2n\pi, n = 2, 3, 4, \ldots$, unstable steady solutions bifurcate from $B \equiv 0$. All steady solutions can be found analytically up to a single real quadrature. In case (24) the steady solutions can even be given explicitly in terms of elliptic functions.

It should be noted that, if B is any solution of (18), (19), then $B\exp(\mathrm{i}\varphi_0)$ also satisfies (18), (19), where φ_0 is a real constant. We restrict this gauge freedom by the constraint

$$B^*(1-z,t) = B(z,t), \tag{27}$$

which means that the real part of B is an even function and the imaginary part of B an odd function of z with respect to the midpoint $z = \frac{1}{2}$ of the considered interval $[0,1]$. Note that, according to (18), (19), the constraint (27) is conserved in course of time if it is initially satisfied. This constraint, however, leaves the freedom $B \to -B$, i.e. for a given solution B, $-B$ is also a solution of our problem. Thus, for $C > 2\pi$, we have exactly two stable steady solutions B_s and $-B_s$. Which of them is attained as the final state of evolution depends on the initial conditions. Numerical investigations of (18), (19) revealed that for large values of the dynamo number C there is also a third possibility for the final state of evolution: a stable oscillating solution B_{osc}. The basin of attraction of this limit cycle increases with growing C while the basins of the two steady solutions shrink. The oscillating solution changes periodically its sign:

$$B_{\text{osc}}(z,t+T) = -B_{\text{osc}}(z,t). \tag{28}$$

The full cycle period is given by $2T$.

3.3 Noise-induced transitions

For large values of κ in case (26) the oscillating solution B_{osc}, at particular time instants $t = t_e$ (turning points), closely approaches the steady solutions $\pm B_s$. This approach is the closer the larger the value of κ. A close approach means that

$$\int_0^1 |B_{\text{osc}}(z,t) \mp B_s|^2 dz \tag{29}$$

is a small quantity for $t = t_e$. Since both solutions B_s and B_{osc} are stable with respect to infinitesimal perturbations there exists a basin boundary between B_s and

$B_{osc}(t_e)$. However, it is clear that a suitable small finite perturbation is sufficient to produce a jump from B_s to B_{osc} and vice versa. Therefore the inclusion of a small noise-term to (18) may lead to an interesting time-behaviour of the resulting solutions. Figure 1 shows solutions of

$$\dot{B} = B'' + \mathrm{i}C[f(B^*B)B]' + AG(z)F(t) \tag{30}$$

with $G(x) = \sin \pi z + \frac{1}{2} \sin 2\pi z$ and F is a stochastic function:

$$F(t) = \sum_{k=1}^{\infty} \epsilon_k \delta(t - k\Delta t), \quad t > 0, \tag{31}$$

where ϵ_k, for each k, can take the values $-1, 0, 1$ with probabilities $p, 1 - 2p, p$, respectively. $A = 0.2$, $p = 10^{-2}$, $\Delta t = 10^{-3}$. The function f was chosen according to (26) and $\kappa = 4$. The steady solutions $\pm B_s$ are characterized by $\mathrm{Re}\, B(z = \frac{1}{2}) \approx \pm 1$. Obviously, the noise is sufficient for $B(z, t)$ to leave the basin of attraction of B_s. The system follows then a part of B_{osc}. In this way it is even possible to reach $-B_s$, i. e. to generate a reversal of the magnetic field. With increasing C the oscillating solution and with decreasing C the steady solutions dominate. (Note that B_{osc} exists as a stable solution for $C \gtrsim 48$ only.)

In general, the noise-level which is necessary to produce jumps between the different attractors depends on the dynamo number C and on the degree of the nonlinearity f, e. g. on κ.

3.4 An illustrative example with a two-dimensional phase-space

The system (18), (19) has an infinite-dimensional phase-space. A discussion of basin boundaries etc. is therefore very difficult. The aim of this section is to present a "toy" system which shows the same coexistence of attractors as (18), (19) but is as simple as possible. A dynamical system of the form

$$\dot{x} = F(x, y), \tag{32}$$

$$\dot{y} = G(x, y) \tag{33}$$

is sufficient for this purpose. We choose

$$F(x, y) = xJ(x, y) - yK(x, y), \tag{34}$$

$$G(x, y) = yJ(x, y) + xK(x, y), \tag{35}$$

with

$$J(x, y) = A - B(x^2 + y^2) + C(x^2 + y^2)^2 - (x^2 + y^2)^3, \tag{36}$$

$$K(x, y) = 2 - (x^2 + y^2) - \tfrac{2}{5}xy, \tag{37}$$

and take

$$A = 15, \quad B = 18.5, \quad C = 7.5. \tag{38}$$

This system has the following attractors:

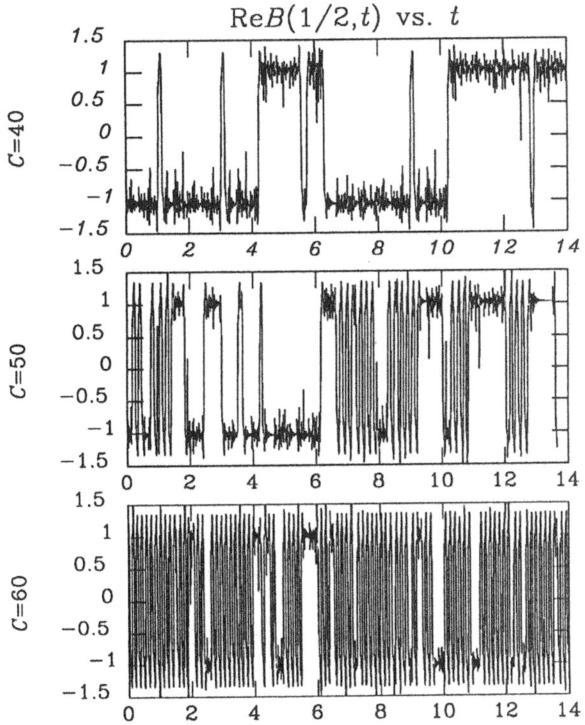

Fig. 1. Time-behaviour of the dynamo model with coexisting two steady and one periodic attractors under the influence of a small-amplitude noise for different values of the dynamo number C (Meinel and Brandenburg, 1990)

$$(i) \quad x = \pm\sqrt{2}, \quad y = 0, \tag{39}$$

$$(ii) \quad x^2 + y^2 = 3, \tag{40}$$

i. e. two fixed points and one limit cycle. Unstable steady solutions are given by $x = y = 0; x = 0, y = \pm\sqrt{2}; x = -y = \pm\sqrt{5/4}$. The two-dimensional phase-space is shown in Fig. 2. It can be seen that the limit cycle, when crossing the x-axis, closely approaches the fixed points. The basin boundary of the limit cycle is given by the circle $x^2 + y^2 = \frac{5}{2}$. All trajectories starting at $x^2 + y^2 > \frac{5}{2}$ reach asymptotically the limit cycle. All trajectories starting at $x^2 + y^2 < \frac{5}{2}$ reach one of the fixed points. The boundary between the basins of the two fixed points can also be seen from Fig. 2 approximately.

The behaviour of our system (32), (33) can easily be understood by introducing polar coordinates:

$$x = r\cos\varphi, \quad y = r\sin\varphi. \tag{41}$$

Together with (34)-(38) this leads to

$$\dot{r} = r(2 - r^2)(\tfrac{5}{2} - r^2)(3 - r^2), \tag{42}$$

$$\dot{\varphi} = 2 - r^2 - \tfrac{1}{5}r^2 \sin 2\varphi. \tag{43}$$

The steady solutions (except $x = y = 0$) are simply given by the crossing points of the isoclinals $\dot{r} = 0$ (circles) and $\dot{\varphi} = 0$ (ellipse). Note that $\dot{\varphi} < 0$ for all φ on the circle $r^2 = 3$ (limit cycle). The basin-boundary property of the circle $r^2 = \frac{5}{2}$ can easily be deduced from (42): for r slightly exceeding $\sqrt{5/2}$ we find $\dot{r} > 0$ while $\dot{r} < 0$ for r slightly less than $\sqrt{5/2}$. In a similar way the stability properties of the steady solutions and the limit cycle can be found.

Looking at Fig. 2 one can easily imagine a time-behaviour completely comparable to that of Fig. 1 if some small noise is added to (32), (33).

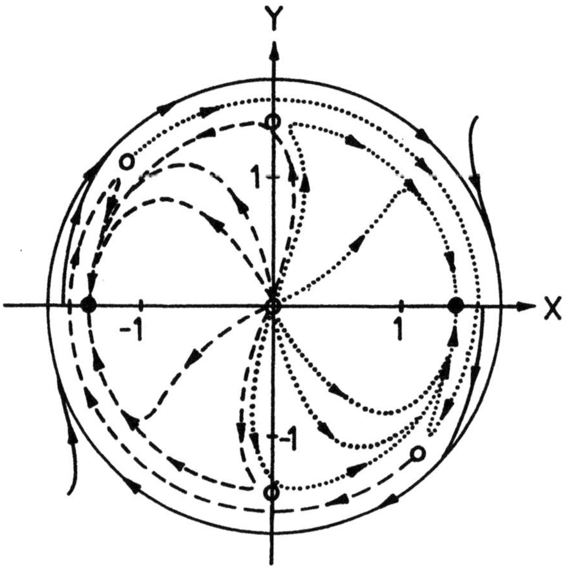

Fig. 2. The trajectories for the "toy" system lead either to the limit cycle $x^2 + y^2 = 3$ (full lines) or to one of the two fixed points $x = \pm\sqrt{2}, y = 0$ (dotted resp. dashed lines). Small empty circles indicate unstable steady solutions

4. Discussion

The coexistence of two stable steady solutions (which differ only in their polarity) and a periodic solution which closely approaches these steady solutions offers an interesting possibility for explaining the irregular time-behaviour of the magnetic field of various astrophysical objects. (Stability means here only stability with respect to infinitesimal perturbations – but not with respect to small finite perturbations.) The assumption of some small noise is quite natural. It is even not necessary to make use of external perturbations to explain it. Within the frame of mean-field theory a small noise term may occur as a correction to the induction equation (5) if the Reynolds relations are not exactly satisfied in the usual two-scale approach (cf. Hoyng, 1988).

Such an explanation of a random-like behaviour of mean-field dynamo models seems to be at least as plausible as an explanation based on deterministic chaos.

In particular, the behaviour of the Earth's magnetic field as obtained from paleomagnetic records can be compared with our solutions, e. g. in the upper panel of Fig. 1. The lower panel of Fig. 1 which corresponds to a larger value of the dynamo number, where the periodic solution dominates, may be of interest for explaining the activity behaviour observed for young and very active stars. Also the solar cycle is not exactly periodic.

Of course, the nonlinear dynamo models under consideration are far away from real astrophysical objects. This is, however, just an argument in favour of simple models. It does not make much sense to study too complicated models of this "quasi-kinematic" kind (three-dimensional models with many free parameters etc.). To be closer to reality the hydrodynamic part of the theory must be treated more thoroughly.

The variational principle discussed in section 2 can certainly be generalized to the full magnetohydrodynamic equations.

References

Brandenburg, A., Krause, F., Meinel, R., Moss, D., Tuominen, I.: 1989, *Astron. Astrophys.* **213**, 411

Glansdorff, P., Prigogine, I.: 1971, *Thermodynamic theory of structure, stability and fluctuations*, Wiley-Interscience, London

Hoyng, P.: 1988, *Astrophys. J.* **332**, 857

Jennings, R., Brandenburg, A., Moss, D., Tuominen, I.: 1990, *Astron. Astrophys.* **230**, 463

Krause, F., Meinel, R.: 1988, *Geophys. Astrophys. Fluid Dyn.* **43**, 95

Krause, F., Rädler, K.-H.: 1980, *Mean-field magnetohydrodynamics and dynamo theory*, Akademie-Verlag, Berlin

Meinel, R., Brandenburg, A.: 1990, *Astron. Astrophys.* **238**, 369

Moss, D., Tuominen, I., Brandenburg, A.: 1990, *Astron. Astrophys.* **228**, 284

Rüdiger, G.: 1974, *Astron. Nachr.* **295**, 275

Schmitt, D., Schüssler, M.: 1989, *Astron. Astrophys.* **223**, 343

Nonlinear Nonaxisymmetric Dynamos for Active Stars

David Moss

Mathematics Department, The University, Manchester M13 9PL, United Kingdom

Abstract: Recent observations seem to have detected large scale nonaxisymmetric structures on active giant stars. These structures are plausibly associated with underlying, dynamo generated, nonaxisymmetric magnetic fields. Such developments have motivated the development of a computer code to solve the nonlinear mean field dynamo equation in spherical geometry with no imposed geometrical symmetries. The nonlinearity is a simple α-quenching. The nature of the stable solutions found depends quite sensitively on the assumed spatial distribution of both α-effect and differential rotation, and also on the degree of supercriticality of the dynamo. Such a dynamo model with stable, purely nonaxisymmetric solutions is described in this paper.

1. Introduction

The first aim of astrophysical dynamo theory has been to understand the solar magnetic cycle. Observations suggest that the mean solar field is approximately axisymmetric. Plausible *kinematic* models, in which large and small scale velocity fields are prescribed in an ad hoc but hopefully physically reasonable manner, can be constructed, but there is as yet no successful dynamically consistent model. These considerations have largely restricted the study of stellar dynamos to axisymmetric models, which does also allow considerable mathematical simplification.

Recently indicators of markedly non-axisymmetric surface structure have been detected in a few late type "active giant" (FK Comae) stars. Piskunov *et al.* (1990) have used surface imaging techniques to produce maps showing large-scale nonaxisymmetric surface temperature inhomogeneities on HD 32918. Jetsu *et al.* (1990) have found a 2.8 year variation in the amplitude of the photometric rotation modulation, superimposed on a 9 year cycle in the mean brightness. Phase seems to have been preserved for more than 13 years. Persistent active longitudes (a "flip-flop" phenomenon) was found in FK Comae (Jetsu *et al.*, in these Proceedings).

RS CVn stars, which are often in a state of spin-orbit synchronization, sometimes appear to have persistent spots at fixed longitudes (e.g. Zeilik, these Proceedings; and Hackman *et al.*, these Proceedings). Assuming an analogy between the origin of sunspots and starspots, such observations can be interpreted as evidence for the presence of large-scale nonaxisymmetric stellar magnetic fields in these stars with deep convective envelopes.

It is known that in spherical dynamo models axisymmetric modes are often more easily excited than nonaxisymmetric. However Brandenburg *et al.* (1989) have demonstrated, for a strictly axisymmetric model, that a knowledge of linear growth rates is of very limited usefulness in determining the nonlinear stability of a system at supercritical dynamo number, and it can be anticipated that a similar situation exists when the restriction to axisymmetric fields is removed. By analogy, it may then be possible to find stable solutions consisting wholly or partly of nonaxisymmetric components.

2. The Model

The standard mean-field dynamo equation in dimensionless form

$$\frac{\partial \boldsymbol{B}}{\partial t} = \nabla \times (\boldsymbol{u} \times \boldsymbol{B} + \alpha \boldsymbol{B}) - \nabla \times (\eta \nabla \times \boldsymbol{B})$$

is solved in the spherical volume $0 \leq x \leq 1$. τ is in units of the global diffusion time, R^2/η, $\alpha = \bar{\alpha}\tilde{\alpha}$, $\Omega = \bar{\Omega}\tilde{\Omega}$, $x = r/R$, where R is the radius of the computational volume, η is the uniform resistivity and $\bar{\alpha}$ and $\bar{\Omega}$ are typical values of α and Ω. The dynamo parameters are given by

$$C_\alpha = R\bar{\alpha}/\eta, \quad C_\omega = R^2\bar{\Omega}/\eta.$$

For the models described here

$$\tilde{\alpha} = f(x)\cos\theta/[1 + \boldsymbol{B}^2(x, \theta, \lambda)],$$

x, θ, λ being spherical polar coordinates, and

$$f(x) = \begin{cases} \frac{15}{16}\frac{1}{d_\alpha}(1 - \xi_\alpha^2)^2, & |\xi_\alpha| \leq 1, \\ 0 & |\xi_\alpha| > 1, \end{cases}$$

$$\tilde{\Omega}(x) = \begin{cases} -1, & \text{for } \xi_\Omega \leq -1, \\ -\frac{1}{2}(1 - \frac{3}{2}\xi_\Omega + \frac{1}{2}\xi_\Omega^3), & \text{for } |\xi_\Omega| \leq 1, \\ 0, & \text{for } \xi_\Omega \geq 1, \end{cases}$$

where $\xi_\alpha = x - x_\alpha/d_\alpha$ and $\xi_\Omega = x - x_\Omega/d_\Omega$. More details are given in Moss *et al.* (1991).

3. Results

The magnetic field can be split into parts that are symmetric (S) or antisymmetric (A) with respect to the rotational equator, and can be Fourier analysed into components proportional to $\exp(im\lambda), m = 0, 1, 2, ...$ (cf Stix, 1971). The gross field properties can be described by the parity parameter

$$P = [E^{(S)} - E^{(A)}]/[E^{(S)} + E^{(A)}]$$

and the symmetry parameter

$$M = 1 - E_0/E,$$

where $E(S)$, $E(A)$, E_0 and E are respectively the energies in the symmetric, antisymmetric, axisymmetric and total fields. For example, $P = +1/-1$ denotes fields of pure even/odd parity and $M = 0$ corresponds to a completely axisymmetric field.

Computations were performed for parameters $x_\omega = 0.55$, $x_\alpha = 0.75$, $d_\omega = 0.25$, $d_\alpha = 0.25$, $C_\omega = -150$. C_α values for the excitation of the lowest axisymmetric and nonaxisymmetric modes are given in Table 1. Supercritical values of C_α were taken, $5 \leq C_\alpha \leq 30$. Computations were started from a field of arbitrary structure, consisting of a mixture of axisymmetric and nonaxisymmetric parts. In each case the field evolved to a purely nonaxisymmetric S-type configuration, consisting of $m = 1, 3...$ components. Fig. 1 shows the evolution of $E_0, E_N = E - E_0, P$ and M with time, and contours of constant surface field strength for a typical model are shown in Fig. 2. The solutions are steady in the sense that the total energy is constant, but the field structure rotates.

Table 1. Approximate critical values of C_α for excitation of linear modes and corresponding frequencies.

Mode	$C_{\alpha,\mathrm{crit}}$	Frequency
A0	4.48	18.4
S0	4.13	13.2
A1	3.98	0.57
S1	3.94	0.68

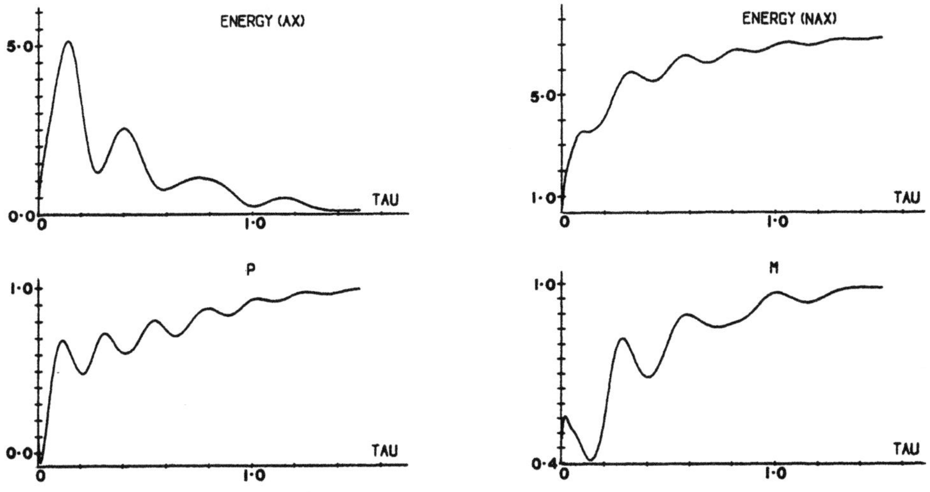

Fig. 1. Evolution with time of the energy in the axisymmetric and nonaxisymmetric parts of the field and aof the parameters P and M for model with $C_\alpha = 30$.

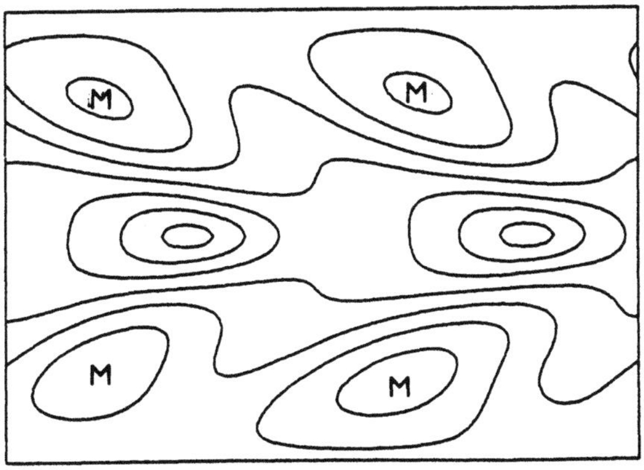

Fig. 2. Contours of equal surface field strength for the model with $C_\alpha = 30$. Longitude runs horizontally and the top and bottom borders correspond to the north and south poles respectively. M denotes regions where the field strength is highest. Slight equatorial asymmetries are because the field has not quite attained its final pure parity state.

4. Discussion

In a real stellar dynamo the nonlinear limitation of the field at finite amplitude may be caused by any (or all) of a number of mechanisms. Buoyancy, the effects of the Lorentz force on the small scale motions or on the differential rotation or meridional circulation are all possibilities. This model, with its very idealized α-quenching mechanism, demonstrates the possibility of the existence of persistent nonaxisymmetric mean-field dynamo solutions. The existence of stable nonaxisymmetric solutions is very sensitive to the assumed source profiles, and also to the value of C_ω. For example, in a series of experiments in which x_α, d_α, d_ω were kept fixed, for $x_\omega = 0.40$, 0.65 and 0.70 no stable nonaxisymmetric solutions were found, whereas for $x_\omega = 0.50$ and 0.75 there are several nonaxisymmetric stable solutions but their properties can be somewhat different to those of the solutions described here with $x_\omega = 0.55$ - see Moss *et al.* (1991) for details. Whether or not suitable α and Ω profiles can coexist as part of a dynamically self-consistent solution is a much deeper and more difficult question that is not tackled here.

It is known that strong differential rotation suppresses nonaxisymmetric modes (which is why the magnitude of C_ω in these models has been taken to be quite small. The nature of the differential rotation in the interior of the "active giants" is quite uncertain. However in the close, tidally interacting components of RS CVn systems it is probable that the differential rotation *is* relatively small. Very speculatively, this might favour the excitation of nonaxisymmetric dynamo fields and so explain the predominance of nonaxisymmetric structure ("active longitudes") in these stars.

Acknowledgement. I am grateful to Axel Brandenburg for performing linear calculations.

References

Brandenburg, A., Krause, F., Meinel, R., Moss, D., Tuominen, I.: 1989, ***Astron. Astrophys.*** **213**, 411

Jetsu, L., Huovelin, J., Tuominen, I., Vilhu, O., Boppp., B.W., Piirola, V.: 1990, ***Astron. Astrophys.*** **236**, 423

Moss, D., Tuominen, I., Brandenburg, A.: 1991, ***Astron. Astrophys.***, in press

Piskunov, N.E., Tuominen, I., Vilhu, O.: 1990, ***Astron. Astrophys.*** **230**, 363

Stix, M.: 1971, ***Astron. Astrophys.*** **13**, 203

Waves of Solar Activity

M.R.E. Proctor[1], E.A. Spiegel[2]

[1] Department of Applied Mathematics and Theoretical Physics, University of Cambridge, Silver Street, Cambridge CB3 9EW, U.K. [2] Department of Astronomy, Columbia University, New York, NY 10027, U.S.A.

Abstract: We develop a theory of the sunspot cycle predicated on the assumption that the observed bands of activity are packets of dynamo waves. An approximate equation is proposed to describe the dynamics of these packets, using standard ideas from bifurcation theory. We show that in a certain limit the system can be described in terms of a slowly-evolving solitary wave, and that periodic behavior, like that of the observed butterfly diagram, can easily be found. Generalizations of the theory are discussed.

1. Motivations

Elsewhere in this volume you can read Brandenburg's description of simulations of hydromagnetic convection, in which descending plumes pull down and stretch out magnetic field lines, defying (at least our) intuition about how magnetic buoyancy might drive the field upward and outward. This will have to be studied hard and understood, but it does seem to suggest that the bulk of the solar convection zone has the capability of performing some serious dynamo action.

On the other hand, participants in the conference also heard from Goode that there is a sharp change in solar angular velocity with radius that can be detected just under the convection zone. This tachycline, as such a feature may be called, may be expected to do a good job of stretching any field that dares to enter it into a strong toroidal field. From there magnetic buoyancy and penetrative convection can cause this enhanced field to intrude back into the zone where it may be swept up to the surface to emerge and form spots.

The interplay of two such mechanisms may be responsible for much of the complexity of the solar cycle and this promises to be a source of concern to stellar physicists for some time. But here, we are interested in studying the second of these two possible processes: subphotospheric dynamo action. Our plan is to study the dynamo waves in a thin fluid layer. We shall first write down the partial differential equations for an α–ω dynamo and recall how they are subject to the onset of overstability for a critical value of the dynamo number at a critical value of the horizontal wave number.

For dynamo numbers slightly in excess of the critical value, for a thin layer of fluid, such as that described by Goode, we expect that a packet of waves will be unstable. This packet will grow to finite amplitude until it causes back reactions on the fluid motions. The nature and the details of such feedbacks were much discussed in the conference that spawned this volume, but it is not clear that any general agreement was reached on what the principal mechanisms are. In this situation, we have decided to let the modern theory of nonlinear instability come to our aid.

To the student of nonlinear stability theory, a band of overstable modes such as we are describing is just a Hopf bifurcation in an extended system. In that case, bifurcation theory reveals what the form of the equation for the envelope of the packet must be, up to certain coefficients that depend on the details of the problem. We shall present this equation here and use it to study the propagation of nonlinear packets of dynamo waves in the manner of the butterfly diagram of solar physics. Of course, what we are describing here is but the beginning of a more extensive modelling process, but we find the results encouraging enough, not only to report them here, but to mention some possible extensions.

2. Equations

As we stated in the introduction, we are concerned with a thin layer of fluid under the convection zone. For this discussion we shall not worry about the sphericity and consider a plane-parallel osculating layer at mid-latitudes, in analogy to the f-plane of meteorology. In that layer we take x to be the North-South coordinate, y as the azimuthal coordinate and z as the vertical coordinate. We confine ourselves to axisymmetric situations.

The magnetic field is

$$\mathbf{B} = b\hat{\mathbf{y}} + \nabla\times(a\hat{\mathbf{y}}) \tag{2.1}$$

where $\hat{\mathbf{y}}$ is a unit vector in the y–direction; that is, $b\hat{\mathbf{y}}$ is the toroidal field and $a\hat{\mathbf{y}}$ is the vector potential for the poloidal field. The equations for $\mathbf{B}$ are these:

$$\partial_t a = \alpha b + \eta\nabla^2 a \tag{2.2}$$

$$\partial_t b = \Omega\partial_x a + \eta\nabla^2 b \tag{2.3}$$

where α measures the ability of helical turbulence to generate poloidal field from toroidal field, Ω is a measure of the solar differential rotation, and $\nabla^2 = \partial_x^2 + \partial_z^2$. We take the differential rotation in the x–direction for illustrative purposes so as to avoid detailed discussions here of what may involve vertical eigenfunctions. By taking this scientific license we are able to avoid technical details. These equations may either be taken as a mathematical statement of the familiar Babcock-Leighton-Parker ideas about solar activity or understood as a direct outcome of the simplest versions of mean field dynamo theory.

Axisymmetry translates into the statement that the equations are invariant with respect to translations in y, so that we may look for solutions of the form $a(x,z,t) = A(x,t)\exp(i\ell z)$, $b(x,z,t) = B(x,t)\exp(i\ell z)$. The equations become

$$\dot{A} = \alpha B + \eta(A'' - \ell^2 A) \tag{2.4}$$

$$\dot{B} = \Omega A' + \eta(B'' - \ell^2 B) \tag{2.5}$$

where the prime denotes derivative with respect to x and the dot indicates time derivative.

To get an idea of the dynamics contained in these equations, we suppose, for the moment, that the coefficients, such as α, are constants. Then we can look for solutions of the form $A(x,t) = A_0 \exp(st + ikx,)$ and $B(x,t) = B_0 \exp(st + ikx)$, where A_0 and B_0 are constants. Then we get two equations for A_0 and B_0. The condition for solvability of these equations is the vanishing of the determinant of the coefficients. We obtain the result

$$s = -\eta\left(k^2 + \ell^2\right) \pm \sqrt{\alpha\Omega k/2}(1+i), \tag{2.6}$$

where we shall consider only cases where $\alpha\Omega > 0$. There is a Hopf bifurcation for wavenumber k when the dynamo number

$$D \equiv \sqrt{\frac{\alpha\Omega}{2\eta^2\ell^3}} \tag{2.7}$$

has the value $(1+K^2)/K^{\frac{1}{2}}$ where $K = k/\ell$. The minimum value of this expression over K is the critical dynamo number $D_c = 4\cdot 3^{-\frac{3}{4}}$ and it occurs when $K = K_c = 1/\sqrt{3}$. When $D \geq D_c$, we have dynamo action for a range of K in a neighborhood of K_c.

3. Wave packets

For K in the neighborhood of the critical value for the onset of instability, we can construct a wave packet that we propose as a model of the band of solar activity delineated in the butterfly diagram. To make this packet, it is convenient to simplify the expression for s for D close to D_c.

Let $D = D_c + \delta$ and $K = K_c + q$. For small δ and q the nondimensional growth rate $\sigma \equiv s/(\eta\ell^2) - 4/3$ is

$$\sigma = 3^{\frac{1}{4}}\delta - 2q^2 + i\left(3^{\frac{1}{4}}\delta + \frac{2}{\sqrt{3}}q - q^2\right). \tag{3.1}$$

Just as in the complete expression for s, the group and the phase velocities of the waves are towards the equator, which is at $x = 0$.

This expression for the growth rate was derived for constant coefficients in (2.4) and (2.5). We shall now assume that these coefficients may vary slowly in x and represent this effect by allowing such a dependence in δ. We can also simplify the expression for σ by factoring out a constant and reabsorbing it into the time. Then we can write down the linear equation for $\Psi(\xi, \tau)$, the amplitude function for a superposition of waves. This takes the form

$$\partial_\tau \Psi = 3^{\frac{1}{4}}(1+i)\delta\Psi + \frac{2}{\sqrt{3}}\partial_\xi \Psi + (2+i)\partial_\xi^2 \Psi. \tag{3.2}$$

Here we have introduced a new spatial coordinate ξ and time variable τ to emphasise the weak dependence of the wave envelope on position and time. Thus, if we replace Ψ by $\exp(i(K_c x + q\xi) + \sigma\tau)$, we recover the dispersion relation (3.1).

In the chosen frame, δ depends weakly on space and time, and the wave packet will deform as it propagates and grows in amplitude. Eventually nonlinear terms become significant, and we have to make allowance for this by *nonlinearizing* (3.2). The possible nonlinear processes arising through the action of the Lorentz force were much discussed at the Symposium. Nevertheless the generic forms of the nonlinear terms that can come into (3.2) are dictated by nonlinear stability theory; the leading one of these is $F(\xi, t)|\Psi|^2\Psi$ for some complex function F.

4. Envelopes of solar activity

Nonlinear waves arising from the modulation of overstability in thin layers are often very robust. Even though they may not arise as completely integrable systems, such waves often have sufficient stability to be considered as solitons. That is the sort of wave that we wish to liken to the waves of solar activity that propagate from mid-latitudes to the equator as the solar cycle unfolds. However the kind of theory for such envelope solitons that we have outlined in §III involves quite a few complications and the solution of the full amplitude evolution equation for Ψ will require numerical integrations. Here we are satisfied to see the qualitative content of such a theory by choosing the parameters of the problem to place the system in *nearly* integrable conditions.

There are two kinds of effects that will spoil the complete integrability of the activity envelope equation: (1) nonconstant coefficients of slow variability and (2) dissipative terms of small amplitude. Let us for the purposes of this discussion suppose that both kinds of smallness can be measured by the same parameter, ϵ. Then the equation formulated in the previous section can conveniently be expressed (in a frame moving with the leading order group velocity) as follows:

$$\partial_\tau \Psi - i\partial_\xi^2 \Psi - i|\Psi|^2\Psi = f(Z, [\Psi]) + \epsilon g(Z, [\Psi]), \tag{4.1}$$

where $Z = \epsilon(\xi - c_0\tau)$, and c_0 is the group velocity, so that Z gives the position of the wave packet. Thus $f(Z)$ represents effects of type (1) while the small amplitude, type (2) effects are summarized by g, and the square brackets indicate dependence on derivatives of Ψ as well as Ψ itself.

In this picture, for $\epsilon = 0$, (4.1) is the cubic Schrödinger equation, which has a solitary wave solution of the form

$$\Psi = \mathrm{Re}(\xi, \tau) e^{i\Theta(\xi,\tau)}, \tag{4.2}$$

where

$$\mathrm{Re} = \sqrt{2} R \operatorname{sech}[R(\xi - \xi_0)], \tag{4.3}$$

and

$$\Theta = U(\xi - \xi_0) + \int (U^2 + R^2) d\tau. \tag{4.4}$$

This Schrödinger soliton has two arbitrary parameters, U and R, with $\xi_0 = 2U\tau$.

The effect of the terms on the right side of (4.1) is to modify the behavior of this nonlinear wave and, when ϵ is small, we can expect to capture the effects of the modification by letting the constants vary slowly. So, to solve (4.1) for small ϵ, we let R and U vary slowly and redefine ξ_0 as $2\int U d\tau$. In short, instead of the simple soliton of the cubic Schrödinger equation with arbitrary values assigned to its basic parameters, we seek a similar object whose basic parameters are allowed to vary according to the dictates of the terms that destroy exact integrability. So we shall use (4.1) to find equations of motion for U, R and $X_0 = \epsilon\xi_0$. These equations control the proposed model for the evolution of the solar cycle. To get them in explicit form, we need to become more specific about f and g. We do that in the next section.

5. Equations of motion of the activity waves

To derive the dynamical equations for the parameters of the solitary waves that we associate to the solar cycle, we first need to give the forms of the terms that cause their variation. Terms of type (2), small dissipative terms, are familiar from studies of the complex Ginzburg-Landau equation for Hopf bifurcation in extended systems. They take the form

$$g = \mu(Z)\Psi + \partial_\xi^2 \Psi - \nu|\Psi|^2\Psi. \tag{5.1}$$

The type (1) terms, formally of order unity though their average effect is of order ϵ, are

$$f = i\kappa(Z)\Psi + c(Z)\partial_\xi \Psi. \tag{5.2}$$

To find the slow evolution of R and U, we use the following integral relations, reminiscent of familiar results from dynamics:

$$\partial_\tau \int_{-\infty}^{\infty} |\Psi|^2 d\xi = 2\epsilon \int_{-\infty}^{\infty} [\mu(Z)|\Psi|^2 - \frac{dc}{dZ}|\Psi|^2 - |\partial_\xi \Psi|^2 - \nu|\Psi|^4] d\xi \tag{5.3}$$

$$\partial_\tau \int_{-\infty}^{\infty} \frac{1}{i}(\Psi^* \partial_\xi \Psi - \Psi \partial_\xi \Psi^*) d\xi = 2\epsilon \int_{-\infty}^{\infty} \frac{d\kappa}{dZ} |\Psi|^2 d\xi +$$

$$+\frac{2\epsilon}{i} \int_{-\infty}^{\infty} [(\mu(Z) - \nu|\Psi|^2)(\Psi^* \partial_\xi \Psi - \Psi \partial_\xi \Psi^*) + (\partial_\xi^2 \Psi^* \partial_\xi \Psi - \partial_\xi^2 \Psi \partial_\xi \Psi^*)]\, d\xi. \tag{5.4}$$

The integrals on the left hand sides of equations (5.3) and (5.4) are easily evaluated in terms of R and U. In fact

$$\int_{-\infty}^{\infty} |\Psi|^2 d\xi = 4R\,, \quad \int_{-\infty}^{\infty} \frac{1}{i}(\Psi^* \partial_\xi \Psi - \Psi \partial_\xi \Psi^*) d\xi = 8UR. \tag{5.5}$$

The terms on the right hand sides can be determined similarly, with the functions of Z considered constant while the integrals over the shorter ξ scale are being performed. Consequently we may now regard the spatial variable Z as giving the centre of the soliton, so that $Z \approx X_0 - \epsilon c_0 \tau$. We have thus reduced the problem of the slow evolution of the wave packet to the determination of the solutions to three simple nonlinear o.d.e.'s. The first two can be obtained from equations (5.4) and, after some obvious manipulations to separate the evolution of R and U, take the form

$$\partial_T R = 2R[\mu(Z) - \omega(Z) - U^2] - \frac{2}{3}(1 + 4\nu)R^3, \tag{5.6}$$

$$\partial_T U = U[2\omega(Z) - \frac{4}{3}R^2] + \lambda(Z), \tag{5.7}$$

while the third comes from the definitions of U and Z and can be written

$$\partial_T Z = 2U - c_0. \tag{5.8}$$

Here $T \equiv \epsilon\tau$ is the slow time on which the wave packet changes its form, and we have written $\omega(Z) \equiv dc/dZ$, $\lambda(Z) \equiv d\kappa/dZ$.

These three equations, then, give a simplified description of the evolution of a single wave packet in a rather special limit. Of course, we could and should elaborate the model by discussing the interaction of two or more packets, but such a program is far from trivial and we plan in conjunction with such a study to solve the full Ginzburg-Landau equation numerically in due time. For now we content ourselves with analysing (5.6)-(5.8) for simple forms of μ, ω, and λ so as to adumbrate the variety of possible behaviors. Before doing this, we may briefly discuss the physics of the system represented by the evolution equations.

Clearly the wave packet is being forced to the 'left' (Z decreasing) by the c_0 term. Were U to be exactly zero, the packet would continue on its merry way until R decreased to exponentially small values. Because of the λ term, however, U cannot be zero and, if circumstances are such that U is forced to be positive, then the packet is stopped in its tracks if U becomes large enough. This in turn is possible if ω is positive, so that U grows exponentially for small R. If μ is supposed to increase with Z, then we can find the following sequence of events: (i) dynamo action (represented by R) arises at some value of Z. Both Z and the amplitude of the packet then decrease, while U begins to grow. Eventually R falls to a small value and the (now almost invisible) disturbance makes its way back to larger Z, where the whole process repeats itself. Thus in this simple model the cyclical behavior of the dynamo is manifested, not in the oscillations of the dynamo waves themselves, but in the periodic motion of their envelope. These ideas are illustrated in the next section for particular forms of the variable coefficients in equations (5.6)-(5.8).

6. Butterfly diagrams and self-renovating envelopes

In this section we present some numerical solutions to equations (5.6)-(5.8). Since we have already made rather drastic simplifying assumptions about the magnitudes of the coefficients so as to treat the system as almost integrable, we do not worry too much about their correspondence with the forms given in earlier sections. In particular, we take ω to be a positive constant (and equal to 1 throughout as this can be accomplished by a simple rescaling). We also take λ to be constant and positive (and $c_0 > 0$ without loss of generality), since only when both ω and λ are positive can we achieve recurrent behavior (in other cases the soliton just moves towards the equator and decays away so that a more complicated model would be required). For purposes of illustration we use two different models for the growth rate $\mu(Z)$. In what follows we envisage that the 'equator' of our model is at some large negative value of Z, while the pole is represented by Z large and positive. For Model I, $\mu = qZ$ and for Model II, $\mu = q - Z^2$, where in each case q is a constant. The dynamics then depends on the four parameters q, ν, λ and c_0. We only investigate the case of positive q since this is the only one with a physical correspondence.

It proves convenient to use, instead of R, the variable $E \equiv R^2/3$. The equations have just one fixed point at

$$(E_0, U_0, Z_0) \equiv \left(\frac{1}{4}[2 + \frac{2\lambda}{c_0}], \frac{1}{2}c_0, \frac{1}{q}[U_0^2 + 1 + E_0(1 + 4\nu)]\right); \tag{5.9}$$

provided $c_0^3 < (1 + 4\nu)\lambda$ this is stable for $0 < q < c_0^{-2}[\lambda + (1 + 4\nu)(c_0 + \lambda)((1 + 4\nu)\lambda - c_0^3)]$; for larger values of q (and for all positive q if the first inequality fails) it is unstable to an oscillatory instability. Depending on the initial conditions the solution may run away to infinity; otherwise it is attracted to a periodic orbit with the general characteristics described in the previous section. We have not been able to find any further bifurcations leading to bounded orbits with more complicated time dependence.

Figure 1 shows a typical orbit of the system, together with time traces of the three dependent variables. It is notable that (seemingly because c_0 is small), there are two distinct timescales; there is a slow progress towards negative Z during which the magnitude of R (equivalently, the magnetic energy) decreases; this timescale must then be identified with the 11 years of the solar cycle during which sunspot activity makes its way from mid-latitudes to the equator. There follows a rapid recovery phase during which the field strength is very small and the disturbance propagates rapidly back towards the pole. We can construct a butterfly diagram by plotting the contours of the soliton amplitude as a function of ξ and time, and this is shown for the same parameters in Figure 2.

For Model II there are two fixed points which appear at a saddle-node bifurcation at the origin as q passes through zero. That with $Z > 0$ is nonstable, while the other, with opposite sign has similar properties to the single fixed point in Model I. For moderate q there is again a stable limit cycle which for small c_0 spends most of its time close to a 'slow manifold'; now, however, this has parabolic shape in

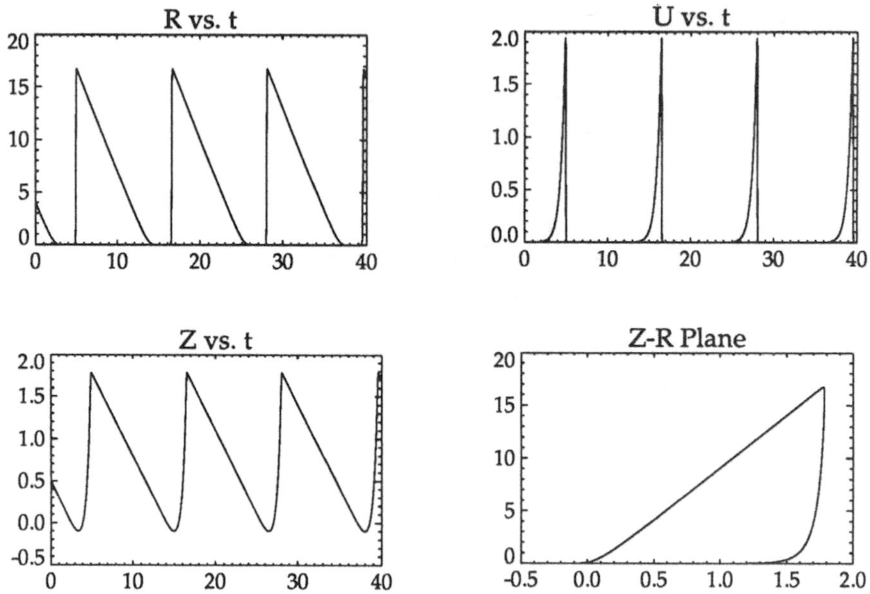

Fig. 1. A periodic solution for Model I with $q = 10$, $\lambda = .0093$, $\nu = 0$, $c_0 = 0.2$. Plots of each of the variables are shown as functions of time, together with a phase portrait of R against Z.

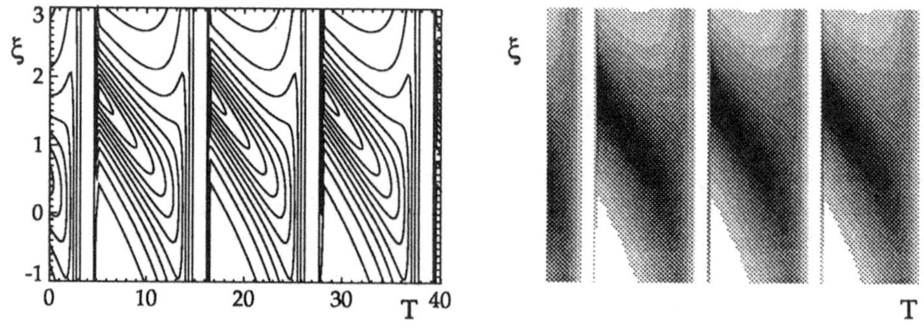

Fig. 2. As Figure 1, but showing a "butterfly diagram" constructed by plotting contours of the quantity Re as a function of ξ and T. A grayscale plot is also shown.

(Z, R) space. Thus Model II gives a better picture of the magnitude of magnetic activity first waxing and then waning as the disturbance propagates towards the equator. Figures 3 and 4 show a typical case, in the same manner as for Figures 1 and 2.

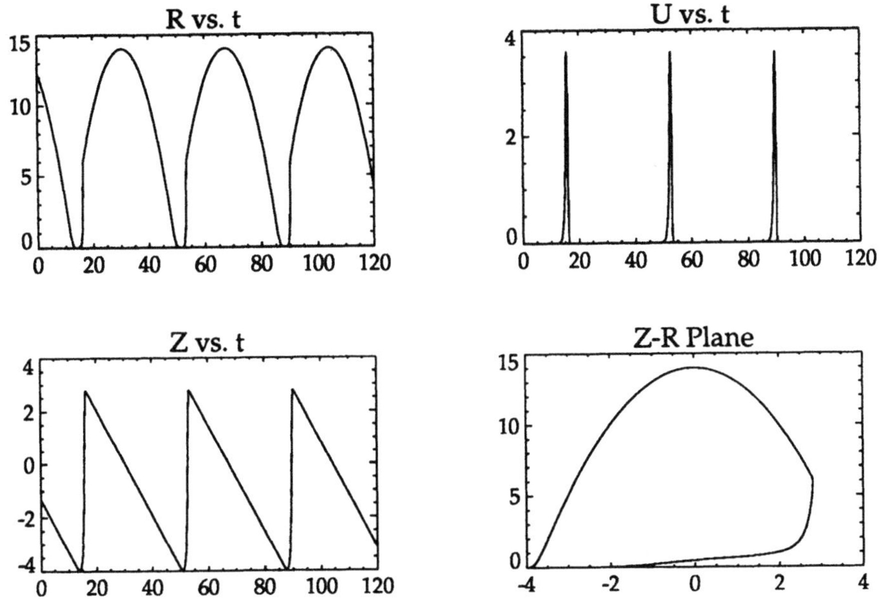

Fig. 3. As for Figure 1, but now for Model II with $q = 15$, $\lambda = .0064$, $\nu = 0$, $c_0 = 0.2$.

In both cases the resemblance to the observed butterfly diagram is striking, although our neglect of the (temporal) width of the wave packet means that the slight overlap (in time) between old waves at low latitude and new ones at high latitude cannot be modelled here.

7. Conclusions

The aim of this work is to make a mathematical model for the solar activity that is represented by the butterfly diagram. To do this, we have had to decide what the diagram actually represents. We are suggesting that the butterfly diagram is a spacetime diagram showing the cyclic appearance of waves of activity forming at ±37° and propagating toward the equator. With this image, there are several features that must be explained. When the waves arrive at the equator what happens to them? Why do they seem to disappear rather than move through or bounce back? Why do the waves in the two hemispheres remain so nearly

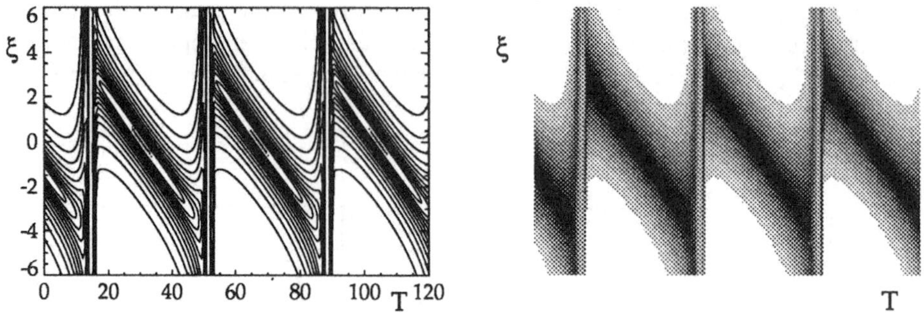

Fig. 4. Butterfly diagram for the calculation of Figure 3 (cf. Figure 2).

synchronous? Is it reasonable that the waves should have such a steady velocity as they move equatorward?

To be able to respond to such questions, we have chosen a particular kind of wave as our model. It is known that in thin-layer dynamo theory there are waves and that, in the standard $\alpha - \omega$ picture for example, these waves are overstable given suitable conditions. Our hypothesis is that the solar activity waves seen in the butterfly diagram are not simple dynamo waves but the envelopes of packets of overstable dynamo waves. The velocity of the observed waves is then not the phase velocity of any dynamo wave process, but the group velocity.

Once this picture is accepted it is not difficult to use qualitative mathematical arguments familiar in modern nonlinear stability theory to write an appropriate general model equation for the waves of solar activity. The main ingredient that we have added that is not usual in the more standard versions of such nonlinear instability theory is the possiblity of weak spatial dependence of the stability characteristics of the system. Given simple choices for such dependence, we have here examined, in a particularly tractable case, the evolution of just one wave packet and found that there do exist periodic solutions that are reminiscent of the observed butterfly diagrams. There are three (to us) pleasing outcomes of this modelling: (a) The velocity of the wave as it progresses toward the equator is reasonably constant; (b) Though the wave is reflected on reaching the equator, the reflected wave has such a small amplitude that we can understand its indetectability to date; and (c) The speed of motion toward the equator is much smaller than that of the return toward the higher latitudes, at least for our choice of parameters. Thus, in this model, a given wave can be thought to appear and move to the equator in eleven years but to make the return trip (to high latitudes) unnoticed in under a year, when it grows again in amplitude to begin a new half-cycle. But there is more to be understood and it remains to be seen whether extensions of the calculations done so far will work to provide the missing features.

Our present theoretical butterfly has only one wing. We need to study a pair of envelope waves, one in each hemisphere, to see if they lock into phase with each other to make a reasonably real butterfly. To do this, we can introduce the two waves in the near integrable case studied here and derive a coupled, sixth-order system for the properties of the two waves. Then we can see how well the two waves get in step with each other. With a sixth order system, we can of course expect at least a little chaos, which would bespeak some north-south asymmetry, but the main question at this stage is whether we can get a full butterfly.

We may also have to allow for two pairs of activity waves, since that is what is seen just at the end of the half-cycle. The possibilities for chaos and intermittency (including minima) in such cases are of course much enhanced, but so are the difficulties of the analysis. Indeed, at this stage we begin to confront harder questions, such as: why are there two pairs of waves active and not some other number? In addressing such questions, we are best advised to solve the evolution equation (4.1) numerically, without relying on the near-integrable approximation. This should yield more disordered behavior due to the interaction of several wave packets. In that case, we can contemplate the extension of the model to incorporate longitudinal dependence of the amplitudes. It is well known that two-dimensional wave patterns can be unstable to three-dimensional disturbances, and the resulting solutions could help us to understand the sector structure that is observed in solar activity. Such a program will of course require extensive computation, but we do hope (and expect) to be heard from again on these matters.

In the meantime, we must not neglect the physical modelling. We have based our derivations on the assumed presence of overstable, simple dynamo waves. These are most accessible in thin-layer dynamos. In line with current thinking, we have assumed that such a thin layer is subconvective. Given such a concrete vision, we ought to be able to derive explicitly the kind of nonlinear wave equation we are here postulating, thus adding some astrophysical input to our astromathematical discussion.

In conclusion, we acknowledge that the collaboration leading this work was made possible by the financial support of our solar research by the Air Force under grant AFOSR89-0012. The work has also been fostered by the N.S.F. under PHY87-04250. The preparation of the manuscript began during the Geophysical Fluid Dynamics School of 1990 at the Woods Hole Oceanographic Institution and was completed during a visit by EAS to the École Normale Supérieure de Lyon.

8. Bibliographical notes

In this informal sketch of our ideas on the mathematics of the solar cycle, we have not given any explicit references since we have assumed that our readers have in mind the basic ideas that we are using. We shall of course make a careful accounting of the provenance of the basic ideas in a more detailed treatment that we shall prepare in the relatively near future. What we shall simply do here is to give some references to basic topics that we have drawn on in this discussion, dynamo theory and nonlinear instability theory.

The main physical assumption of this work is that there is lurking somewhere inside the sun, but accessible to convective upwelling, a thin-layer dynamo. The theory of such dynamos has been elaborated in terms of the 'α-effect' or two-scale model by

Zel'dovich, Ya.B., Ruzmaikin, A.A. & Sokoloff, D.D.: 1983, *Magnetic Fields in Astrophysics*, Gordon & Breach.

Moffatt, H.K.: 1978, *Magnetic Field Generation in Electrically Conducting Fluids*, Cambridge University Press.

In particular, we relied on the property of these models of leading to instabilities in the form of propagating dynamo waves, according to the original ideas of

Parker, E. N.: 1980, *Cosmical Magnetic Fields: their Origin and Activity*, Oxford University Press.

Given the known result that such waves are overstable we studied the nonlinear propagation of a packet of them following the known procedures of nonlinear stability theory as described for instance in

Manneville, P.: 1990, *Dissipative Structures and Weak Turbulence*, Academic Press.

Then we applied standard nonlinear perturbation theory to the solitary envelope waves, a subject introduced in

Lamb, G.L, Jr.: 1980, *Elements of Soliton Theory*, J. Wiley and Sons.

That the solutions to the Ginzburg-Landau equations for which we have derived the governing dynamical system may indeed be prone to the development of more than one activity wave is a danger that is foreshadowed in

Bretherton, C.S. and Spiegel, E.A.: 1983, " Intermittency Through Modulational Instability," *Phys. Lett.* **A96**, 152

Non-steady Global Magnetic Fields in Kinematic Theory

J.H.G.M. van Geffen[1,2], P. Hoyng[1], C. Zwaan[2]

[1] Laboratory for Space Research, Sorbonnelaan 2, 3584 CA Utrecht
[2] Astronomical Institute, P.O. Box 80000, 3508 TA Utrecht

Abstract: The dynamo equation for the mean field $\langle \boldsymbol{B} \rangle$ contains a random forcing term of unknown magnitude, which is therefore always omitted. The influence of this term is potentially large. To evaluate its effect, we employ *ensemble averaging.* If an ensemble average is used, there is no random forcing term in the dynamo equation. The effect of fluctuations is that the ensemble members get out of phase, so that $\langle \boldsymbol{B} \rangle \rightarrow 0$. The damping time of $\langle \boldsymbol{B} \rangle$ can be found by requiring that the mean energy $\langle \boldsymbol{BB} \rangle$ remains finite. The eigenvalues κ of the dynamo equation then all have negative real parts. Imκ determines the period, and -Reκ/Imκ the relative period stability of the dynamo. We have developed a code to solve the equation for $\langle \boldsymbol{BB} \rangle$ in a spherical shell (the convection zone), assuming axisymmetry. We report our first results, which do not yet include differential rotation. Using spherically symmetric boundary conditions, we reproduce the well known α^2-dynamo, whose behaviour is known analytically. For instance, for an α^2-dynamo located in a shell with inner boundary at $R/2$, we find that $\langle \boldsymbol{BB} \rangle$ remains finite for $R^2\gamma/\beta = 1.48$, where β represents turbulent diffusion and γ turbulent vorticity. Taking $\alpha = {}^1/_4(\beta\gamma)^{1/2}$ — a factor of four below maximum helicity — implies that we have a dynamo number $C_\alpha \equiv R\alpha/\beta = 0.30$. Using this value we find a damping time of $6 \times 10^{-2}\, R^2/\beta$ for $\langle \boldsymbol{B} \rangle$, which is a measure for the coherence time of $\boldsymbol{B}$ in a single ensemble member. This result implies that the large-scale field of this particular α^2-dynamo reorganizes its structure completely on a time scale of only about one year (for solar values of R and β), and it shows the enormous influence of random forcing in general.

1. Introduction

The effects of turbulence in the convection zone of stars on the magnetic field $\boldsymbol{B}$ are described by the dynamo equation:

$$\partial_t \langle \boldsymbol{B} \rangle = \alpha \nabla \times \langle \boldsymbol{B} \rangle + \beta \nabla^2 \langle \boldsymbol{B} \rangle \ , \tag{1}$$

where $\langle \boldsymbol{B} \rangle$ is some average of the magnetic field $\boldsymbol{B}$, which we discuss presently. We have assumed that there is no differential rotation. The first term in (1) is the turbulent helicity term, and the second is the turbulent diffusion term. Their coefficients are defined as:

$$\alpha \approx -\langle \boldsymbol{u}_1 \cdot \nabla \times \boldsymbol{u}_1 \rangle \tau_c / 3 \ ; \qquad \beta \approx \langle u_1^2 \rangle \tau_c / 3 \ , \tag{2}$$

which we assume to be constant; $\boldsymbol{u}_1(\boldsymbol{r}, t)$ is the turbulent velocity, and τ_c its correlation time.

All finite solutions of the dynamo equation are *strictly* periodic, and in our case this period is infinite, since we assume no differential rotation. *Strict* periodicity is very strange, since the underlying mechanisms are of a *stochastic* nature, so we would expect to find at best quasi-periodic solutions. Yet, the traditional dynamo equation suggests that the dynamo has an infinite memory. The origin of this feature lies in the derivation of (1). The averaging operator $\langle\cdot\rangle$ must satisfy the Reynolds relations (e.g., Krause and Rädler, 1980):

$$\langle f+g\rangle = \langle f\rangle + \langle g\rangle\ ; \quad \langle f\langle g\rangle\rangle = \langle f\rangle\langle g\rangle\ ; \quad \langle c\rangle = c\ ; \tag{3a, b, c}$$

$$\langle\cdot\rangle \text{ commutes with } \nabla,\ \partial_t \text{ and all integrals}\ , \tag{3d}$$

where $f(\boldsymbol{r},t)$ and $g(\boldsymbol{r},t)$ are arbitrary functions and c is an arbitrary constant.

The average over space or time, generally used in (1), does not satisfy (3b) and (3d), which means that there is actually an extra term on the r.h.s. of (1) to account for this discrepancy. This extra term depends on the turbulent velocity $\boldsymbol{u}_1$ and has therefore the nature of a random forcing term, henceforth abbreviated as *r.f.t.* Even if $\langle\cdot\rangle$ is defined as a longitudinal average, i.e. $\langle\bullet\rangle = (2\pi)^{-1}\oint\langle\bullet\rangle d\phi$, which satisfies (3) exactly, there still is random forcing, because according to their definition α and β will now have a fluctuation component according to their definition (2).

The *r.f.t.* is usually omitted, and that explains the (unphysical) infinite phase memory of the dynamo. We wish to investigate the magnitude of this *r.f.t.*, but since it is very difficult to obtain a manageable expression for it, we adopt another approach.

2. Ensemble average

We interpret $\langle\cdot\rangle$ in (1) as an *ensemble average*, which does obey (3) exactly. In this case α and β are constant, meaning that (1) holds as it stands: there is *no r.f.t.* if an ensemble average is used. One has to take the ensemble literally: an infinite set of identical systems, each having a different realization of $\boldsymbol{u}_1(\boldsymbol{r},t)$. This, however, implies that $\langle\boldsymbol{B}\rangle$ no longer refers to a specific dynamo, or in other words: we don't know which of the ensemble members is *our* dynamo. Nevertheless, the ensemble averaging turns out to be a useful approach for estimating the order of magnitude of the *r.f.t.*

Our line of thought is as follows: Since each of the ensemble members has a different realization of fluctuations, the ensemble members gradually get out of phase. In other words: eventually $\langle\boldsymbol{B}\rangle \rightarrow 0$. If we can determine the damping time of this process, we know the *correlation time* of $\boldsymbol{B}$ itself in a given ensemble member.

3. Mean magnetic energy

In order to find the damping time of $\langle\boldsymbol{B}\rangle$, we analyse the *mean magnetic energy* $\langle\boldsymbol{BB}\rangle$. On physical grounds we must require that this mean magnetic energy remains *finite*. For if $\langle\boldsymbol{BB}\rangle \rightarrow 0$, then $\boldsymbol{B}=0$ in the end, contrary to the assumption that there is a dynamo. And if $\langle\boldsymbol{BB}\rangle$ would increase, the energy would grow indefinitely, which is physically not acceptable either. Therefore we must require that $\langle\boldsymbol{BB}\rangle$ remains finite. We can now find the damping time of $\langle\boldsymbol{B}\rangle$, because the requirement "$\langle\boldsymbol{BB}\rangle$ *remains finite*" fixes a combination of parameters also present in the dynamo equation.

The equation for the tensor $T_{ij} = \langle B_iB_j\rangle$ is (Knobloch, 1978; Hoyng, 1987):

$$\partial_t T_{ij} = \alpha\big(\epsilon_{ikl}\nabla_k T_{lj} + \epsilon_{jkl}\nabla_k T_{li}\big) + \beta\nabla^2 T_{ij} + \frac{2}{5}\gamma\big(2T_{kk}\delta_{ij} - T_{ij}\big), \tag{4}$$

where i, j represent Cartesian coordinates; recall that differential rotation is omitted. The first and second terms in (4) are turbulent helicity and diffusion, also present in the dynamo equation (1). The third term depends on the turbulent vorticity, and it represents generation of magnetic energy. The coefficient is defined as:

$$\gamma \approx \langle|\nabla \times \boldsymbol{u}_1|^2\rangle\tau_c/3\ , \tag{5}$$

which we also assume to be constant. Note that α, β and γ must fulfill Schwarz' inequality:

$$\alpha^2 \leq \beta\gamma \quad \text{or} \quad C_\alpha^2 \leq C_\gamma, \tag{6}$$

where:

$$C_\alpha \equiv R\alpha/\beta\ ; \qquad C_\gamma \equiv R^2\gamma/\beta\ , \tag{7}$$

with R the outer radius of the dynamo. C_α is the usual dynamo number, important in both (1) and (4); C_γ is a second "dynamo number" relevant for (4) only. The definition of these two numbers follows upon the introduction of a dimensionless time $t\beta/R^2$ in both (1) and (4).

The damping time of $\langle \boldsymbol{B}\rangle$ is found as follows. We choose a value of C_α and determine, subject to condition (6), the value of C_γ for which eq. (4) has a solution that remains finite. With the adopted value of C_α we determine the eigenvalues of (1), from which we obtain the damping time of $\langle \boldsymbol{B}\rangle$. This is expounded in the following sections.

4. Results

We have developed a code to determine the eigenvalues of (4) in a spherical shell, using spherical coordinates r, θ, ϕ, and assuming axisymmetry. Eq. (4) is in fact a set of six coupled differential equations, and therefore six boundary conditions are needed, both on the inner boundary $r = r_c$ and on the outer boundary $r = R$; for completeness they are summarized in the Appendix.

Let us write the eigenvalues of (4) as $\Lambda_k = \lambda_k + \mu_k \mathrm{i}$ ($k = 0, 1, 2, \ldots$). The real parts λ_k of the eigenvalues form a decreasing series (because (4) is a parabolic equation). The requirement that $\langle \boldsymbol{BB}\rangle$ remains finite implies that the largest of the real parts must equal zero, i.e. $\lambda_0 = 0$. The eigenfunction $T_{ij,0}$ then describes the corresponding energy distribution within the dynamo; $T_{ij,0}$ depends on the value of C_α, but λ_0 does not (Van Geffen and Hoyng, 1990). Figure 1 shows two examples:

a) *Spherically symmetric* boundary conditions, which reproduces the spherically symmetric α^2-dynamo, whose behaviour is known analytically (Van Geffen and Hoyng, 1990), and can serve as a check on the computations.

b) *Axisymmetric* boundary conditions. In this case there is a preferred direction (the vertical axis).

In both figures, the top row shows the distribution of magnetic energy along the radial direction $\langle B_r^2\rangle$, the bottom row along the ϕ direction $\langle B_\phi^2\rangle$. Since there is no magnetic field within the core, we have $B_r = 0$ at the inner boundary, implying that $\langle B_r^2 \rangle$ equals zero at $r = r_c$, and increases outward. On the other hand, $\langle B_\phi^2 \rangle$ is maximal on the inner boundary and decreases outward.

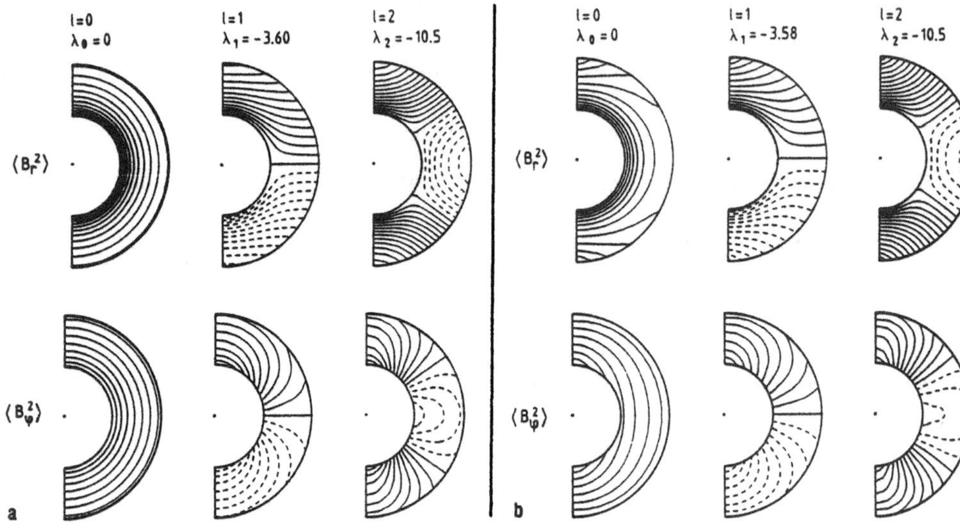

Fig. 1. Contour lines of the distribution of magnetic energy in a spherical shell, using **a)** spherically symmetric and **b)** axisymmetric boundary conditions. Solid lines represent positive contours, broken lines negative contours. In both cases $C_\gamma = R^2\gamma/\beta = 1.48$, $C_\alpha = {}^1\!/\!_4\sqrt{C_\gamma} = 0.30$. See the text for an explanation.

The left column shows the eigenfunction $T_{ij,0}$ belonging to the eigenvalue with the largest real part $\lambda_0 = 0$. The second and third column show the eigenfunctions belonging to the states with the next two largest real parts: λ_1 and λ_2. The solutions shown in Figure 1 are non-periodic, i.e. they have $\mu_k = 0$. The states with negative energy ($k = 1, 2, \ldots$) are damped and will play only a transient role, as the solution of (4) relaxes from arbitrary initial conditions to the stationary solution. Some of the higher eigenfunctions are periodic (that is: they have $\mu_k \neq 0$) in addition to being damped; we found, however, that μ_0 always equals zero.

We can assign quantum numbers to the eigenmodes. First of all m equals zero because of axisymmetry. The quantum number l, representing the number of zeros along the θ direction, takes values 0, 1 and 2 as indicated in Figure 1. The radial quantum number n equals zero in the cases shown, but going to larger values of $|\lambda_k|$ there will also be states with $n = 1, 2, \ldots$.

5. Damping time of $\langle B \rangle$

It turns out that $\lambda_0 = 0$ is reached if $C_\gamma = 1.48$, independent of C_α (see Van Geffen and Hoyng, 1990). In the figure we have taken $C_\alpha = {}^1\!/\!_4\sqrt{C_\gamma} = 0.30$, so that α equals ${}^1\!/\!_4$ of its maximum value according (6). With this value we solve (1) analytically by extending the method of Krause and Rädler (1980, Ch. 14), to allow for the existence of an inner boundary (where $B_r = 0$). The eigenvalues κ_k of (1) are real, since there is no differential rotation. For $C_\alpha = 0.30$ it turns out that $\kappa_k < 0$ for *all* k, thus confirming that *all* solutions of the dynamo equation are damped (cf. Hoyng, 1987). The longest living mode is determined by the the smallest $|\kappa_k|$, which we found to be equal to $16\,\beta/R^2$. Hence the damping time of $\langle B\rangle$ is $(16\,\beta/R^2)^{-1} \approx 6\times 10^{-2}\,R^2/\beta$.

Our interpretation of this result is as follows. The dynamo equation (1) with $\langle\cdot\rangle$ is a spatial or time average contains a *r.f.t.* and may be written as follows:

$$\partial_t\langle B\rangle = \alpha\nabla\times\langle B\rangle + \beta\nabla^2\langle B\rangle + \underbrace{\cdots\langle B\rangle}_{r.f.t} \,. \tag{8}$$

The diffusion term is of the order of β/R^2, so we must conclude that in this case the *r.f.t.* is about 16 times *larger* than the diffusion term. Although this is only a rough estimate, it shows that the *r.f.t.* is certainly not small, as has been assumed so far (e.g., Krause and Rädler, 1980). We have analyzed values of C_α ranging from 1 to $^1/_{128}$ times the maximum value $\sqrt{C_\gamma}$, and we found the damping time to be of the same order of magnitude as in the above example.

6. Concluding remarks

We have shown above that by using ensemble averaging and requiring that $\langle BB\rangle$ remains finite, the damping time of $\langle B\rangle$ can be found, and the magnitude of the random forcing term in the dynamo equation can be estimated. We have also shown that this random forcing dominates in (8), rather than being negligible. This result raises several issues:

- First of all the role of non-linearities. Our main result is that random forcing is an essential ingredient in the dynamo equation (1), if an average over space or time is used. The use of an ensemble average to estimate the magnitude of the *r.f.t.* is merely a technical matter. Apparently the *r.f.t.* is very large and this makes the dynamo unsteady and erratic. At present it is not clear whether non-linearities (which are of course important) will further destabilize the dynamo. The traditional approach, however, where non-linear effects are included in (1) — for example by making α a function of $|\langle B\rangle|$ — while random forcing is ignored, seems rather pointless in the light of the present results.
- We have assumed isotropic, homogeneous turbulence and infinite resistivity. These assumptions are not very realistic and have to be removed, but we believe that this will not drastically change the outcome of our study.
- What happens if we adopt a larger value of C_α? We found the dipole solution of (1) to be marginally stable only at $C_\alpha = 6.57$. We have verified that for this large value of C_α, eq. (4) always has growing solutions (i.e. $\lambda_0 > 0$). A full discussion will appear later.
- The aim of our work is to study the effect of random forcing in the solar dynamo, in order to find out whether the observed period stability of the 22-year cycle of the sun ($\delta P/P \approx 0.1$) can be explained within linear theory.

Acknowledgements: We thank Dr. K.-H. Rädler and Dr. A. Brandenburg for many stimulating discussions. This research has been supported by the Netherlands Foundations for

Astronomical Research (ASTRON) and the Space Research Organization of the Netherlands (SRON). Both organizations are supported by the Netherlands Organizations for Scientific Research (NWO). The computations are financed by the Foundation SURF from the National Fund for the use of Supercomputers (NFS).

Appendix: Boundary conditions

We consider a dynamo in a spherical shell: $0 < r_c \leq r \leq R$, and assume axisymmetry. The first column shows the boundary conditions and quantities specified on the inner boundary, and the second column those on the outer boundary. The first two lines follow directly from physics, where $\epsilon \equiv 2gem$ is the mean magnetic energy density, and where σ, τ are related to the distribution of magnetic energy over spatial scales; these are discussed in more detail in Van Geffen and Hoyng (1990). The bottom three lines represent cross-correlation functions at the boundaries. On the axis: $T_{r\theta} = T_{r\phi} = T_{\theta\phi} = 0$, $T_{\theta\theta} = T_{\phi\phi}$ because of axisymmetry.

$r = r_c$	$r = R$
$\partial\epsilon/\partial r = 0$	$\partial\epsilon/\partial r + \epsilon/\varrho = 0$
$T_{rr} = T_{r\theta} = T_{r\phi} = 0$	
	$\sigma = (T_{\theta\theta} + T_{\phi\phi})/T_{rr}$
$\tau^0 = T_{\theta\theta}/T_{\phi\phi}$	$\tau = T_{\theta\theta}/T_{\phi\phi}$
	$C_{r\theta} = T_{r\theta}/\sqrt{T_{rr}T_{\theta\theta}}$
	$C_{r\phi} = T_{r\phi}/\sqrt{T_{rr}T_{\phi\phi}}$
$C^0_{\theta\phi} = T_{\theta\phi}/\sqrt{T_{\theta\theta}T_{\phi\phi}}$	$C_{\theta\phi} = T_{\theta\phi}/\sqrt{T_{\theta\theta}T_{\phi\phi}}$

The distribution and correlation functions chosen for the graphs in Figure 1 are:

	spherically sym.	axisymmetric
σ	1/2	$1/2+\sin\theta$
τ^0, τ	1	$1+\sin\theta$
$C^0_{\theta\phi}, C_{\theta\phi}$	0	$\sin\theta$
$C_{r\theta}, C_{r\phi}$	0	0
R/ϱ	1	1

References

Hoyng, P.: 1987, *Astron. Astrophys.* **171**, 357

Knobloch, E.: 1978, *Astron. J.* **220**, 330

Krause, F., Rädler, K.-H.: 1980, *Mean Field Magnetohydrodynamics and Dynamo Theory*, Pergamon Press, London

Moffatt, H.K.: 1978, *Magnetic Field Generation in Electrically Conducting Fluids*, Cambridge University Press, Cambridge

Van Geffen, J.H.G.M., Hoyng, P.: 1990, *Geophys. Astrophys. Fluid Dynam.* **53**, 109

Asymptotic Methods in the Nonlinear Mean-field Dynamo

D.D. Sokoloff[1], A. Shukurov[2] and A.A. Ruzmaikin[2]

[1]Physics Department, Moscow State University, Moscow, USSR
[2]IZMIRAN, USSR Academy of Sciences, Troitsk, Moscow Region, USSR

Abstract: We discuss the methods and results of analysis of nonlinear mean-field dynamo models based on α-quenching in two asymptotic regimes, namely for weakly and highly supercritical excitation. In the former case the spatial distribution of the steady-state magnetic field is close to that given by the neutrally stable eigenfunction of the corresponding kinematic dynamo. In the latter case the magnetic field distribution within the main part of the dynamo volume is presumably determined by the balance between the Lorentz and Coriolis forces while near the boundaries boundary layers arise in which the field adjusts itself to the boundary conditions. The asymptotic behaviour of the highly supercritical $\alpha\omega$-dynamos is sensitive to the particular form of dependence of the mean helicity on magnetic field while α^2-dynamos are less sensitive to this dependence.

1. Introduction

One of widely employed approaches to the analysis of the steady states of magnetic field generated by hydromagnetic mean-field dynamos is based on introduction of the mean helicity as an explicit function of the large-scale magnetic field. In fact, this approach is an extrapolation to nonlinear regime of ideas of the kinematic dynamo theory in which the velocity field is prescribed as a known function of position whose form is justified by simple physical arguments of a general character, rather than follows from the Navier-Stokes equations. Such approach is much less complicated than solution of the coupled induction and Navier-Stokes equations and allows the analysis of general types of evolution in nonlinear dynamo systems and also a qualitative comparison of the theory with observations.

Two basic concepts of the steady-state distribution of magnetic field are usually discussed in the literature. It is often presumed that this distribution is close to the neutrally stable eigenfunction of the kinematic dynamo based on the observed (or presumed) velocity field. On the other hand, comparisons with observational results sometimes rely on the assumption either of equipartition of kinetic and magnetic energies or of local balance between the Lorentz and Coriolis forces. Our results discussed below allow the reconciliation of these apparently contradictory ideas.

2. The nature of nonlinear solutions

A steady state of a dynamo arises when the generated magnetic field modifies the velocity field in such way that the generation efficiency becomes marginally weak due to, e.g., a local balance of amplification and dissipation. In this state the effective dynamo number (or any other relevant measure of the generation efficiency) is presumed to decrease down to the critical value D_0 which corresponds to vanishing of the field growth rate in the allied kinematic dynamo. These arguments contain an implicit contradiction: on the one hand, it is presumed that modification of the velocity field has a *local* character and, on the other hand, the criterion of such balance is based on a *global* parameter, the dynamo number.

A resolution of this contradiction follows from the fact that the generated magnetic field is inhomogeneous and, therefore, the velocity field is modified to different extent at different positions. Meanwhile, the critical dynamo number is rather sensitive to the form of the velocity field and, thereby, the effective critical dynamo number cannot be determined basing on the initial velocity field and remains unknown in advance. It is clear that the steady-state magnetic field is stronger in those regions where the fluid density or kinetic energy density is larger. Correspondingly, the velocity field is modified more strongly in these regions.

The arguments of the first paragraph in this section bring us to the conclusion that the steady-state distribution of magnetic field is close to the neutrally stable eigenfunction of the corresponding kinematic dynamo. Meanwhile, the arguments of the second paragraph imply that spatial distributions of both the velocity field and the fluid density affect spatial distribution of the steady-state magnetic field, i.e. one comes to an idea of a balance between kinetic and magnetic energies or between Lorentz and Coriolis forces in the steady state of the dynamo. Although these two ideas on the nature of the steady state – the neutrally stable eigenfunction and the dynamic equilibrium – seem to be inconsistent with each other, they both exist independently in the current literature.

The aim of this contribution is to demonstrate that both nonlinear regimes can be realized in dynamo systems. In the case of weakly supercritical excitation, $|D - D_0| \ll |D_0|$ with D the dynamo number, the steady-state distribution of magnetic field is asymptotically close to the neutrally stable eigenfunction while the maximal field strength is proportional to $|D - D_0|^{1/2}$. In the other limiting case of highly supercritical excitation, $|D| \gg |D_0|$, the magnetic field is distributed in accordance with the local balance between the Lorentz and Coriolis forces.

For definiteness, we employ the terminology of the mean-field $\alpha\omega$-dynamos although our arguments are applicable in a general case. For illustration, we consider two particular forms of the helicity coefficient $\alpha(\boldsymbol{r})$ as a function of magnetic field:

$$\alpha(\boldsymbol{r}, \boldsymbol{B}) = \alpha_0(\boldsymbol{r}) \left[1 - g(\boldsymbol{r})|\boldsymbol{B}|^2\right], \tag{1}$$

and

$$\alpha(\boldsymbol{r}, \boldsymbol{B}) = \frac{\alpha_0(\boldsymbol{r})}{1 + g(\boldsymbol{r})|\boldsymbol{B}|^2}, \tag{2}$$

where $\boldsymbol{B}$ is the large-scale magnetic field and $\alpha_0(\boldsymbol{r})$ and $g(\boldsymbol{r})$ are certain functions of position. We consider the α-quenching as the only nonlinearity although ω-quenching can be considered in the same way.

The physical meaning of the function $g(\boldsymbol{r})$ is the following. This function is connected with the characteristic strength of magnetic field which is determined, in turn, by a specific physical mechanism of saturation of the dynamo. For example, such saturation in the mean-field dynamos can be shown to be associated with a balance of the Lorentz and Coriolis forces which is formally similar to the magnetostrophic balance typical of convective dynamos:

$$\frac{\tilde{B}_\phi \tilde{B}_r}{4\pi L} \simeq \rho v \Omega, \tag{3}$$

where $\tilde{B}_\phi$ and $\tilde{B}_r$ are the characteristic azimuthal and radial cylindrical components of the large-scale magnetic field, respectively, L is the field scale, v is the r.m.s. turbulent velocity and Ω is the overall angular velocity of rotation. Then the function $g(\boldsymbol{r})$ is defined as $g(\boldsymbol{r}) = |\tilde{\boldsymbol{B}}|^{-2}$, where $\tilde{\boldsymbol{B}}$ follows from the balance relation (3).

3. The form of nonlinear solutions

In this section we discuss the method of derivation and the explicit form of nonlinear solutions of the mean-field dynamo problem in the two limiting cases mentioned above.

3.1 Weakly supercritical dynamos

In the case when the dynamo number D only slightly exceeds the critical value D_0, i.e. $|D - D_0| \ll |D_0|$, the steady-state strength of magnetic field is rather small and the mean helicity is modified only weakly. Therefore, the steady-state distribution of the helicity can be expected to be close to that corresponding to $\boldsymbol{B} = 0$, i.e. $\alpha(\boldsymbol{r}) \approx \alpha_0(\boldsymbol{r})$. As a result, the spatial distribution of magnetic field is asymptotically close to that given by the neutrally stable eigenfunction of the kinematic dynamo with $\alpha(\boldsymbol{r}) = \alpha_0(\boldsymbol{r})$. The corresponding nonlinear asymptotic solutions have been obtained by Rüdiger (1973) for the simplest case of the α^2-dynamo and by Kvasz *et al.* (1990) for the $\alpha\omega$-dynamo, both for a nonlinearity of the form given by eq. (1) and for a thin-disk geometry. For $|D - D)| \ll |D_0|$ we have $|\boldsymbol{B}| = O(|D - D_0|^{1/2}) \ll 1$ and, therefore, any dependence of α on $\boldsymbol{B}$ (e.g., that given by eq. 2) can be expanded in power series in B^2 and the form (1) is a general one. Kleeorin and Ruzmaikin (1981) have considered nonlinear weakly supercritical solutions for α described by a differential equation rather that by an algebraic function.

3.2. Highly supercritical dynamos

One-dimensional highly supercritical α^2-dynamos have been recently investigated numerically by Meinel and Brandenburg (1990) (see also Baryshnikova and Shukurov, 1987). An analytical asymptotic analysis of a one-dimensional $\alpha\omega$-dynamo with nonlinearity of the form (1) has been performed by Kvasz *et al.* (1990) whose results and their generalizations are discussed below.

The stationary $\alpha\omega$-dynamo equations in a slab surrounded by vacuum can be reduced to the following equation for B, the y-component (or the azimuthal one in cylindrical coordinates) of the large-scale magnetic field (it is presumed that $|\boldsymbol{B}| \approx B$ which is typical of $\alpha\omega$-dynamos):

$$\frac{\mathrm{d}^3 B}{\mathrm{d}z^3} + D\alpha_0(z)\left[1 - g(z)B^2\right]B = 0, \tag{4}$$

with the boundary conditions $B(1) = \mathrm{d}^2B(1)/\mathrm{d}z^2 = \mathrm{d}B(0)/\mathrm{d}z = 0$, where z is the coordinate measured across the slab (Vainshtein and Ruzmaikin, 1972) and the nonlinearity is chosen to be of the form (1).

For $|D| \gg |D_0|$ eq. (4) represents a singularly perturbed equation and its solution reduces to a combination of boundary layers situated near $|z| = 1$ (the slab surface) and, possibly, $z = 0$ (the midplane) and the smooth (*degenerate*) field distribution in the main part of the slab, for which the field derivative can be neglected to the leading order. The structure of the boundary layers is considered in detail by Kvasz *et al.* (1990). Here we note that their thickness is as small as $O(|D|^{-1/3})$ and they hardly can be observed in real dynamo objects. Therefore, here we concentrate on the degenerate solution.

For $|D| \gg 1$ we have $B = O(1)$ and $\mathrm{d}^3B/\mathrm{d}z^3 = O(1)$ outside the boundary layers and eq. (4) can be satisfied when $1 - g(z)B^2 = O(|D|^{-1})$. Thus, the degenerate solution is given by $B(z) = [g(z)]^{-1/2} + O(|D|^{-1})$. The terms of the order of $|D|^{-1}$ and higher ones can be easily obtained using the regular perturbation techniques (Kvasz *et al.*, 1990). Thus, for a nonlinearity of the form (1) the field distribution outside the boundary layers does not depend, to the leading order, on the magnetic diffusivity.

For $\alpha(\boldsymbol{r}, \boldsymbol{B})$ of the form (2), the asymptotic solution has somewhat different properties. In this case eq. (4) is replaced by

$$\frac{\mathrm{d}^3 B}{\mathrm{d}z^3} + D\frac{\alpha_0(z)}{1 + g(z)B^2}B = 0. \tag{5}$$

In contrast to eq. (4), the second term on the left-hand side of eq. (5) cannot be identically zero. Therefore, the field strength in the steady state is large, $B \gg 1$ (in contrast to eq. (4) where $B = O(1)$). As a result, the balance of the field generation and diffusion is maintained even at the leading order and the derivative in eq. (5) cannot be neglected even outside the boundary layers. One can easily verify that in this case $B = O(|D|^{1/2})$. The leading-order solution is given by $B(z) = |D|^{1/2} b(z)$ where $b(z)$ obeys the following equation:

$$b\frac{\mathrm{d}^3 b}{\mathrm{d}z^3} + \frac{\alpha_0(z)}{g(z)} = 0. \tag{6}$$

Notice that this solution depends on magnetic diffusivity through the dynamo number D.

Thus, properties of highly supercritical $\alpha\omega$-dynamos are very sensitive to the particular form of nonlinearity. The case of the α^2-dynamo is much simpler in this respect because the highest order derivative can be always neglected outside the boundary layers in equation

$$\frac{\mathrm{d}^2 B}{\mathrm{d}z^2} - \mathrm{i}R_\alpha \frac{\mathrm{d}}{\mathrm{d}z}\left[\alpha(z, B)B\right] = 0, \tag{7}$$

which corresponds to eqs (4) or (5), and the degenerate solution governed by $\alpha(z, B)B = \text{const}$ always exist. This implies that, in contrast to $\alpha\omega$-dynamos, solutions of highly supercritical α^2-dynamos depend only weakly on the magnetic diffusivity.

References

Baryshnikova, Yu.S., Shukurov, A.: 1987, *Astron. Nachr.* **308**, 89
Kleeorin, N.I., Ruzmaikin, A.A.: 1981, *Geophys. Astrophys. Fluid Dyn.* **17**, 281
Kvasz, L., Sokoloff, D., Shukurov, A.: 1990, *Geophys. Astrophys. Fluid Dyn.* in press
Meinel, R., Brandenburg, A.: 1990, *Astron. Astrophys.*, in press
Rüdiger, G.: 1973, *Astron. Nachr.* **294**, 183
Vainshtein, S.I., Ruzmaikin, A.A.: 1972, *Sov. Astron.* **16**, 365

Fractal Flux Tubes of the Solar Magnetic Field

Alexander Ruzmaikin[1], Dmitry Sokoloff[2], Theodore Tarbell[3]

[1]IZMIRAN, Troitsk, Moscow Region 142092, USSR
[2]Moscow State University, Moscow, USSR
[3]Lockheed Palo Alto Research Labs, Palo Alto, CA 94304, USA

Abstract: The small-scale solar magnetic field exceeding a given threshold forms a fractal set. A dimension of this fractal is found from magnetograms with varying linear resolution. The dimension depends on the value of the threshold magnetic field (multifractality). A simple dynamo model explaining the origin of the fractal magnetic structure is considered. The dynamo produces a magnetic field in the form of flux tubes with a fractal distribution of magnetic field across the tube. The observed dimension gives a possibility of estimating a degree of structuredness of the solar velocity field.

1. Introduction

Analysis of many spectroscopic observations suggests that a great part of the solar magnetic field, except for a weak mean field, is concentrated in kilogauss flux tubes too small to be observed directly (they have diameters of order or less than 1000 km), see a review by Stenflo (1989) and a recent paper by Tarbell *et al.* (1990b).

Dynamo theory predicts that a random flow of highly conducting fluid can lead to self-excitation of an initially weak magnetic field (Molchanov *et al.* 1983). However, the field grows highly inhomogeneously (intermittently): ropes and layers of intense field are prominent against a background of relatively slowly growing field. This was also demonstrated by numerical simulations (Meneguzzi *et al.* 1981). These ropes and layers form a network which randomly changes in space and time.

The solar magnetic field can be divided into a mean field, fluctuations of the mean field and a random magnetic field independently excited by random convective motions. In this paper we will be concerned with the interpretation of the random magnetic field, leaving aside the weaker and larger-scale mean field and its fluctuations which are important in the context of the solar cycle. Of course, the fluctuations of the mean field also may have a fractal structure. They probably tend to be vertical because the mean field emerges at the solar surface due to buoyancy. In a model approximation the fluctuating field can be considered as vertical and therefore be transported like a passive scalar which is stretched and subdivided by fine-scale turbulent motions into a fractal subset. At present it is difficult to distinguish these fluctuations from the random field excited independently. In principle it can be done by studying the solar cycle phase dependence and the correlation with the sunspots.

The fraction of the total volume occupied by the field structures in which the field strength exceeds a given level can serve as a quantitative measure of such intermittent field distribution. However, this part of the total volume at large magnetic Reynolds numbers can turn not to be a volume (even a small one). Due to the strong intermittency, the magnetic structures can occupy a manifold of a non-integer (fractal) dimension. Strictly speaking, at large but finite magnetic Reynolds numbers this manifold has the form of a volume. However in the solar conditions a characteristic scale for these ropes is too small (of the order of the electromagnetic skin-depth) to be observable so it is justified to consider this manifold as fractal. An adequate measure of the field distribution is the dimension of this manifold rather than the fraction of the total volume occupied by magnetic structures (which depends on the chosen resolution).

2. The observed fractal dimension

When some surface is observed at the linear resolution ϵ, the fraction of the surface area occupied by the concentrations of magnetic field is given by

$$S(\epsilon) = S_{\mathfrak{f}}\epsilon^{2-d}, \tag{1}$$

where d is known as the fractal dimension. It is clear that for $d = 2$, i.e. when the concentrations occupy the whole surface, the area (1) does not depend on the resolution. In the other limiting case, when $d = 0$ and the concentrations are actually points, we have $S(\epsilon) \propto \epsilon^2$, i.e. circles of radius ϵ, which cover individual concentrations, do not overlap. Instead of S one can use the number of pixels covering the fractal set, $N = S/\epsilon^2 = S_{\mathfrak{f}}\epsilon^{-d}$.

The value $S_{\mathfrak{f}}$ we may call "a fractal area". For a given solar box under observation $S_{\mathfrak{f}}$ can be excluded by normalization. However it may appear that by observing the different boxes one will find different $S_{\mathfrak{f}}$'s. We guess that this value is proportional to the number of flux tubes filling the observational box. In principle the box can be filled up by one or two or more flux tubes having a complicated fractal form. The observational determination of $S_{\mathfrak{f}}$ is thus meaningful.

Observations of solar magnetic fields at the scale of a single supergranulation cell with progressively higher resolution were first presented at the IAU Symposium No. 138 by Tarbell *et al.* (1990a) and later by Tarbell *et al.* (1990b). These first results, based on Kitt Peak magnetograms obtained at resolutions below 2.5 arc seconds and La Palma magnetograms at resolutions below 0.6 arc seconds, give only preliminary results for the fractal dimension d. The estimated value is about 1.6 over the range 0.6–10 arc second considered, which indicates that the distribution of small-scale solar magnetic fields has fractal properties. An example of the

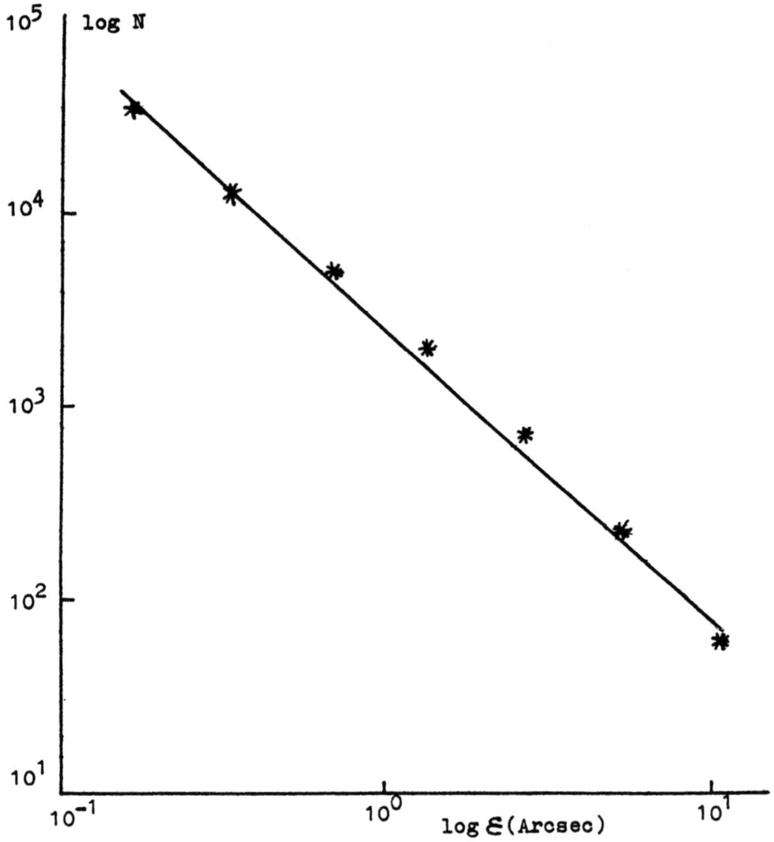

Fig. 1. The number of pixels covering the fractal set versus a pixel size. Threshold of the magnetograms used is 800 G. The resulting dimension is 1.49 (Tarbell *et al.*, 1990b).

dependence of the number of pixels covering the area occupied by the fields whose strength exceeds a certain given value H_{th} is given in Fig. 1. This dependence is linear when plotted in double logarithmic scale. The tangent of this line yields the desired dimension. Of course, the value of d depends on the chosen threshold value of magnetic field strength. In that sense the magnetic field distribution can be described as a multifractal (the idea of multifractality is discussed by Frisch (1983) and recently by Zeldovich *et al.*, 1990)

It can be expected that the fractal dimension decreases when the threshold field strength increases (Figs. 2 and 3) since the stronger the concentrations, the more rarely they are encountered.

Note that the fractal structure could hardly be revealed by Fourier power analysis, losing the important information about the phase correlations between different Fourier modes. Tarbell *et al.* (1990b) showed that the solar magnetic field spectrum is consistent with a power law $k^{-\sigma}$ where the exponent is close to -3/2 in the inertial range of the wavenumber space. This 3/2 power law is exactly the

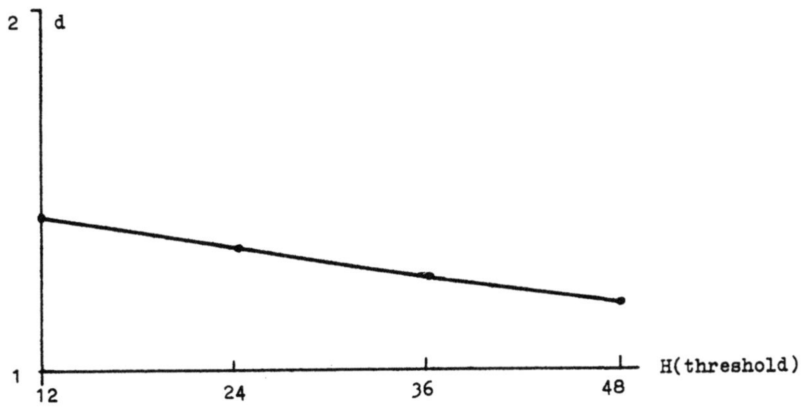

Fig. 2. Dimension of magnetic fractal set versus the threshold magnetic field for the Kitt Peak magnetograms.

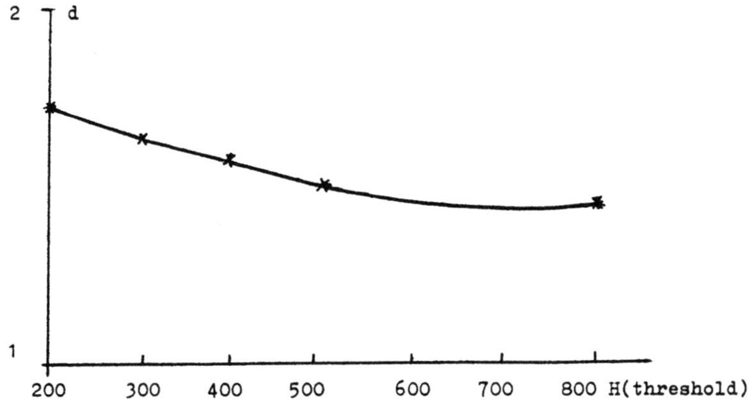

Fig. 3. d versus H (threshold) for the La Palma magnetograms.

known Kraichnan spectrum in the MHD turbulence of uncorrelated Alfvén waves. To catch the intermittency in this way it is necessary to find deviations from the 3/2-law as accurately as possible.

3. A dynamo model of the fractal distribution of magnetic field

Twisting and folding of magnetic ropes form the basis of the dynamo in a random flow (see, e.g., Zeldovich *et al.*, 1983). In the simplest case, the rope is stretched homogeneously to twice its size, which transformation is accompanied by a two-fold decrease of its cross-section because the magnetic flux is conserved. Consequent twisting and folding of the rope restores the cross-section and, thus, the magnetic field and flux are finally doubled. As noted by Finn and Ott (1988), the stretching can be inhomogeneous: some part of the rope can be stretched strongly (by the factor α^{-1}, $\alpha < 1/2$) while the remaining part is stretched only weakly (by the factor $(1-\alpha)^{-1}$, see Fig. 4). Obviously, in this case the field is stronger in that part of the rope which is stretched most strongly.

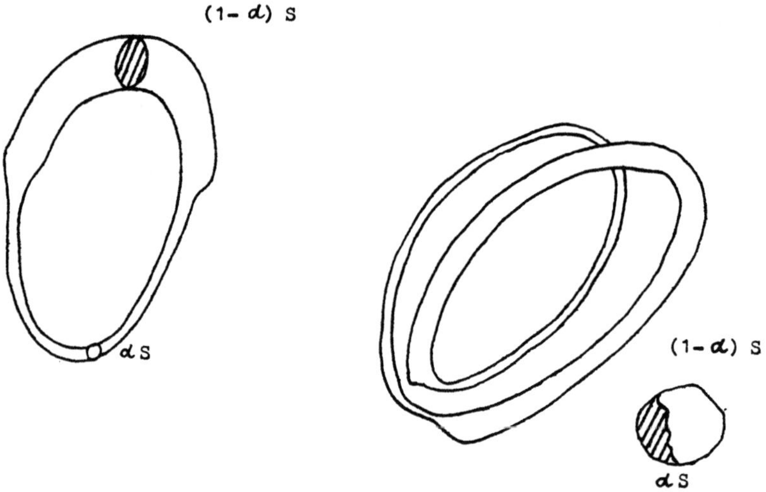

Fig. 4. The inhomogeneous stretching of magnetic rope results in a fractal distribution of the magnetic field over the cross-section of the rope.

Let us clarify the physical meaning of α. This quantity characterizes a ratio of a turbulent cell size to an element of its structure. In that sense one can imagine the turbulent flow considered not as a structureless gaussian flow but as the structured random flow consisting of vortex lines with α characterizing a scale of these vortices.

After multiple repetitions of this transformation, the field concentrates within a small fraction of the initial loop cross-section. According to Finn and Ott (1988), the dimension of this part of the cross-section is given by

$$d = 1 + \frac{\ln 2}{\ln\left[1/\sqrt{\alpha(1-\alpha)}\right]}. \tag{2}$$

The simplest case of the Zeldovich rope dynamo corresponds to $\alpha = 1/2$ and $d = 2$. In a real flow the value of α varies from one loop to another whilst remaining between 0 and 1/2. Thus, it can be expected that d can vary from one position to another. In this sence, it would be correct to introduce an average dimension d and consider deviations from this average value. However, as shown by equation (2), even strong variations in α only weakly affect the value of d.

4. Confrontation with observations and discussion

The degree of inhomogeneity of the loop stretching can be determined from the observed value of d (see Fig. 5).

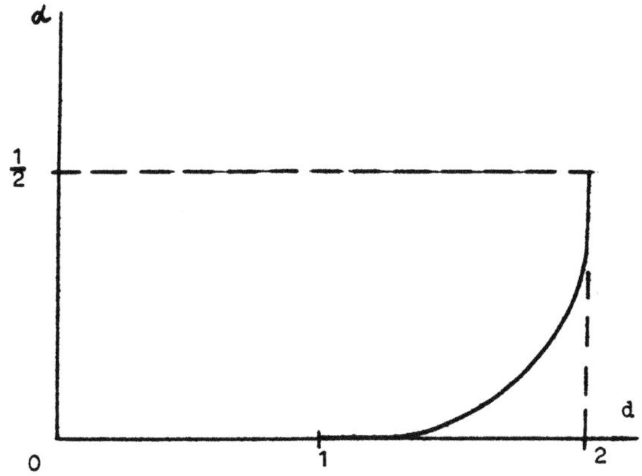

Fig. 5. Degree of structuredness of the flow versus the dimension of magnetic set.

In particular, for $d = 1.2$ we have $\alpha \cong 10^{-3}$ and for $d = 1.73$ we obtain $\alpha \cong 0.2$. Thus, the observed picture implies a highly inhomogeneous stretching of magnetic ropes, i.e. a high inhomogeneity of the velocity field.

While at this stage it is dangerous to make general conclusions it is clear that α is essentially smaller than $1/2$ corresponding to the simple Zeldovich model, which can be considered as a dynamo created by the structureless gaussian flow. The small value of α is an indication that the solar convective flow is highly structured in the observed cell. This deduction is well supported by observations of the solar velocity field. "Inside" the supergranule space under consideration there are, as it is well known, mesogranules and granules. In that sense the analysis of magnetic data may provide some evidence concerning the velocity field. The characteristic sizes of supergranule, mesogranule, and granule are of the order 3×10^4, 8×10^3, and 10^3 km correspondingly. Putting $\alpha = 8\times 10^3/3\times 10^4 = 0.27$ or $10^3/3\times 10^4 = 0.03$ one has $d = 1.87$ or 1.5. Thus the structuredness predicted from the observed dimension and the dynamo model is comparable with the structuredness of independently observed velocity fields.

The fractal distribution of the magnetic field over the cross-section of the tube causes a breakdown of the classical conception of the flux tube. We can introduce the concept of "a fractal flux tube" which has one aspect in common with the classical flux tube: there is no magnetic field perpendicular to the surface of the tube. Note however that in the dynamo model considered here it is assumed that the flux tube has a well defined surface. In the general case the boundary probably has a fractal form. That is evident from the study of the correlation properties of typical realizations of the magnetic field in a random flow (Galeeva *et al.*, 1989).

References

Galeeva, R., Ruzmaikin, A.A., Sokoloff, D.D.: 1989, *Magnitnaya Gidrodinamika* No 3, 3

Finn, J., Ott, E.: 1988, *Phys. Rev. Lett.* **60**, 760

Frisch, U.: 1983, in *Chaotic Behaviour of Deterministic Systems*, Les Houches Session, North Holland

Molchanov, S.A., Ruzmaikin, A.A., Sokoloff, D.D.: 1983, *Geophys. Astrophys. Fluid Dyn.* **30**, 241

Meneguzzi, M., Frisch, U., Pouquet, A.: 1981, *Phys. Rev. Lett.* **47**, 1060

Stenflo, J.O.: 1989, *Astron. Astrophys. Rev.* **1**, 3

Tarbell, T., Ferguson, S., Frank, Z., Shine, R., Title, A., Topka, K.: 1990a, in *Solar Photosphere: Structure, Convection and Magnetic Fields*, ed. J.O. Stenflo, IAU Symp. No 138, Kluwer Academic Publishers, p. 147

Tarbell, T., Acton, D.S., Topka, K., Title, A., Schmidt, W., Scharmer, G.: 1990b, submitted to *Phys. Fluids*

Zeldovich, Ya.B., Ruzmaikin, A.A., Sokoloff, D.D.: 1983, *Magnetic Fields in Astrophysics*, Gordon and Breach

Zeldovich, Ya.B., Ruzmaikin, A.A., Sokoloff, D.D.: 1990, *The Almighty Chance*, World Sci. Pub., Singapore

$\alpha\Lambda$-dynamos

A. Brandenburg[1,2], D. Moss[3], M. Rieutord[4],

G. Rüdiger[5], I. Tuominen[1]

[1] Observatory and Astrophysics Laboratory, University of Helsinki Tähtitorninmäki, SF-00130 Helsinki, Finland
[2] NORDITA, Blegdamsvej 17, DK-2100 Copenhagen, Denmark
[3] Dept. of Mathematics, The University, Manchester M13 9PL, UK
[4] Observatoire de Toulouse, 14 av. E. Belin, F-31400 Toulouse, France
[5] Astrophysikalisches Observatorium Potsdam, O-1560 Potsdam, FRG

Abstract: In contrast to $\alpha\omega$-dynamos, where the angular velocity is arbitrarily prescribed, we consider here $\alpha\Lambda$-dynamos, for which the differential rotation and meridional circulation are solutions of the momentum equation. The non-diffusive parts of the Reynolds stress tensor are parameterized by the Λ-effect. In earlier investigations we have shown that the turbulent magnetic diffusivity has to be much smaller than the eddy viscosity, otherwise the dynamo is not oscillatory or else the contours of constant angular velocity are cylindrical, contrary to observations. In the present paper we investigate the effects of compressibility.

1. The basic equations

We consider here the initial value problem of a conducting fluid in a rotating sphere with radius R. We solve the hydromagnetic equations, starting with a rigid rotation with angular velocity, Ω. A weak magnetic seed-field is present initially. We are interested in the evolution of the velocity and magnetic fields on time scales that are much longer than the sound travel time and can therefore adopt the anelastic approximation. The equations governing the generation of magnetic field and differential rotation are

$$\frac{\partial \boldsymbol{B}}{\partial t} = \mathrm{curl}\,(\boldsymbol{u} \times \boldsymbol{B} + \alpha \boldsymbol{B} - \eta_t \mu_0 \boldsymbol{J}), \tag{1}$$

$$\rho \frac{\mathrm{D}\boldsymbol{u}}{\mathrm{D}t} = -\nabla p + \boldsymbol{J} \times \boldsymbol{B} - \rho \boldsymbol{g} - \mathrm{Div}\,(\rho \mathcal{Q} - \mathcal{B}), \tag{2}$$

$$\rho T \frac{\mathrm{D}s}{\mathrm{D}t} = -\mathrm{div}\, \boldsymbol{F}, \tag{3}$$

with $\mathrm{div}\, \rho \boldsymbol{u} = \mathrm{div}\, \boldsymbol{B} = 0$. $\mathrm{D}/\mathrm{D}t = \partial/\partial t + \boldsymbol{u} \cdot \nabla$ denotes the advective derivative, $\boldsymbol{J} = \mathrm{curl}\, \boldsymbol{B}/\mu_0$ is the electric current, where μ_0 is the induction constant, η_t is the turbulent magnetic diffusivity, and $\mathcal{Q}$ and $\mathcal{B}$ are respectively the Reynolds and Maxwell stress tensors. Axisymmetry is assumed and, as in Paper I (Brandenburg *et al.*, 1990), we neglect $\mathcal{B}$.

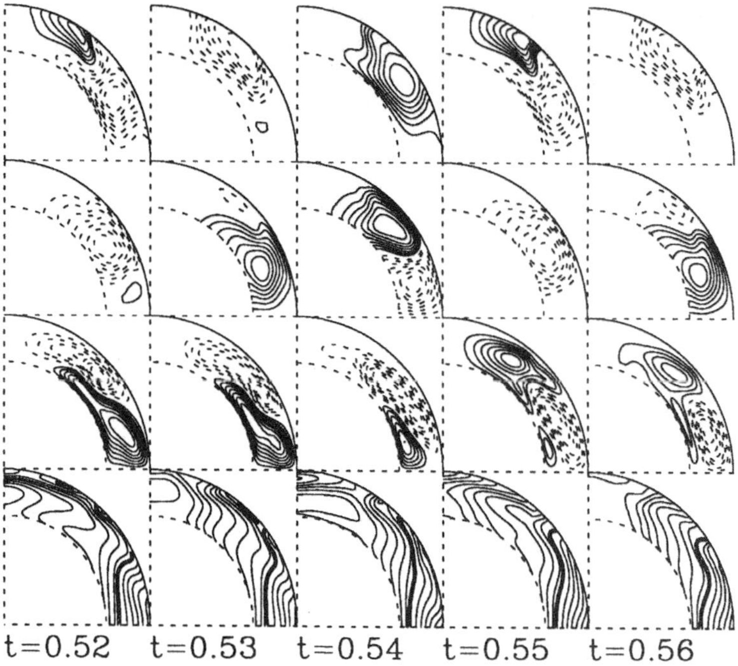

Fig. 1. The effect of compressibility on the magnetic cycle: Snapshots showing the variation of magnetic and velocity field through approximately one magnetic cycle period. In the upper row are the field lines of the poloidal magnetic field, in the second row contours of constant toroidal field strength, in the third row stream lines of the meridional motion, and in the last row contours of constant angular velocity.

2. Results

The Taylor-Proudman theorem shows that for idealized rotating flows the surfaces of constant angular velocity are cylinders parallel to the rotation axis. In many of our preliminary models we found this property to hold approximately at Taylor numbers $\mathrm{Ta} \gtrsim 10^5$, even though the strict conditions of the theorem did not apply. It is possible that stratification may give rise to deviations from cylindrical Ω-contours even for high Taylor numbers.

We are mainly interested in oscillatory dynamo solutions of $\alpha\omega$-type, because such solutions are of interest for understanding the solar cycle. Previous investigations have shown that for such solutions $|C_\Omega|$ must be sufficiently large ($|C_\Omega| \gtrsim 10^3$). This can be achieved only when $\mathrm{Pr}_M^2 \mathrm{Ta}$ is larger than $10^{7...8}$ ($\mathrm{Pr}_M = \nu_t/\eta_t$). We assume here the simplest form for the Λ-effect, taking $V^{(0)} = 1$ and $V^{(1)} = H^{(1)} = 0$ (see Paper I for the notation). The compressible case is characterized by two more dimensionless numbers: ξ, and $\tilde{\xi}$, where ξ is the surface pressure scale height relative to the stellar radius, and $\tilde{\xi}$ is the superadiabatic gradient at the surface. We take $\xi = 0.01$, and $\tilde{\xi} = 2 \times 10^{-6}$. Approximately twelve pressure scale heights are covered in the model. The field geometry for $C_\alpha = 10$ and $\mathrm{Ta} = 10^8$ is displayed in Fig. 1. When the magnetic field is included, the Ω contours differ substantially from the non-magnetic case, in particular at the surface where the density becomes small and $\boldsymbol{J} \times \boldsymbol{B}/\rho$ is large. The meridional circulation is also strongly affected by the presence of the field. We note that our Ω-contours are still basically parallel to the rotation axis through much of the shell. The influence of the magnetic field on the mean surface temperature is shown in Fig. 2.

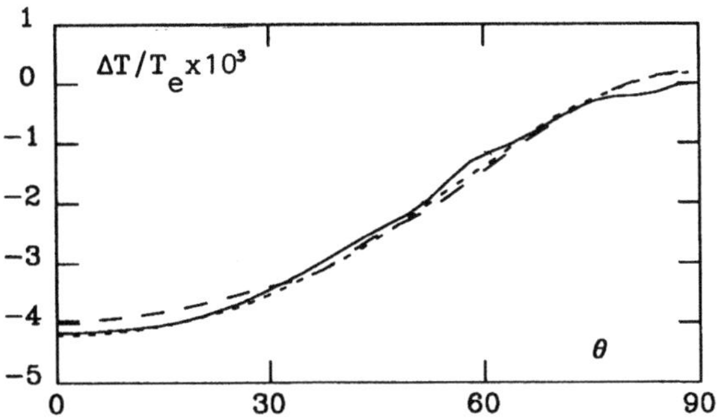

Fig. 2. Temperature difference at the surface relative to the equator temperature for the same model as in Fig. 1 at different instants during the magnetic cycle. The cyclic modulation of the temperature is about 0.03%, whilst the pole-equator difference is about 0.4%.

We have compared this model with an equivalent one where stratification is absent. The corresponding field geometry for such a model is displayed in Fig. 3. Clearly, cyclic variations of the differential rotation and the meridional circulation are much weaker than in the compressible, stratified case. It is surprising that the dynamo period is longer when stratification is absent. This is in contrast to results of Glatzmaier (1985), who found longer cycle periods for his models with stratification than for the incompressible models of Gilman and Miller (1981). Usually, periods become shorter as nonlinear feedbacks grow. In fact, in our compressible case the feedback is very strong, in particular in the upper layers, where the density is low. The disagreement with Glatzmaier's result may be due to the fact that the model discussed here has a highly supercritical dynamo number.

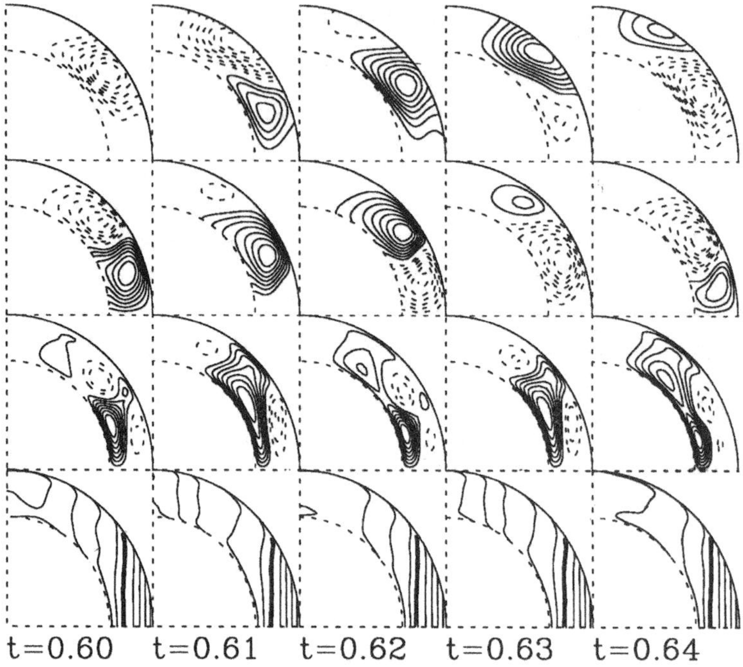

Fig. 3. Same as Fig. 1, but without thermodynamics and without stratification. Note that the modulation of meridional flow and angular velocity is much less than in the compressible case. (The heavy Ω-contour denotes the basic rotation velocity.) Also the cycle period is longer than in the compressible case. For the present choice of parameters ($V^{(0)} > 0$ and $C_\alpha > 0$), dynamo waves migrate polewards.

3. Conclusion

The present investigation demonstrates the problems of dynamo models where the differential rotation is a selfconsistent solution of the dynamo. Other effects might be included to make the model more realistic, e.g. a decrease in eddy viscosity at the base of the convective layer, or including an anisotropy in the eddy viscosity of α-effect.

References

Brandenburg, A., Moss, D., Rüdiger G., Tuominen, I.: 1990, *Solar Physics* **128**, 243 (Paper I)

Gilman, P. A., Miller, J.: 1981, *Astrophys. J. Suppl.* **46**, 211

Glatzmaier, G. A.: 1985, *Astrophys. J.* **291**, 300

Rüdiger, G.: 1989, *Differential rotation and Stellar Convection: Sun and solar-type stars*, Gordon and Breach, New York

Fossil Magnetic Fields and Activity of Young Stars

A.E. Dudorov, E.E. Gorbenko

Astronomical Department of Moscow State University,
Sternberg State Astronomical Institute, Universitetskij Pr., 13,
119899 Moscow V-234, USSR

Abstract: The strength of fossil magnetic fields in young stars is estimated. The interaction of a fossil magnetic field with convection in stellar envelopes is investigated. The influence of such a magnetic field, transformed to a small scale, on the activity of young stars is discussed.

1. The theory of the fossil magnetic field of stars has been developed recently with the help of numerical simulations of star formation in interstellar magnetic molecular clouds (see Dudorov, 1988, 1990). In the framework of this theory the magnetic field of protostellar clouds is transformed during star formation into the fossil magnetic field of young stars. The energy of the stellar fossil magnetic field is $10 - 10^5$ times lower than that of the corresponding frozen-in field. It depends on the efficiency of magnetic ambipolar and Ohmic diffusion, processes that occur in the protostellar cloud cores, with ionization ratio $x < 10^{-12}$, and densities $n = (10^4 - 10^{10})\, n_o$, where n_o is the initial number density of the cloud.

Magnetic diffusion leads to the fragmentation of molecular cloud cores. In the process of magnetic fragmentation, magnetic protostars are formed with masses $M > (0.1 - 0.3)\, M_\odot$, with the initial ratio of magnetic energy to gravitational $e_m = 0.1 - 0.3$, and with the corresponding ratio for rotational energy, $e_\omega = 0.01$–0.1. The oblate magnetic protostellar cloud continues to flatten during the accretion stage under the influence of electromagnetic and centrifugal forces and as the result of the action of magnetic torques and flow along the rotational axis.

In the protostars and in newborn stars that are still accreting mass, magnetic flux is pumped from the envelope of the protostellar cloud. After magnetosphere formation, the pumping of magnetic field from the accretion disk weakens. The interaction of the magnetosphere and accretion disk leads to outflows and jets along the lines of force of the magnetic field of the molecular cloud. In the boundary region between the magnetosphere and the inner edge of the accretion disk, the additional acceleration of charged particles along open magnetic lines may occur. The ejection of 30 per cent of the incoming mass flow from the disk is sufficient to maintain the angular velocity of such young stars (accretion stage of T Tauri stars) at about of 20 – 30 km/s.

2. Analyses of early stages of stellar evolution leads to the conclusion that efficient convection is absent from the envelopes of stars with masses above $2\,M_\odot$. These stars can retain a strong fossil magnetic field in their envelopes. For this to happen, strong sources of ionization, SN or OB stars, must be present nearby. Intense cosmic rays, the depletion of dust or an excess of heavy elements in protostellar matter, all help to produce a fossil field with a surface intensity $B_s \geq 500 - 1000\,\mathrm{G}$ (Dudorov and Tutukov, 1990).

The results of calculations show that the magnetic field strength inside cool active stars with masses $M < 1.5 - 2.0\,M_\odot$ in the T Tau stage is a hundred times lower than that of the frozen-in field. After the Hayashi stage the internal magnetic field may be of order $10^6 - 10^7$ G at the core boundary and $10^4 - 10^6$ G near the base of the convection zone (Dudorov, 1986; Dudorov *et al.*, 1989). The surface fossil magnetic field of normal stars increases along the ZAMS from $B_{fs} = 1 - 10\,\mathrm{G}$ (M-stars) to $B_{fs} = 100 - 200\,\mathrm{G}$ (B-stars). The magnetic energy of these stars is $E_m = 10^{-1} - 10^{-2} E_g$ (E_g – gravitational energy).

3. The subsequent evolution of the fossil magnetic field depends on hydromagnetic or resistive instabilities, buoyancy, Ohmic and turbulent diffusion. The turbulent convection in newborn stars on the Hayashi stage converts large scale magnetic field into a small scale, because the convective energy may be as large as the thermal. The buoyancy of magnetic flux tubes can cause the activity of the pre-main sequence stars. On the other hand turbulent convection together with the differential rotation enables the operation of an $\alpha\omega$-dynamo, which turns on after sufficient turbulent dissipation of fossil magnetic field. The differential rotation is generated by the gravitational compression of protostars (Dudorov *et al.*, 1989).

4. The results of numerical calculations and estimates show that the strength of the fossil magnetic field inside the Sun in the T Tauri stage (Dudorov *et al.*, 1989) exceeds the strength of the dynamo magnetic field by a factor of more than 10 – 100. The difference of energies of these types of fields may appear as an energy of stellar activity which is a result of convective conversion of large-scale fossil field into turbulent magnetic field and a result of the evolution of small-scale magnetic field in convective and radiative stellar envelopes. The slender magnetic flux tubes, rising to stellar surface, cause the activity of young stars. Numerical experiments (Gorbenko and Dudorov, 1989) show that the basic contribution to the energetics of the activity of young stars is made by the largest existing magnetic flux tubes. This explains the attenuation of T Tauri star activity with age.

In newborn stars magnetic flux tubes with the dimensions $d = 0.05 - 0.1\,R$ rise to the stellar surface (R the stellar radius). Their energies are $10^{30} - 10^{32}$ erg.

In the post T Tauri stars magnetic flux tubes with dimensions of order of pressure scale height are rising to the surface. The behaviour of the dynamo depends on the convective intensity and the differential rotation. The maximum effect is reached in G-stars.

5. In K- and M-stars, magnetic ambipolar diffusion slows down the motion of slender magnetic flux tubes in the regions of partial hydrogen ionization (Dudorov, 1990).

In a newborn star, the magnetic diffusion zone is situated in the region of temperature minimum. A buoyancy mechanism may concentrate the magnetic flux tubes of medium dimensions beneath the temperature minimum, causing an increase of the radius of the star and also of its luminosity, giving rise to the mechanism of Gershberg (1983) as well as that hypothesised by Appenzeller and Dearborn (1984). In a such situation luminosity variations may arise, with amplitudes less than $2^m - 3^m$.

References

Appenzeller, R.I., Dearborn, D.S.P.: 1984, *Astrophys. J.* **278**, 689

Dudorov, A.E.: 1986, *Astron. Tsirk.*, No 1446, 3

Dudorov, A.E.: 1989, *Astron. Tsirk.*, No 1535, 25

Dudorov, A.E.: 1988, in *Magnetic stars*, eds. I.M.Kopylov, Yu.V.Glagolevskij, Nauka, Leningrad, p. 226

Dudorov, A.E.: 1990, *Pis'ma Astron. Zh.*, in press.

Dudorov, A.E., Sazonov Yu.V.: 1987, *Nauchn. Inf.* **63**, 68

Dudorov, A.E., Krivodubskij, V.N., Ruzmaikina, T.V., Ruzmaikin, A.A.: 1989, *Astron. Zh.* **66**, No 4

Dudorov, A.E., Tutukov, A.V.: 1990, *Astron. Zh.* **67**, 342

Gershberg, R.E.: 1983, *Flare stars*, Nauka, Moscow

Gorbenko, E.E., Dudorov, A.E.: 1989, *Astron. Tsirk.*, No 1536, 27

III

Solar magnetism and large-scale flows

– Solar rotation from helioseismology

– The theory of differential rotation

– Evolution of large-scale and small-scale magnetic fields

– The solar dynamo

– Indicators of solar activity

– How typical is the Sun among solar type stars

The Sun's Internal Differential Rotation From Helioseismology

Philip R. Goode

Department of Physics
New Jersey Institute of Technology, Newark NJ, USA

Abstract: Well-confirmed helioseismic data from several groups using various observational techniques at different sites have allowed us to determine the differential rotation in the outer half of the Sun's interior. The resulting rotation law is simple – the surface differential rotation persists through much of the convection zone with a transition toward solid body rotation beneath. To date there is no appealing evidence for a rapidly rotating core. There is however, weak evidence for a solar cycle dependence of the Sun's internal rotation.

1. Introduction

Knowledge of the sun's internal rotation is a key to unlocking secrets of both solar evolution and the cause of surface activity. From helioseismology we have learned what we know about the Sun's internal rotation. Our seismic information comes from observations of intensity fluctuations in white light or Doppler shifts in a particular line. The oscillations discussed here are observed as the surface manifestation of trapped standing sound waves which sample the solar interior. Goldreich and Kumar (1989) have shown that these acoustic vibrations are apparently excited by turbulence in the outer most layers of the convection zone. The oscillations are global in nature and have a period of about five minutes. Global means that the lifetime of a mode is longer than the time it takes for the mode to circuit the part of the solar cavity it samples. Each normal mode samples its particular region of the interior and the seismology to be reviewed here is from oscillations which best sample between 0.5 and 0.9 R, where R is the solar radius. This choice is made because those data represent the only confirmed seismic data.

2. The nature of the oscillations

The solar oscillations are described by

$$\xi_{nlm}(r,\theta,\phi) = r\,[y(r), z(r)\frac{\partial}{\partial\theta}, z(r)\frac{1}{sin\theta}\frac{\partial}{\partial\phi}\,]\,Y_l^m(\theta,\phi), \tag{1}$$

where the y_{nl} *and* z_{nl} are the radial and horizontal fluid displacement and ω is the angular frequency of the oscillation. The n, l and m quantum numbers are the radial order, angular degree and angular order respectively. The well-confirmed splitting data under consideration here are for l-values roughly between 10 and 100. In a high frequency asymptotic limit which is sufficiently valid in the five minute regime for us to gain insight to the oscillations, we can describe a unitless oscillation f by

$$f'' + f[\frac{\omega^2 r^2}{c^2} - l(l+1) + O(|\frac{dln\rho}{dlnr}|^2, 1)] = 0, \tag{2}$$

where c is the local speed of sound. The inner turning point for the sound waves occurs when the first two terms in brackets are comparable. This means that the lower the degree of the oscillation the deeper it samples. On the other hand, near the surface where the speed of sound is low, the outer turning point is reached – for the intermediate l-values under consideration here – where the first term is comparable to the logarithmic derivative of the density. Thus the sampling depends on frequency rather than l. Since the five minute band is rather narrow, the oscillations sample the outer regions in the same way. This leads to the curious result that we know the rotation rate better at the base of the convection zone than near its top. If the Sun lacked perturbing forces like those due to rotation and magnetism then each (nl)-multiplet would be $(2l+1)$-fold degenerate in m. The observers tell us that it is best to view the available data in (nl)-multiplets with a fine structure in m so that

$$\nu_{nlm} - \nu_{nlo} = L\sum_{i=1}^{N} a_{i,nl}P_i(\frac{m}{L}), \tag{3}$$

where P_i is a Legendre polynomial and $L = l$ or $\sqrt{l(l+1)}$ depending on the choice of the observer and $N = 5$ or 6 for now. The fact that equation (3) works implies that the perturbations manifest in the a-coefficients have the rotation axis as their axis of symmetry. If for instance, the Sun possessed an intense inclined magnetic field in its core, there would be difficulty in using equation (3) to describe the fine structure especially in multiplets which sample this region ($l < 5$). In detail, each multiplet would be characterized by at least $(2l+1)^2$ peaks per multiplet which is inconsistent with equation (3). The antisymmetric a-coefficients in equation (2) arise from the linear effect of rotation and the symmetric ones from shape distorting forces like the centrifugal force. The a_1-, a_3- and a_5- terms are precisely correlated with a rotation law of the form

$$\Omega(r,\theta) = \Omega_0(r) + \Omega_1(r)\cos^2\theta + \Omega_2(r)\cos^4\theta, \tag{4}$$

where θ is the co-latitude. The inverse problem to determine $\Omega(r,\theta)$ follows from formally accounting for the effects of the Coriolis force and advection on the fine structure in the oscillation spectrum. In general, the three equations to determine $\Omega_0(r)$, $\Omega_1(r)$, and $\Omega_2(r)$ are coupled. However, the coupling is so weak that at the current level of accuracy in the five minute oscillation data the coupling can be safely ignored and

$$\frac{1}{2\pi}\int K_{nl}(r)\Omega_0(r)dr = a_{1,nl} + H_{03}(L)a_{3,nl} + H_{05}(L)a_{5,nl} \longrightarrow a_{1,nl} + a_{3,nl} + a_{5,nl} \tag{5}$$

$$\frac{1}{2\pi}\int K_{nl}(r)\Omega_1(r)dr = H_{13}(L)a_{3,nl} + H_{15}(L)a_{5,nl} \longrightarrow -5a_{3,nl} - 14a_{5,nl} \tag{6}$$

$$\frac{1}{2\pi}\int K_{nl}(r)\Omega_2(r)dr = H_{25}(L)a_{5,nl} \longrightarrow 21a_{5,nl} \tag{7}$$

where the $H(L)$ and angular integral coupling coefficients and the arrows indicate $L^2 >> 1$. Examples of the $H(L)$ are given in Table 1. If $L^2 = l(l+1)$ is chosen instead of $L = l$ then asymptotic limit is reached for lower l. The kernel, K_{nl} is

$$K_{nl}(r) = [y_{nl}^2(r) + l^2 z_{nl}^2(r) - 2y_{nl}(r)z_{nl}(r)]\rho r^4, \tag{8}$$

where ρ is the density.

Table 1. The $H(L)$–coefficients

l	$H_{25}(L)$	$H_{13}(L)$	$H_{15}(L)$	$H_{03}(L)$	$H_{05}(L)$
			$L = l$		
10	24.38	-5.46	-17.88	1.24	1.65
20	22.90	-5.24	-16.12	1.12	1.34
30	22.31	-5.16	-15.25	1.08	1.23
40	22.00	-5.12	-15.10	1.06	1.17
50	21.81	-5.10	-14.89	1.05	1.14
60	21.68	-5.08	-14.74	1.04	1.12
70	21.58	-5.07	-14.64	1.04	1.10
80	21.51	-5.06	-14.56	1.03	1.09
90	21.46	-5.05	-14.50	1.03	1.08
100	21.41	-5.05	-14.45	1.02	1.07
∞	21.00	-5.00	-14.00	1.00	1.00
			$L = \sqrt{l(l+1)}$		
10	20.15	-4.97	-13.19	0.99	0.90
20	20.78	-4.99	-13.19	1.00	0.97
30	20.90	-5.00	-13.90	1.00	0.99
40	20.94	-5.00	-13.95	1.00	0.99

50	20.96	-5.00	-13.97	1.00	0.99
60	20.97	-5.00	-13.98	1.00	1.00
70	20.98	-5.00	-13.98	1.00	1.00
80	20.99	-5.00	-13.99	1.00	1.00
90	20.99	-5.00	-13.99	1.00	1.00
∞	21.00	-5.00	-14.00	1.00	1.00

3. The data

The resolved disk observational data have been used to determine the confirmed rotation laws. In these observations as many as 106 elements are used to detect oscillations. From these data splittings for l between 0 and 1000 can be determined in principle. The analysis proceeds by reducing the data into subspectra from which splittings can be identified. Even these projected spectra are quite complicated. Major complications arise from spatial and temporal sidebands which are a consequence of the fact that we observe only one side of the Sun and, to date, observe from a single site. To simplify the data reduction, many groups average their data over n. Libbrecht (1989) has shown that intrinsic linewidths of the peaks are sufficiently narrow that more accurate data will come from longer observing runs. Further, global networks like Global Oscillations Network Group (GONG) should provide data from multiple sites by the mid-1990s effectively eliminating the day-night temporal sidelobe problem. This network should yield data strings 10 time longer than the longest available now ($\sim$ 100 days). Nonetheless, there are several data sets currently available from which the internal rotation rate of the Sun has been determined. Some of the properties of these sets are given in Table 2. Each set provides data from somewhat different parts of the five minute spectrum – different in both l-range and frequency range. The data are from various observational groups using different instruments at several sites. Further, among

Table 2. The splitting data sets

	Epoch	Frequency Range(mHz)	Range in l	Averaged over n
Duvall *et al.* (1986)	1982.0	2.4 – 4.0	20–98	yes
Duvall and Harvey (1984)	1983.4	1.5 – 4.0	1–131	no
Tomczyk (1988)	1984.5	2.57–3.74	10–120	yes
Rhodes *et al.* (1987)	1984.6	2.2 – 3.6	3–89	yes
Brown and Morrow (1988)	1984.9	2.5 – 4.0	15–99	yes
Libbrecht (1989)	1986.5	1.5 – 4.0	10–60	no

the groups of observers, intensity and four different kinds of Doppler measurements were used as well as different reduction methods. Yet all the data are remarkably similar. The longest data string is that due to Libbrecht (1989). Further, this set is not averaged over n. As a result the most accurate inversions follow from these data and extra attention is given to that set in this review. Splitting data results from that set are shown in Fig. 1.

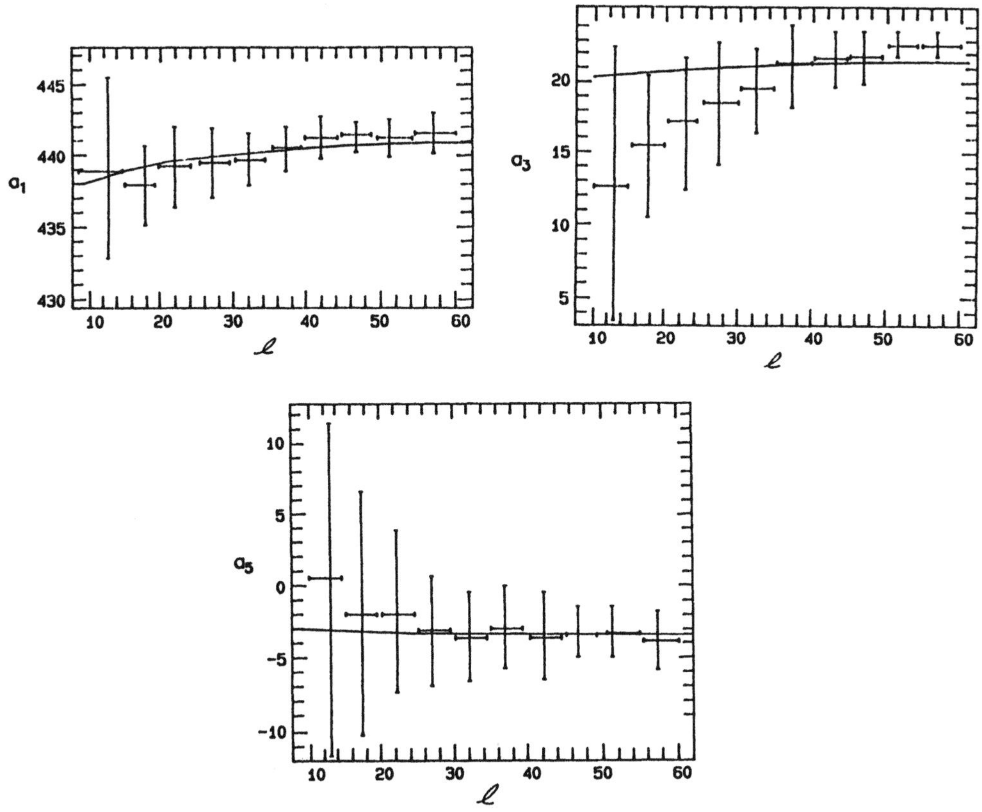

Fig. 1. Odd a-splitting coefficients (nHz) vs. l from Libbrecht's (1989) data. The solid line represents the determination of the coefficients from the mean rotation law of Eq. (9).

We have averaged the data over n and binned it in l for ease of viewing. We see that the a_1 are largely independent of l and have an average of about 440 nHz. The solid line through the a_1 data follows from assuming the mean rotation law from Libbrecht's (1989) data,

$$\Omega(r,\theta) = 460.2 \pm 0.2 - (58.3 \pm 1.3)\cos^2\theta - (73.1 \pm 2.6)\cos^4\theta. \qquad (9)$$

This rotation law is quite close to that for the surface differential rotation. If we assumed rotation constant on cylinders instead, then the calculated a_1 varies between 420 and 430 nHz. The 420 nHz value applies for $l \sim 40$ where we know a_1 best. There the discrepancy is $\sim 10\sigma$. The a_3 coefficients average about 20 nHz . Their tendency to decrease with l results in rotation laws which show a decreased differential rotation with depth. The solid line is from a differential rotation which does not decrease with depth. The a_5-coefficient averages about $-3\,\mathrm{nHz}$ and shows, perhaps, a trend for a_5 to increase toward zero with decreasing l. If we assume rotation constant on cylinders, the calculated a_3 and a_5 values would not be inconsistent with those from Eq. (9). From Fig. 1 it is clear that we know a_1 better than a_3 and a_3 better than a_5. With this and Eqs. (5)–(7), we will know $\Omega_0(r)$ better than $\Omega_1(r)$ and $\Omega_1(r)$ better than $\Omega_2(r)$. Another way to understand why we know the rotation rate best in the Sun's equatorial plane is to recall that sectoral modes ($m = \pm l$) are the ones which are confined to that region of the solar interior. Thus, we could determine the equatorial rate by observing the spacing between $m = +l$ and $m = -l$ modes rather than the spacing between each of the $(2l + 1) - m$ peaks. The errors for each l would be reduced by a factor of $2l$ – the spacing between the two sectoral components.

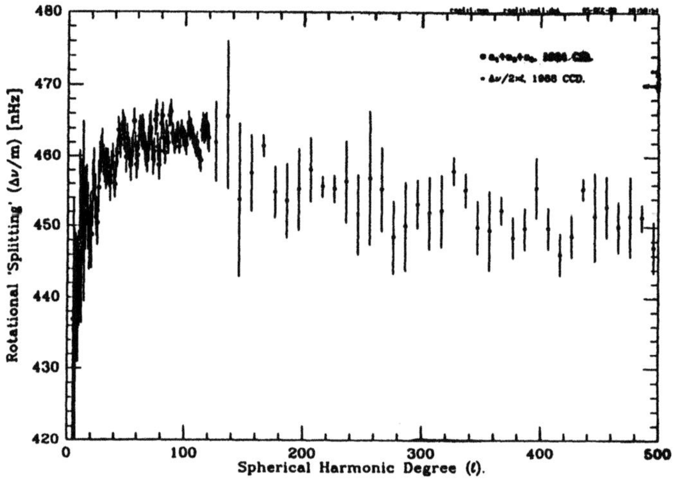

Fig. 2. $a_1 + a_3 + a_5$(nHz) vs. l from the data of Rhodes *et al.* (1990b) for the l=3–500 range.

The inner turning points for modes with $l = 10$ and 60 are about 0.3 R and 0.8 R, respectively. Thus, it is clear that modes in the l=10–60 range cleanly sample the region near the base of the convection zone. Considering the range of turning points, the asymptotic limits of Eqs. (5)–(7) and most importantly the data in Fig. 1, we anticipate the transition away from differential rotation with increasing depth. The detailed results are presented in the next section.

There are small differences between a_1-, a_3- and a_5-coefficients from data set to data set. However the $a_1 + a_3 + a_5$ values are more robust from set to set suggesting that there is some residual correlation between the a-coefficients. This is also reflected in the fact that the first splitting data for the $l = 1 - 500$ range are for $a_1 + a_3 + a_5$ only. These data are due to Rhodes *et al.* (1990b) and are those above l=120 are from 1988. The new high-l splitting data are consistent with but more accurate than the earlier sets due to Hill *et al.* (1988) and Deubner *et al.* (1979). The new high-l data are especially valuable in determining structure in the rotation near the surface in the equatorial plane. The primary feature they reveal is a $\sim$ 5% bump in Ω_0 centered at 0.93 R and having a width of about 0.1 R (Korzennik, 1990). However, this result must be regarded with caution because the bump in the splitting data centered near l=80 occurs where the sidelobes are very close to the "true" peaks.

For $l < 5$ whole disk observations should be useful. To date however, the limited results are contradictory. For instance, the reported rotational splitting for $l = 1$ by Jefferies *et al.* (1988) for the period of 1981–1984 is about 750 nHz implying a rapidly rotating core. On the other hand, Woodard (1984) reports an upper limit consistent with a core rotating more slowly than the surface in its equatorial region. Other results range between these two. Owing to this absence of consistency in the low-l splittings we do not use them in the calculation of the internal rotation.

Three groups have used whole disk observations to report the discovery of rotational splittings in gravity modes. Gravity modes are confined to the interior and such a discovery would enable a much more accurate seismology. Even though consistent splittings are reported by three groups indicating a rapidly rotating core, the groups also report significantly different identifications and frequencies for the g-modes. Until these discrepancies are cleared up, the identification of g-mode splittings should be regarded as unconfirmed. This situation is quite different than that for the acoustic modes for which the observers listed in Table 2 agree on the mode identifications and frequencies and their splittings. The SCLERA group reports rotational splittings which are about 3–4 times larger than any of the aforementioned splittings (Hill, 1984). This group observes fluctuations in the limb darkening function near the Sun's equator. The spectra are considerably more complicated than any of the spectra from aforementioned data.

We have well-confirmed rotational splitting data for acoustic modes in the five minute band for l-values in the 10–120 range. These facilitate the determination of the differential rotation between 0.4 R and 1.0 R.

4. The inversions

The typical approach to solving the inverse problem posed by equations (5)–(7) is a least squares one in which χ^2 per degree of freedom is minimized to determine the rotation rate. The problem here is that the data have a finite resolution width, and a gridding which is too fine will yield artificial short wavelength oscillations in the calculated rotation. The resolution width shown in Fig. 3 is for 0.7 R from Libbrecht's (1989) data using the Backus-Gilbert (1970) method. This least squares approach circumvents the problem of short wavelength oscillations but the cost is that the calculations require a large amount of time on a sizeable computer. The method employed here is regularization in which one minimizes α_s^2, where

$$\alpha_s^2 = \chi_s^2 + \eta_s M_s P_s, \tag{10}$$

and where η_s is the regularization parameter and M_s is the number of (nl)-multiplets and

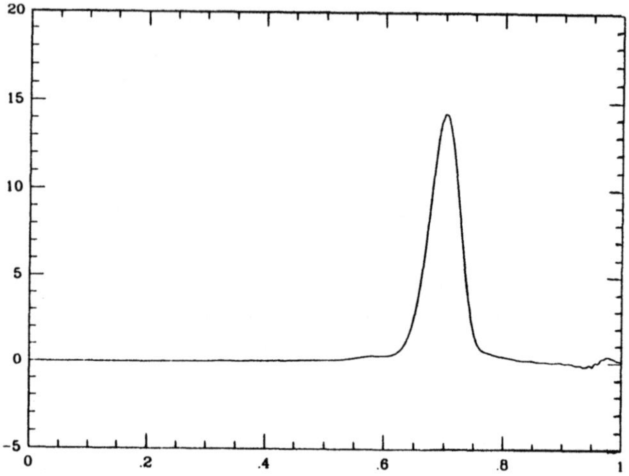

Fig. 3. Unitless Backus-Gilbert (1970) kernel for 0.70 R vs. fractional radius using the data of Libbrecht (1989).

$$P_s = \int |\frac{d\Omega_s}{dr}|^2 dr \tag{11}$$

and $s = 0$, 1 or 2 as in Eqs. (5)–(7). However, η_s is not a free parameter. The constraint invoked here reflects our *a priori* view that the rotation rate is a slowly varying function of radius. In the minimization of α_s^2 we vary η_s until χ_s^2 per degree of freedom is 1.

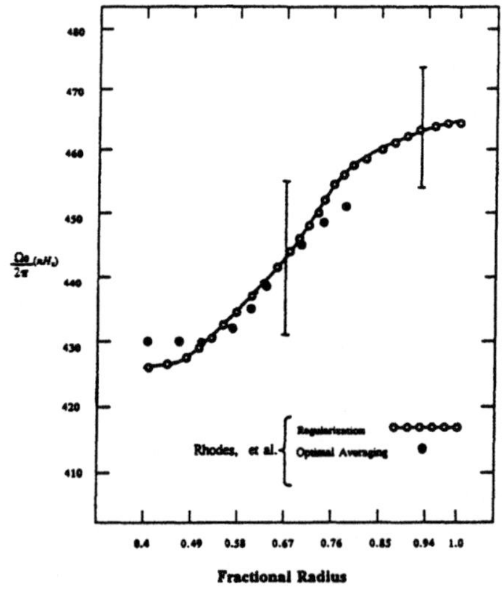

Fig. 4. $\frac{\Omega_0}{2\pi}$ vs. fractional radius from the data of Rhodes *et al.* (1987, 1990a).

Then the 1σ least square errors from the inversion are consistent with the 1σ errors in the data. In practice, the χ_s^2-space is fairly flat and we further weaken η_s

until short wavelength oscillations almost begin to appear in the rotation. In Fig. 4, we see that $\Omega_0(r)$ calculated using the Backus-Gilbert (1970) approach compares favorably to that from regularization.

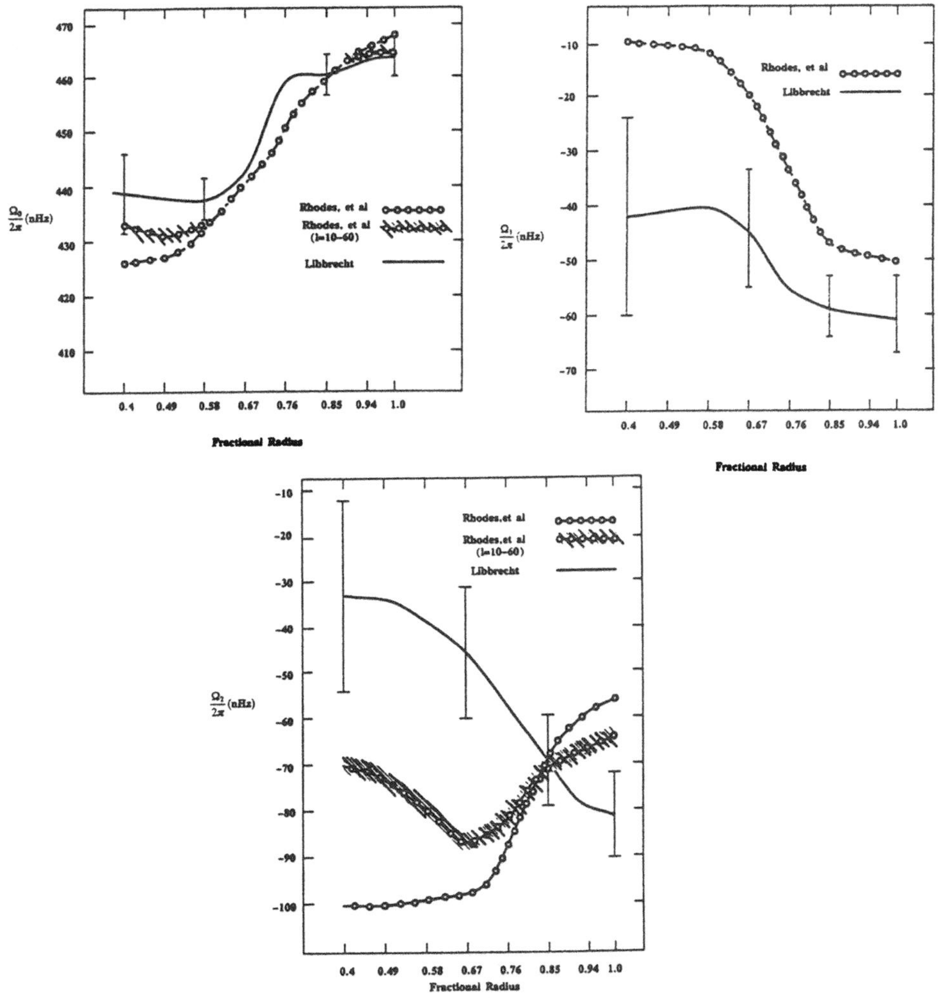

Fig. 5. $\frac{\Omega_0}{2\pi}$, $\frac{\Omega_1}{2\pi}$ and $\frac{\Omega_2}{2\pi}$ vs. fractional radius from Libbrecht's (1989) data. The errors from the Rhodes *et al.* (1987, 1990a) are about twice as large as those from Libbrecht's (1989) data. The shaded functions appear where the two rates from the Rhodes *et al.* (1987, 1990a) data differ.

In Fig. 5, we see the overall agreement for $\Omega_0(r)$, $\Omega_1(r)$, $\Omega_2(r)$ from Libbrecht's (1989) data and those from Rhodes *et al.* (1987, 1990a) data for l=3–89 and its l=10–60 subset. The error bars on the full Rhodes *et al.* (1987, 1990a) data set are about twice those from Libbrecht's reflecting the fact that Rhodes *et al.* averaged over n and their observation run has fewer days in it. The $\Omega_0(r)$ functions tend to be flat through the convection zone with a sharp gradient below. It seems though that the sharpness of the gradient depends on which data are excluded. Near the surface, as mentioned earlier, the Rhodes *et al.* (1990b) data yielded a bump in $\Omega_0(r)$ near 0.93R. However that rate is fairly flat throughout the rest of the convection zone showing a decreasing rate going inwards like that in Fig. 5 from the data of Libbrecht. The decrease beneath the bump starts near the base of the convection zone. The low-l data are not so secure so it is not clear whether or not $\Omega_0(r)$ continues to decline beneath 0.6 R. For $\Omega_1(r)$ both sets show a strong tendency away from differential rotation beneath the convection zone. For $\Omega_2(r)$ the results are consistent at the 1σ level in the least square error bars, while showing opposite trends. The function $\Omega_2(r)$ is not yet well determined.

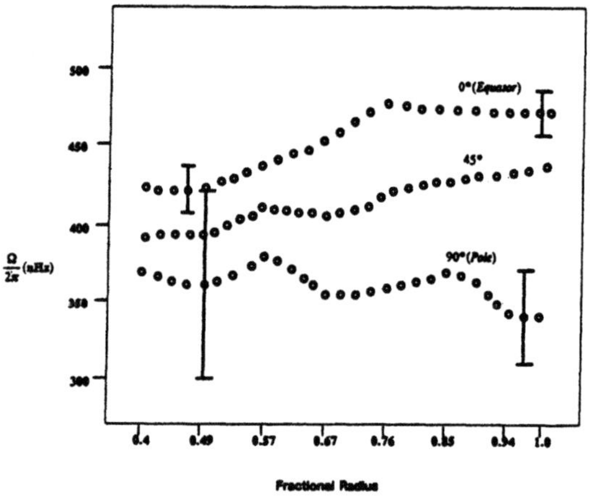

Fig. 6. $\frac{\Omega}{2\pi}$ at three latitudes vs. fractional radius from the data of Tomczyk (1988).

Nonetheless, the net tendency away from the surface like differential rotation is clear and is arbitrarily illustrated with the l=10–120 splitting data of Tomczyk (1988). Regularization oversmooths the rotation rate and so the gradients in $\Omega_0(r)$ and $\Omega_1(r)$ could be even sharper than they appear in Fig. 4. Fig. 7 shows the results from Libbrecht's (1989) data of allowing a discontinuity at 0.73 R, near the base of the convection zone. The striking result is surface-like differential rotation throughout the convection zone with an abrupt transition to solid body rotation beneath at a rate close to mid-latitude surface rate. Moving the arbitrary discontinuity either closer to or further from the surface results in a reduction in magnitude of the jump. In fact, the jump essentially vanishes if the arbitrary discontinuity is moved by 0.1R in either direction. Of course Fig. 7 is only a curiosity because of the limited resolving power of the data as illustrated in Fig. 2. Owing to the resolving power we do not pay a high price for the oversmoothing of regularization. To further illustrate the accuracy to which we know the rotation we try a second constraint

$$P_s = \int |\frac{d^2\Omega_s}{dr^2}|^2 dr \tag{12}$$

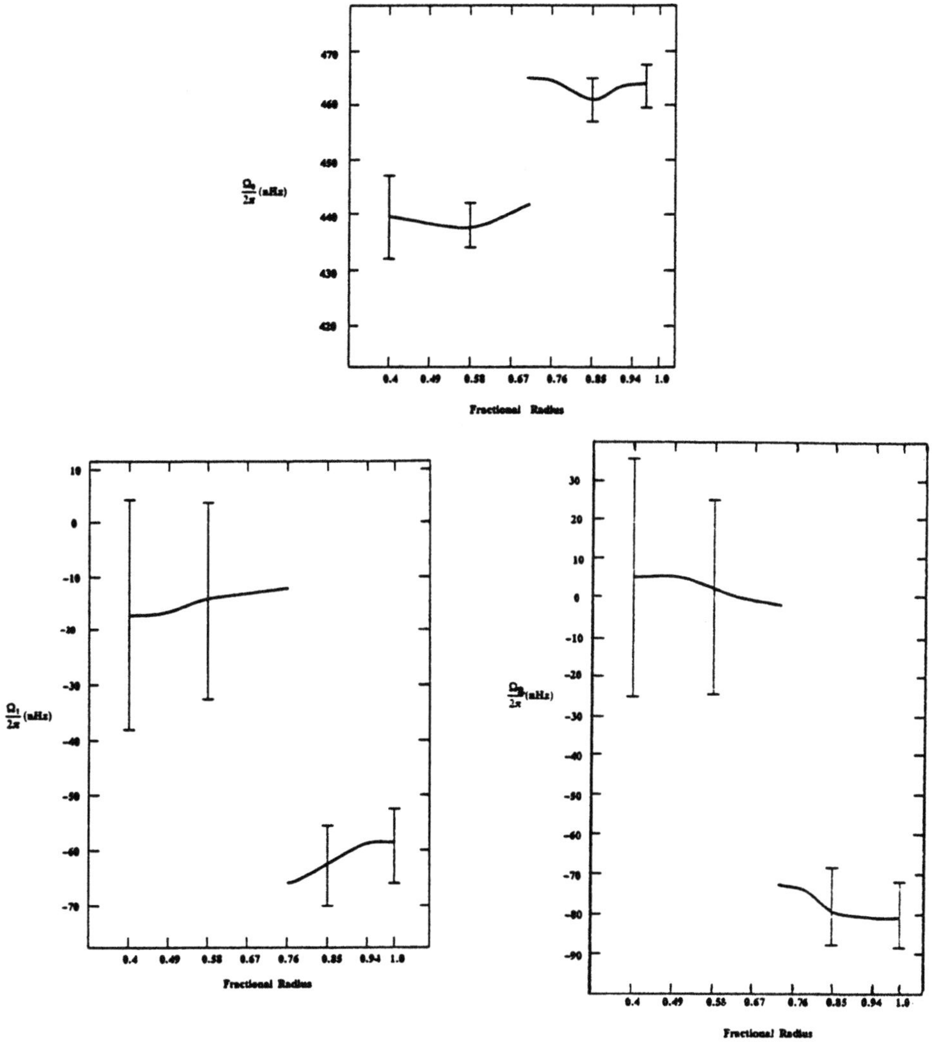

Fig. 7. $\frac{\Omega_0}{2\pi}$, $\frac{\Omega_1}{2\pi}$ and $\frac{\Omega_2}{2\pi}$ vs. fractional radius from Libbrecht's (1989) data allowing a discontinuity at 0.73 R.

In Fig. 8, the results of this constraint are compared to that of equation (11) for the data of Libbrecht. For $\Omega_0(r)$, the results are the same except that, not surprisingly, the gradient in Ω_0 is larger near the base of the convection zone. For the two $\Omega_1(r)$ results the trends are the same but the role of the constraint is larger here. The $\Omega_2(r)$ functions are controlled by the constraints. The first determination of the Sun's internal rotation was made by Duvall *et al.* (1984). That inversion for the rotation near the equatorial plane is largely consistent with the $\Omega_0(r)$ values here. However, the results also suggested the hint of a rapidly rotating core. That hint is probably an artifact of the previously discussed short wavelength oscillations in rotation resulting from too fine a grid in $\Omega_0(r)$. In Fig. 9, regularized inversions are compared with η_0 for Figures 4–7 to that from a much smaller η_0 with a step size of 0.05 R. The latter solution roughly shows the form given by Duvall *et al.* (1984) including the dip near 0.3R and the rapidly rotating core. As η_0 is increased from the too small value to a reasonable one, first the oscillations near the center disappear then those near the surface disappear.

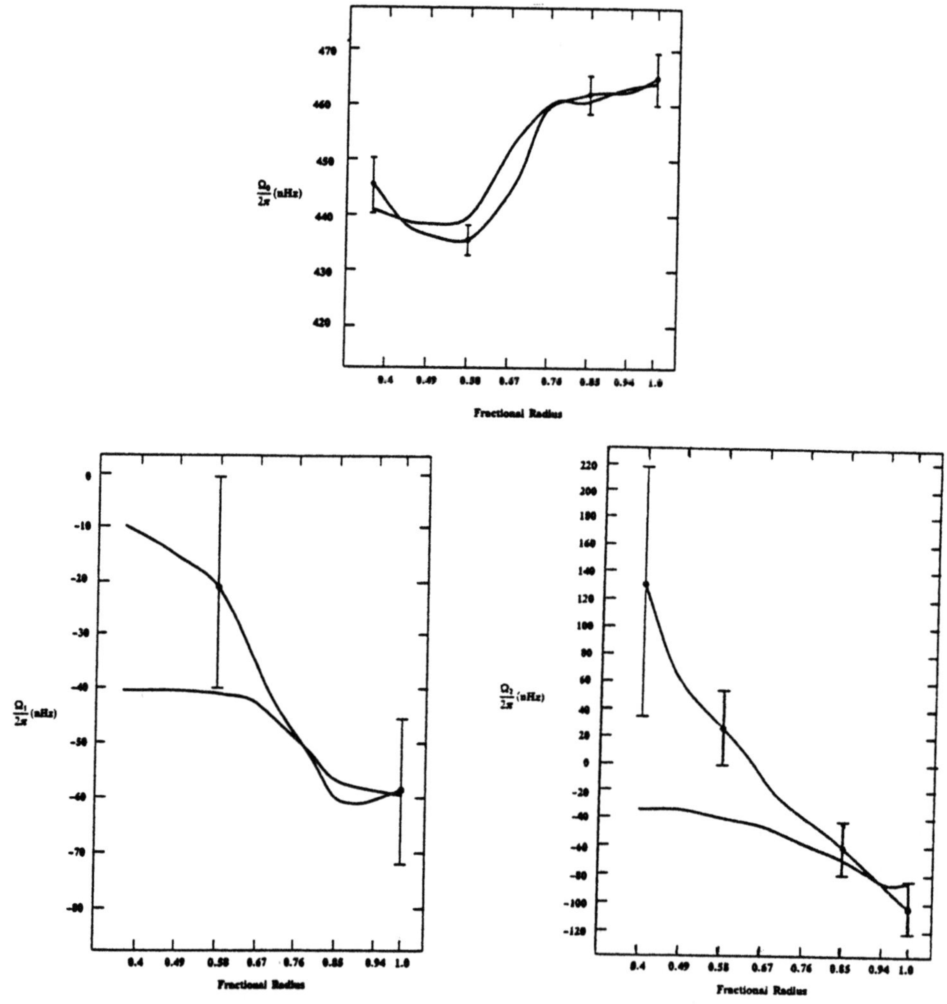

Fig. 8. $\frac{\Omega_0}{2\pi}$, $\frac{\Omega_1}{2\pi}$ and $\frac{\Omega_2}{2\pi}$ vs. fractional radius from the data of Libbrecht (1989). The error bars are on the inversion using the constraint of Eqs. (12).

5. Is the internal rotation time dependent?

We have rotation splitting data from the last solar maximum until now. The only change in rotation that appears to be marginally significant is in Ω_0 in the radiative interior (Goode *et al.*, 1991; Goode and Dziembowski, 1991) near 0.4 R. The rate at that location appears to be anticorrelated with surface activity as shown in Fig. 10. In detail, that rate was maximal at solar minimum.

Further, Ω_0 near the base of the convection zone does not appear to change at all over the cycle. Dziembowski and Goode (1991) inverted the symmetric part of preliminary oscillation data of Libbrecht and Woodard (1990) from the summer of 1986 and 1988 to determine that there appears to be a persistent megagauss quadrupole toroidal field near the base of the convection zone. Such a field is too intense to vary over the activity cycle. In an effort to rationalize these apparently disparate and marginally significant effects of rotation and magnetism, Goode and Dziembowski (1991) suggested the possibility of a torsional oscillation vaguely like that suggested years ago by Walen (1947).

The rationalization is a picture in which the megagauss field is maintained by a dynamo action on an "invisible" constant kilogauss poloidal field near the base of the convection zone. The equation of motion for the torsional oscillation is

$$\rho r^2 \sin^2\theta \frac{\partial^2 \Omega}{\partial t^2} = \frac{1}{4\pi}(\mathbf{B_p}\cdot\nabla)[r^2\sin^2\theta(\mathbf{B_p}\cdot\nabla)\Omega] \tag{13}$$

where $\mathbf{B_p}$ is the poloidal field and its kilogauss magnitude is chosen so that the torsional oscillation has a period comparable to that of the activity cycle. The dynamo action generating the toroidal field, $\mathbf{B}_\phi$, is described by

$$\frac{\partial B_\phi}{\partial t} = r\sin\theta(\mathbf{B_p}\cdot\nabla)\Omega. \tag{14}$$

Since we expect the time dependent dynamo to be amplitude limited, the larger B_p is the smaller amplitude variation in Ω over time. Thus, we would expect that the rotation rate near the base of the convection zone would change much less than the rate deep down where B_p would be considerably smaller. Of course there are problems with this picture, like how would the rotational energy stored deep down, even after conversion to magnetic energy, be transferred to the surface over 11 years.

6. Conclusions

Various groups of observers have collected several sets of consistent splitting data spanning overlapping regions of the five minute period acoustic band. These data have been used to infer that the surface rate at any point persists through the convection zone going inward along the radius from the chosen point. Near the base of the convection zone there is a fairly abrupt transition towards solid body

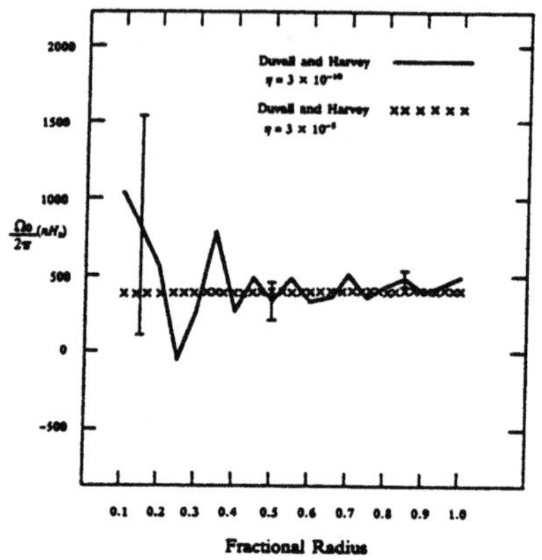

Fig. 9. $\frac{\Omega_0}{2\pi}$ vs. fractional radius from the data of Duvall and Harvey (1984) mimicking the calculation of Duvall *et al.* (1984). For the smooth function the error bars are comparable to the width of the x's.

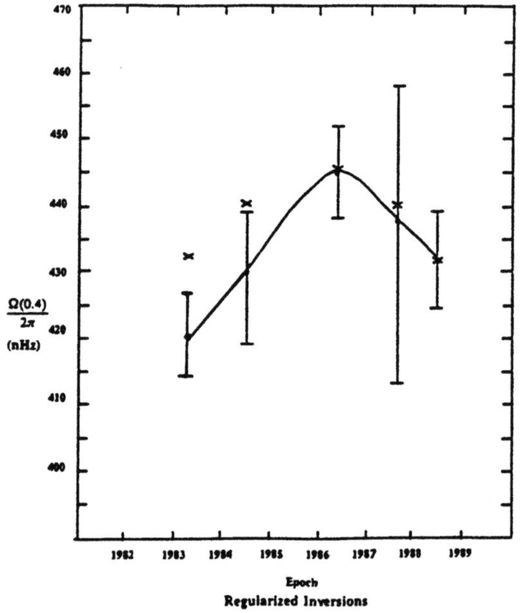

Fig. 10. $\frac{\Omega_0(r=0.4R)}{2\pi}$ vs. epoch. The x's represent the inverse sunspot number calibrated between 1986.5 and 1988.5.

rotation beneath. The solid body rate is close to that of an intermediate surface latitude. Brown *et al.* (1989) and Dziembowski *et al.* (1989) used the confluence of the radial and latitudinal gradients to suggest that the dynamo driving solar activity is seated near the base of the convection zone. The rotation law in the convection zone is not consistent with rotation on cylinders as suggested by detailed treatments of the consequences of turbulent convection on solar activity. Nonetheless, we should be encouraged because the rotation law is simple. We need to search for the dynamical origin of this simple law. In the near future we expect to know the Sun's internal differential rotation about ten times more accurately than we know it now because of projects like GONG as well as observations from space.

P.R.G is partly supported by AFOSR 89-0048.

References

Backus, G., Gilbert, F.: 1970, *Phil. Trans. Roy. Soc. London, A* **266**, 123

Brown, T.M., Christensen-Dalsgaard, J., Dziembowski, W.A., Goode, P.R., Gough, D.O., Morrow, C.A.: 1989, *Astrophys. J.* **343**, 526

Brown, T.M., Morrow, C.A.: 1987, *Astrophys. J. Letters* **314**, 121

Deubner, F.L., Ulrich, R.K., Rhodes, E.J.: 1979, *Astron. Astrophys.* **72**, 177

Duvall, T.L., Dziembowski, W.A., Goode, P.R., Gough, D.O., Harvey, J.W., Leibacher, J.W.: 1984, *Nature* **310**, 22

Duvall, T.L., Harvey, J.W.: 1984, *Nature* **310**, 19

Duvall, T.L., Harvey, J.L., Pomerantz, M.A.: 1986, *Nature* **321**, 500

Dziembowski, W.A., Goode, P.R.: 1991, *Astrophys. J.*, in press.

Dziembowski, W.A., Goode, P.R. and Libbrecht, K.G.: 1989, *Astrophys. J. Letters* **337**, L53

Goldreich, P., Kumar, P.: 1988, *Astrophys. J.* **326**, 462

Goode, P.R., Dziembowski, W.A.: 1991, *Nature*, in press.

Goode, P.R., Dziembowski, W.A., Korzennik, S.G., Rhodes, E.J.: 1991, *Astrophys. J.*, in press.

Hill, F., Gough, D.O., Toomre, J. and Haber, D.A.: 1988, in *Advances in Helio- and Asteroseismology*, IAU Symposium No. 123, ed. J. Christensen Dalsgaard and S. Frandsen, Dordrecht, Reidel, p. 45

Hill, H.A.: 1984, SCLERA Monograph Series, No. 1.

Jefferies, S.M., McLeod, C.P., Van der Raay, H.B., Palle, P.L., Roca Cortes, T.: 1988, in *Advances in Helio-and Asteroseismology*, IAU Colloquium 123, ed. J. Christensen-Dalsgaard and S. Frandsen, Dordrecht, Reidel, p. 25

Korzennik, S.G.: 1990, Ph.D. dissertation, University of California at Los Angeles, in preparation.

Libbrecht, K.G.: 1989, *Astrophys. J.* **336**, 1092

Libbrecht, K.G., Woodard, M.: 1990, *Nature* **345**, 779 and private communication

Rhodes, E.J., Cacciani, A., Korzennik, S., Tomczyk, S., Ulrich, R.K., Woodard, M.F.: 1990a, *Astrophys. J.*, in press.

Rhodes, E.J., Cacciani, A., Woodard,. M.F., Tomczyk, S., Korzennik, S., Ulrich, R.K. 1987, in *The Internal Solar Angular Velocity*, eds. B.R. Durney and S. Sofia, Dordrecht, Reidel, p. 75

Rhodes, E.J., Korzennik, S., Ulrich, R.K.: 1990b, in preparation.

Tomczyk, S.: 1988, Ph.D. dissertation, University of California at Los Angeles.

Walen, C.: 1947, *Ark. Mat. astr. Fys.* **33a**, 1

Woodard, M.F.: 1984, *Nature* **309**, 530

Λ-effect, Meridional Flow and the Differential Solar Rotation

G. Rüdiger[1] and I. Tuominen[2]

[1]Universitäts-Sternwarte Göttingen, Göttingen, F.R.G.
[2]Observatory and Astrophysics Laboratory, University of Helsinki, Helsinki, Finland

Abstract: Rotation laws for the solar convection zone are produced by the Λ-effect in rotating anisotropic turbulence fields. In this paper we use the structure of the turbulence parameters introduced by Rüdiger and Kichatinov (1990), based on a simplified turbulence model. When we neglect the meridional circulation, for small inverse Rossby numbers the angular velocity isolines are spherical, while for increasing inverse Rossby number they approach more and more the helioseismologically derived shape. This simple picture becomes more complicated if the meridional circulation is allowed to act as an angular momentum transporter.

1. Introduction

In a rotating turbulence without any preferred direction apart from the angular velocity vector, $\boldsymbol{\Omega}$, angular momentum is transported only according to the well-known Boussinesq relations

$$Q_{\mathrm{r}\varphi} = -\nu_{\mathrm{T}}\, r \frac{\partial \Omega}{\partial \mathrm{r}} \sin\theta, \quad Q_{\theta\varphi} = -\nu_{\mathrm{T}} \frac{\partial \Omega}{\partial \theta} \sin\theta \tag{1}$$

for the zonal off-diagonal components of the correlation tensor Q_{ij}

$$Q_{ij} = \langle u_{\mathrm{i}}'(x,t) u_{\mathrm{j}}'(x,t) \rangle. \tag{2}$$

ν_{T} is the eddy viscosity. For anisotropic turbulence, however, with the radial unit vector $\mathring{\boldsymbol{g}}=\boldsymbol{r}/r$ as an additional preferred direction, new terms

$$\begin{aligned} Q_{\mathrm{r}\varphi} &= \nu_{\mathrm{T}} \left\{ -\frac{r\partial\Omega}{\Omega\partial \mathrm{r}} + V^{(0)} + \sin^2\theta\, V^{(1)} \right\} \sin\theta\, \Omega, \\ Q_{\theta\varphi} &= \nu_{\mathrm{T}} \left\{ -\frac{\partial\Omega}{\Omega\partial\theta} \frac{\sin\theta}{\cos\theta} + H^{(0)} + \sin^2\theta\, H^{(1)} \right\} \cos\theta\, \Omega \end{aligned} \tag{3}$$

appear – known as the Λ-effect. They act in a non-diffusive way as they do not vanish for $\Omega = \text{const.}$, thus preventing the existence of uniform rotation. For given coefficients $V^{(0)}, \ldots, H^{(1)}$ a mean-field flow system $\langle \boldsymbol{u} \rangle$ develops whose azimuthal (φ-) component reflects the differential rotation.

2. The turbulence model

The general theory of the Λ-effect is presented in Rüdiger (1989) using the so-called second-order correlation approximation. It works with the simplest truncation procedure, i.e. the restriction to a basically linear treatment of the coupling of all small-scale modes. The interaction of the laminar large-scale modes with the random small-scale modes is non-linear. As Vainshtein and Kichatinov (1983) have pointed out, the influence of higher-order correlations can be modelled by

$$\partial u_i'/\partial t - \langle (\boldsymbol{u}'\nabla) u_i' \rangle + (\boldsymbol{u}'\nabla) u_i' = u_i'/\tau \tag{4}$$

with τ as the relaxation time of the correlations (Orszag, 1970). It is tempting to apply this model to the general expressions for the turbulent angular momentum transport coefficients $V^{(0)}, \cdots, H^{(1)}$. One has then simply to use the relations

$$q(\boldsymbol{k}, w) = q(\boldsymbol{k})\, \delta(w)/\tau, \quad \nu k^2 = 1/\tau \tag{5}$$

for the dependence of the spectral function q on the wave-number $\boldsymbol{k}$ and the frequency w of the turbulence field.

3. The Λ-terms

Rüdiger and Kichatinov (1990) presented the Λ-coefficients $V^{(0)}, \ldots, H^{(1)}$ for the turbulence model (5):

$$\begin{aligned} V^{(0)} &= \frac{1}{\nu_{\mathrm{T}}} \left(\frac{8}{15} - \frac{64}{35} \tau^2 \Omega^2 \right) \int k^2 q(k)\, dk \\ V^{(1)} &= H^{(1)} = -\frac{32}{35} \frac{\tau^2 \Omega^2}{\nu_{\mathrm{T}}} \int k^2 q(k)\, dk, \end{aligned} \tag{6}$$

$H^{(0)}$ being zero. It is easy to show that

$$\int k^2 q(k)\, dk = \frac{\tau}{3} \left(\langle u_\theta'^2 \rangle - \langle u_r'^2 \rangle \right), \tag{7}$$

which gives the relation for the anisotropy in the turbulence field. Thus, ***positive*** q implies dominance of the ***horizontal*** motions and *v.v.* All our micro-scale expressions are correct to order Ω^3. That is a minimal constraint as the inverse Rossby number $\mathrm{Ro}^{-1} = 2\tau\Omega$ is expected to be of order unity at the bottom of the convection zone. According to (6), a strongly reduced $V^{(0)}$ and maximal values of $V^{(1)}$ and $H^{(1)}$ are the consequence of high Ro^{-1} values. At the solar surface, however, we have the opposite situation: because of the very small Ro^{-1}, maximal

$V^{(0)}$ occurs with minimal $V^{(1)}$ and $H^{(1)}$. From the observations of the horizontal motions of sunspots we obtain $H^{(1)} \cong 1$. If this is correct, Eq. (6) leads directly to $q < 0$, so that a preference for vertical turbulent motions must be assumed.

With the anisotropy parameter δI,

$$\delta I = -\frac{1}{2\nu_{\rm T}} \int k^2 q(k) dk, \tag{8}$$

we write

$$V^{(0)} = -2\left(\frac{8}{15} - \frac{64}{35}\tau^2 \Omega^2\right) \delta I, \qquad V^{(1)} = H^{(1)} = \frac{64}{35}\tau^2 \Omega^2 \, \delta I. \tag{9}$$

We consider δI as *depth-independent*, while, on the other hand, the correlation time τ can be parameterized by

$$\tau = \tau_i (x_i/x)^{\lambda/2} \tag{10}$$

with $x = r/R$, so that simply

$$V^{(0)} \cong \left(-\frac{16}{15} + \frac{32}{35}(x_i/x)^\lambda \, {\rm Ro_i}^{-2}\right) \delta I, \; V^{(1)} = H^{(1)} \cong \frac{16}{35}(x_i/x)^\lambda \, {\rm Ro_i}^{-2} \delta I, \tag{11}$$

with the inner rotation parameter

$${\rm Ro_i}^{-1} = 2\tau_i \Omega. \tag{12}$$

According to its construction, the expression in brackets in Eq. (11) is not allowed to change sign. For smaller Rossby numbers only $V^{(0)}$ is quenched. In principle, three parameters describe the non-diffusive part of the turbulent angular momentum transport: Rossby number, Ro, anisotropy parameter, δI, and the radial dependence of the correlation time, λ. The observations are the surface profile of the angular velocity, $\Omega(\theta)$, the surface value of $H^{(1)}$ and the surface characteristics of the meridional circulation.

We ask whether there is a range for these parameters and the Taylor number Ta for which the helioseismologically derived smooth rotation law in the convection zone appears and whether these values are in agreement with usual estimates for the Sun. Using a representation

$$\Omega = \Omega_0 \sum \omega_{n-1} P_n^1(\cos\theta)/\sin\theta, \tag{13}$$

the "observed rotation law" corresponds to

$$\omega_0(1) = 1, \quad \omega_2(1) = -0.033, \qquad \omega_0(x_i) \cong 1, \quad \omega_2(x_i) \cong 0. \tag{14}$$

The Taylor number, Ta$= \Omega^2 R^4/\nu_{\rm T}^2$ (with $\nu_{\rm T} = C_\nu L^2/\tau$), can be rewritten as

$${\rm Ta} = \frac{{\rm Ro_i}^{-2}}{C_\nu^2 \xi_i^4}, \tag{15}$$

so that additionally the (normalized) mixing length $\xi_i = L_i/R$ at the base of the convection zone enters our formulation. With $\xi = 0.1$

$$\mathrm{Ta} \cong 10^{5\ldots6}\mathrm{Ro}_i^{-2}. \tag{16}$$

It remains to determine the rotationally created terms in the eddy heat transport tensor. Again we restrict ourselves to the isotropic part of the turbulence and find

$$VV^{(1)} = HV^{(1)} = Q/\chi_\mathrm{T}, \tag{18}$$

with Q taken from Rüdiger and Tuominen (1989), so that

$$VV^{(1)} = HV^{(1)} = \frac{1}{5}(x_i/x)^\lambda\,\mathrm{Ro}_i^{-2}. \tag{19}$$

4. Models

The aim of our calculations is to explain the results for the angular velocity in the solar convection zone presented by Stenflo (these Proceedings, Fig. 8), derived from helioseismology and from the surface pattern of magnetic fields. The given isolines (strictly surfaces) of the angular velocity, Ω = const., are cylindrical or radial in the equatorial region, and rather disk-like in the polar region, while the isolines are radially directed in middle latitudes.

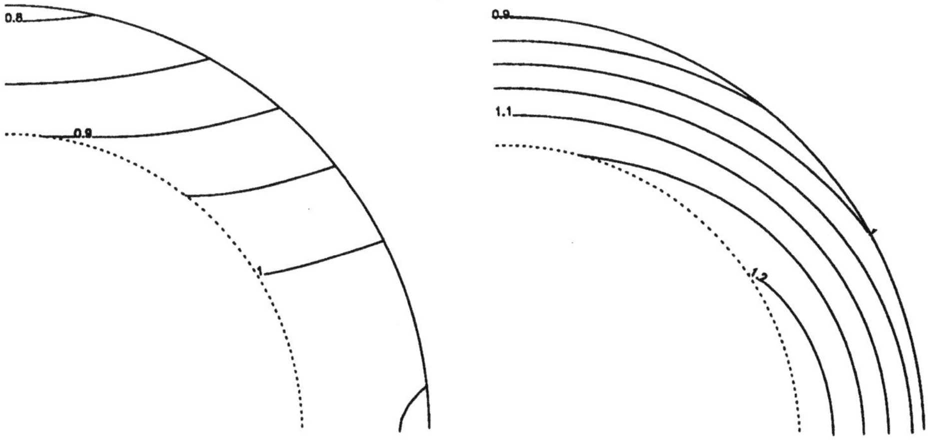

Fig. 1. Ω-contours for models in which the meridional circulation has been set to zero. a) Uniform Λ-effect ($\lambda = 0$). b) Λ-effect concentrated to the base of the convection zone ($\lambda = 5$).

Let us first focus on the angular momentum balance. Neglect of the meridional flow gives the simplest possible model. The Λ-effect alone then produces a characteristic profile of the angular velocity $\Omega(r,\theta)$. We present here models with the

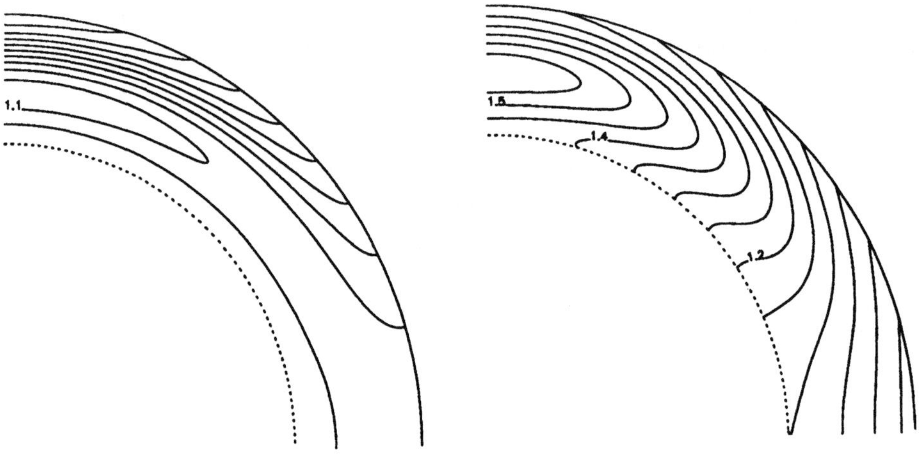

Fig. 2. Ω-contours for models in which the meridional circulation acts. The cases a) and b) as in Fig. 1.

parameter values Ta $= 10^5$ and Pr $= 0.33$, and with a uniform Λ-effect ($\lambda = 0$) and with Λ-effect more concentrated to the base of the convection zone ($\lambda = 5$).

Fig. 1 describes the consequences of the pure Λ-effect. We find already here essential differences. With increasing Ro_i^{-1}, the Ω-contours change from spherical to cylindrical, being close to the observations with $\mathrm{Ro}_i^{-1} = 0.90$ (Fig. 1a). The case of non-uniform Λ differs drastically (Fig. 1b). If the Λ-effect is concentrated at the bottom of the convection zone, then any differential rotation at the surface can only follow from a strong pole-equator difference of Ω deep in the SCZ. The observed radially aligned isolines in middle latitudes are then always missing. If this result is of some generality then the Λ-effect must be depth-independent which, however, is not compatible with the mixing-length theory.

Let us now consider the effects of a meridional flow. Its inclusion requires the simultaneous solution of the temperature equation. It is, of course, formulated for a stratified medium, but simplifying the behaviour of the eddies, the medium is assumed to be adiabatic. Fig. 2 gives the solutions. For the uniform Λ-effect the angular velocity contours are less modified. The particular turbulence model adopted here does not give a reasonable solution for the non-uniform Λ-effect: there is an equatorial deceleration. The two circulation drivers – non-conservative centrifugal force and buoyancy – change the simple picture, given in Fig. 1. Clearly a proper turbulence model and the appropriate Λ-parameters may be sought, combining the "inverse" method, attempted in Tuominen and Rüdiger (1989), and their derivation by a direct numerical simulation, as in Pulkkinen *et al.* (these Proceedings).

5. The key question

Our numerical results look convincing despite the relatively simple turbulence model used. The free parameters are Prandtl number and Taylor number. It is, of course, an interesting task to study their influence on the resulting flow pattern. Doing so, we find a very unexpected behaviour of the solutions since they do not depend continuously on Ta or Pr. The pole-equator differences of the angular velocity and/or the temperature do not remain finite for all values of the dimensionless parameters. If one of them is fixed, critical values of the other exist for which the solution loses its relevance.

Mathematically speaking we have established an inhomogeneous system as a natural description of the over-all problem. That is in contrast to the dynamo theory, which essentially deals with a self-excitation problem. We are here confronted with the numerical fact that for certain values of Pr or Ta the characteristic determinant vanishes so that at these points self-excitation occurs.

Similar results have already been obtained by Gierasch (1974), Schmidt (1982) and Chan *et al.* (1987). It simply means that the full system of equations, in which the influence of the turbulence only enters via the eddy diffusivities, possesses an eigensolution for certain eigenvalues. Chan *et al.* (1987) propose that these solutions – which already appear for very simplified equations – give the desired answer for the problem of the maintenance of the solar differential rotation. Nonlinear calculations are necesssary to clarify the reality of this important phenomenon. Without the answer to this "key question of the theory of differential rotation", all results presented from linearized equations must be considered with care.

References

Chan, K.L., Sofia, S., Mayr, H.G.: 1987, in *The internal solar angular velocity*, eds. B.R. Durney, S. Sofia, Reidel, p. 347

Gierasch, P.J.: 1974, *Astrophys. J.* **190**, 199

Orszag, S.A.: 1970, *J. Fluid Mech.* **41**, 363

Rüdiger, G.: 1989, *Differential rotation and stellar convection, Sun and solar-type stars*, Gordon and Breach

Rüdiger, G., Kichatinov, L.L.: 1990, *Astron. Astrophys.* **236**, 503

Rüdiger, G., Tuominen, I.: 1990, in *Solar Photosphere: structure, convection and magnetic fields*, ed. J.O. Stenflo, Kluwer, p. 315

Schmidt, W.: 1982, *Geophys. Astrophys. Fluid Dyn.* **21**, 27

Tuominen, I., Rüdiger, G.: 1989, *Astron. Astrophys.* **217**, 217

Vainshtein, S.I., Kichatinov, L.L.: 1983, *Geophys. Astrophys. Fluid Dyn.* **24**, 273

Differential Rotation as an Axisymmetric Resonant Mode of Convection

Kwing L. Chan[1], Hans G. Mayr[2]

[1] Applied Research Corporation, Landover, MD, USA
[2] Goddard Space Flight Center, Greenbelt, MD, USA

Recent results from helioseismology (see Goode, these Proceedings) have shown that the inferred contours of the solar angular velocity are more or less radial in the convection region, and the rotation becomes uniform below. These observations contradict the prevailing numerical models of Taylor columns which predict angular velocity contours parallel to the rotation axis of the Sun. Thus, an alternative explanation of solar differential rotation is called for.

Presently, it is not feasible to construct a thermally-relaxed, dynamically self-consistent numerical model of the solar convection zone (see Chan and Serizawa, these Proceedings). It is then appropriate to explore simplified models that may shed some light. A number of analytical models have been proposed for the solar differential rotation, and the reader is referred to the book by Rüdiger (1989) for a comprehensive review of this subject. Here, we report on some recent development on the convective resonance model proposed by Chan *et al.* (1987; hereafter referred as CSM).

In the convective resonance model, differential rotation is described in terms of the zonally symmetric velocity field. The model is linear, axisymmetric, and eddy diffusion describes the Reynolds stresses that dissipate energy and momentum. Latitudinal variations of convective heat transfer (Weiss, 1965; Durney and Roxburgh, 1971; Belvedere and Paternò, 1977) are invoked to drive the motions. An important aspect of the model is that the superadiabatic gradient is dynamically important; and a resonant convective eigenmode then develops which is rather insensitive to the magnitude and latitude dependence of the excitation source.

Using scale analysis, we illustrate how such a resonance develops. With standard notations, the linearized conservation equations, for the zonally averaged temperature perturbation, δT, meridional velocity, V, and zonal velocity, U, are written in the simplified form

$$\frac{\Delta T}{\mathrm{Pr}\tau_d} - \frac{T\delta\nabla}{\lambda H}V = q, \tag{1}$$

$$\frac{gd}{\lambda T}\Delta T - \Omega U - \frac{V}{\tau_d} = 0, \tag{2}$$

$$\Omega V - \frac{U}{\tau_d} = 0, \tag{3}$$

where q is the excitation source, Ω the mean rate of rotation, λ and d the horizontal and vertical length scales, H the pressure scale height, $\tau_d = d^2/\nu$ the dissipation time constant, and Pr the Prandtl number; the hydrostatic approximation $\Delta \ln p = \Delta T d/TH$ is used. The solution then is

$$V = \frac{\mathrm{Pr} g d}{T\lambda[\Omega^2 + \tau_d^{-2} - \frac{g \mathrm{Pr}\delta\nabla d^2}{\lambda^2 H}]} q, \tag{4}$$

$$U = \Omega \tau_d V = \sqrt{\mathrm{Ta}} V. \tag{5}$$

Since the Taylor number is large, Ta$\gg$ 1, and the solar envelope is convectively unstable, i.e. $\delta\nabla > 0$, the denominator in Eq. (4) can vanish. A resonance then develops such that the wavelength

$$\lambda = \frac{d}{\Omega}\sqrt{\frac{g\mathrm{Pr}\delta\nabla}{H}} \tag{6}$$

is preferentially excited. When the conservation equations are formulated for a spherical envelope, using spherical harmonics, only discrete wavelengths can be excited.

For the simplified one-layer Boussinesq model of CSM it was shown that the fundamental resonance mode matches the surface differential rotation of the sun. In the following, we present a compressible model which accounts for stratification; the resulting differential rotation in the solar interior is discussed and compared with observations.

Our stratified model is spectral in terms of zonal spherical harmonics and is constructed with the following specifications: (i) The linearized hydrodynamic equations are solved for the response to a small source (described with the spherical harmonic, P_2) in the energy equation. (ii) The domain of computation extends from 0.57 to 0.93 solar radii, and the bottom of the convection zone is at 0.69. (iii) The distribution of the superadiabatic gradient $\delta\nabla$ in the convection zone is determined by $\nu\rho C_p T\delta\nabla/\mathrm{Pr}H$ = flux; where ν, the diffusivity is constant, and Pr = 0.3; ρ is the density; C_p is the specific heat under constant pressure; i.e. $\delta\nabla \propto \rho^{-1}$. (iv) The stratification of the stable layer below the convection zone is based on a standard solar model. (v) No-slip boundary conditions are adopted at the bottom and stress-free conditions at the top; (vi) The degree of spherical harmonics is up to 6. (vii) The number of vertical grid points is 100.

A search for resonance was performed by varying $\delta\nabla_0$, the value of the superadiabatic gradient at the top of the convection layer. The lowest order resonance occurs at $\delta\nabla_0 = 3.7 \times 10^{-4}$ (with $\delta\nabla \sim 5 \times 10^{-5}$ in the middle of the convective layer, $\nu = 1.2 \times 10^{10}$ cm^{2}s^{-1}).

We show the computed angular velocity distribution as a surface plot in Fig. 1 and the observations (Dziembowski *et al.*, 1989) in Fig. 2. Latitude and depth are represented by horizontal coordinates, and the angular velocity by the vertical

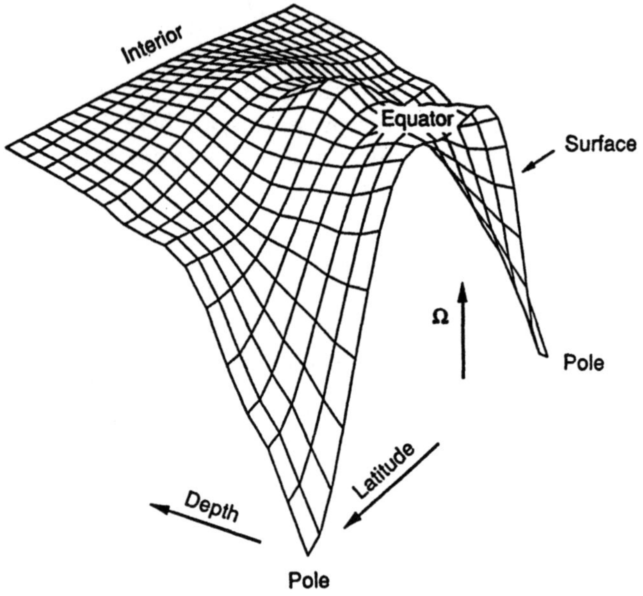

Fig. 1. Computed differential rotation

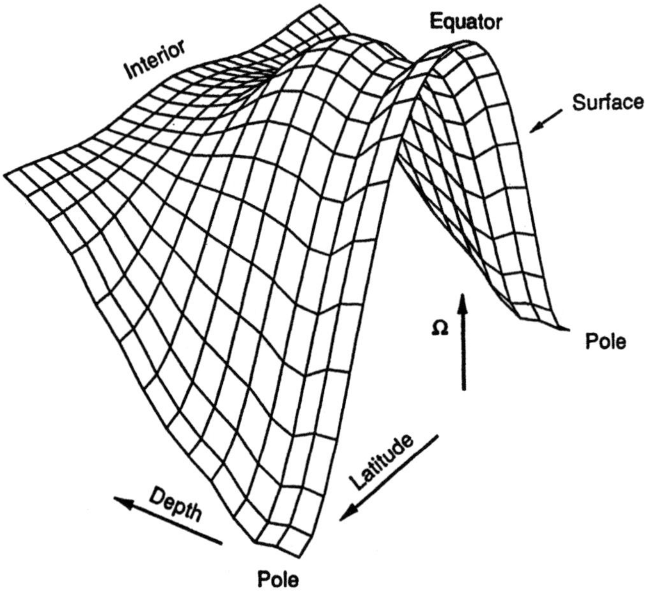

Fig. 2. Observed solar differential rotation (Dziembowski *et al.*, 1989)

coordinate. The velocity distribution is very similar to that obtained from helioseismology, and when its magnitude is matched to that observed at the surface, the maximum meridional velocity is only about 10 cm/sec. There are also some disagreements. The small extraneous dip in the surface angular velocity at the equator is not observed, and the magnitude of the differential rotation decreases somewhat faster towards the bottom of the convection zone. However, these discrepancies are not serious, considering that the model has been constructed with a simple set of assumptions without any fine tuning. The model needs to be explored further by varying the distributions of $\delta\nabla$ and ν in the convection zone.

Acknowledgement: KLC thanks NSF for support (AST-8815457)

References

Belvedere, G., Paternò, L.: 1977, *Solar Phys.* **54**, 289
Chan, K.L., Sofia, S., Mayr, H.G.: 1987, In *The internal solar angular velocity*, eds. B.R. Durney and S. Sofia, Reidel, Dordrecht, p.347
Durney, B.R., Roxburgh, I.W.: 1971, *Solar Phys.* **16**, 3
Dziembowski, W.A., Goode, P.R., Libbrecht, K.G.: 1989, *Astrophys. J.* **336**, 53
Rüdiger, G.: 1989, *Differential rotation and stellar convection: Sun and solar-type stars*, Gordon and Breach, New York
Weiss, N.O.: 1965, *Observatory* **85**, 37

The Solar Internal Rotation and its Implications

Lucio Paternò

Istituto di Astronomia, Università di Catania, Italy
CNR - Gruppo Nazionale di Astronomia, UdR di Catania, Italy

Abstract: The present internal rotation of the Sun as deduced from helioseismological data has many implications which concern the location of the solar cycle dynamo mechanism, core magnetism and history of the angular momentum. These aspects and their mutual connections are briefly discussed.

1. The present internal rotation

Recent measurement of p-mode rotational splittings (Morrow, 1988; Rhodes *et al.*, 1988; Woodard and Libbrecht, 1988) have pointed out that the surface rotation persists through the solar convection zone, at the base of which there is a tendency towards a rigid rotation. The inversion of helioseismological data indicates that the change of the rotational behaviour occurs in a transition layer located between $r = 0.7\,R_\odot$ and $r = 0.5\,R_\odot$. The optimization of data (Morrow, 1988) leads to a rotational behaviour that can approximately be described by the following expressions:

For $r_2 \le r \le R_\odot$ (standard c.z.) $\omega(r,\varphi) = \omega(R_\odot,\varphi) = a + b\sin^2\varphi + c\sin^4\varphi$, (1a)

for $r_1 \le r \le r_2$ (transition layer) $\omega(r,\varphi) = \omega + f(\varphi)[(r - r_1)/(r_2 - r_1)]$, (1b)

for $r \le r_1$ (radiative interior) $\omega(r,\varphi) = \omega_c$, (1c)

where φ is the latitude, $f(\varphi) = \omega(R_\odot,\varphi) - \omega_c$ and a = 2.87μrad s^{-1}, b = – 0.31 μrad s^{-1}, c = – 0.503 μrad s^{-1}, $\omega_c = 2.74$ μrad s^{-1}. The latitude at which $\omega(R_\odot,\varphi)$ = ω_c is $\varphi_0 = 32.35°$ as deduced from (1a). The radial behaviour of the angular velocity for four latitudes ($\varphi = 0°, 30°, 60°, 90°$) and the isorotation surfaces, as deduced from expressions (1), are shown in Fig. 1 (respectively top and bottom). The inversion of the p-mode rotational splittings gives very uncertain results below $0.3\,R_\odot$.

2. The location of the dynamo

For an $\alpha - \omega$ dynamo it is well known that in order to obtain the correct field migration at the solar surface, since the dynamo wave migrates along the isorotation surfaces and the migration direction is given by the vector $\alpha \nabla \omega \times i_\varphi$ (Yoshimura, 1975), it is necessary that $\partial\omega/\partial r < 0$, as in the solar convection zone $\alpha > 0$ in the north hemisphere and $\alpha < 0$ in the south hemisphere (Steenbeck and Krause, 1969).

However the picture which emerges from the analysis of helioseismological data allows for ω isoplanes displaced almost radially (Fig. 1 bottom) and therefore for $\partial\omega/\partial r \simeq 0$. This situation seems to preclude an $\alpha - \omega$ dynamo from operating in the solar convection zone.

The behaviour of the angular velocity in the transition layer indicates the possibility of a consistent overshooting below the standard convection zone. In this case the sign of α is reversed in both the hemispheres because the convective motions are mainly directed towards the interior. In this region we have $\partial\omega/\partial r > 0$ for $\varphi < \varphi_0$ and $\partial\omega/\partial r < 0$ for $\varphi > \varphi_0$. This picture is consistent with the migration of the sunspots towards the equator at the lower latitudes and migration of other magnetic features towards the poles at the higher latitudes. In the weak buoyancy transition layer it is possible to intensify the magnetic field to exceed greatly the equipartition value over spatial scales comparable with those of the active regions, as it has been shown by an instability analysis of Spruit and Van Ballegooijen (1982). For high order modes, or small spatial scales, the limit field strength is of the order of 1 MGauss (Paternò, 1990).

Attemps to model non linear α–ω dynamos in the transition layer (Belvedere *et al.* these Proceedings) are not yet conclusive about the possibility of a dynamo operating in this layer, as they show that magnetic field is preferentially intensified at high latitudes. This probably depends on the radial shear $\partial\omega/\partial r = f(\varphi)/(r_2 - r_1)$ which is much stronger at high latitudes.

3. The angular momentum evolution

The possibility of tracing back the rotation of the Sun is offered by the observations of the stars with about the same mass of the Sun in earlier evolutionary stages. In Fig. 2 (top) typical rotation rates of a sample of younger suns are shown as functions of the age.

The spinup phase from T Tau ($t \simeq 10^6$ ys, $\Omega \simeq 3\Omega_\odot$) to α Per ($t \simeq 5 \cdot 10^7$ ys, $\Omega \simeq 26\Omega_\odot$) is consistent with the conservation of angular momentum and rigid body rotation. The hypothesis of rigid rotation for almost fully convective stars is consistent with a rapid redistribution of angular momentum due to the convective motions. On taking into account the slight contraction which affects the mass distribution inside, we obtain:

$$\Omega_2/\Omega_1 = (h_1 R_1/h_2 R_2)^2 \simeq 10, \qquad (2)$$

where the suffices 1 and 2 refer respectively to the initial and final conditions and $h_1^2 \simeq 0.2$, $h_2^2 \simeq 0.1$ are the radii of gyration, in units of R^2, of the two configurations, assuming two polytropic structures respectively with $n = 1.5$ and $n = 2.5$.

The spin-down phase from α Per through Pleiades and Hyades to the present Sun can be discussed in terms of magnetic wind torques considering the field to be a potential field inside the Alfvénic surface and radial thereafter (Schatzman, 1962; Mestel, 1968). In this situation it is possible to have a strong braking with a small mass loss provided that the Alfvén radius is sufficiently large. In the absence of strong mass loss, which on the other hand is not observed during these evolutionary phases, the rate of braking is given by

$$\frac{\partial \Omega}{\partial t} \simeq \frac{2}{3h^2} \frac{dm/dt}{M} \left[\frac{r_A}{R}\right]^2 \Omega, \tag{3}$$

where dm/dt is the rate of mass loss, r_A the Alfvénic radius, M the mass of the layers in which the braking occurs, and expression (3) has been deduced from the relevant equations (Castellani and Paternò, 1984). The slow spin-down phase (Pleiades-Sun) is quite consistent with the slowing down of the whole Sun (Paternò, 1984), while the fast spin-down phase (α Per - Pleiades) can only be explained if the convection zone spins-down independently of the radiative interior.

This picture is not consistent with helioseismological data which suggest for rigid rotation at least until to a depth of $0.3\,R_\odot$. A very effective transport mechanism must have been active during the fast spin-down phase for transferring the internal angular momentum to the outer layers.

In Fig. 2 (bottom) we compare the marginal gradient of Ω for the two hydrodynamical instabilities GSF and ABCD, as calculated by Dziembowski *et al.* (1986), with the mean equatorial velocity of the Sun as deduced by Libbrecht and Woodard (1990) analyzing the 1986 and 1988 helioseismological data. Since the ABCD instability, which has the lowest threshold among the hydrodynamical instabilities, does not explain the observed behaviour of Ω, we are induced to think that some hydromagnetic mechanism is at work in the Sun's core (Dziembowski *et al.*, 1985).

References

Castellani, V., Paternò, L.: 1984, *Astron. Nachr.* **305**, 251

Dziembowski, W.A., Paternò, L., Ventura, R.: 1985, *Astron. Astrophys.* **151**, 47

Dziembowski, W.A., Paternò, L., Ventura, R.: 1986, in *Seismology of the Sun and Distant Stars*, ed. D.O. Gough, Reidel, Dordrecht, p. 257

Libbrecht, K.G., Woodard, M.F.: 1990, *Nature* **345**, 779

Mestel, L.: 1968, *Monthly Notices Roy. Astron. Soc.* **138**, 359

Morrow, C.A.: 1988, ESA SP-286, p. 91

Paternò, L.: 1984, in *Space Research Prospects in Stellar Activity and Variability*, eds. A. Mangeney and F. Praderie, Obs. Paris-Meudon, France, p. 327

Paternò, L.: 1990, in *Progress of Seismology of the Sun and Stars*, eds. Y.Osaki and H. Shibahashi, Lecture Notes in Physics, Springer-Verlag, Berlin, in press

Rhodes Jr., E., Cacciani, A., Korzennik, S., Tomczyk, S., Ulrich, R.K., and Woodard, M.F.: 1988, ESA SP-286, p.73
Schatzman, E.: 1962, *Ann. Astrophys.* **25**, 1
Spruit, H.C., Van Ballegooijen, A.A.: 1982, *Astron. Astrophys.* **106**, 58
Steenbeck, M., Krause, F.: 1969, *Astron. Nachr.* **291**, 49
Woodard, M.F., Libbrecht, K.G.: 1988, ESA SP-286, p. 67
Yoshimura, H.: 1975, *Astrophys. J.* **201**, 740

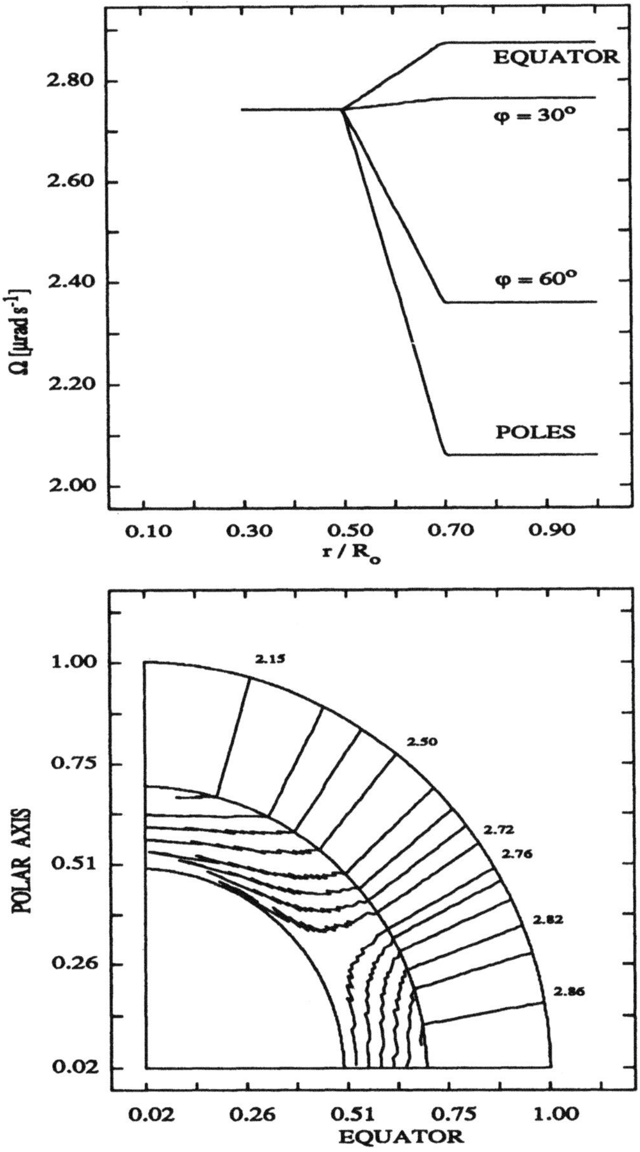

Fig. 1. The radial behaviour of the solar angular velocity at four latitudes (top) and the corresponding isorotation surfaces (bottom) as deduced form a best fit with helioseismological data (Morrow, 1988).

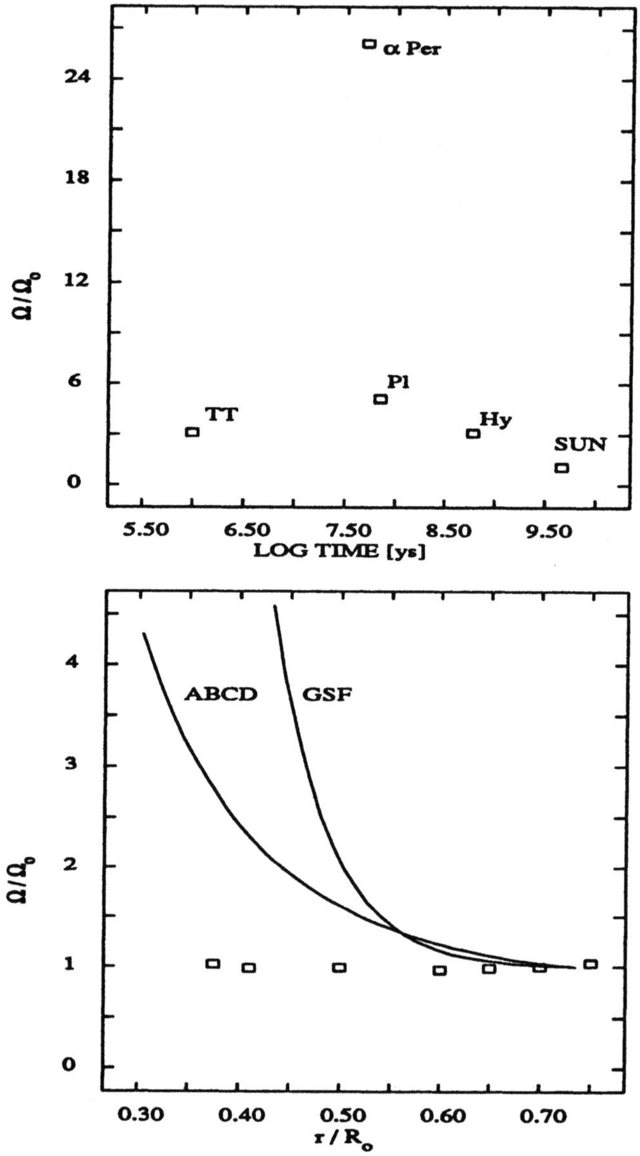

Fig. 2. Observed mean rotation rates of a sample of stars of about 1 $M_\odot$ in earlier evolutionary stages than the Sun as functions of the time (top) and the marginal gradient of the solar angular velocity for the GSF and ABCD instabilities compared with the mean equatorial angular velocity, as deduced from the 1986 and 1988 helioseismological data (Libbrecht and Woodard, 1990), (bottom).

The Toroidal Magnetic Field inside the Sun

V.N. Krivodubskij[1], A.E. Dudorov[2], A.A. Ruzmaikin[3], T.V. Ruzmaikina[4]

[1] Kiev University Astronomical Observatory, Kiev, 252053, U.S.S.R.
[2] Moscow University Astronomy Department, Moscow, 117899, U.S.S.R.
[3] IZMIRAN, Troitsk, Moscow region, 142092, U.S.S.R.
[4] Institute of Physics of Earth, Moscow, 123810, U.S.S.R.

Analysis of the fine structure of the solar oscillations has enabled us to determine the internal rotation of the Sun and to estimate the magnitude of the large-scale magnetic field inside the Sun. According to the data of Duvall *et al.* (1984), the core of the Sun rotates about twice as fast as the solar surface. Recently Dziembowski *et al.* (1989) have showed that there is a sharp radial gradient in the Sun's rotation at the base of the convection zone, near the boundary with the radiative interior. It seems to us that the sharp radial gradients of the angular velocity near the core of the Sun and at the base of the convection zone, acting on the relict poloidal magnetic field B_r, must excite an intense toroidal field B_ϕ, that can compensate for the loss of the magnetic field due to magnetic buoyancy.

Magnetic buoyancy plays the main role in constraining the amplitude of the magnetic induction of the toroidal field generated at the present stage of solar evolution (Dudorov *et al.*, 1989; Krivodubskij, 1990). There, from the condition of stationarity, $\partial B_\phi/\partial t = 0$, neglecting ohmic dissipation, we obtained the following expression for the maximum value of the established stationary toroidal field (Dudorov *et al.*, 1989; Krivodubskij, 1990):

$$\max|B_\phi| \equiv B_\phi^0 \simeq \left[r\frac{\Delta\omega}{\Delta r} B_r \frac{8\pi P_e L}{u_T}\left(\frac{a}{\lambda}\right)^2 \right]^{1/3} . \tag{1}$$

Here $\Delta\omega$ is the increament in the angular velocity over the step Δr along the solar radius; P_e is the electron pressure; u_T is the mean velocity of upward transport of thermal energy; L is the space scale of the magnetic field; a is the transverse radius of the magnetic flux tube; λ is the temperature scale height; r is the distance from the centre of the Sun.

Krivodubskij (1990) performed an analysis of the energetic aspects of the problem. The total magnetic energy of the excited toroidal field, $E_H^* = E_H V_H$ ($E_H = B_\phi^2/8\pi$ is the magnetic energy density), cannot exceed the total kinetic energy of the rapidly rotating core, E_ω^*. To make $E_H^* \leq E_\omega^*$, toroidal field with intensity B_ϕ must be concentrated in the volume $V_H \leq E^*_\omega / E_H = 8\pi E^*_\omega / B^2_\phi$. In the generation mechanism considered the toroidal magnetic flux tubes are localized in the region of the radial jump of the angular velocity and are oriented along the parallels of solar latitude, with opposite signs on either side of the equatorial plane. Then $V_H = 2V_t = 2(2\pi r_0 S_t)$ where $S_t = d_t^2/4$ is the cross-sectional area of the torus with radius r_0 and volume V_t. The cross-sectional diameter $d_t = (4S_t/\pi)^{1/2}$ of the toroidal magnetic flux tube must not exceed the value $(8/\pi E^*_\omega / r_0 B^2_\phi)^{1/2}$.

The expression (1) together with the helioseismological data of Duvall *et al.* (1984), yields a maximum estimate of the toroidal field near the core of the Sun in the range $1 \times 10^7 - 2 \times 10^8$ G. Energetic constraints suggest that this toroidal field is concentrated in relatively thin magnetic flux tubes with diameters $d_t \leq 16\,000 - 800$ km, respectively (Krivodubskij, 1990).

Dziembowski *et al.* (1989) used the oscillation data of Libbrecht (1989), and found that there is a sharp radial jump of the angular velocity, $\Delta\omega \simeq 18$ nHz $\simeq 1.1 \times 10^{-7}$ rad/s at the transition from the convection zone to the radiative zone ($r = 0.7\,R_\odot, \Delta\omega/\Delta r \simeq 1.6 \times 10^{-17}$ rad/s cm). If for the radial field B_r we take a value of about 1 G, which to order of magnitude is in agreement with the observed poloidal field on the surface of the Sun, then from the expression (1) we obtain an estimate of the maximum magnetic induction of the resulting stationary toroidal field of about 2×10^6 G near the bottom of the convection zone ($d_t \leq 50\,000$ km).

Dziembowski and Goode (1989) determined the value of the toroidal magnetic field inside the Sun directly from the analysis of the helioseismological data of Libbrecht (1989), using for this an asymptotic formulation of the inverse problem analogous to that from which the internal rotation of the Sun was determined (Dziembowski *et al.*, 1989). They investigated the effect of a magnetic field on the 5-min oscillations and found evidence for an axisymmetric quadrupole toroidal field of about $(2 \pm 1) \times 10^6$ G, centred near the base of the convection zone. This value clearly agrees with our estimate of the toroidal magnetic field.

As we pointed out, the origin of the strong toroidal field in deep layers of the Sun is the action of the differential rotation on the relict poloidal field. This has been noted also by Dicke (1979) (the excitation of the toroidal field near the boundary of the fast rotating core) as well as by Dziembowski and Goode (1989) (the toroidal field excitation of the boundary of the convection zone and the radiative zone). However, they neglected the effect of magnetic buoyancy on the magnetic field regeneration processes. In our investigation the effect of the magnetic buoyancy is taken into account and as a result it is shown that it is the magnetic buoyancy that is the factor which limits the maximum intensity of the toroidal field to about $(1 - 2) \times 10^6$ G near the base of the convection zone, and to about $10^7 - 2 \times 10^8$ G near the solar core.

The magnetic field intensities obtained by us are sufficient to induce the observed splitting effect in the 5-min oscillation frequencies. Therefore, to interpret the fine structure of the solar oscillations, the effect of the magnetic field and the effect of rotation must be taken equally into account. Considering the spatial structure of the excited toroidal field it should be remembered that the contribution of the magnetic field to the splitting of oscillations will depend on its distribution both with depth and latitude.

References

Dicke, R.H.: 1979, *Astrophys. J.* **228**, 898

Dudorov, A.E., Krivodubskij, V.N., Ruzmaikina, T.V., Ruzmaikin, A.A.: 1989, *Astron. Zh.* **66**, 809 (*Sov. Astron.* **33**, 420)

Duvall Jr., T.L., Dziembowski, W.A., Goode, P.R., Gough, D.O., Harvey, J.W., Leibacher, J.W.: 1984, *Nature* **310**, 22

Dziembowski, W.A., Goode, P.R.: 1988, IAU Symp. No 123, p. 171

Dziembowski, W.A., Goode, P.R.: 1989, *Astrophys. J.* **347**, 540

Dziembowski, W.A., Goode, P.R., Libbrecht, K.G.: 1989, *Astrophys. J. Letters* **337**, L53

Krivodubskij, V.N.: 1990, *Vestn. Kiev. Univ., Astron.* **32**, 3

Libbrecht, K.G.: 1989, *Astrophys. J.* **336**, 1092

The Transfer of Large-scale Magnetic Field by Radial Inhomogeneity of the Material Density in the Rotating Convection Zone

V.N. Krivodubskij[1], L.L. Kichatinov[2]

[1] Kiev University Astronomical Observatory, Observatorna Street 3, Kiev - 53, 252053, Ukraine, USSR
[2] Sib IZMIR, P.O. Box 4026, Irkutsk - 33, 664033, Russia, USSR

Abstract: The influence of rotation on the transfer of the mean magnetic field of the Sun, caused by the radial inhomogeneity of the solar turbulent plasma density, is investigated. It turns out that the transfer directions of the poloidal and toroidal magnetic fields do not coincide.

Random turbulent hydrodynamic motions (with velocities u') in a medium of high conductivity generate random fluctuating magnetic fields (h'). Such small-scale magnetic fluctuations cause the mean large-scale magnetic field to rise with speed proportional to the density gradient of the medium (∇Q) (Kichatinov, 1982). In the nonlinear formulation of the problem the magnetic field transfer takes place in the direction of increasing plasma density, independent of the magnetic field configuration (Kichatinov, 1982; Vainshtein, 1983). The expulsion velocity of the horizontal large-scale magnetic field of the Sun in the convective zone has been calculated by Krivodubskij (1987). Here the calculation of the transfer effect of the solar magnetic field is carried out for a convection zone with density stratification, taking into account the anisotropy in the turbulence generated by the solar rotation.

The calculation of the mean electromotive force in the turbulent medium is carried out in the quasilinear approximation under the condition of equipartition of the kinetic and magnetic energies of small-scale fluctuations (Kichatinov, 1990). It is found that the transfer directions of poloidal ($\boldsymbol{V}^a$) and toroidal ($\boldsymbol{V}^b$) magnetic fields do not coincide in the highly turbulent rotating convective envelope of the Sun. We write

$$\begin{aligned} \boldsymbol{V}^a &= -\boldsymbol{V} + \boldsymbol{V}_m + \boldsymbol{V}'_m, \\ \boldsymbol{V}^b &= +\boldsymbol{V} + \boldsymbol{V}_m, \end{aligned} \tag{1}$$

where $\boldsymbol{V}$ is the transfer velocity of the mean magnetic field caused by the turbulent motions (the kinematic contribution); $\boldsymbol{V}_m$ is the transfer velocity of the mean field caused by the small-scale magnetic fluctuations (the nonlinear contribution); $\boldsymbol{V}'_m$

is the additional transfer velocity of the poloidal magnetic field (resulting from rotational and magnetic fluctuations). These components of the transfer velocity are given by

$$\begin{aligned} \boldsymbol{V} &= -\boldsymbol{\lambda}_{\perp}\tau u'^2 \Phi_1(\beta), \\ \boldsymbol{V}_m &= +\boldsymbol{\lambda}\tau u'^2 \Phi_2(\beta), \\ \boldsymbol{V}'_m &= -\boldsymbol{\lambda}_{\perp}\tau u'^2 \Phi_3(\beta), \end{aligned} \tag{2}$$

where $\boldsymbol{\lambda} = \nabla Q/Q$; $\tau \simeq l/u'^2$ is the timescale of the nonlinear relaxation of fluctuations which is equal, to order of magnitudes, to the correlation time; l is the characteristic size of small-scale velocity (u') and magnetic field (h') fluctuations; $\Phi_n(\beta)$ are functions of the Coriolis number $\beta = 2\tau\Omega$; Ω is the angular velocity.

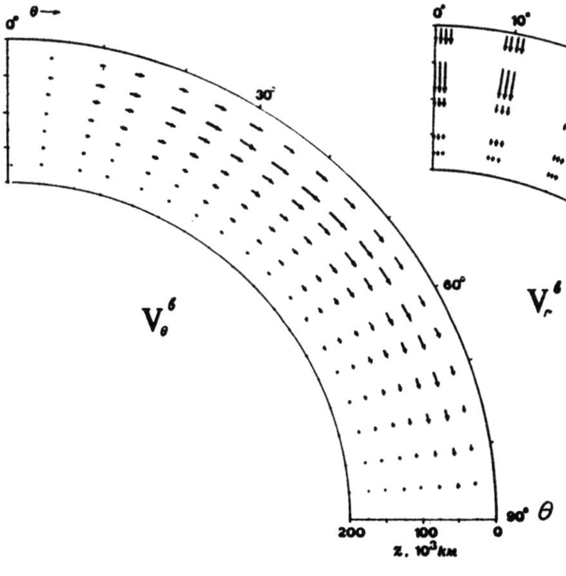

Fig. 1.

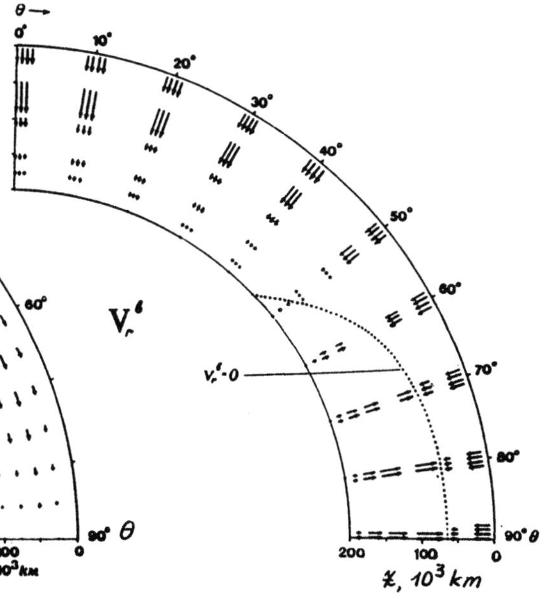

Fig. 2.

The expressions for the meridional (θ) and radial (r) transfer velocity components of poloidal (V^a, V_r^a) and toroidal (V^b, V_r^b) large-scale magnetic fields in the usual spherical coordinate system r, θ, ϕ (θ is the co-latitude) take the form (Kichatinov, 1990)

$$
\begin{aligned}
V^a_\theta &= -\frac{3\kappa g}{(\gamma-1)C_pT}\cos\theta\sin\theta[\Phi_1(\beta)-\Phi_3(\beta)],\\
V^a_r &= -\frac{3\kappa g}{(\gamma-1)C_pT}\{\Phi_2(\beta)+\sin^2\theta[\Phi_1(\beta)-\Phi_3(\beta)]\},\\
V^b_\theta &= +\frac{3\kappa g}{(\gamma-1)C_pT}\cos\theta\sin\theta\Phi_1(\beta),\\
V^b_r &= -\frac{3\kappa g}{(\gamma-1)C_pT}[\Phi_2(\beta)-\Phi_1(\beta)\sin^2\theta],
\end{aligned}
\tag{3}
$$

where $\kappa \simeq \frac{1}{3}\tau u'^2$ is the turbulent diffusion coefficient, C_p is the specific heat capacity at constant pressure, T is the temperature, g is the acceleration due to gravity and γ is the adiabatic index.

The values of the different components of the transfer velocities of the magnetic fields for the convection zone model of Spruit (1977) are calculated. The poloidal field is transferred deep into the convection zone (the maximum transfer velocity V^a_r is about 2×10^3 cm/s) and mainly to the pole ($V^a_{\theta\,\max} \simeq 10^2$ cm/s). The horizontal component of the toroidal field V^b_θ is directed to the equator ($V^b_{\theta\,\max} \simeq 6\times10^2$ cm/s) (Fig. 1), in accordance with the observed picture of the active region migrations.

The radial component of this velocity (V^b_r) depends on the latitude (Fig. 2). At high latitudes it is directed inwards ($V^b_{r\,\max} \simeq -2\times10^3$ cm/s) whilst near the equator it is directed towards the surface ($V^b_{r\,\max} \simeq +2\times10^2$ cm/s), with the exception of the upper layers of the convection zone where it is directed deep into the Sun. Therefore, the toroidal magnetic fields in the near-pole regions may be supposed to be blocked in deep layers of the convective zone, due to the transfer mechanism considered. In the near-equator region the toroidal fields in the lower part of the convection zone migrate upward reaching levels where the magnetic buoyancy starts to play the essential role (i.e. the effect of buoyancy exceeds the effect of blocking) and as a result it causes the active regions to rise.

References

Kichatinov, L.L.: 1982, *Magnitnaya Gidrodinamika* **2**, 67

Kichatinov, L.L.: 1990, *Astron. Astrophys.* (in press)

Krivodubskij, V.N.: 1987, *Pis'ma Astron. Zh.* **9**, 803 (*Sov. Astron. Lett.* **9**, 338)

Spruit, H.C.: 1977, Thesis, University of Utrecht, p.17

Vainshtein S.I.: 1983, *Magnetic Fields in Outer Space* (in Russian), Nauka, Moscow

Diagnostics of the Solar Dynamo Using the Observed Pattern of Surface Magnetic Fields

J.O. Stenflo

Institute of Astronomy, ETH-Zentrum, CH-8092 Zurich, Switzerland

Abstract: The solar surface represents for the solar dynamo an outer boundary that is directly accessible to observations. The evolutionary and rotational properties of the magnetic fields at this boundary can be empirically determined using existing synoptic magnetograph data. The derived properties provide detailed constraints on the underlying theory, such that an inversion approach to the dynamo problem becomes feasible. Ambiguities in the interpretation may be removed using the independent constraints from helioseismology.

1. Introduction

Most solar dynamo work of the past has aimed at qualitatively reproducing the butterfly diagram of sunspots. The sunspots are however only secondary manifestations of the primary dynamo parameter, the magnetic field, they reflect only a certain aspect of the magnetic field (intense concentrations of toroidal flux), and they occur only at certain discrete locations on the sun. The magnetic fields, on the other hand, have been observed over the whole solar disk for three decades now, since 1959, and provide us with rich empirical constraints that may be used for dynamo inversions. The detailed magnetograph data however need to be converted into forms more suitable for direct diagnostic purposes, as will be described in Sects. 2–4.

Besides revealing the evolutionary properties and global wave modes of the solar dynamo, the surface magnetic flux can also be used as a tracer for determinations of the solar rotation. The rotational phase velocity of the magnetic field pattern depends however not only on processes in the surface layers, but also on the processes in the deep layers of the solar interior, from which the surface fields are being replenished. In Sect. 5 we will indicate how these various effects may be untangled.

2. Data base

Magnetograms (maps of the line-of-sight component of the magnetic field) have been recorded daily at the Mount Wilson Observatory since August 1959 and at the Kitt Peak Observatory since 1976. We do not have information on the vector magnetic field, but for investigations of the evolution of the field pattern it is a good approximation to assume that the true orientation of the field is on the average in the radial direction in the layers, where the field is being measured (the photosphere). The justification for this approximation is that the photospheric magnetic flux is extremely fragmented in individual fluxtubes, and the buoyancy forces acting on these fluxtubes in the radial direction are very strong due to the rapid exponential density decrease in the photosphere. The magnetograms may then be converted into maps of the radial component of the magnetic flux. When we in the following talk about the surface magnetic field B, we mean the radial magnetic flux, although we omit index r for brevity of notation.

For a global mode analysis we like to expand the surface magnetic field pattern in the spherical harmonics $Y_\ell^m(\vartheta,\varphi)$, where ϑ is the colatitude, φ the longitude:

$$B(\vartheta,\varphi,t)=\sum_{\ell=0}^{\infty}\sum_{m=-\ell}^{\ell} c_\ell^m(t)\,Y_\ell^m(\vartheta,\varphi)\,. \tag{1}$$

The dependence of B on time t is then accounted for by the time dependence of the complex harmonic coefficients c_ℓ^m. Using the orthonormal properties of the spherical harmonics, we can solve for the coefficients:

$$c_\ell^m(t)=\int B(\vartheta,\varphi,t)Y_\ell^{m*}(\vartheta,\varphi)\,\mathrm{d}\Omega\,. \tag{2}$$

The harmonic decomposition cannot however be carried out entirely as described by Eqs. (1) and (2), since the sun is not observed from all sides at the same time. Observational sampling of the magnetic field over all longitudes can only be accomplished over the course of one solar rotation, i.e., about 27 days. Since the magnetic field pattern evolves significantly over this period, the longitude φ and time t coordinates are no longer independent of each other. Moreover, as the sun rotates differentially, there is no unique way to define a longitude system on the sun (since any longitude system rotates rigidly). We will return to these questions later.

A condensed synoptic data set that is easier to work with can be extracted from the full-disk daily magnetograms by sampling the magnetic field in one longitude band around the central meridian in each magnetogram. As the sun rotates, different longitudes (decreasing from 360° to 0°) pass the central meridian, so during one month all longitudes get sampled, and a synoptic map can be constructed, giving the radial magnetic field over all latitudes and longitudes of the sun. The longitude system normally adopted for such maps is based on the Carrington rotation period $P_C = 27.2753$ days. Let us denote the corresponding rotation frequency by $\nu_C = 1/P_C$. Then we have the following relation between the Carrington longitude φ and time t:

$$\varphi = -2\pi(\nu_C t - n)\,. \tag{3}$$

n is an integer that ensures that φ always stays in the interval $-\pi < \varphi \leq \pi$. For convenience we have chosen the zero point of the time scale such that $\varphi = 0$ when $t = 0$.

The choice of the longitude system is arbitrary and does not affect the analysis that will be described in the rest of the present review. If we want we can always replace the longitude coordinate φ by time t, using Eq. (3), and consider the magnetic field B as a function of latitude and time only. It is also useful to replace the colatitude ϑ by $x = \cos\vartheta$, since x represents the projected distance on the sun (when the axis of rotation is perpendicular to the line of sight), and the synoptic maps are digitized at points equidistant in x. The magnetic field is thus basically a function of x and t, and may be written $B(x,t)$.

The Kitt Peak data have been recorded with much higher spatial resolution than the Mount Wilson data. To obtain a homogeneous data base extending all the way back to 1959 we have therefore spatially averaged the Kitt Peak data and used them from 1976 onwards. The selected common format for the whole synoptic data set is defined by a grid of 30 zones equidistant in x (the sine of the latitude), and 36 sectors equidistant in longitude. In the following we will describe the results based on such a synoptic data base extending over a period of 26 yr.

3. Evolution of the axisymmetric field component

3.1 "Butterfly diagram" of the radial magnetic flux

The axisymmetric component of the magnetic field pattern may be obtained by averaging over all longitudes:

$$\bar{B}(x,t) = \frac{1}{2\pi}\int_{-\pi}^{\pi} B(x,\varphi,t)\,\mathrm{d}\varphi\,. \tag{4}$$

This longitude averaging may also be regarded as a smoothing of the data using a rectangular time window with a width of one Carrington rotation. The results of this smoothing process are not sensitive to the exact choice of width for the time window, except that a wider window is useful to suppress noise and short-term fluctuations. For the harmonic analysis we will use a window of $1P_C$ as in Eq. (4), but for a smoother representation of "butterfly diagrams" we will apply a time window of $16P_C$ (corresponding to 1.2 yr).

The time smoothed $\bar{B}$ can be graphically represented as isocontours in latitude – time space, analogous to the representation of sunspots in the form of "butterfly diagrams". Figure 1 gives $\bar{B}$, smoothed with a running $16P_C$ window, vs. x (sine of the latitude) and time t. The pattern is clearly predominantly of odd parity, i.e., anti-symmetric with respect to reflections in the equatorial plane. It exhibits the well-known features of polarity reversals every 11 yr, equatorward migration of the low-latitude zones, and steep poleward migration of the higher-latitude zones.

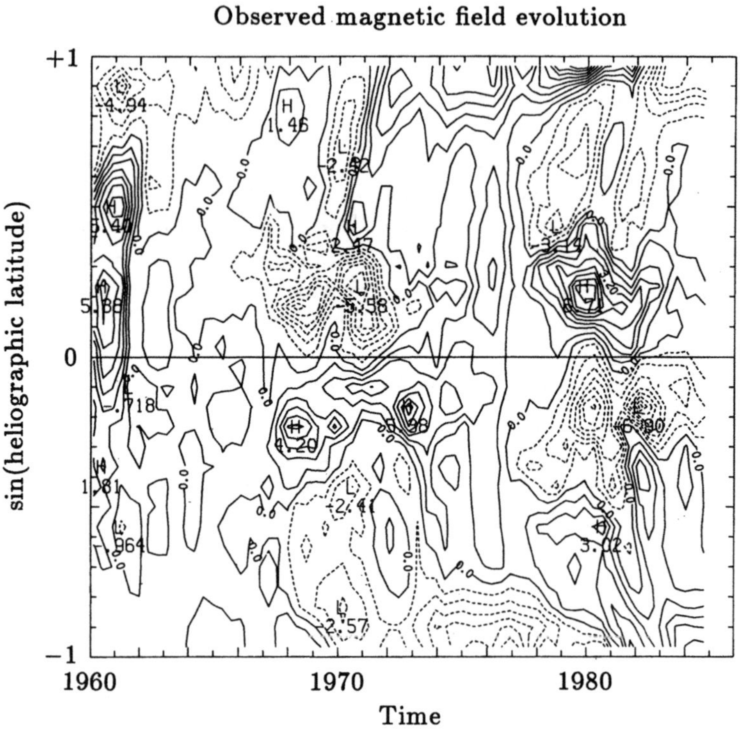

Figure 1. Isocontours of the axially symmetric component of the radial magnetic field, vs. time and sine of the latitude. The solid (dashed) contours represent fields of positive (negative) polarity. From Stenflo (1988).

3.2 Power spectrum analysis of the axisymmetric modes

We can explore whether the pattern of Fig. 1 can be explained in terms of globally resonant waves by making a harmonic decomposition of the pattern and a power spectrum analysis of the harmonic coefficients. The power spectrum $Pc_\ell^m(\nu)$ is given by

$$Pc_\ell^m(\nu) = \frac{1}{T_e} |\tilde{c}_\ell^m(\nu)|^2 , \tag{5}$$

where T_e is the effective length of the time series (taking apodization and data gaps into account), and

$$\tilde{c}_\ell^m(\nu) = \int_{-\infty}^{\infty} c_\ell^m(t)\, e^{-i2\pi\nu t}\, \mathrm{d}t \tag{6}$$

is the Fourier transform of the harmonic coefficients. The axisymmetric harmonic coefficients (with $m = 0$) are obtained from the Legendre transform of $\bar{B}$, as follows from Eqs. (2) and (4):

$$c_\ell^0(t) = \sqrt{\pi(2\ell+1)} \int_{-1}^{1} \bar{B}(x,t) P_\ell(x)\, \mathrm{d}x . \tag{7}$$

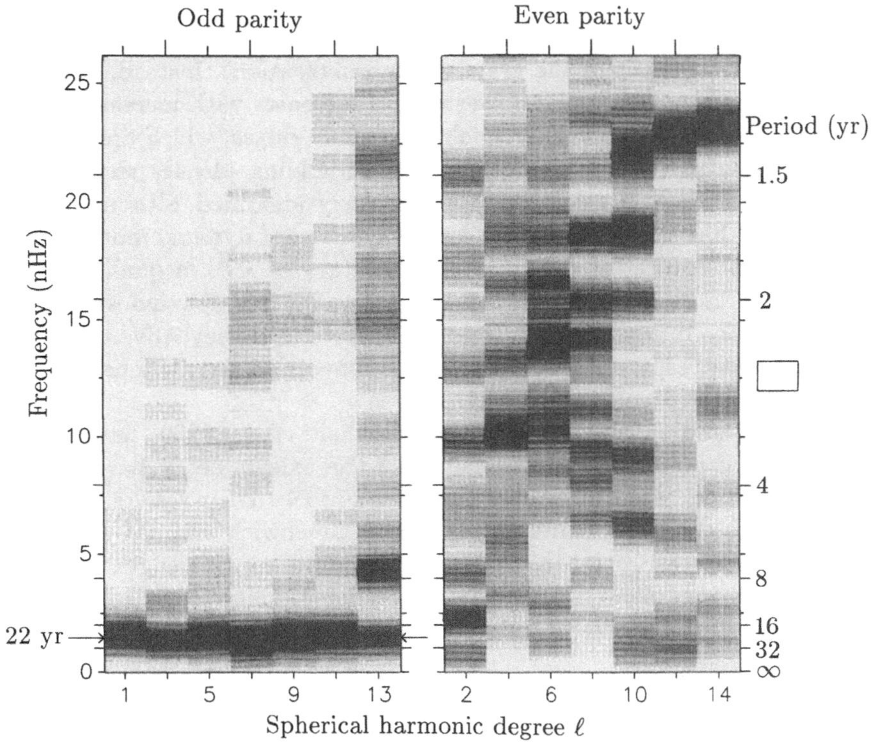

Figure 2. Power spectra of the axisymmetric harmonic coefficients of the radial magnetic field, as functions of frequency and ℓ. The white areas correspond to zero power, the darkest areas to the maximum spectral density for the given ℓ value. The resolution is indicated by the small box to the right of the diagram. From Stenflo (1988).

Power spectra $Pc_\ell^0(\nu)$ for the axisymmetric modes were first obtained by Stenflo and Vogel (1986). Figure 2 shows Pc_ℓ^0 for the present 26 year data set. Odd and even parity refer to odd and even values of ℓ, and correspond to patterns that are anti-symmetric and symmetric, respectively, with respect to reflections in the equatorial plane.

We see that the power is not distributed continuously over the diagram, but is strongly concentrated around one frequency for each value of ℓ. The power amplitudes of the even modes are 5–10 times smaller as compared with the odd modes, but they have been made visible by normalizing the gray scale in Fig. 2 independently for each ℓ value to the maximum power in the respective ℓ band. The low amplitudes of the even modes explain why the background noise is more prominent in the even-parity diagram as compared with the odd-parity one.

The most striking feature of Fig. 2 is the entirely different behaviour of the odd and even modes, giving evidence for the operation of a strict parity selection rule on the sun. For the odd modes the power is concentrated around the 22 yr resonance, with hardly any sign of a second harmonic at 11 yr. In contrast the even modes do not show any trace of the 22 yr resonance (the peak of the quadrupole $\ell = 2$ mode occurs at a period of 14 yr and is of very small amplitude — it appears prominent as a result of the gray-scale normalization). Instead, the frequency at which maximum even-mode power occurs increases with increasing ℓ value, in a way reminiscent of the helioseismology p-mode ridges, which are due to standing global waves in the sun's interior. The underlying physics responsible for this dispersion relation has not yet been definitely identified, although Hoyng (1987, 1988, 1990) has shown that stochastic excitation of dynamo modes would lead to a dispersion relation of the type $\omega \sim \sqrt{k}$ or $\nu \sim \sqrt{\ell}$, in qualitative agreement with Fig. 2. $\omega \sim \sqrt{k}$ is the dispersion relation for free dynamo waves in the $\alpha - \omega$ dynamo if one omits the diffusion terms. Note also that Alfvén waves along kG toroidal fluxropes in the lower part of the convection zone will have periods in the observed range (Stenflo and Vogel, 1986).

Gokhale and Javaraiah (1989) and Gokhale *et al.* (1990) have performed the same type of power spectrum analysis of the axisymmetric harmonic modes for the distribution of sunspots over the solar surface. Their results, which are based on the more than one hundred years long Greenwich sunspot data set, confirm and extend our results for the magnetic field pattern. The length of the Greenwich data set has also allowed them to explore the stability of the relative amplitudes and phases of the modes of different ℓ values when moving in time from cycle to cycle.

3.3 "Modal cleaning" of the butterfly diagram

The circumstance that the power spectrum diagram of Fig. 2 has the character of an emission-line spectrum with a weak and noisy continuous background suggests that the sun's magnetic field can to a good approximation be described as a superposition of discrete global modes, which may be the eigenmodes of an underlying linear wave equation.

Assuming that the observed pattern can be represented by one discrete global mode per ℓ value, the axisymmetric part of Eq. (1) can be written as

$$B(x,t) = \mathrm{Re} \sum_{\ell=1}^{N} a_\ell \, e^{i2\pi\nu_\ell t} \, P_\ell(x) , \tag{8}$$

where N is the maximum ℓ value that we have spatially resolved ($N = 14$ in our case), and a_ℓ are the complex mode amplitudes. Eq. (8) implies that we approximate the power spectrum of Fig. 2 as a superposition of δ-function peaks, one for each ℓ value.

If we factorize each complex amplitude in an absolute value and a phase factor, Eq. (8) can be written as

$$B(x,t) = \sum_{\ell=1}^{N} |a_\ell| \, \cos[2\pi\nu_\ell(t - t_\ell)] \, P_\ell(x) , \tag{9}$$

where we have introduced a time lag t_ℓ related to the phase ϕ_ℓ by

$$\phi_\ell = -2\pi\nu_\ell t_\ell . \tag{10}$$

Equation (9) contains three unknowns for each ℓ value: the absolute amplitude $|a_\ell|$, the time lag t_ℓ, and the frequency ν_ℓ. The frequency can be fixed by choosing the value at which the power spectrum maximum occurs for each ℓ value in Fig. 2. The two remaining free parameters for each of the 14 modes can be determined by an iterative least squares fit to the "butterfly diagram" of Fig. 1 (Stenflo, 1988). This gives us the amplitudes and phases of the modes, and allows us to reconstruct "cleaned-up" butterfly diagrams through superposition of selected discrete modes, using Eq. (9).

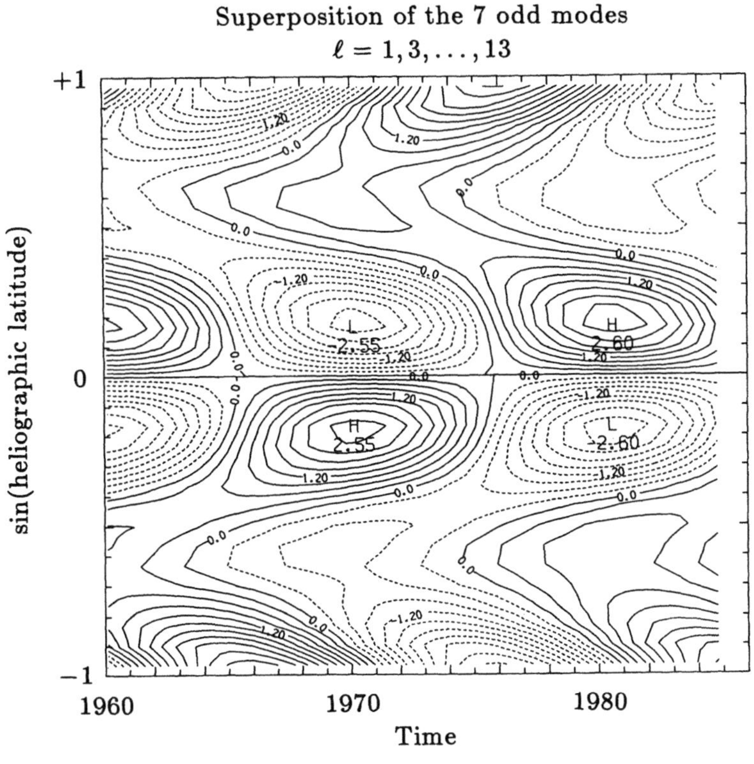

Figure 3. The axisymmetric component of the radial magnetic field described as a superposition of 7 discrete harmonic modes of odd parity, each varying sinusoidally with a 22 yr period. From Stenflo (1988).

As the odd modes are the dominating ones and are all represented by one and the same frequency, $(22\,\text{yr})^{-1}$, it is particularly useful to reconstruct the "modally cleaned" butterfly diagram as a superposition of the 7 odd modes alone. This has been done in Fig. 3, which nicely brings out the equatorward and poleward migrations of the anti-symmetric pattern, as well as the 11 yr polarity reversals. This is the pattern that dynamo models of the sun need to reproduce.

Note that each linearly superposed mode is strictly sinusoidal and therefore time symmetric, whereas the composite diagram of Fig. 3 exhibits a clear arrow of time as a consequence of the definite sense of latitude migration of the pattern. As all the superposed sinusoidal modes have the same frequency, the time direction is exclusively determined by the systematic, relative phase relations between the modes of different ℓ values, as will be seen below.

3.4 Amplitudes and phases, and their use for dynamo inversions

The amplitudes and phases determined by the iterative least squares fit to the pattern of Fig. 1 are given in Fig. 4. The phases in the diagram to the lower right have been obtained from the time lags via Eq. (10), adding multiples of 2π to some of the ϕ_ℓ values to bring out the systematic shift of the phases with increasing ℓ. It is this systematic shift that establishes the arrow of time in Fig. 3.

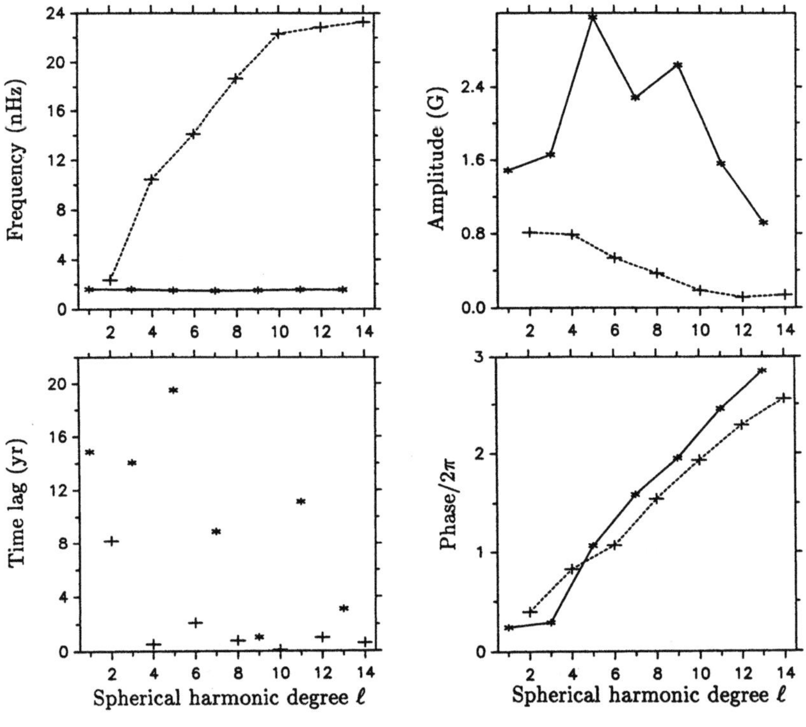

Figure 4. Frequencies, amplitudes, and phases of the odd (stars and solid lines) and even (pluses and dashed lines) modes. Adapted from Stenflo (1988).

We notice in Fig. 4 that the maximum mode amplitude does not occur for the dipole ($\ell = 1$) mode, as is often assumed, but for the mode with $\ell = 5$. The even-mode amplitudes are smaller than the odd-mode ones by typically a factor of five, and the amplitude declines monotonically with ℓ.

The amplitudes and phases of the odd modes represent 14 observables that may be used as observed boundary constraints for inversions of the dynamo problem. Let us give an example of how such an inversion can be done. The main parameters governing the operation of the solar dynamo are the helicity, the turbulent diffusivity, and the angular velocity gradient. As the angular velocity of rotation can be obtained from helioseismology, it need not be regarded as an unknown parameter for the dynamo problem. What remains is therefore the depth and latitude variation of the helicity and diffusivity. If this variation is parameterized in terms of a sufficiently small number of free parameters, then an inversion using iterative least squares fitting of the observables will give us the values of the free model parameters with error bars.

This type of inversion approach has proven very useful even for highly non-linear problems, e.g. the inversion of Stokes line profiles to derive the temperature stratification in the interior of small-scale magnetic fluxtubes (Keller *et al.*, 1990). It is important to choose the free parameters of the model carefully to make the problem numerically well conditioned, and test for the uniqueness of the solution by trying iterations with widely different starting values. The goodness of the fit is given by the resulting value of χ^2, which tells us how well the observables can be reproduced within the general framework of the chosen theoretical model.

4. The non-axisymmetric modes

4.1 Effect of differential rotation

As we cannot observe the sun simultaneously from all sides and thus not sample all longitudes simultaneously, the longitude and time coordinates are not independent of each other, but are related according to Eq. (3). This relation is somewhat arbitrary, since it depends on the choice of a rigidly rotating coordinate system, and we know that the sun in fact rotates differentially (and there is even a coexistence of different rotation laws, as we will see in Sect. 5). In contrast to the axisymmetric modes, for which the longitude information is simply averaged away, differential rotation has a profound effect on the non-axisymmetric modes (with $m \neq 0$).

To obtain information on possible periodicities and resonant behaviour of the non-axisymmetric modes we need to derive the Fourier transform $\tilde{c}_\ell^m(\nu)$ of the time series of the harmonic coefficients. These coefficients are formally determined by Eq. (2), which can be expressed more explicitly, using the associated Legendre polynomials $P_\ell^m(x)$, as

$$c_\ell^m(t) = f_\ell^m \int_{-1}^{1} \mathrm{d}x \; P_\ell^m(x) \int_{-\pi}^{\pi} \mathrm{d}\varphi \; B(x,\varphi,t)\, e^{-im\varphi}\,, \tag{11}$$

where

$$f_\ell^m = (-1)^m \sqrt{\frac{2\ell+1}{4\pi}\,\frac{(\ell-m)!}{(\ell+m)!}}\;. \tag{12}$$

The second integral in Eq. (11) constitutes a Fourier transform over a circular longitude interval, while the first integral makes a Legendre transform of the second integral.

Due to the coupling between time and longitude, the longitude interval of 2π used in Eq. (12) can be regarded as a time interval of length P_C, the Carrington period. We can thus replace the Fourier transform in longitude by a Fourier transform over an infinite time interval, if we also introduce a time window function w_C of width P_C, with a value of unity inside the window, zero outside. Accordingly

$$c_\ell^m(t) = 2\pi\nu_C \int_{-\infty}^{\infty} D_\ell^m(t')\, w_C(t'-t)\, e^{i2\pi m\nu_C t'}\, \mathrm{d}t' \,, \tag{13}$$

where

$$D_\ell^m(t) = f_\ell^m \int_{-1}^{1} B(x,t) P_\ell^m(x)\, \mathrm{d}x \tag{14}$$

is the Legendre transform of the radial magnetic field for a given longitude and time t.

The needed Fourier transform, defined by Eq. (6), of the time series described by Eq. (13), then becomes

$$\tilde{c}_\ell^m(\nu) = 2\pi \tilde{D}_\ell^m(\nu - m\nu_C)\, \mathrm{sinc}(\nu/\nu_C) \,, \tag{15}$$

where the sinc function, defined by

$$\mathrm{sinc}(\nu/\nu_C) = \frac{\sin(\pi\nu/\nu_C)}{\pi\nu/\nu_C} \,, \tag{16}$$

arises from Fourier transforming the window function.

We can now see how differential rotation affects the frequency spectrum of the non-axisymmetric modes. The Fourier transform $\tilde{D}_\ell^m(\nu)$ of the Legendre transform time series $D_\ell^m(t)$ is dominated by the effect of the pattern recurrence after an integer number of rotation periods. This recurrence causes "rotation peaks" in the frequency spectrum to occur at frequencies $\nu = m\nu_R$, where ν_R is the rotation frequency. ν_R lies in the neighbourhood of ν_C, but it varies with latitude, due to the sun's differential rotation. The Legendre transform superposes the contributions from different latitudes with different weights. Therefore, instead of ideally having δ-function peaks at $m\nu_R$, we get a distribution or "forest" of subpeaks around $m\nu_C$.

When forming the Fourier transform $\tilde{c}_\ell^m(\nu)$ of the harmonic coefficients, the "rotational forest" of the given m value is translated down to the region around zero frequency, as seen by Eq. (15). The sinc function in Eq. (15) serves as a fixed low-pass filter, which transmits the low frequencies while suppressing the rotational peaks of the other, not wanted m values. The problem is that the rotational distribution around zero frequency covers the whole frequency range where evolutionary effects of the non-axisymmetric modes are expected. We need to somehow dig out the subtle evolutionary information from the "rotational jungle". This is not possible by direct inspection of the power spectra as was done for the axisymmetric modes, but requires the procedure that will be described next.

4.2 Extraction of the "evolutionary splitting"

Evolutionary periodicities or resonances will cause "evolutionary splitting" of the rotational peaks. The systematic splitting pattern of each peak can be extracted by making an autocorrelation analysis of the Fourier transform of the harmonic coefficients. The complex autocorrelation function is defined by

$$A_\ell^m(\tau) = \langle\, \tilde{c}_\ell^m(\nu)\, \tilde{c}_\ell^{m*}(\nu+\tau) \,\rangle \,/\, \langle\, |\tilde{c}_\ell^m(\nu)|^2 \,\rangle\,, \qquad (17)$$

where τ is the frequency lag.

An evolutionary resonance, like the Hale frequency $\nu_H = (22\,\mathrm{yr})^{-1}$, gives rise to doublets with a total peak separation of $2\nu_H$, which results in an autocorrelation peak at $\tau = 2\nu_H$. Due to this frequency doubling we will in the following diagrams give the autocorrelation results vs. $\tau/2$ instead of vs. τ, to allow us to read off the relevant frequencies directly.

The spatial resolution of our data set allows us to compute the harmonic coefficients for m values from zero through 17 without aliasing. In Fig. 5 we have averaged the complex autocorrelation functions for all the 17 m values $1 \le m \le 17$ for each given, fixed value of $\ell - m$. The moduli of these averaged autocorrelation functions have then been plotted in a grey-scale representation (with darker regions representing higher autocorrelation) as functions of $\tau/2$. As odd and even values of $\ell - m$ correspond to patterns anti-symmetric and symmetric with respect to reflections in the equatorial plane, respectively, Fig. 5 allows us to look for any possible breaking of parity symmetry, which was such a prominent feature in the case of the axisymmetric modes. We also recall that $\ell - m + 1$ represents the number of zones of the associated Legendre polynomial P_ℓ^m, so $\ell - m$ can also be regarded as representative of a zonal number. The global ($\ell - m$-independent) grey scale cuts used in Fig. 5 will be defined in Fig. 6 below.

In contrast to the axisymmetric modes, the non-axisymmetric modes in Fig. 5 do not show any sign of parity dependence, and seem to have no $\ell - m$-dependent dispersion relation. Allowing for the noise fluctuations, the diagrams of Fig. 5 are instead characterized by horizontal bands, indicating a resonant but $\ell - m$-independent evolution of the non-axisymmetric modes. The 22 yr resonance dominates, but the grey-scale cuts have been set low in Fig. 5 to bring out the other low-amplitude resonances. We see for instance one band with a period of 4–5 yr, with the next band occurring around the second harmonic of this frequency.

As there seems to be no ℓ and m dependence of the resonant pattern, we may average the complex autocorrelation functions for *all* the spatially resolved ℓ and m modes to better bring out the low-amplitude resonances from the background noise. This has been done separately for odd and even parity ($\ell - m$) in Fig. 6. Thus 119 ($= 7 \times 17$) autocorrelation functions contribute to the odd mode curve, 136 ($= 8 \times 17$) functions to the even mode curve. The right diagram of Fig. 6 gives a close-up look at the lower-frequency region of the left diagram.

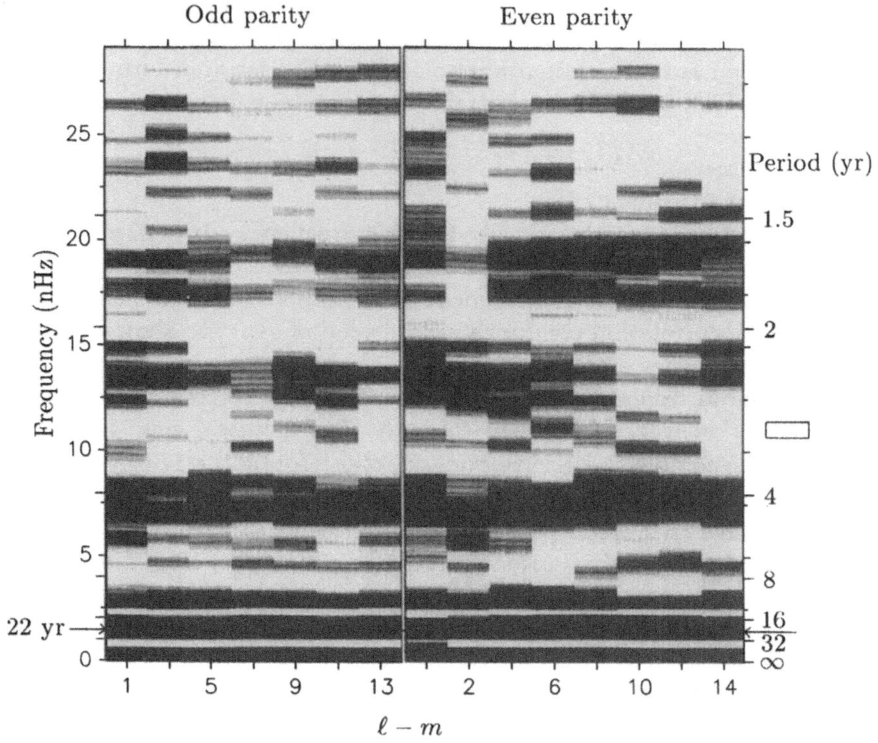

Figure 5. Autocorrelation amplitudes of the non-axisymmetric modes, as functions of frequency ($\tau/2$) and zonal number. The results are given in separate diagrams for odd and even $\ell - m$, and contain the averaged contributions from all the m values from 1 through 17. Global grey-scale cuts, independent of the value of $\ell - m$, have been used (as defined in Fig. 6). From Stenflo and Güdel (1988).

4.3 Interpretation of the non-axisymmetric modes

The most striking feature of Fig. 6 is the almost perfect agreement between the independently derived curves representing odd and even parity. The amplitudes of the dominating 22 yr peak are for instance equal to within 2 %. Even the high-frequency, low-amplitude wiggles are extremely well matched between the two parities. This strongly suggests that practically all of these wiggles are real and of solar origin, and may therefore be subject to a serious analysis.

The horizontal band with its second harmonic seen in Fig. 5 appears in Fig. 6 with an apparent triplet structure, with a central frequency of 4.6 or 2.3 yr, surrounded by side peaks shifted by $\pm(22\,\mathrm{yr})^{-1}$. At higher frequencies the diagram gets very crowded with apparently overlapping peak patterns. This part of the diagram cannot be interpreted without a much more elaborate analysis, which has not yet been carried out.

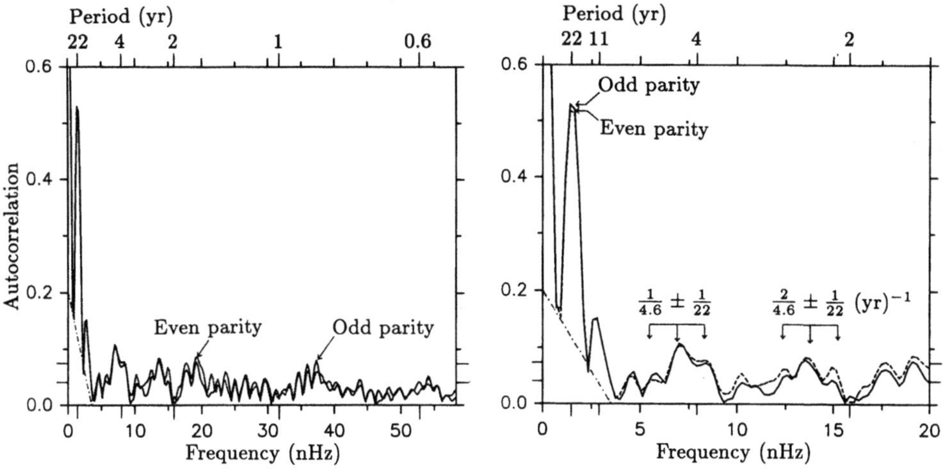

Figure 6. Autocorrelation amplitude as a function of frequency $\tau/2$, for odd (solid curve) and even (dashed curve) values of $\ell - m$. The curves represent averages of the contributions from all the ℓ and m values with the given parity. A possible triplet structure with its second harmonic is indicated in the right diagram. The cuts used for the grey scale of Fig. 5 are marked by the two longest tick-marks on the vertical axis. From the lower to the upper tick-mark the grey scale varies from white to black. Before applying the cuts, however, the lower, straight-line envelop marked by the dashed-dotted line in the lower left-hand corner has been subtracted, to bring the minima surrounding the 22 yr peak down to the range of the cuts. From Stenflo and Güdel (1988).

The following scenario may provide a general, qualitative framework for understanding the results for the non-axisymmetric modes. The dominant 22 yr mode represents the periodicity in the dynamo production of new toroidal flux inside the sun. This flux is made visible when part of it emerges to the surface from the production region. Flux emergence is triggered by non-axisymmetric deformations of the toroidal flux ropes. As these deformations may propagate along the closed toroidal flux as waves, they may develop a resonant structure through constructive and destructive interference. These interferences will then appear as periodicities in the emergence of new bipolar flux and get revealed by our analysis. The frequencies that we find then represent the eigenmodes of the underlying waves. Wave periods of the order of a few years are expected if the waves travel with the Alfvén velocity along kG fields at the bottom of the convection zone.

When a localized bipolar magnetic region is described by an expansion in terms of spherical harmonics, it will contribute to the harmonic modes of all ℓ and m values up to a spatial frequency corresponding to the inverse size of the bipolar region. Therefore the periodicities in the flux emergence will appear with the same frequencies for all ℓ and m values, which explains the horizontal band structure in Fig. 5. The available amount of flux is modulated by dynamo action with a 22 yr period, and this modulation combines with the various frequencies of flux emergence to produce the triplet splitting patterns seen in Fig. 6 (Stenflo and Güdel, 1988).

If this general scenario is adopted, then the various low-amplitude resonant frequencies may be used to provide diagnostic constraints on the distribution of magnetic flux and on the field strengths in the solar interior.

5. Pattern phase velocity and the internal angular velocity

5.1 Coexistence of two magnetic field rotation laws

The observed pattern of surface magnetic flux may be used as a tracer for determinations of the sun's differential rotation. Two different methods have been used: (1) The "longitude displacement method" (Snodgrass, 1983), and (2) the "pattern recurrence method" (Stenflo, 1989). Although the two methods have been applied to the same magnetograph data, and both measure the *phase* velocity of the magnetic pattern, they produce entirely different rotation laws.

The first method determines the proper motion in the longitude direction of the pattern by comparing consecutive full-disk magnetograms, separated in time by 1–4 days. The longitude displacement of the pattern is determined by cross-correlation of the flux pattern in the different magnetograms, without any identification of individual flux features. The second method uses only the flux sampled at the central meridian. For each latitude zone we get the field as a function of time, and the time series that has been used is 26 yr long (1959–1985), as described above in Sect. 2. Autocorrelation or power spectrum analysis can be directly applied to the 26 yr time series for each of the 30 latitude zones. The recurrence of the pattern at the central meridian after an integer number of solar rotations gives rise to equidistant peaks in the autocorrelation function as well as in the power spectrum. The peak locations determine the pattern phase velocity.

Figure 7 compares the resulting rotation laws. The four different types of symbols plotted refer to rotation rates determined from the center of gravity of each of the four first autocorrelation peaks (with lags from one to four months). They show that the four peaks give consistently very similar rotation rates. The synodic rotation period reaches a maximum of about 29.5 days around a latitude of about 55°, with a tendency for a small polar spin-up. In the case of the Snodgrass law (solid curve), however, the rotation period increases steeply with latitude, reaching a value of about 38 days at the poles. The Snodgrass law on the other hand closely agrees with the rotation law determined from Doppler measurements (Howard *et al.*, 1983), represented by the dashed-dott ed curve.

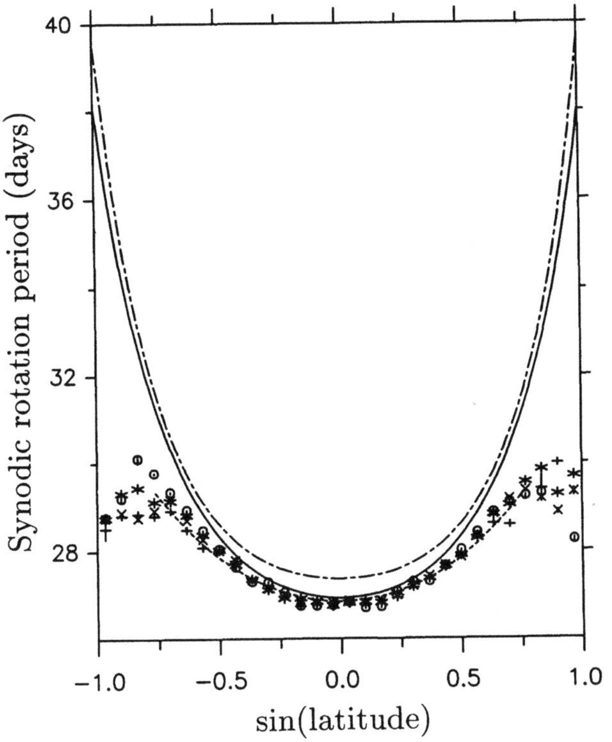

Figure 7. Synodic period of rotation of the magnetic field pattern. The four different symbols represent the rotation rate determined with the recurrence method, using lags of one (open circles), two (stars), three (pluses), and four (crosses) rotation periods. The solid curve represents the rotation law of Snodgrass (1983), while the dashed-dotted curve has been obtained from Doppler line shifts (Howard *et al.*, 1983). From Stenflo (1989).

5.2 "Active longitudes" and pattern regeneration

We note that the longitude displacement and recurrence methods both refer to the *phase* velocity of the magnetic pattern. Both use a rather coarse spatial grid (Snodgrass averages over about one arcmin square). The only essential difference between them is the time lag used in the correlation analysis, 1–4 days for the longitude displacement method, 27 days or more for the recurrence method.

The close agreement between the longitude displacement and the Doppler results shows that the pattern phase velocity determined from small correlation lags agrees with the *plasma* velocity in the photosphere. It has in the past been popular to invoke surface flux redistribution processes (like turbulent diffusion and meridional circulation) to explain how the phase velocity will deviate with time more and more from the plasma velocity, such that a quasi-rigid pattern rotation is approached (Sheeley *et al.*, 1987). Such processes can however be ruled out as an explanation of the two coexisting rotation laws, even if the time scale for such a redistribution would be arbitrarily short, e.g. less than a month (Stenflo, 1989). The reason is the following:

In the redistribution models (e.g. Sheeley *et al.*, 1987) a quasi-rigid pattern phase velocity develops due to reshuffling of existing, old surface flux. To clarify the consequences of such models, let us assume that the flux pattern initially rotates with the plasma velocity but within some short time, say a month, develops a quasi-rigid phase velocity. If then the longitude displacement method of Snodgrass is applied, this quasi-rigid pattern phase velocity will be picked up, in strong contradiction with what is being observed for the real sun. Once a quasi-rigid pattern phase velocity has been established by reshuffling of the old flux, assuming no significant contributions from new emerging flux, then correlation analysis in the limit of small lags, like the longitude displacement method, will find the quasi-rigid phase velocity, *not* the plasma velocity. In contrast Snodgrass (1983) finds a steep differential rotation law that agrees with the plasma rotation. This law is moreover found to be time invariant, applying to the whole 15 yr period that he studied.

The only process that has been found to account for the coexistence of the two pattern phase velocities is pattern renewal by flux emergence from the solar interior over a time scale of less than 27 days, but more than a few days. In this scenario the surface magnetic pattern observed at any given time represents new flux that has emerged within the past month, and does *not* consist of old fluxes that have their origin in redistribution of active-region fluxes. New flux that arrives at the surface will initially drift with the "local winds", and thus have a phase velocity that agrees with the plasma velocity, as found by the longitude displacement method. As the flux turnover time is less than a month, however, the recurrence method will not measure something related to the photospheric plasma velocity, but will correlate the regions of flux emergence, the "active longitudes", with each other. They rotate quasi-rigidly, not because of some redistribution processes, but because of the rotation properties of the source in the solar interior, from which the surface pattern is constantly being replenished.

This explanation revives and generalizes the old concept of "active longitudes" (cf. Sawyer, 1968; Bogart, 1982). The concept is generalized because it is used here not so much for active-region flux, but primarily for flux emerging far from the active-region belts, at high latitudes (where the discrepancy between the two rotation laws is the most pronounced). It may appear surprising that flux emergence would be so pervasive and occur at such an enormous rate. Video magnetograph observations of flux emergence and cancellation rates (Martin, 1990) however indicate that the turnover time scale for the flux pattern is really this extremely short. No other viable explanation of the two coexisting rotation laws has been proposed.

5.3 Location of the magnetic source region inside the sun

The recurrence method has also been used to explore if the quasi-rigid rotation law varies with the phase of the 11 yr cycle. Such variations could occur if for instance the depth distribution of the sources of magnetic flux varies with the cycle (and the angular velocity of rotation varies with depth), or if there are torsional oscillations in the source region. For this purpose the 26 yr time series have been divided into 21 time windows of length 16 Carrington rotations (≈ 1.2 yr), and the rotation rate has been determined by power spectrum analysis for each of the $21 \times 30 = 630$ time series.

The quasi-rigid rotation law determined by the recurrence method is thereby found to be time invariant within tight limits (Stenflo, 1990). Thus the random rms fluctuations are $\lesssim 0.3\,\%$ at low latitudes, increasing to 2–3 % at high latitudes (where the signal-to-noise ratio is smaller due to the smaller amplitudes of the line-of-sight magnetic flux there). Translated to linear velocities for a source region at the bottom of the convection zone these limits correspond to $\lesssim 4\,\mathrm{m\,s^{-1}}$ at low latitudes, $\lesssim 10$–$20\,\mathrm{m\,s^{-1}}$ at high latitudes. In contrast the Snodgrass rotation law deviates from the recurrence rotation law by as much as about 20 % near the poles.

These results speak in favour of flux storage at the bottom of the convection zone throughout the course of the solar cycle. If the storage depth would vary with the cycle, then any radial gradient of the angular velocity of rotation would show up as an apparent cycle-dependent variation of the phase velocity of the surface pattern, which is not observed. Now results from helioseismology (Morrow, 1988; Libbrecht, 1988; Brown *et al.*, 1989; Dziembowski *et al.*, 1989) suggest that the radia l gradient of the angular velocity is small throughout the bulk of the convection zone, but that there is a transition to a rigidly rotating core near the bottom of the convection zone. As however the bulk of the convection zone rotates nearly with the surface rate according to this picture, the flux storage region with its quasi-rigid rotation can hardly be located there. Thus the results from helioseismology when combined with the magnetic-field data support the conclusion that the magnetic flux is stored near the bottom of the convection zone. Flux storage at this location, which has been advocated in the past on theoretical grounds (e.g. Spiegel and Weiss, 1980; Spruit and van Ballegooijen, 1982; DeLuca, 1987; Schüssler, 1987), has thus been given empirical support.

5.4 Angular velocity structure of the solar interior

The apparent absence of time variations in the observed phase velocity of the source pattern is easier to understand if the phase velocity approximately agrees with the plasma velocity in the source region. Any deviation between the phase and plasma velocities would be expected to vary with the solar cycle, since the dynamo evolves with time, and the sources of flux migrate in latitude. The absence of observed variations suggests that the phase velocity closely follows the plasma velocity, which should be time invariant (as long as the kinetic energy densities at the bottom of the convection zone are much larger than the magnetic energy densities, which is the case if the field strengths are much less than megagauss).

It thus appears that the longitude displacement method of Snodgrass gives the rotation law for the surface layers (the photosphere), whereas the recurrence method gives the law for the source region, which has a likely location at the bottom of the convection zone. Our analysis of the magnetic-field data however has not provided us with any information on the angular velocity of the intervening layers of the convection zone. If we use the mathematically simplest procedure (without any physical justification) of linear interpolation between these two regions (top and bottom of the convection zone), then we obtain the isocontour plot of the angular-velocity structure of the sun given to the right in Fig. 8. It is compared with the rotational structure obtained from helioseismology (Libbrecht, 1988), given to the left in Fig. 8. Such a comparison has also been provided by Brandenburg (1990).

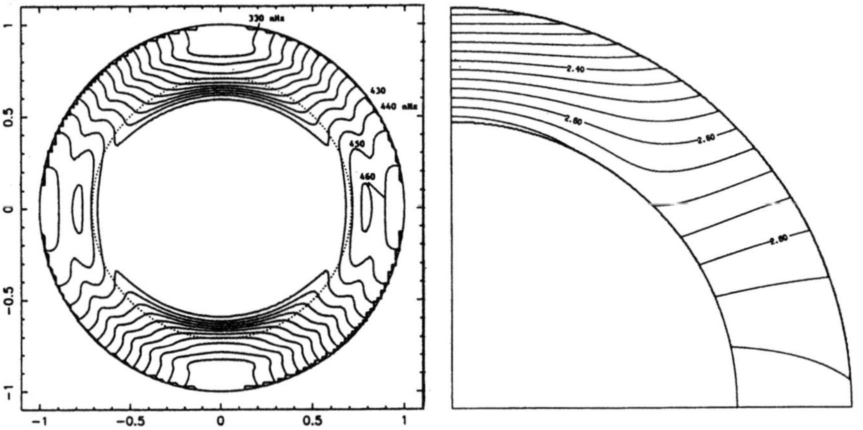

Figure 8. Isocontours of the sun's angular velocity of rotation, derived from helioseismology (left diagram, from Libbrecht (1988)) and from the observed pattern of surface magnetic fields (right diagram, from Stenflo (1989)). The two rotation laws for the pattern phase velocity have been taken to represent the top and bottom of the convection zone, with linear interpolation in between.

In spite of the simplistic use of linear interpolation between the top and bottom of the convection zone there are considerable similarities between the two diagrams of Fig. 8. Thus while the contours are predominantly radial at low latitudes, they become increasingly perpendicular to the sun's rotational axis at higher latitudes. This qualitative result is in direct contradiction with predictions from numerical simulations that the contours should be largely parallel to the rotational axis (Glatzmaier, 1987), but it agrees with other theories of the sun's differential rotation (e.g. Rüdiger and Tuominen, 1990).

Closer agreement with the helioseismology diagram would be obtained if we instead of linear interpolation between the top and bottom of the convection zone would let the surface rotation law prevail throughout most of the convection zone, and do our interpolation within a transition zone near the bottom of the convection zone. This implies that the isocontours in the right diagram of Fig. 8 would be strictly radial from the surface down to the top of the transition zone, and that the isocontour

structure within the transition zone would look like a compressed version of the present right-hand diagram.

The only significant disagreement with helioseismology occurs at low latitudes, in the active-region belts, where helioseismology finds an angular velocity at the bottom of the convection zone that is *slower* as compared with the surface rate, whereas our low-latitude rates instead are very similar to the surface rates, with a tendency of being slightly *faster* (although by less than one percent). The cause for this discrepancy has not been identified yet, but it may be that the scenario of rapid regeneration of the flux pattern over a time scale of less than a month

is less applicable at active region-latitudes, since many of the individual flux elements (e.g. sunspots) there are much larger than at high latitudes and have much longer lifetimes, which means that their old fluxes will dominate the pattern for much longer periods than the flux elements at high latitudes.

The regeneration scenario was introduced as the necessary and apparently only possibility for explaining the coexistence of highly different pattern phase velocities. The two rotation laws rapidly diverge from each other at high latitudes, but at low latitudes the difference is below one percent and not very significant. Therefore, if the low latitudes were considered in isolation, there would be practically only one single magnetic rotation law, and thus no problem requiring an explanation.

These questions may be clarified, on the one hand by direct observations of the flux regeneration rate as a function of latitude, on the other hand by more definite observations of the p-mode rotational splittings. The regeneration problem is particularly difficult, since most of the "background" magnetic flux far from active regions is expected to emerge and disappear over very small spatial scales, possibly beyond the resolution capabilities of currently used magnetographs. It is the accumulated effect of many such small-scale emergence and disappearance events that produces the large-scale flux pattern that we use for our correlation analyses. The key to an understanding of the large scales thus lies in the observation of the smallest scales, which requires powerful telescopes. The planned, international LEST (Large Earth-based Solar Telescope) has a design (Engvold and Andersen, 1990) that is optimized for solving this problem.

Acknowledgements. The Mount Wilson data used throughout this review has been recorded under the direction of Robert Howard and been subsequently organized in the 30×36 array synoptic format at the High Altitude Observatory. The Kitt Peak synoptic data set has been compiled and made available by J.W. Harvey.

References

Bogart, R.S.: 1982, *Solar Phys.* **76**, 155

Brandenburg, A.: 1990, Ph.D. Thesis, Observatory and Astrophysics Laboratory Rep. 2/1990, Univ. of Helsinki

Brown, T.M., Christensen-Dalsgaard, J., Dziembowski, W.A., Goode, P., Gough, D.O., Morrow, C.A.: 1989, *Astrophys. J.* **343**, 526

DeLuca, E.E.: 1987, Ph.D. Thesis, Univ. of Colorado (available as NCAR Cooperative Thesis No. 104)

Dziembowski, W.A., Goode, P.R., Libbrecht, K.G.: 1989, *Astrophys. J.* **337**, L53

Engvold, O., Andersen, T. (eds.): 1990, *LEST Design*, LEST Foundation

Glatzmaier, G.A.: 1987, in *The Internal Solar Angular Velocity*, eds. B.R. Durney, S.Sofia, *Astrophys. Space Sci. Library* **137**, 263

Gokhale, M.H., Javaraiah, J.: 1989, *Monthly Not. Royal Astron. Soc.*, in press

Gokhale, M.H., Javaraiah, J., Hiremath, K.M.: 1990, in *Solar Photosphere: Structure, Convection, and Magnetic Fields*, ed. J.O. Stenflo, *IAU Symp.* **138**, 375

Howard, R., Adkins, J.M., Boyden, J.E., Cragg, T.A., Gregory, T.S., LaBonte, B.J., Padilla, S.P., Webster, L.: 1983, *Solar Phys.* **83**, 321

Hoyng, P.: 1987, *Astron. Astrophys.* **171**, 357

Hoyng, P.: 1988, *Astrophys. J.* **332**, 857

Hoyng, P.: 1990, in *Solar Photosphere: Structure, Convection, and Magnetic Fields*, ed. J.O. Stenflo, *IAU Symp.* **138**, 359

Keller, C.U., Solanki, S.K., Steiner, O., Stenflo, J.O.: 1990, *Astron. Astrophys.* **233**, 583

Libbrecht, K.G.: 1988, in *Seismology of the Sun and Sun-Like Stars*, eds. V. Domingo, E.J. Rolfe, ESA SP-286, p. 131

Martin, S.F.: 1990, in *Solar Photosphere: Structure, Convection, and Magnetic Fields*, ed. J.O. Stenflo, *IAU Symp.* **138**, 129

Morrow, C.A.: 1988, Ph.D. Thesis, Univ. of Colorado (available as NCAR Cooperative Thesis No. 116)

Rüdiger, G., Tuominen, I.: 1990, in *Solar Photosphere: Structure, Convection, and Magnetic Fields*, ed. J.O. Stenflo, *IAU Symp.* **138**, 315

Sawyer, C.: 1968, *Ann. Rev. Astron. Astrophys.* **6**, 115

Schüssler, M.: 1987, in *The Internal Solar Angular Velocity*, eds. B.R. Durney, S. Sofia, *Astrophys. Space Sci. Library* **137**, 303

Sheeley, N.R., Jr., Nash, A.G., Wang, Y.-M.: 1987, *Astrophys. J.* **319**, 481

Snodgrass, H.: 1983, *Astrophys. J.* **270**, 288

Spiegel, E.A., Weiss, N.O.: 1980, *Nature* **287**, 616

Spruit, H.C., van Ballegooijen, A.A.: 1982, *Astron. Astrophys.* **106**, 58

Stenflo, J.O.: 1988, *Astrophys. Space Sci.* **144**, 321

Stenflo, J.O.: 1989, *Astron. Astrophys.* **210**, 403

Stenflo, J.O.: 1990, *Astron. Astrophys.* **233**, 220

Stenflo, J.O., Güdel, M.: 1988, *Astron. Astrophys.* **191**, 137

Stenflo, J.O., Vogel, M.: 1986, *Nature* **319**, 285

The Large-scale Magnetic Field in the Global Solar Cycle: Observational Aspects

V.I. Makarov[1] and K.R. Sivaraman[2]

[1]Kislovodsk Solar Station, Kislovodsk 357741, USSR
[2]Indian Institute of Astrophysics, Bangalore 560034, India

Abstract: The global solar cycle is considered as an interaction of 3 types of activity: at low-latitude (sunspots), at high-latitude (polar faculae) and the weak magnetic field. The properties of single and 3-fold reversals of the polar magnetic field are considered. The variation spectrum of the large-scale magnetic field of the Sun is analyzed in the range of 1–30 nHz. A dependence between the rate of a poleward meridional flow and phase of the global cycle is discussed.

1. Introduction

The studies of the magnetic field in polar faculae, their latitude distribution with the phase of the cycle (Sheeley, 1976, Makarov *et al.*, 1987, Wang *et al.*, 1989), of the ephemeral active regions and X-ray bright points (Martin and Harvey, 1979, Golub *et al.*, 1977), of the latitude distribution of the coronal brightness in the 5303Å line (Leroy and Noëns, 1983; Makarov *et al.*, 1987; Wilson *et al.*, 1988) and of the migration of the magnetic neutral lines (Makarov and Sivaraman, 1983, 1986, 1989a,b) show that a toroidal component of the magnetic field B_φ is present over all latitudes on the Sun. The synoptic view of the solar cycle (Makarov and Sivaraman, 1983, 1989a,b) shows that the epoch of the reversal of the polar magnetic field heralds the beginning of the global solar activity processes. Starting from this epoch the first type of magnetic activity, in the form of high-latitude B_φ, begins to show up at latitudes from 40° to 70°, as polar faculae migrate poleward. The second type of magnetic activity (sunspots) starts when activity of the first type is at its peak, and drifts equatorwards.

The third type of activity is connected with the dynamics of the radial component of a weak magnetic field. This type of activity shows up in dynamics of the zonal and sector structures of the solar field and also the interplanetary magnetic field (Makarov and Sivaraman, 1989b).

2. Interaction of three types of magnetic activity of the sun

Each type of activity is connected with one each other and has its own peculiarities. The entire magnetic flux on the Sun varies by a factor 2–3 from minimum to maximum of solar activity (Howard and La Bonte, 1981). This means that a latitude redistribution of the magnetic flux with the phase of the global cycle takes place. The cycle of the background magnetic field is connected with a sunspot cycle and the polar faculae cycle .

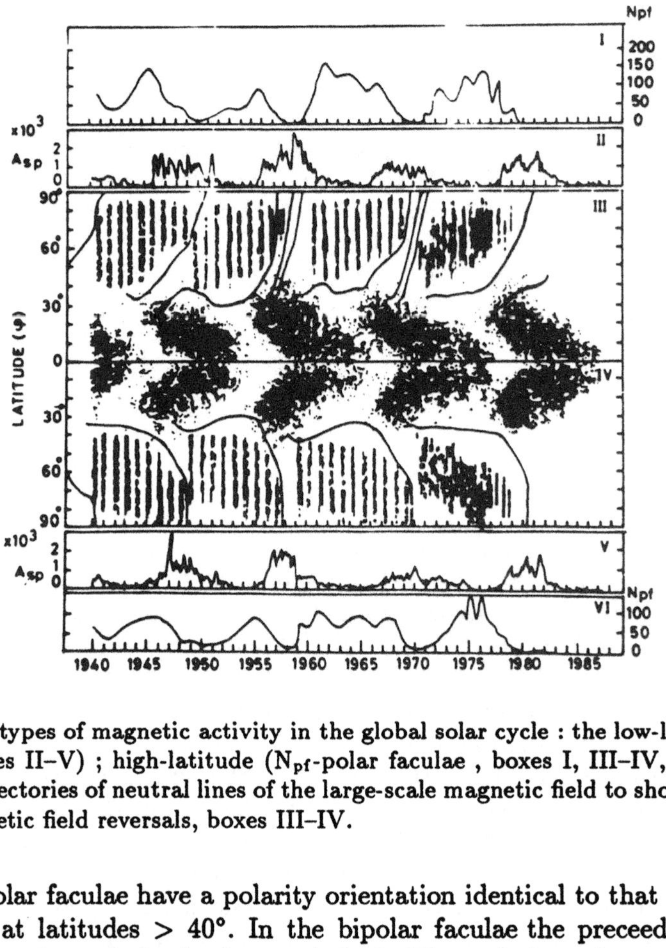

Fig. 1. Three types of magnetic activity in the global solar cycle : the low-latitude (A_{sp}-sunspots, boxes II–V) ; high-latitude (N_{pf}-polar faculae , boxes I, III–IV, VI) and the migration trajectories of neutral lines of the large-scale magnetic field to show the epochs of polar magnetic field reversals, boxes III–IV.

The unipolar faculae have a polarity orientation identical to that of the background field at latitudes > 40°. In the bipolar faculae the preceeding polarity is identical to that of the background field. The polarity orientation of those bipoles is opposite to that of the spots of the same cycle, but identical to the polarity orientation for bipolar spots of the next following cycle (Martin and Harvey, 1979; Makarov and Makarova, 1987; Wang *et al.*, 1989). In other words, while faculae of the (N+1) cycle make their appearance at high latitudes, activity of the preceding cycle (N) is still present at lower latitudes in the sunspot phase. This is similar to the behaviour of ephemeral active regions .

So, the new cycle shows up first as faculae at high latitudes and leads the sunspot phenomenon by 5–6 years. Each has a duration of 11 years, but occurs at separate latitudes, displaced from each other by 5–6 years within a 22 year magnetic cycle . The duration of this global cycle turns out to be 16–18 years (Wilson *et al.*, 1988, Makarov and Sivaraman, 1989b).

3. Single and three-fold field reversal cycles

During the last years considerable information has been derived both from the full disc magnetograms as well as the Hα synoptic charts (Makarov and Sivaraman, 1989a). On the Hα charts one can see that the polemost filament reaches the pole first and causes the reversal of the polar field. This is the picture when a single fold reversal takes place in both hemispheres. It was observed after the maximum activity in odd sunspot cycles Nos 11(1872.3), 13(1895.0), 15(1918.7), 17(1940.1), and 21(1981.8). There are instances when a three-fold reversal occurs in one of the hemispheres. In such cases all three filament bands travel to the respective poles one after the other and cause a three-fold reversal. This took place in the even cycles in the Southern hemisphere Nos. 12(1928.5), 18(1949.0), and 20(1970.6). This means that the 22-year solar magnetic cycles begin with a cycle in which a three-fold reversal is observed in one of the hemispheres. A single reversal takes place in the succeeding 11-year cycle . The phenomenon of the three-fold reversal in both hemispheres has not been observed during the last 115 years (see the paper by Benevolenskaya, these Proceedings).

One can note that the timescale of latitude zone evolution in a single field reversal is about 22 years, but in the case of a three-fold field reversal it is about 10–12 years. The velocity of the poleward motion is increased from 5 ms^{-1} to 40 ms^{-1} with a steep rise in the sunspot activity in the case of a three-fold field reversal.

4. Spectrum of solar large-scale magnetic field variations in the frequency range 1–30 nHz

From the Hα charts for the period 1910–1985 we have calculated the distribution of the dominant polarity area over every 10° latitude zone. Figure 2 shows the complete power spectra for all latitudes. Besides Hale's magnetic cycle, whose duration is 20 years there are additional quasi-periods of 7.0, 3.4, 2.5 and 1.7 yrs. The last period of 1.7 years, which is of greatest interest (Makarov *et al.*, 1985), was detected from quasi-periodic oscillations in the latitude zones from Hα charts.

The periods 3.4 and 7.0 yrs corresponds to the amplitude modulation of the principal period 1.7 yrs, which in the evolution of background magnetic fields corresponds to a time interval between the zones of alternating polarity of the magnetic field. It determines the quasi-period of the high-frequency component. This enabled us to show topologically that single and three-fold polarity reversals of the solar magnetic fields can result from an interaction of two types of

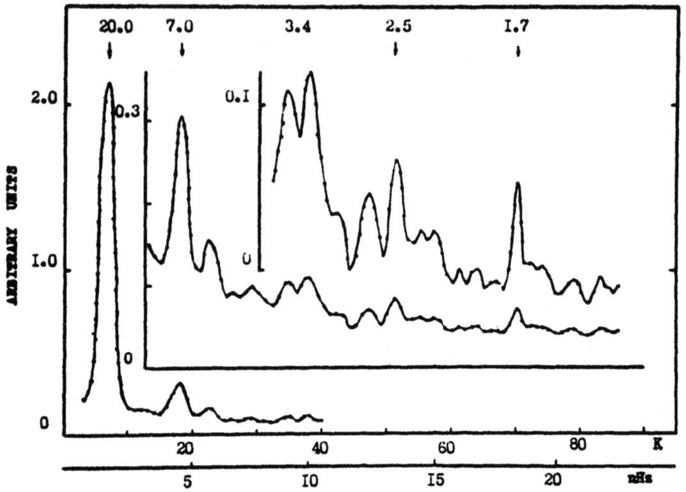

Fig. 2. The power spectra of solar large-scale magnetic field variations for the period 1910–1985.

magnetic field: a low-frequency component with period of the order of 20 yrs and a high frequency component, period of the order of 1.7 yrs (Belevolenskaya and Makarov, 1990).

5. On the correlation between the sunspot cycle and polar-faculae cycle at the current cycle 22

For the period 1940–1985 we have found a high degree of correlation (0.76 to 0.85) between the variations of monthly numbers of polar faculae N(t) and sunspot areas S(t) (Makarov *et al.*, 1989). The correlation was maximal for a value Δt ranging from 5.2 to 6.2 yrs. This time is the interval between the finish of the polar magnetic field reversal and the beginning of a new sunspot cycle . We have compared the sunspot activity with that of polar faculae when $\Delta t = 6.6$ yrs for current cycle 22, see Fig. 3. It is seen that the activity fluctuations in polar zones are repeated in 6 years in sunspot fluctuations. The most powerful activity burst may be expected at the end of 1991 or beginning of 1992. Regrettably the connection between the 1st and 2nd type of activity has not been explained so far.

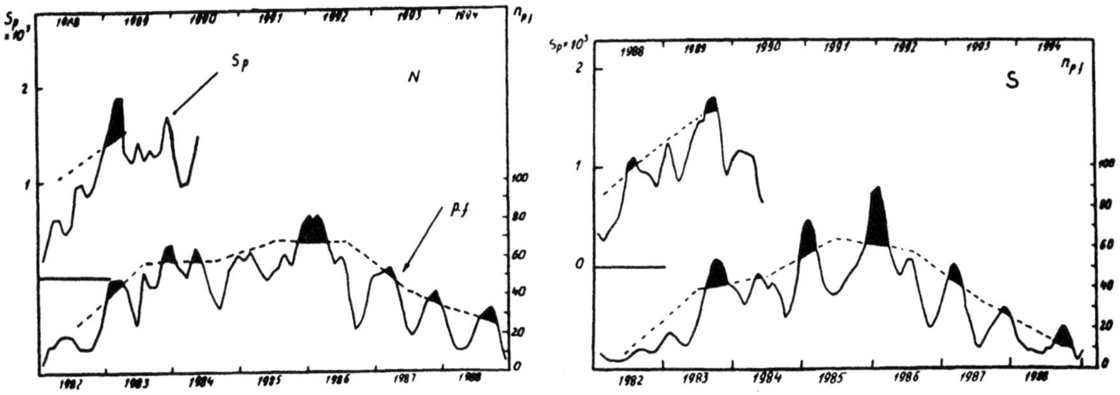

Fig. 3. Relation of the polar faculae number N_{pf} to the sunspot area cycle S_p in the current 22nd global solar cycle from the Kislovodsk solar station data.

References

Belevonkaya, E.E. and Makarov, V.I.: 1990, *Soln. Dannye* No 5, 75

Golub, L.K., Krieger, A.S., Harvey, J.W. and Vaiana, J.S.: 1977, *Solar Phys.* **53**, 111

Howard R. and LaBonte, B.J.: 1981, *Solar Phys.* **74**, 131

Leroy, J.L. and Noëns, J.C.: 1983, *Astron. Ap.* **120**, L1

Makarov, V.I. and Sivaraman K.R.: 1983, *Solar Phys.* **85**, 227

Makarov, V.I., Tavastsherna, K.S. and Petrova, N.S.: 1985, *Soln. Dannye* No 6, 69

Makarov, V.I. and Sivaraman K.R.: 1986, *Bull. Astron. Soc. India* **14**, 163

Makarov, V.I., Makarova V.V. and Sivaraman K.R.: 1987, *Soln. Dannye* No 4, 62

Makarov, V.I., Leroy, J.C. and Noëns, J.C.: 1987, *Astron. Zh.* **64**, 1072

Makarov, V.I. and Makarova, V.V.: 1987, *Soln. Dannye* No 3, 62

Makarov, V.I., Makarova, V.V. and Sivaraman K.R.: 1989, *Solar Phys.* **119**, 45

Makarov, V.I. and Sivaraman K.R.: 1989a, *Solar Phys.* **119**, 35

Makarov, V.I. and Sivaraman K.R.: 1989b, *Solar Phys.* **123**, 367

Martin, S.F. and Harvey, K.L.: 1979, *Solar Phys.* **64**, 93

Sheeley, N.R.,Jr.: 1976, *J. Geophys. Res.* **81**, 3462

Wang, Y.M., Nash, A.G. and Sheeley, N.R.,Jr.: 1989, *Ap. J.* **347**, 529

Wilson, P.R., Altrock, R.C., Harvey, K.L., Martin, S.F. and Snodgrass, H.B.: 1988, *Nature* **333**, No 6175, 748

Evolution of Large and Small Scale Magnetic Fields in the Sun

Peter A. Fox, Michael L. Theobald and Sabatino Sofia

Center for Solar and Space Research, Yale University, P.O. Box 6666, New Haven, CT 06511 USA

Abstract: This paper will discuss issues relating to the detailed numerical simulation of solar magnetic fields, those on the small scale which are directly observable on the surface, and those on larger scales whose properties must be deduced indirectly from phenomena such as the sunspot cycle. Results of simulations using the ADISM technique will be presented to demonstrate the importance of the treatment of Alfvén waves, the boundary conditions, and the statistical evolution of small scale convection with magnetic fields. To study the large scale fields and their time dependence, the magnetic resistivity plays an important role; its use will be discussed in the paper.

1. Introduction

In attempting to understand the magnetic activity on stars, it is inevitable that we begin by searching for an understanding of the Sun. In turn, the Sun presents us with a very complex array of features related to magnetic activity. From a theoretical viewpoint, these features separate naturally into a large scale component and a small scale component. Both components are important in their own right, but the key to a realistic understanding of magnetic activity almost certainly lies in the interaction between the two scales.

The evolution of the large scale solar magnetic field divides into two timescales: that of the solar cycle, and one of a much longer (and presently unknown) timescale. Current theoretical models of the large scale field are necessarily simplified, often drastically, in order to make reasonable progress. On the other hand the small scale field, which can be constrained by observations, can be modeled in some detail. In particular, we can do both qualitative and quantitative studies in an attempt to understand how the magnetic field and the surrounding flow interact.

An understanding of both scales, even perhaps an incomplete one, will be important in analyzing the components of the solar dynamo process. Current dynamic models do not give a satisfactory explanation of even the simplest solar cycle features for a number of reasons (see Chan, these Proceedings). Unfortunately, since the kinematic models lack predictive capabilities, further investigation is required. This paper will discuss some limited aspects of the large scale field but will mainly discuss the small scale fields.

2. Large scale magnetic fields

Because of the tremendous uncertainty in the interior structure of the solar magnetic field, much closer attention should be given to calculating the evolutionary aspects of those fields. Since the understanding of angular momentum transport in stars is increasing rapidly (Endal and Sofia, 1976; Pinsonneault *et al.*, 1989), the possibility of constraining the magnetic component becomes a reasonable next step. There is strong evidence (observational and implied theoretical) for the existence of internal differential rotation and large scale magnetic fields in solar-type stars.

The implications of certain magnetic field structures, for example at the base of the solar convection zone, are crucial to popular scenarios for the solar dynamo. In addition, it is important to understand what magnetic flux (i.e. in the mean sense) the solar convection zone may have had during its evolution which can consequently be compared to solar–type stars.

One constraint on interior magnetic fields that is yet to be fully developed is helioseismology. Except for very high field strengths (of order MGauss) the direct impact on oscillation mode frequencies is small. However changes in the solar structure (which may be position dependent) due to magnetic fields are likely to produce measurable changes in the oscillation frequencies. In addition there are, as yet still only suggestive, variations of oscillation frequencies (at particular wavelengths) over the solar cycle (Libbrecht and Woodard, 1990).

3. Small scale magnetic fields

Studies of the small scale interaction of convection and magnetic fields require the solution of the fully compressible form of the Navier–Stokes equations coupled with Maxwell's equations (see, for example, Priest, 1982). We will not attempt to review the present state of knowledge in this area but suffice to say that rapid progress has been made in the last few years.

The numerical solution of the governing equations can be thought of as a two stage procedure, one of a qualitative nature using calculations that have two spatial dimensions (with or without axisymmetry), and the other of a quantitative nature that includes all three dimensions. Both approaches are useful but we will make a few comments about the former in this paper. In particular, we know from existing work that many magnetic field related features are evident in two dimensional calculations (compared to three dimensions) even though some details are wrong (e.g. reconnection).

It is important to have a good qualitative understanding of the physical conditions, as well as of the limitations of the numerical and physical approximations, before progressing to more complex models. Part of this procedure should lead to better approximations or scaling relationships when investigating larger scale fields (and flows).

The results presented here are from 2D simulations in cartesian geometry using the ADISM (Chan and Wolff, 1982) method. More details on these and related calculations can be found in Fox *et al.* (1990). The usual features such as expulsion of magnetic flux, concentration of flux which is stronger in downflows and weaker, but present, in upflows (depends on the magnetic resistivity, η which we will emphasize shortly), and magnetic reconnection are all evident in the simulations. In addition, beyond a threshold magnetic flux (here measured as 5% of the surface gas pressure), there is a time dependence manifested as a continuous transformation of weak (but not necessarily uniformly distributed) fields into small regions of intense fields (which implies the ubiquity of fibril-type fields) which in turn decay, etc.

These small scale interactions show "complex" exchange between kinetic and magnetic energies which, when analyzed, should give direct information on wave interaction and perhaps indirect information on dynamos if the features actually can be extended to larger scales.

One quantity which is very uncertain in simulations of magnetohydrodynamics is the magnetic resistivity. It limits the concentration of magnetic flux, smooths the fibril field, influences the decay of large scale fields, and most importantly limits the time steps in numerical schemes, usually to satisfy some numerical stability criterion. Because of the important role that the resistivity plays, we would like to encourage detailed investigations to determine the answers to a number of questions: Is there turbulent enhancement of resistivity? What is its magnitude and form? Is the Alfvén hypothesis correct, i.e. is the dissipation of magnetic and kinetic energy the same on small scales?

Figure 1 shows the effect of changing the magnetic resistivity in a 2-D simulation. The vector potential, which in this case gives the lines of magnetic force, is overlayed with the velocity vector field. There are obvious differences between the field structures that develop.

As the strength of the magnetic field is increased there is strong evidence of overstable oscillations and a preference for convection cells of small aspect ratio (width to depth) to develop. This is not a new result but there is a subtle dependence on other quantities, such as magnetic resistivity and boundary conditions, that also needs further investigation (see Fox *et al.*, 1990).

Another feature of these simulations is the presence of waves, i.e Alfvén, sound and MAG. Figure 2 shows the magnetic energy spectra for two simulations. The low frequency peaks are magnetic and the high frequency are from sound wave exchange.

One important result for this type of modeling is that the initial distribution of magnetic field does not determine the final configuration once the simulation has been allowed to "relax" (see Chan and Sofia, 1986). The main external constraint on the magnetic field arises from the boundary conditions. Unfortunately this is a very difficult problem to treat accurately and we are left with compatibility relations that are only partially realistic, for example, imposed radial field component at the surface (but non-zero horizontal component).

The statistical relaxation time of simulations of compressible convection depends on the mass within a layer and the number of pressure scale heights. So far no attempts have been made to determine the influence of magnetic fields on this process, except that for small magnetic fields there should be no difference. Preliminary indications are that the relaxation time can be shortened by strong magnetic fields, essentially because of a combination of the Poynting flux carrying a fraction of the flux, and the smaller aspect ratio cells mentioned previously allow more rapid communication between the lower and upper layers. Unfortunately we have not yet fully investigated the influence of a large number of pressure scale heights, and so some of our conclusions are preliminary.

4. Connecting the two scales

There are a number of important physical features that occur in large scale magnetic fields, or are implied by our primitive understanding of the solar dynamo. One of these is to determine the region of influence of the solar dynamo, i.e. how important is it to know about the sub-convection zone field? What interaction occurs at the boundary and do we have to worry about generating a cyclic field there? In addition, can we evaluate the concept of the α-effect, either on small scales or large scales using detailed models?

5. Summary

In this paper we have attempted to promote the use of numerical simulations as a tool, not only for modeling surface features, but for determining more global properties associated with the interaction of convection and magnetic fields. Since there is *less* adjustability (in terms of parameters) in current numerical models, compared to analytical or simplified numerical approximations, we should be able to identify structures and effects that can be modeled on the larger scales and also determine which features are unimportant and neglect them. In essence we view this approach as the development of a numerical "mean field" theory.

We consider the statistical evolution of magnetic fields, the influence of boundary conditions, and the role of magnetic resistivity in determining magnetic field structures to be crucial to the understanding of the solar dynamo. Finally, it is essential that we develop ideas about the importance of "waves" (sound, Alfvén, MAG, etc.) as a contributor to both large scale fields and the solar dynamo. These

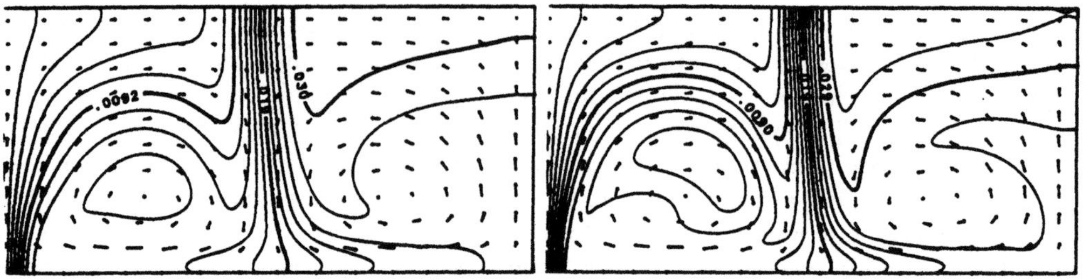

Fig. 1. Comparison of two simulations of two dimensional compressible convection interacting with magnetic fields. Using the same initial magnetic flux but different magnetic resistivities (factor of two greater in the left panel). Only a subset of the domain is shown.

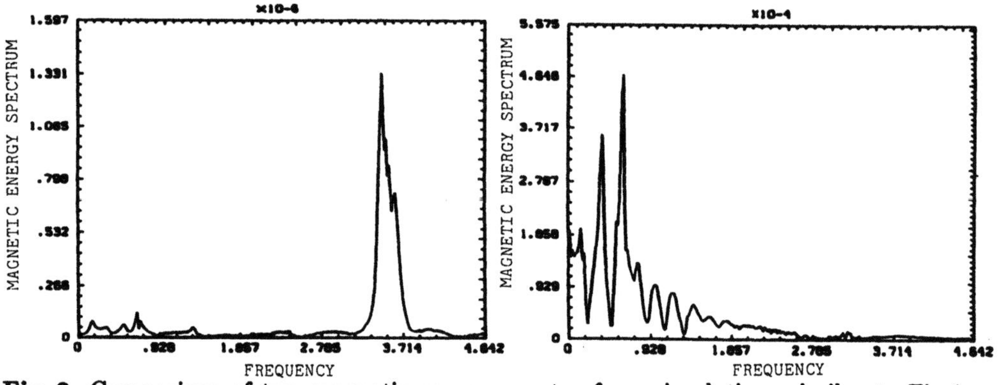

Fig. 2. Comparison of two magnetic energy spectra from simulations similar to Fig.1. The left panel is for weak initial field and the right is for strong field, using the same magnetic magnetic resistivities.

waves presently constitute a significant barrier to numerical simulations on solar cycle timescales and so a "non-wave" approximation would be extremely useful.
This work is supported by grants from NASA (NAGW-777) and the USAF (AFOSR-88-0054).

References

Chan, K. L. , Sofia, S.: 1986, *Astrophys. J.* **307**, 222
Chan, K. L. , Wolff, C. L.: 1982, *J. Comp. Phys.* **47**, 109
Endal, A. S. , Sofia, S.: 1976, *Astrophys. J.* **210**, 184
Fox, P. A., Theobald, M. L. , Sofia, S.: 1990, submitted to *Astrophys. J.*
Libbrecht, K. G. , Woodard, M. F.: 1990, *Nature* **345**, 779
Pinsonneault, M. H. *et al.*: 1989, *Astrophys. J.* **338**, 424
Priest, E. R.: 1982, *Solar Magnetohydrodynamics*, (Reidel, Dordrecht)

The Solar Dynamo

Axel Brandenburg [1,2] and Ilkka Tuominen [1]

[1] Observatory and Astrophysics Laboratory, University of Helsinki
Tähtitorninmäki, SF-00130 Helsinki, Finland
[2] NORDITA, Blegdamsvej 17, DK-2100 Copenhagen, Denmark

Abstract: The traditional $\alpha\Omega$-dynamo as a model for the solar cycle has been successful in explaining the butterfly diagram, phase relations between poloidal and toroidal field, and polar branch migration features. Observational and theoretical achievements in recent years have however shaken this picture. The current trend is towards dynamos operating in the overshoot region of the convection zone. Nevertheless, there are many open questions and a consistent picture has not been established. In this paper we compare recent approaches and discuss remaining problems.

1. Introduction

Magnetic fields are the engine of solar and stellar activity. The most prominent activity phenomenon is the 11 year sunspot cycle. At the beginning of this century Hale showed that sunspot pairs involve strong magnetic fields which reverse orientation from one cycle to another, and thus the original field orientation is recovered after two cycles, making the magnetic cycle period 22 years. There are variations of activity on a much smaller timescale as well, for example those related to prominences, faculae or flares, which occur in a more irregular fashion. Also, long-term variations are known, such as Grand Minima, which do not seem to occur very regularly either. Of course, the period of the solar cycle is not exactly 11 years, but may vary between 7 and 17 years, and the cycle's amplitude also varies. However, there is a clear peak in the power spectrum computed from the time sequence of the sunspot number. Also, the migration of sunspot activity belts is very systematic.

The overall geometry of the solar cycle magnetic fields has been successfully explained by the $\alpha\Omega$-dynamo models of Steenbeck and Krause (1969). Twenty years ago it was hoped that minor disagreements between models and reality could be ironed out with the development of more realistic turbulence models ($\rightarrow$ α-effect) and with improving observations of solar differential rotation ($\rightarrow$ Ω-effect).

2. The solar field geometry

Before we discuss different approaches in more detail let us first consider the solar observations that are of direct relevance for mean-field dynamos. Perhaps the most important observation is Hale's polarity law. According to Parker's (1955) interpretation sunspot pairs are formed when toroidal flux ropes rise and break through the solar surface. The two footpoints of an emerging field loop then correspond to a sunspot pair. With remarkably small statistical scatter the orientation of sunspot pairs and bipolar magnetic regions are opposite on opposite sides of the equator, reversing after 11 years (Wang and Sheeley, 1989). From Hale's polarity law we learn that the Sun has a systematic azimuthal magnetic field B_ϕ which is *antisymmetric* about the equator. Measurements of the radial field component B_r can be obtained from the Mt. Wilson and Kitt Peak magnetograms. Yoshimura (1976) and Stix (1976) have shown that B_r and B_ϕ vary approximately in antiphase ($B_r B_\phi < 0$ for most of the time). The radial field at the pole is of special interest, because there the observations of B_r are not contaminated by a B_ϕ-field. The curve $B_r = 0$ corresponds to the location of faculae (see Stix, 1974). For latitudes above 60° this curve shows a poleward migration. This feature is often called this *polar branch*.

3. The traditional $\alpha\Omega$-dynamo

The governing equation in the theory of $\alpha\Omega$-dynamos (Steenbeck and Krause, 1969; Roberts and Stix, 1972) is the induction equation for the mean field $\langle \boldsymbol{B} \rangle$:

$$\frac{\partial}{\partial t}\langle \boldsymbol{B} \rangle = \mathrm{curl}\left(\langle \boldsymbol{u} \rangle \times \langle \boldsymbol{B} \rangle + \alpha \langle \boldsymbol{B} \rangle - \eta_t \mathrm{curl}\, \langle \boldsymbol{B} \rangle\right). \tag{1}$$

Induction effects are due to the α-effect and to gradients in the angular velocity $\Omega = \langle u_\phi \rangle / r \sin\theta$. In the following, by a *traditional* $\alpha\Omega$-dynamo we mean that

$$\alpha \cos\theta > 0, \quad \partial\Omega/\partial r < 0. \tag{2}$$

In the model of Steenbeck and Krause (1969), simple profiles for α and Ω were used. If Eq.(1) is solved inside a sphere of radius R, and if a vacuum outside is assumed ($\mathrm{curl}\,\langle \boldsymbol{B} \rangle = 0$ for $r > R$), then there is a marginal stable oscillatory solution which is antisymmetric about the equator (odd parity, in accordance with Hale's law) with dynamo waves migrating equatorwards. It should be stressed that the preference for an antisymmetric field is sensitive to boundary conditions and to the thickness of the convection zone (see e.g. Table 5 in Roberts, 1972). The magnetic cycle period is $T \approx 0.20\tau_{\mathrm{diff}}$, where $\tau_{\mathrm{diff}} = R^2/\eta_t$ is the global diffusion timescale. For $\eta_t = 10^{13}\mathrm{cm}^2/\mathrm{s}$ we have $\tau_{\mathrm{diff}} \approx 16$ years, i.e. T=3 years, which is too short for the 22 year solar period by a factor of 7. A polar branch is not present in model 1 of Steenbeck and Krause, but models by Köhler (1973) and Yoshimura (1975), which include a gradient $\partial\Omega/\partial\theta$, do indeed show a poleward migration at higher latitudes. Also a more complicated θ-dependence in α (Schmitt, 1987) or in $\partial\Omega/\partial r$ (Makarov *et al.*, 1988; Belvedere, 1990) can produce a polar branch.

4. The parity problem

We mentioned in the previous section that the parity of the generated magnetic field depends on details of the dynamo model (e.g. boundary conditions and thickness of the convection shell). In addition, the degree of nonlinearity also can have an influence on the parity. The parity of the dynamo can depend critically on the dynamo number D. Models show that there is a range of D, where the odd parity solution loses stability to a mixed parity solution (Brandenburg *et al.*, 1989a). The importance of studying the stability properties of nonlinear dynamos has been stressed by Krause and Meinel (1988). The most detailed investigation has been presented by Jennings (1991).

The parity problem is relevant to understanding details of the solar dynamo. Deviations of the solar field from the pure dipole-type symmetry can be deduced from the observed North-South asymmetry of sunspot numbers. This has been investigated in detail by Vizoso and Ballester (1990). The degree of asymmetry of the toroidal field is an observable quantity which can in principle be predicted from dynamo models (cf. Brandenburg *et al.*, 1989b).

The phenomenon of active longitudes is surprisingly regular (Tuominen, 1962) and may be due to non-axisymmetric contributions to the dynamo generated field (Stix, 1974). It is known that in the presence of strong differential rotation non-axisymmetric fields are much harder to excite than axisymmetric ones (Rädler, 1986). However, in the highly nonlinear regime secondary bifurcations to mixed parity solutions with nonaxisymmetric contributions can occur. In a simplified two-dimensional model Jennings *et al.* (1990) found an example, where the A0-type solution can lose stability to a mixed mode with A0 and S1 contributions. Certainly, more realistic three-dimensional models are required to make conclusive statements concerning active solar longitudes.

5. Transport coefficients

A severe uncertainty in the theory of solar mean-field dynamos concerns transport coefficients such as α and η_t. These coefficients have often been determined analytically using a simple model for isotropic turbulence and considering rotation and stratification as small perturbations. A closure is achieved by taking only second order correlations into account (first order smoothing). In their original paper Steenbeck *et al.* (1966) also obtained the turbulent diamagnetic effect, and anisotropies in α.

In a similar manner coefficients for the eddy viscosity and the turbulent heat conductivity can be obtained. Latitudinal dependences and anisotropies of these coefficients are important here, because they can drive differential rotation. Another, perhaps more important, driver of differential rotation is Rüdiger's (1977) Λ-effect – a hydrodynamic analogue to the α-effect. Similar ideas have been developed by Frisch *et al.* (1987), who called this the AKA-effect (anisotropic kinetic alpha-effect).

A useful complement to analytical theories are numerical simulations. An early example of this type is the two-dimensional model by Moss (1971, unpubl. report), who found $\eta_t \approx 0.2u_t\ell$, where u_t is the turbulent rms-velocity and ℓ is a correlation length. Kraichnan (1977) found examples where η_t can even be negative. Pulkkinen *et al.* (these Proceedings) determined the latitude dependence of the Λ-effect and found agreement between observations and theory (see Tuominen, 1990). A numerical determination of the eddy heat conductivity and other stellar mixing length parameters has been carried out by Chan and Sofia (1989). Recent simulations of magnetoconvection in the presence of rotation have indicated that the α-effect in the vertical direction can be of opposite sign to the horizontal components of the α-tensor (Brandenburg *et al.*, 1990a).

As discussed by Krause (these Proceedings), simulations even at a resolution of 63^3 gridpoints can hardly be expected to represent the circumstances under which mean-field theory is valid. The correlation length in these models is comparable with the size of the simulated domain. Simulations of stellar convection by Stein and Nordlund (1989) do indicate that the correlation length in the vertical direction is much longer than in the horizontal directions (for a review see Spruit *et al.*, 1990). On the other hand, mean-field theory applies only if the correlation length is short compared with global dimensions. In addition, first order smoothing is applicable only if the correlation time is short compared with the turnover time and rotation period. For the Sun this is not the case. It seems therefore that simulations and analytical theories may approach the solar case from opposite directions.

6. Nonlinear feedbacks

An important property of stellar dynamos is nonlinearity. Although the induction equation (1) is at first glance linear, there can be nonlinear feedbacks if the average velocity $\langle \boldsymbol{u} \rangle$ or the α-coefficient depend on $\langle \boldsymbol{B} \rangle$. One may refer to these nonlinearities as micro- and macro-feedback. In both cases the feedback originates from the Lorentz force in the momentum equation and it is the length scale associated with the Lorentz force which leads to this distinction. These aspects have been examined by Gilbert and Sulem (1990), and in the context of $\alpha\Lambda$-dynamos by Brandenburg *et al.* (1990b). Properties of macro-feedback and the relevance of the Elsasser number have been stressed by Roberts (these Proceedings).

Another important feedback is that from magnetic buoyancy (Noyes *et al.* 1984; see also Rädler, 1990; Moss *et al.*, 1990a). Here the magnetic pressure term, as part of the Lorentz force, provides the feedback. The length scale associated with the buoyancy force is usually short compared with the thickness of the convection zone and we may classify this as a micro-feedback. However, convective dynamo simulations indicate that the main feedback is from the magnetic curvature force, rather than from the magnetic pressure gradient (Brandenburg *et al.*, 1990c).

The dynamo has various properties which arise from the presence of nonlinearities. We mentioned already the stability problem (Sect. 4), which makes sense only if nonlinearity is involved. Feedbacks on the mean motions caused by the mean magnetic fields are observed in the form of torsional waves and cyclic variations of the meridional circulation (Tuominen and Virtanen, 1987; 1988). Quite another aspect is chaos and irregularity, which is usually studied using ordinary differential equations derived from the three-dimensional dynamo equations (for reviews see Weiss, 1989, 1990).

7. The solar angular velocity

Contours of constant angular velocity in the Sun are not cylindrical, as early theories of differential rotation predicted (Durney, 1976), but rather are "disk shaped" (Rüdiger, 1989). Realistic numerical models show this property if the Rossby number is of order unity (Tuominen and Rüdiger, 1989) and if the Taylor number is not too large (Ta $\lesssim 10^6$, i.e. $\nu_t \gtrsim 3 \times 10^{13}$). There is a remarkable consequence for $\alpha\Omega$-type dynamo models: oscillatory solutions can only be obtained if C_Ω is large enough ($\gtrsim 10^3$), i.e. $\eta_t \lesssim 10^{12}$. This is a puzzle, because ν_t and η_t are *turbulent* kinematic and magnetic diffusivities which are not expected to differ substantially from each other. Future work will show how much this Taylor number puzzle (Brandenburg *et al.* 1990b) is model dependent.

8. The $\partial\Omega/\partial r > 0$ problem

Differential rotation is a good candidate not only for causing the cyclic behavior, but also the migration of the solar magnetic field pattern. However, strong anisotropies of the α-effect can produce solar-like butterfly diagrams as well (Weisshaar, 1982). One reason why such approaches have not been considered further is that the degree of anisotropy needed is rather large.

Latitudinal differential rotation can also yield oscillatory dynamos, but such models do not exhibit any significant field migration, only periodic field emergence at mid-latitudes followed by diffusion both to higher and lower latitudes (Köhler, 1973). Only with a sufficient amount of radial differential rotation is there equatorward migration at low latitudes and poleward migration at high latitudes (polar branch). Such models have solar-like field geometry if the inductive effects at low latitudes are as in Eq.(2). Parker (1989) suggested that, in addition to a purely latitudinal differential rotation, suitable nonuniform distributions of the α- and Ω-effects might result in an equatorward migration. For example, an enhanced α at low latitudes (cf. Schmitt, 1987) might modify latitudinal diffusion in some way. However, satisfactory models have not been presented yet.

The most promising mechanism for explaining (i) the cyclic behavior and (ii) the migration properties as well as (iii) the strength of the toroidal field relative to the poloidal field seems to be a dynamo of $\alpha\Omega$-type with $\partial\Omega/\partial r$ being important. Recent results of helioseismology (e.g. Goode, these Proceedings) seem to exclude the possibility of a negative radial Ω-gradient at the equator. The consequences for current $\alpha\Omega$-dynamo models are either a poleward migration of dynamo waves (if $\alpha \cos\theta > 0$) or an incorrect phase relation between poloidal and toroidal field ($B_r B_\phi > 0$ if $\alpha \cos\theta < 0$).

In recent years various $\alpha\Omega$-dynamo models with $\partial\Omega/\partial r > 0$ have been presented (e.g. Belvedere *et al.*, 1990a), where the problem with the phase relation has obviously been ignored. These models are mainly designed for the lower convection zone or for the overshoot layer below. The hope is perhaps that the upper convection zone might lead to a systematic phase flip. However, even dynamo models with complicated radial profiles of α and $\partial\Omega/\partial r$ do not give deviations from the expected phase relation. The possibility that the observations are not reliable enough does not seem to be justified, although measurements of B_r and B_ϕ are tricky, cf. Stix (1981).

The model presented by Wilson (1988) tries to explain the dynamo dilemma in terms of a two component fluid with magnetic and non-magnetic constituents. He suggests that the dynamo does not "feel" the average angular velocity $\langle\partial\Omega/\partial r\rangle$, but rather the angular velocity $(\partial\Omega/\partial r)_B$, associated with magnetic flux tubes. A negative $(\partial\Omega/\partial r)_B$ seems plausible, if flux tube motion is considered as being predominantly governed by the Coriolis force. That these gradients are different is supported by the observation that the rotation rate of young sunspots (Tuominen and Virtanen, 1988) and magnetic tracers (Stenflo, 1989a) is larger than the value obtained from helioseismology.

9. Dynamos for the overshoot layer

The overshoot layer of the Sun is of considerable interest for the dynamo for two reasons: (i) magnetic flux losses due to magnetic buoyancy are small here, and (ii) there is evidence that α changes sign at the bottom of the convection zone, which when combined with $\partial\Omega/\partial r > 0$ gives equatorward migration. A change of sign of α in the overshoot layer follows from a formula due to Krause (1967). However, this formula is perhaps invalid at this location where the turbulence is expected to be far from homogeneous. One should note that the argument for a reversal of α, often quoted in the literature, comes from results of Boussinesq convection!

DeLuca and Gilman (1986) considered a hydromagnetic, two-dimensional, Cartesian model for the interface between the convection zone and the radiative interior. An interesting property of their model is that magnetic energy considerably exceeds kinetic energy. The importance of this for understanding the observed magnetic fluxes in the Sun has been stressed by Durney *et al.* (1990).

There has been considerable interest in models without explicit radial dependence, with the idea that the dynamo works in a thin layer. However, it is not at all evident that radial structure is unimportant. A model without vertical extent was presented by Leighton (1969), without, however, attributing it to a "thin shell". Nonlinear dynamics of such models have been investigated by Schmitt and Schüssler (1989), Jennings and Weiss (1990), and Belvedere *et al.* (1990b). There is evidence that such dynamos are of $\alpha^2\Omega$-type (Gilman *et al.*, 1989). In order to produce the correct period and number of field belts the correlation length should be two hundred times shorter than the local pressure scale height (Choudhuri, 1990). This result is based on a Cartesian approximation. Dynamos in thin spherical shells, however, seem to have many belts, or otherwise they become steady (Moss *et al.*, 1990b).

10. Flux tubes and magnetic buoyancy

The solar magnetic field is highly intermittent (see review by Stenflo, 1989b). This must have consequences for the traditional $\alpha\Omega$-dynamo. Is it possible, that the mean-field concept is still applicable (Schüssler, 1980), and only that the transport coefficients (e.g. α, η_t) are reduced (Childress, 1979))?

Another consequence is magnetic buoyancy experienced by horizontal field (see Hughes and Proctor, 1988). The idea often presented is that the dynamo generates a diffusive field which gets concentrated into flux tubes by doubly diffusive instabilities (Schmitt and Rosner, 1983), dynamical fragmentation (Schüssler, 1977, 1979), or flux expulsion (Galloway *et al.*, 1977). Magnetic buoyancy acting on these tubes can then rapidly remove field from the dynamo region (Parker, 1975). However, it is unclear how efficient this effect is, or whether it even negates the mechanism of a solar dynamo.

Meanwhile several arguments have been accumulated against the importance of buoyancy, for example turbulent diamagnetism and topological pumping as well as drag forces (Schüssler, 1984). On the other hand, Petrovay (these Proceedings) presented new arguments at this conference as to why topological pumping might actually not work. Many other interesting ideas have been proposed, for example Parker (1987) estimated that thermal shadows could push flux tubes down whereas Choudhuri (1989) argues that flux tube motion is dominated by the Coriolis force and hence tubes rise parallel to the rotation axis. In this case one might expect sunspots to emerge at rather high latitudes. Schüssler (1983) concluded from Hale's polarity law that bipolar regions can only be produced by strong flux tubes which can resist the irregular turbulent motions and prevent the field from being "brainwashed".

11. Fast and convective dynamos

Much attention has been devoted to the investigation of fast dynamos (Vainshtein and Zeldovich, 1972). The relevant timescale here is not the global diffusion time $\tau_{\rm diff} = L^2/\eta$ (where η is the non-turbulent value!), but the convective turnover time $\tau_{\rm conv} = L/u_{\rm turb}$. Thus dissipation acts on the length scale of small magnetic flux concentrations. Fast dynamos are of astrophysical relevance, because the global diffusion timescale for the Sun is comparable to the age of the universe.

Kinematic fast dynamos have been investigated using prescribed ABC-flows, with and without the presence of diffusion (Galloway and Frisch, 1986; Gilbert and Sulem, 1990). Recently there have been a number of *direct simulations* of the dynamo without parametrization of diffusivities and viscosities. Meneguzzi and Pouquet (1989) find dynamo action on a convective timescale, which may be called a fast dynamo. Mean-field dynamos are also fast, because the timescale is essentially independent of the (non-turbulent) diffusivities (cf. Moffatt, 1990). The first dynamo simulations in spherical geometry were those of Gilman and Miller (1981). Valdettaro and Meneguzzi (these Proceedings) presented the first spherical dynamos with compressible flow.

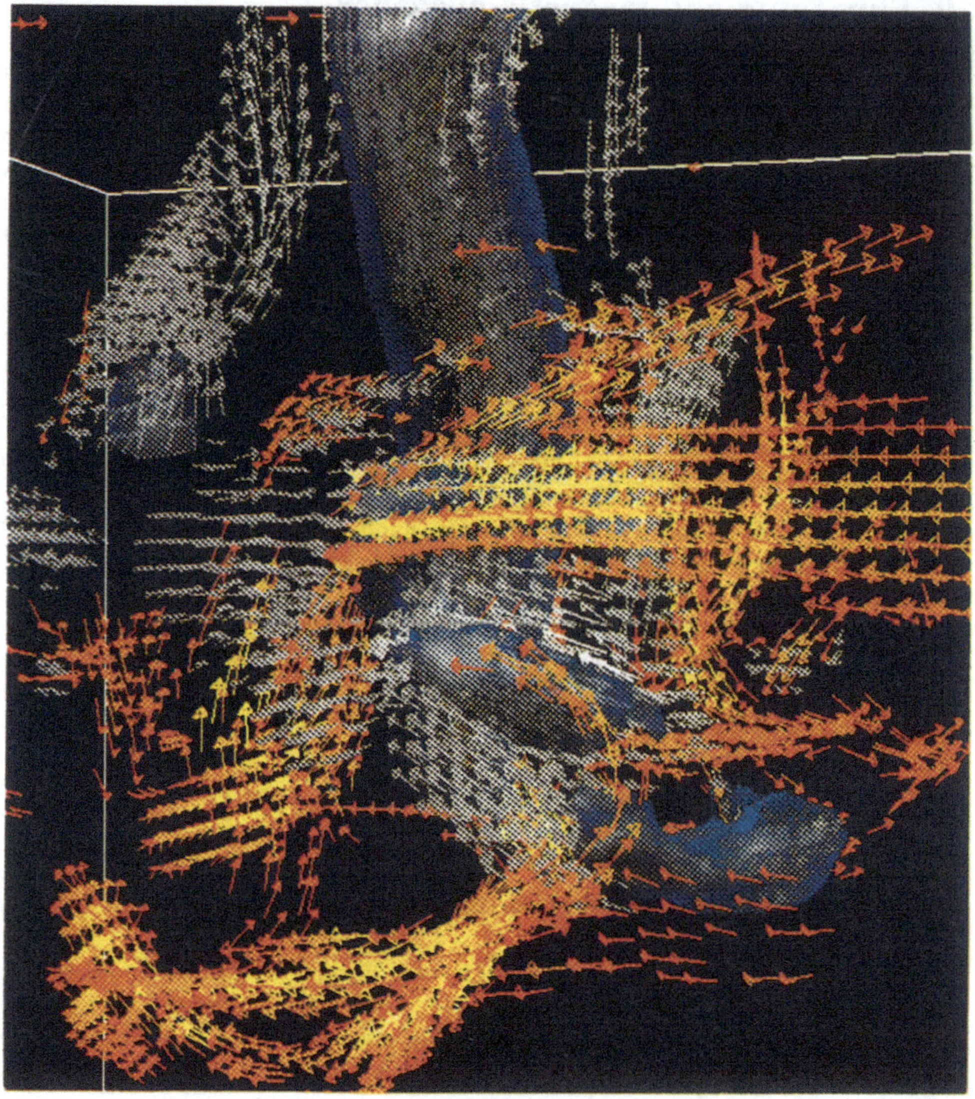

Fig. 1. Snapshot from a video animation showing magnetic field vectors in yellow (the strongest) and red (less strong) and those of vorticity in white. Transparent surfaces of constant negative pressure fluctuation are shown in blue. Note that the vectors of magnetic field form flux tubes which become wound up around a tornado-like vortex.

Simulations including overshoot (Brandenburg *et al.*, 1990c) give the impression that magnetic buoyancy may not be a problem for the dynamo, because it is overwhelmed by other dynamical forces. Indeed, such dynamical forces are responsible for the formation of flux tubes in the first place. The picture of diffuse field generation is therefore not appropriate.

Figure 1 shows a snapshot from the video animation presented at the conference. This video shows the evolution of magnetic field, generated and maintained by the turbulent motions, in a Cartesian box located at the bottom of the convection zone. There is a strong tendency for field to be sucked by the downdrafts whilst upward motions of buoyant flux tubes were barely detectable.

12. Conclusions

The solar dynamo problem has many aspects – only some of them have been highlighted in this review. It is clear that at present there is no good dynamo model for the Sun! Many details are uncertain. Is the solar dynamo really of $\alpha\Omega$-type? Is the traditional mean-field concept appropriate? Is magnetic buoyancy a problem? Recent simulations indicate that flux tube dynamics are governed by suction of field in the vicinity of strong cyclonic downdrafts. Flux tubes are formed and deformed by the flow, and actively participate in the dynamo process. However, these simulations ignore the global nature of the solar dynamo. It remains a challenge to combine these separate pieces of information into a coherent picture of the solar dynamo, which may perhaps then be describable in terms of mean fields after all.

References

Belvedere, G.: 1990, in *Inside the Sun*, eds. G. Berthomieu and M. Cribier, Kluwer, 371

Belvedere, G., Proctor, M. R. E., Lanzafame, G.: 1990a, in *Progress of Seismology of the Sun and stars*, eds. Y. Osaki and H. Shibahashi, Lecture Notes in Physics

Belvedere, G., Pidatella, R. M., Proctor, M. R. E.: 1990b, *Geophys. Astrophys. Fluid Dyn.* **51**, 263

Brandenburg, A., Tuominen, I., Moss, D.: 1989a, *Geophys. Astrophys. Fluid Dyn.* **49**, 129

Brandenburg, A., Krause, F., Tuominen, I.: 1989b, in *Turbulence and Nonlinear Dynamics in MHD Flows*, eds. M. Meneguzzi *et al.*, Elsevier Science Publ., p. 35

Brandenburg, A., Nordlund, Å., Pulkkinen, P., Stein, R.F., Tuominen, I.: 1990a, *Astron. Astrophys.* **232**, 277

Brandenburg, A., Moss, D., Rüdiger G., Tuominen, I.: 1990b, *Solar Phys.* **128**, 243

Brandenburg, A., Jennings, R. L., Nordlund Å., Rieutord, M., Ruokolainen, J., Stein, R. F., Tuominen, I.: 1990c, *Nordita preprint* 90/67

Chan, K. L., Sofia, S.: 1989, *Astrophys. J.* **336**, 1022

Childress, S.: 1979, *Phys. Earth Pl. Int.* **20**, 172

Choudhuri, A. R.: 1989, *Solar Phys.* **123**, 217

Choudhuri, A. R.: 1990, *Astrophys. J.* **355**, 733

DeLuca, E. E., Gilman, P. A.: 1986, *Geophys. Astrophys. Fluid Dyn.* **37**, 85
Durney, B. R.: 1976, *Astrophys. J.* **204**, 589
Durney, B. R., De Young, D. S., Passot, T. P.: 1990, *Astrophys. J.* **362**, 709
Frisch, U., She, Z. S., Sulem, P. L.: 1987, *Physica* **28D**, 382
Galloway, D. J., Proctor, M. R. E. and Weiss, N. O.: 1977, *Nature* **266**, 686
Galloway, D. J., Frisch, U.: 1986, *Geophys. Astrophys. Fluid Dyn.* **36**, 53
Gilbert, A. D., Childress, S.: 1990, *Phys. Rev. Lett.* (in press)
Gilbert, A. D., Sulem, P.-L.: 1990, *Geophys. Astrophys. Fluid Dyn.* **51**, 243
Gilman, P. A., Miller, J.: 1981, *Astrophys. J. Suppl.* **46**, 211
Gilman, P. A., Morrow, C. A., DeLuca, E. E.: 1989, *Astrophys. J.* **338**, 528
Hughes, D. W., Proctor, M. R. E.: 1988, *Ann. Rev. Fluid Mech.* **20**, 187
Jennings, R.: 1991, *Geophys. Astrophys. Fluid Dyn.* (in press)
Jennings, R., Weiss, N.O.: 1990, in *Solar photosphere: structure, convection and magnetic fields*, ed. J. O. Stenflo, Kluwer Acad. Publ., Dordrecht, p. 355
Jennings, R., Brandenburg, A., Moss, D., Tuominen, I.: 1990, *Astron. Astrophys.* **230**, 463
Köhler, H.: 1973, *Astron. Astrophys.* **25**, 467
Kraichnan, R. H.: 1976, *J. Fluid Mech.* **77**, 753
Krause, F.: 1967, Habilationsschrift., Univ. Jena
Krause, F., Meinel, R.: 1988, *Geophys. Astrophys. Fluid Dyn.* **43**, 95
Leighton, R. B.: 1969, *Astrophys. J.* **156**, 1
Makarov, V. I., Ruzmaikin, A. A., Starchenko, S. V.: 1988, *Solar Phys.* **111**, 267
Meneguzzi, M., Pouquet, A.: 1989, *J. Fluid Mech.* **205**, 297
Moffatt, H. K.: 1989, *Nature* **341**, 285
Moss, D., Tuominen, I., Brandenburg, A.: 1990a, *Astron. Astrophys.* **228**, 284
Moss, D., Tuominen, I., Brandenburg, A.: 1990b, *Astron. Astrophys.* **240**, 142
Noyes, R. W., Weiss, N. O., Vaughan, A. H.: 1984, *Astrophys. J.* **287**, 769
Parker, E. N.: 1955, *Astrophys. J.* **121**,
Parker, E. N.: 1975, *Astrophys. J.* **198**, 205
Parker, E. N.: 1987, *Astrophys. J.* **321**, 984
Parker, E. N.: 1989, *Solar Phys.* **121**, 271
Rädler, K.-H.: 1986, *Plasma Physics*, ESA SP-251, 569
Rädler, K.-H.: 1990, in *Inside the Sun*, eds. G. Berthomieu and M. Cribier, Kluwer, 385
Roberts, P. H.: 1972, *Phil. Trans. Roy. Soc.* **A274**, 663
Roberts, P. H., Stix, M.: 1972, *Astron. Astrophys.* **18**, 453
Rüdiger, G.: 1977, *Astron. Nachr.* **298**, 245
Rüdiger, G.: 1989, *Differential rotation and stellar convection: Sun and solar-type stars*, Gordon and Breach, New York
Schmitt, D.: 1987, *Astron. Astrophys.* **174**, 281
Schmitt, D., Schüssler, M.: 1989, *Astron. Astrophys.* **223**, 343
Schmitt, J. H. M. M., Rosner, R.: 1983, *Astrophys. J.* **265**, 901
Schüssler, M.: 1977, *Astron. Astrophys.* **56**, 439
Schüssler, M.: 1979, *Astron. Astrophys.* **71**, 79
Schüssler, M.: 1980, *Nature* **288**, 150
Schüssler, M.: 1983, in *Solar and stellar magnetic fields: origins and coronal effects*, ed. J. O. Stenflo, Reidel, p. 213
Schüssler, M.: 1984, in *The hydromagnetics of the Sun*, eds. T. D. Guyenne and J. J. Hunt, ESA SP-220, p. 67
Spruit, H. C., Nordlund, Å, Title, A. M.: 1990, *Ann. Rev. Astron. Astrophys.* **28**, 263

Steenbeck, M., Krause, F., Rädler, K.-H.: 1966, *Z. Naturforsch.* **21a**, 369
Steenbeck, M., Krause, F.: 1969, *Astron. Nachr.* **291**, 49
Stein, R. F., Nordlund, Å.: 1989, *Astrophys. J. Letters* **342**, L95
Stenflo, J.O.: 1989a, *Astron. Astrophys.* **210**, 403
Stenflo, J.O.: 1989b, *Ann. Rev. Astron. Astrophys.* **1**, 3
Stix, M.: 1974, *Astron. Astrophys.* **37**, 121
Stix, M.: 1976, *Astron. Astrophys.* **47**, 243
Stix, M.: 1981, *Solar Phys.* **74**, 79
Tuominen, I.: 1990, in *The dynamic Sun*, ed. Dezsõ, Publ. Debr. Heliophys. Obs., p. 27
Tuominen, I., Virtanen, H.: 1987, in *The internal solar angular velocity*, eds. B. R. Durney and S. Sofia, D. Reidel, Dordrecht, p. 83
Tuominen, I., Virtanen, H.: 1988, *Adv. Space Sci.* **8**, (7)141
Tuominen, I., Rüdiger, G.: 1989, *Astron. Astrophys.* **217**, 217
Tuominen, J.: 1962, *Z. Astrophys.* **55**, 110
Vainshtein, S. I., Zeldovich, Ya. B.: 1972, *Sov. Phys. Usp.*, **15**, 159
Van Ballegooijen, A. A., Choudhuri, A.,R.: 1988, *Astrophys. J.* **333**, 965
Vizoso, G., Ballester, J. L.: 1990, *Astron. Astrophys.* **229**, 540
Wang, Y.-M., Sheeley, Jr., N. R.: 1989, *Solar Phys.* **124**, 81
Weisshaar, E.: 1982, *Geophys. Astrophys. Fluid Dyn.* **21**, 285
Weiss, N. O.: 1989, in *Accretion disks and magnetic fields in astrophysics*, ed. G. Belvedere, Kluwer Acad. Publ., Dordrecht, p. 11
Weiss, N. O.: 1990, *Phil. Trans. Roy. Soc.* **A 330**, 617
Wilson, P. R.: 1988, *Solar Phys.* **117**, 217
Yoshimura, H.: 1975, *Astrophys. J. Suppl.* **29**, 467
Yoshimura, H.: 1976, *Solar Phys.* **50**, 3

A Topological Model of the Solar Magnetic Field Reversals

E.E. Benevolenskaya

Pulkovo Astronomical Observatory
196140, Leningrad, USSR

The phenomenon of a three-fold reversal of the solar polar magnetic field in both hemispheres has not been observed during the last 115 years. Such three-fold reversals took place in the southern hemisphere alone in the even cycles Nos 12 (1885.8), 14 (1908.4) and in the northern hemisphere alone in solar cycles Nos 16 (1928.5), 18 (1949.0), 20 (1970.6). The single reversal took place in the odd cycles, the only exception is the solar cycle No 19 (Fig. 1).

CYCLES	10–11	12–13	14–15	16–17	18–19	20–21
N	+/-	-/+ +/-	-/+ +/-	-/+/-/+ +/-	-/+/-/+ +/-/+/-	-/+/-/+ +/-
S	-/+	+/-/+/- -/+	+/-/+/- -/+	+/- -/+	+/- -/+	+/- -/+

Fig. 1. Hale's 22-year cycles and the reversal of the Sun's magnetic fields in the odd and even 11-year cycles in the northern (N) and southern (S) hemispheres.

There are periods of 1.7–2.5 years in the variation of background magnetic fields (Makarov *et al.*, 1985). It determines the quasi-period of the high-frequency component and corresponds to a time interval between the zones of alternating polarity of the magnetic field. This enables us to show topologically that single and three-fold polarity reversals of the solar magnetic fields can result from interaction of two types of magnetic fields: a low-frequency component with period of the order of 20 years and a high frequency component with period of order of 1.7–2.5 years (Benevolenskaya and Makarov, 1990).

According to modern concepts, solar magnetic fields are generated by helicity and the differential rotation of the solar convection zone (Parker, 1979). The dynamo process in the linear approximation is instrumental in generating one carrier frequency ω_c. Within the frame of linear stochastic theory, a spectrum of frequencies is generated for the dynamo numbers $D > D_c$. In the non-linear theory, stable frequencies with a period shorter than that of a cycle may occur (Hoyng, 1990). The toroidal fields (B_φ-component) formed in the convection zone show up as sunspots on the surface. The background magnetic fields (B_r-component) in their turn are transferred poleward due to diffusion and meridional circulation (Wang *et al.*, 1989). Let us assume that

$$B_r = B_r' + \delta B_r, \tag{1}$$

where B_r' is a low-frequency component with a period of the order of 22 years and δB_r-is a high-frequency component.

According to Makarov *et al.* (1987) and Kleeorin and Ruzmaikin (1988) the B_r field can be represented as a dynamo wave

$$B_r' = b_r(r,\theta)\cos(\omega_c t + g(r,\theta) - \frac{\pi}{2}), \tag{2}$$

where $b_r(r,\theta)$ is the amplitude of the radial field in the convection zone; $g(r,\theta)$ is the phase function, $g(R_\odot, 0) = 0$; $\omega_c = \frac{2\pi}{T_c}$ is the frequency of the solar cycle. Similarly, the δB_r field will be represented as

$$\delta B_r = \delta b_r(r,\theta)\cos(\omega_\delta(t)t + g_1(r,\theta) - \frac{\pi}{2}). \tag{3}$$

If $\omega_\delta(t)$ is a slowly varying with time and $\omega_\delta(t) = \omega_c$=const and $g_1(r,\theta)$=$g(r,\theta)$+φ, $b_r(r,\theta) = A \cdot \delta\, B_r(r,\theta)$. At some fixed latitude θ_c :

$$B_r = b_r(r,\theta_c)\sin\omega_c t + \delta b_r(r,\theta_c)\sin(\omega_\delta t + \varphi).$$

Consider the domain $B_r = 0$. The function

$$B_r = b_r(r,\theta)\sin\omega_c t + \frac{b_r(r,\theta)}{A}\sin(\omega_\delta t + \varphi)$$

becomes zero depending on the relation between the phase φ, frequencies ω and ω_δ, and the relation of the amplitudes A. If the frequencies are multiple, i.e. $\omega_\delta = K\omega_c$ and K integer, e. g. K=14, A=2.5. This corresponds to 18 months for real T_c=20.6 years. For K=8 and A=2.5 the behaviour of the B_r' field (solid line) is shown in Fig. 2a, the δB_r field in Fig. 2b and B_r in Fig. 2c. The polarity distribution of the B_r field on the solar surface given in Fig. 2d.

When the polarity of the background field low-frequency component reverses from − to + the zones of alternating polarities are formed. Such a situation takes place in the northern hemisphere, beginning with solar cycle No 16. The presence of such zones depend on the relationship of amplitudes and phases of the high-frequency and low-frequency component of the background magnetic field. The combination of the phase and growth of the amplitude δB_r can result in a situation in which such zones form with the reversal of the B_r' field from + to −, which actually took place in solar cycle 19. The width of the zones of the alternating polarities approximately corresponds to the period of the field (Fig. 2). It is of the order of 1.5–2.0 years.

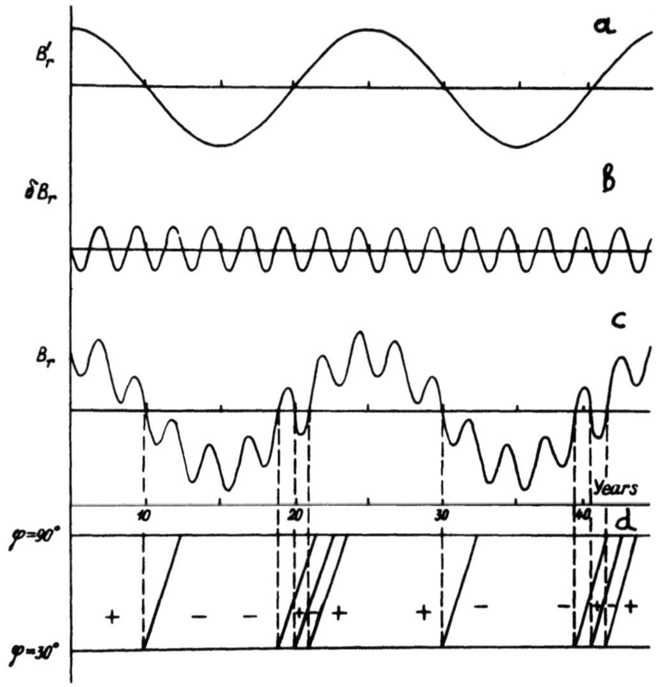

Fig. 2. Schematic representation of the behaviour of B'_r, δB_r and B_r.

Thus, three-fold reversals of the high latitude magnetic field of the Sun can be interpreted as a result of an interaction of the low-frequency component of the field, describing the behaviour of mean background fields during the 22-year period, with a high-frequency component with a slowly-varying phase.

References

Benevolenskaya, E.E. and Makarov, V.I.: 1990, *Soln. Dann.* No 5, 75

Hoyng, P.: 1990 in *Solar Photosphere: Structure, Convection and Magnetc Fields*, ed. Stenflo, O.J., IAU Symp. No 138, p. 359

Kleeorin, N.I. and Ruzmaikin, A.A: 1988, Preprint *IZMIRAN*, No 7(833), p. 3

Makarov, V.I., Tavastsherna, K.S. and Petrova, N.S.: 1985, *Soln. Dann.* No 6, 69

Parker, E.N.: 1979, *Cosmical Magnetic Fields*, Oxford University Press.

Wang, J.W., Nash, A.G. and Sheeley, N.R.,Jr.: 1989, *Astrophys. J.* **347**, 529

Solar Internal Rotation, the Boundary Layer Dynamo and Latitude Distribution of Activity Belts

G. Belvedere[1], M.R.E. Proctor[2] and G. Lanzafame[1]

[1]Istituto di Astronomia, Università di Catania, Italy
[2]Department of Applied Mathematics and Theoretical Physics, University of Cambridge, England

Abstract: We suggest that the latitude distribution of solar activity belts and the related equatorward or poleward migration of different tracers of the solar cycle are a natural consequence of the internal radial profile of angular velocity via the working of a dynamo in the boundary layer beneath the convection zone. This has been confirmed by the results of a non-linear dynamo model in a very thin spherical shell which show that dynamo action may reasonably take place in the boundary layer and reproduce the observed surface phenomenology.

Extending the argument to late main-sequence stars, it is reasonable to think that observations of the latitude distribution and migration of stellar active regions by current sophisticated techniques may make it possible to infer their internal rotation profile in a simple and direct way.

1. Belts as a consequence of internal rotation

The location of dynamo action in the boundary layer at the bottom of the convection zone is indirectly supported by the most recent helioseismological results (Belvedere, 1990). Indeed helioseismic data (e.g. Harvey, 1988; Brown *et al.*, 1989; Dziembowski *et al.* 1989; Libbrecht, 1989) show that the surface differential rotation persists throughout the convection zone ($1\,R_\odot \rightarrow 0.7\,R_\odot$), where the radial gradient of angular velocity vanishes, while rigid rotation dominates beneath the c.z. ($0.65\,R_\odot$), with angular velocity $\omega_\circ = 2.7 \cdot 10^{-6}$ rad s^{-1}, which coincides with the surface value at latitude $\lambda = 37°$.

In the light of these data, radial shear driven dynamos operating in the c.z. might be unrealistic, whereas a boundary layer dynamo is strongly supported by them, on the basis of the following argument (referring to the northern hemisphere).

Since considerations of the helicity of convective overshooting motions show that the dynamo parameter α should be negative in the boundary layer (Yoshimura, 1975a; Glatzmaier, 1985a,b), and, interpolating the helioseismological results in the boundary layer itself, the radial angular velocity gradient $\partial\omega/\partial r$ should be positive for $\lambda < 37°$ and negative for $\lambda > 37°$, the dynamo criterion (Parker 1955; Yoshimura 1975b), based on the sign of the product $\Gamma = \alpha\,\partial\omega/\partial r$, allows equatorward propagation of dynamo waves ($\Gamma < 0$) at lower latitudes ($\lambda < 37°$) and poleward propagation at higher latitudes, in agreement with the observational evidence shown by different tracers of the solar cycle.

Indeed spots and most faculae (equatorial activity belt; $\lambda < 35° - 40°$) do migrate towards the equator during the solar cycle, while polar faculae, filaments and large scale magnetic flux (polar belt; $\lambda > 40° - 50°$) do evidence poleward migration.

We want to stress here that all this has a profound significance for the interpretation of solar activity global features:

(1) the latitude distribution of solar activity belts and the related migration are a natural and direct consequence of the structure of the velocity field in the solar interior;

(2) dynamo action in the boundary layer can still (and better) account for the observed phenomenological evolution in the course of the solar cycle, and remains the basic mechanism to understand solar (and stellar) magnetic activity.

2. The working of boundary layer dynamo

The present picture has been given substantial support by the results of a nonlinear dynamo model in a very thin ($0.05\,R_\odot$) spherical shell, representing the boundary layer, with full time and latitudinal resolution, and an integrated representation only in the radial direction (the so-called radial truncation).

The mean field dynamo equations for the time evolution of the magnetic field $\boldsymbol{B} = B(r,\theta)\boldsymbol{\phi} + \nabla \times (A(r,\theta)\boldsymbol{\phi})$, where r, θ are spherical polar coordinates, $\boldsymbol{\phi}$ is the unit vector in the azimuthal direction, $B\boldsymbol{\phi}$ is the toroidal and $\boldsymbol{B}_p = \nabla \times A\boldsymbol{\phi}$ the poloidal part of $\boldsymbol{B}$, are:

$$\frac{\partial A}{\partial t} = \alpha F(r,\theta)B + \eta_T\left[\nabla^2 - \frac{1}{r^2\sin^2\theta}\right]A, \tag{1}$$

$$\frac{\partial B}{\partial t} = r\sin\theta\boldsymbol{B}_p\cdot\nabla\left[\frac{U(r,\theta)}{r\sin\theta}\boldsymbol{\phi}\right] + \eta_T\left[\nabla^2 - \frac{1}{r^2\sin^2\theta}\right]B. \tag{2}$$

Here αF is the usual α-effect, with F representing its spatial structure and α its magnitude. The quantity η_T is a turbulent diffusivity.

The dynamical influence of the magnetic field enters the model through its effect on the differential rotation $U(r,\theta) = u_\circ + u$, where $u_\circ$ is a prescribed velocity field and u is a perturbation driven by the mean Lorentz force and subject to viscous damping.

The simplest equation that encompasses these features is:

$$\rho\frac{\partial \boldsymbol{u}}{\partial t} = \frac{1}{\mu_\circ}\left[(\nabla \times \boldsymbol{B}) \times \boldsymbol{B}\right]_\phi + \rho\nu_T\left[\nabla^2 - \frac{1}{r^2 \sin^2\theta}\right]\boldsymbol{u} \tag{3}$$

where ν_T is a turbulent viscosity.

The radially truncated model equations have been solved numerically with an explicit time stepping method of DuFort-Frankel type with suitable boundary conditions. For full details we address the reader to the extensive paper by Belvedere, Pidatella and Proctor (1990) who calculated a similar model in the solar convection zone.

Assuming the differential rotation profile $u_\circ$ in the boundary layer as given by interpolating the most recent helioseismological data, the results show the existence of periodic (dynamo wave-like) stable solutions with both equatorward and poleward migrating branches, as shown by the associated butterfly diagram (Fig. 1).

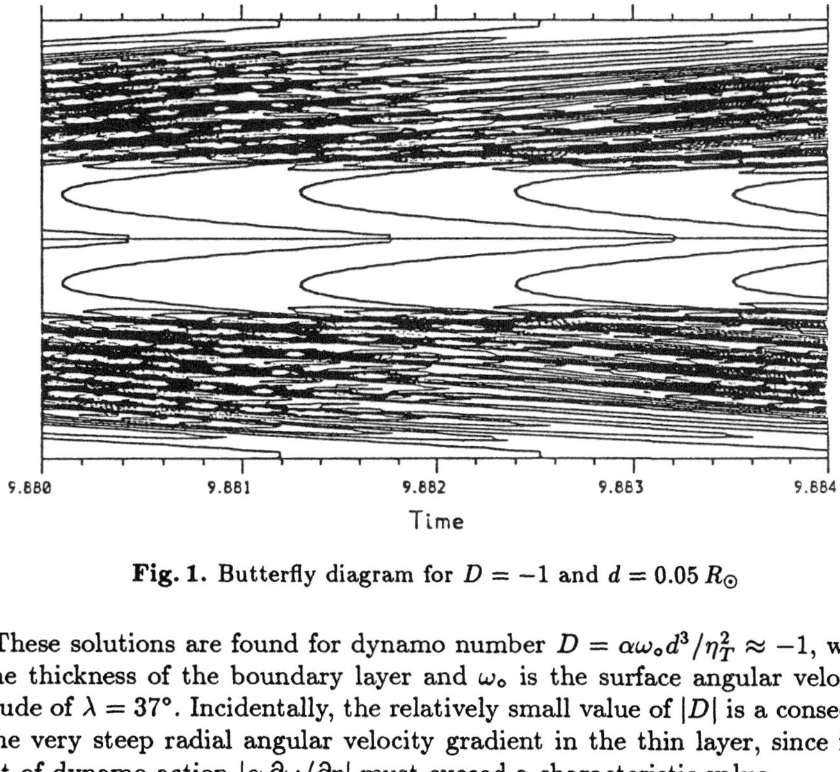

Fig. 1. Butterfly diagram for $D = -1$ and $d = 0.05\, R_\odot$

These solutions are found for dynamo number $D = \alpha\omega_\circ d^3/\eta_T^2 \approx -1$, where d is the thickness of the boundary layer and $\omega_\circ$ is the surface angular velocity at latitude of $\lambda = 37°$. Incidentally, the relatively small value of $|D|$ is a consequence of the very steep radial angular velocity gradient in the thin layer, since for the onset of dynamo action $|\alpha\, \partial\omega/\partial r|$ must exceed a characteristic value.

The relative latitude extent and magnetic field intensity of the polar and equatorial branches shown in Fig. 1 is somewhat artificial, as it depends on the free choice of the latitudinal variation of the model parameter α, which has only to satisfy the requirement of antisymmetry with respect to the equator. Here we adopted $\alpha(\theta) \approx \sin\theta\cos\theta$, where θ is the polar angle, this giving more emphasis to the polar branch.

3. Conclusions

The basic result of our numerical simulation is that horizontally propagating dynamo waves, with equatorial and polar branches, can exist even in a very thin layer, such as the boundary layer at the bottom of the convection zone, where the radial angular velocity gradient changes its sign at latitude $\lambda = 37°$.

Therefore this gives support to our idea that the observed latitude distribution of solar activity belts is a natural consequence of the profile of angular velocity inside the Sun, namely of the fact that the rigid rotation angular velocity beneath the convection zone has approximately the same value as the surface (and c.z.) angular velocity at latitude $\lambda = 37°$.

Of course it remains to be understood why the internal rotation is rigid and its angular velocity coincides with that at the surface just at latitude $\lambda = 37°$; and why do spots occur only for $\lambda < 35° - 40°$. A great amount of detailed study of the interaction of rotation, convection and magnetic field in the solar interior is clearly needed to try to explain these facts.

Nevertheless the present framework allows us to reasonably suggest that observation of latitude distribution and migration of stellar active regions on late main sequence (G,K,M) slowly rotating stars, by the available sophisticated photometric and spectroscopic methods, may conversely give us the possibility to infer their internal rotation profile and angular momentum distribution in a simple and direct way, offering a powerful tool to investigate stellar internal rotation.

Any results obtained by a methodology based on the principles outlined here, should be compared with future stellar oscillations data from space, presumably in the next decade, in order to test the validity of the boundary layer dynamo hypothesis on a large sample of objects.

References

Belvedere, G.: 1990, in *Inside the Sun*, Proc. IAU Coll. 121, G. Berthomieu and M. Cribier eds., Kluwer, Dordrecht, p. 371

Belvedere, G., Pidatella, R.M., Proctor, M.R.E.: 1990, *Geophys. Astrophys. Fluid Dyn.* **51**, 263

Brown, T.M., Christensen-Dalsgaard, J., Dziembowski, W., Goode, P.R., Gough, D.O., Morrow, C.A.: 1989, *Astrophys. J.* **343**, 526

Dziembowski, W.A., Goode, P.R., Libbrecht K.G.: 1989, *Astrophys. J.* **337**, L53

Glatzmaier, G.A.: 1985a, *Geophys. Astrophys. Fluid Dyn.* **31**, 137

Glatzmaier, G.A.: 1985b, *Astrophys. J.* **291**, 300

Harvey, J.W.: 1988, in *Seismology of the Sun and Sun-like Stars*, E.J. Rolfe ed., ESA-SP **286**, 55

Libbrecht, K.G.: 1989, *Astrophys. J.* **336**, 1092

Parker, E.N.: 1955, *Astrophys. J.* **122**, 293

Yoshimura, H.: 1975a, *Astrophys. J. Suppl.* **29**, 467

Yoshimura, H.: 1975b, *Astrophys. J.* **201**, 740

Solar Rotation over Solar Cycle 21

Elisabeth Ribes [1], Istvan Vince [2], Pierre Mein [1],

Eduardo Neto Ferreira [1]

[1]URA 326, Observatoire de Paris, 92195 Meudon, France
[2]Osservatorio Beograd, Beograd, Yugoslavia

Abstract: Having measured the rotation rate of sunspots through solar cycle 21, from 1977 to 1983, we have found that the mean differential rotation averaged over this seven year record is similar to the grand average differential rotation determined by Howard *et al.* (1984) over the period 1921–1982. However, the rotation rate does change from year to year. These changes are evidenced by a steepening or a flattening of the mean differential rotation profile, as well as significant changes in the equatorial rate. The presence of a time-dependent pattern of azimuthal rolls inferred from the meridional circulation pattern of the sunspots offers a qualitative explanation of the observed rotation rates. The amplitude of the changes is almost one order of magnitude larger than that of the torsional oscillations found by Howard and LaBonte (1981).

1. Introduction

The rotation of the solar surface has been thoroughly explored by means of spectroscopic measurements of the photospheric plasma (Howard and LaBonte, 1980), and it has been found that the surface rotation exhibits changes as solar cycle proceeds. The mean differential rotation sometimes steepens and sometimes flattens out. This phenomenon has been described as a "torsional oscillation". Magnetic tracers have also been used, e.g., from measurements of individual sunspot positions on the Mt. Wilson white-light plate collection for the period 1921–1982, grand average rates have been determined that indicate significant changes in the solar rotation (Gilman and Howard, 1984), but which the authors felt were different from the torsional oscillation detected on the spectroscopic measurements.

In order to investigate these changes, we have made use of the Meudon collection of spectroheliograms which exhibit various magnetic features. A digital analysis of the spectroheliograms taken in the violet wing of the chromospheric Ca II line (393.3 nm) provides a more accurate detection of sunspots than by the conventional methods.

2. Methods and results

In most studies involving sunspots as tracers of the surface rotation, the accuracy of individual sunspot velocity field observations is usually no better than $0.^{\circ}2$ per day, i.e. $25\,\mathrm{ms}^{-1}$. Sunspot rotation rates exhibit a considerable scatter, so substantial time and spatial averaging are required to obtain reliable results (Howard *et al.*, 1984; Balthasar *et al.*, 1986). If the changes in rotation rate are of solar origin and time-dependent, it is important to locate sunspot positions with greater accuracy, if we expect to achieve any better sensitivity in the detection of velocity fields. This is the approach used at the Meudon Observatory. To localize a sunspot, we digitize the whole spectroheliogram (1151 by 1151 pixels, each pixel corresponding to the spatial resolution of the spectroheliogram) and apply an image-processing program to correct the solar image for geometric and photometric distorsions. This achieves a $15\,\mathrm{m\,s}^{-1}$ accuracy for each individual sunspot, and is limited only by the spatial resolution of the Meudon spectroheliograms (Mein and Ribes, 1990). We have computed the rotation rates of sunspots observed in summer (from April to October), from 1977 to 1983, and have derived rotation rates from one year to the next.

Figure 1 shows the mean sunspot rotation, using a polynomial fit, averaged over the seven-year sequence from 1977 to 1983, based on observations of 2900 sunspots. The mean synodic rotation laws for both hemispheres are

$$\text{for the northern hemisphere: } \Omega = 13.^{\circ}48 - 2.^{\circ}28 \sin^2\theta + 0.^{\circ}97 \sin^4\theta,$$

$$\text{for the southern hemisphere: } \Omega = 13.^{\circ}53 - 2.^{\circ}05 \sin^2\theta - 1.^{\circ}71 \sin^4\theta,$$

where θ is the latitude. As a few sunspots are observed at high latitudes, the rotation rate derived for these high-latitude belts is not significant. However, within a latitude range of $\pm 30^{\circ}$, the rotation rate derived from our data set is similar to that obtained by Howard *et al.* (1984).

There are noticeable differences in rotation rates from one year to the next. The mean rotation law for each year, showing the difference from the average rotation, is also given in Fig. 1. These departures appear as either a flattening or a steepening of the mean differential rotation profile. The observational effect was first discovered by Howard and LaBonte (1981) on the photospheric rotation rate, and has been referred to as a torsional oscillation. Their finding was interpreted by Ribes *et al.* (1985) and Ribes (1986), as a result of the Coriolis force which accelerates or decelerates the surface flow, in the vicinity of azimuthal convective rolls. The existence of azimuthal rolls has been inferred from a zonal meridional circulation pattern discovered by Ribes and Mein (1984). Such a pattern implies the existence of counter-rotating rolls, with boundary regions of upflows and downflows. These regions coincide with the large scale magnetic field pattern (Fig. 2) measured at Stanford (Hoeksema, 1984). In the present paper, we make use of the differences from two consecutive yearly rotation rates to reconstruct the azimuthal roll pattern, assuming that the change in the rotation from one year to the next is due to the Coriolis forces near the border of converging/diverging rolls.

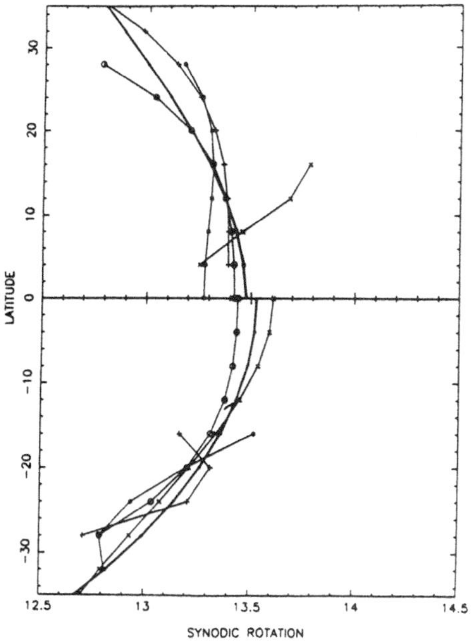

Fig. 1. The polynomial fits of rotation profiles derived from the digital analysis of 2900 sunspots from 1977 to 1983: solid thick curve refer to the mean differential rotation derived for the whole sample, while +, *, 0, and x refer to yearly rotation rate of 1977, 1979, 1981 and 1983 respectively. Errors of the yearly means for 5° latitude intervals are generally about 0.1°/day (from 0.05° to 0.2°).

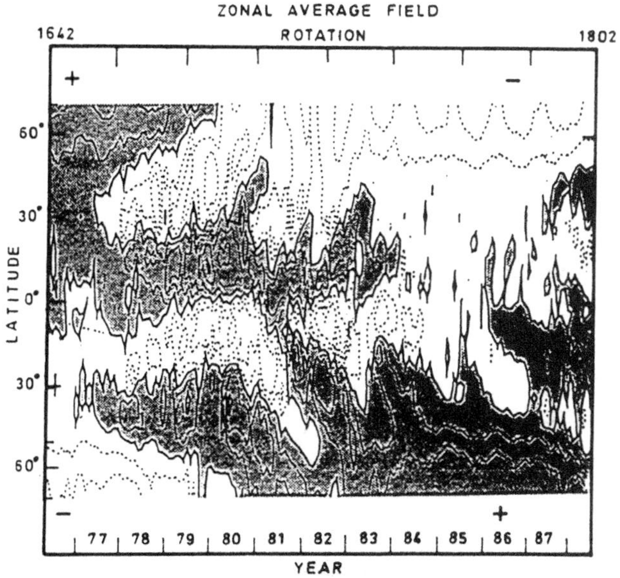

Fig. 2. Large-scale magnetic pattern measured at Stanford. The large-scale magnetic pattern coincides fairly well with the direct large-scale circulation of sunspots shown in Fig. 4.

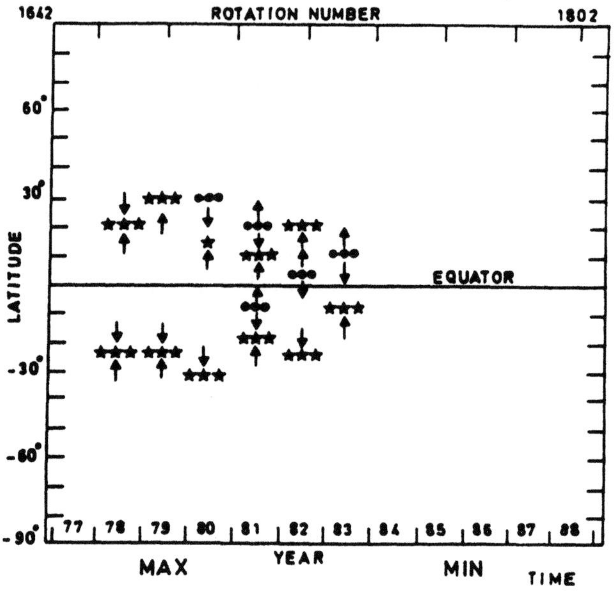

Fig. 3. Meridional circulation pattern deduced from the yearly changes in the differential rotation from one year to the next.

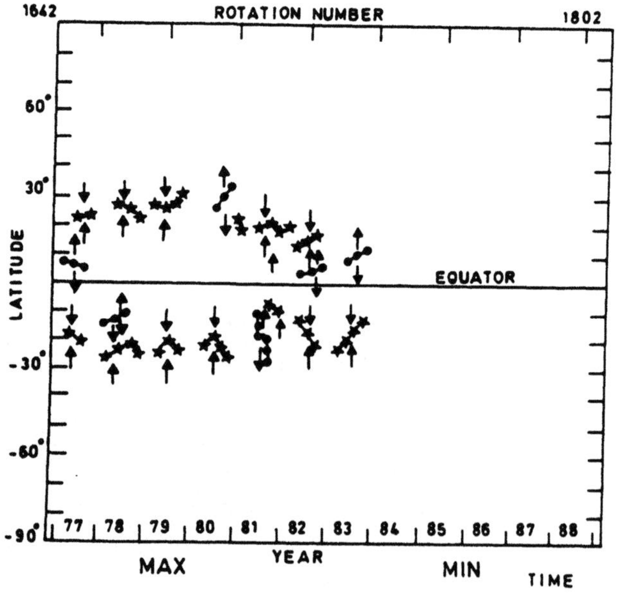

Fig. 4. Direct meridional circulation measurements (north-south drift of sunspots) for the period 1977–1983.

Results are shown in Fig. 3. On the other hand, the zonal meridional circulation (north-south motions) for the whole data set is a direct measurement (Ribes and Bonnefond, 1990), see Fig. 4. It is clear that the deformation of the mean rotation profile (flattening and steepening) can be explained by the presence of a pattern of converging rolls. Some discrepancies do exist, but these can be explained by the fact that our data cover half years rather than the whole years. The amplitude of the changes in the rotation rate obtained in our analysis is almost one order of magnitude greater than the amplitude of the torsional oscillation. A possible explanation for this is that sunspots probe deep into the convective layers while spectroscopic measurements refer to the photospheric plasma. It should also be pointed out that the accuracy of the rotation rate as determined by the a digital analysis of sunspots is significantly greater than by the Doppler measurements.

The poleward propagation of the rolls, as well as their direction of rotation, redistributes angular momentum at the surface and along the radius. This probably accounts for the rotation variations (two peaks per 11-year cycle) reported by Gilman and Howard (1984) on sunspots. Further comparison with Gilman and Howard's results will be published in a separate paper.

References

Balthasar, H., Vasquez, M., Wöhl, H.: 1986, *Astron. Astrophys.* **155**, 87

Hoeksema, T.: 1984, *thesis*, Stanford University

Howard, R.H., Labonte, B.J.:1980, *Astrophys. J. Letter* **239**, L33

Gilman, P.A., Howard, R.H.: 1984, *Astrophys. J.* **283**, 385

Howard, R.H., Gilman, P.A., Gilman, P.I.: 1984, *Astrophys. J.* **284**, 373

Mein, P., Ribes, E.: 1990, *Astron. Astrophys.* **227**, 577

Ribes, E., Mein, P.: 1984, in *Proc. Eur. Astron. Meeting*, ed. R. Muller, Toulouse, p. 283

Ribes, E., Mein, P., Mangeney, A.: 1985, *Nature* **318**, 170

Ribes, E., Bonnefond, F.: 1990, *Geophys. Astrophys. Fluid Dyn.*, in press

Sun-as-a-star: Its Convective Signature and the Activity Cycle

W. Livingston

National Solar Observatory, NOAO
P.O. Box 26732, Tucson, AZ 85726

Abstract: From May 1980 to April 1990 we have observed several times a year the solar irradiance spectrum from 5000Å to 6300Å using the 1-m Fourier Transform Spectrometer. During the ascending phase of the solar cycle, spectrum line asymmetry diminishes about 15% as the total magnetic field encreases. At some epoch near the time of total magnetic field maximum, line asymmetry ceases to track surface magnetism and abruptly assumes a value appropriate to solar minimum.

1. Introduction

At the photospheric level the Sun's surface is covered by the pattern of granulation. High resolution spectra (e.g. Kirk and Livingston, 1968) show that the bright components of granulation are Doppler shifted to the blue while the dark are shifted to the red. These granular motions are attributed to convection driven by the outward transport of heat (Bray *et al.*, 1984). Spatially averaged this brightness-velocity correlation imparts an asymmetry to Fraunhofer lines formed in the high photosphere (e.g. Dravins *et al.*, 1981).

Apparently granular convection is slightly inhibited in the presence of surface magnetic fields so that the resulting line asymmetry is reduced (Livingston, 1982; Cavallini*et al.*, 1985; Immerschitt and Schröter, 1989; Brandt and Solanki, 1990). One explanation for this inhibition is that the presence of magnetic flux tubes, which have an observed preference for inter-granule lanes (Muller, 1989), crowds or confines the convective cells, forcing them to be smaller and convectively less vigorous.

Magnetograph observations indicate that the total amount of magnetic flux, i.e. $|F_+|+|F_-|$, crossing the solar surface varies through the activity cycle (Bruning and LaBonte, 1985). The measured magnitude of the change is sensitive to spatial resolution and the magnetograph zero level. Recently K. Harvey (1990) has derived a time history of the total field variation from Kitt Peak full disk magnetograms.

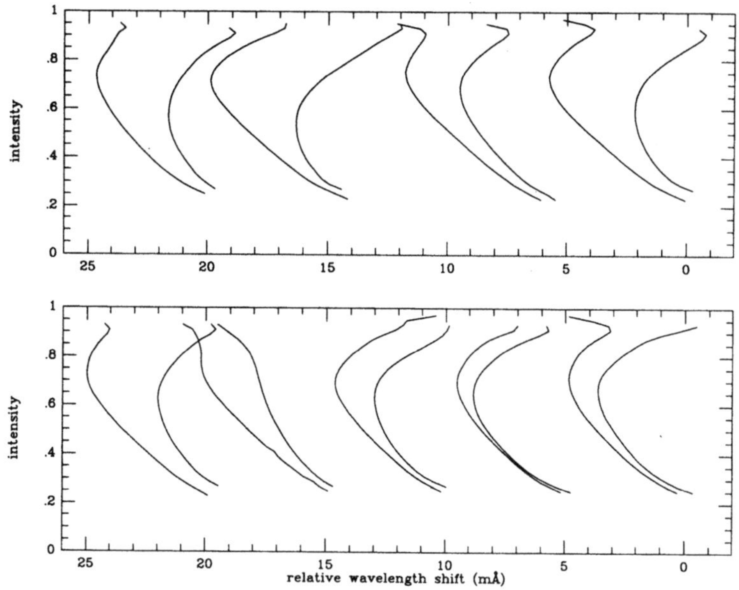

Fig. 1. Samples from the resolved disk with a resolution of about 2 arc-min comparing line bisectors for Fe 5217.4 Å in quiet Sun (left side of each pair) with plage. Although there is considerable variation, in part due to filling factor, in all cases the plage is relatively red shifted and displays less curvature.

She finds that the total field crossing the surface encreased by a factor of about 5 from minimum to maximum in cycle 21 (1976–1986).

The question we pose is this: Given the empirical evidence from the resolved disk that surface magnetism locally reduces line asymmetry, does the activity cycle variation in total field modulate full disk line asymmetry? If the answer is yes, then by observing changes in line asymmetry for other solar-type stars their state of magnetism might be deduced independent of the hard to measure Zeeman effect.

Observations aimed at the above question began in May 1980. Preliminary findings were presented in 1987 (Livingston, 1987), but at that time the temporal record was puzzling to interpret. The main feature was a dip in line asymmetry mid-cycle. Now we have an additional 3 years of data and the situation is clarified. The previous dip remains, of course, but now a new one has developed presumably at about the same phase as in the last cycle. Additionally a sensible relation is also found at the beginning of the cycle: At that initial phase, the magnitude of line asymmetry appears to depend on the total magnetic field.

2. Observations

Spectra in the wavelength interval 5000Å to 6300Å have been obtained with the 1-m FTS in integrated sunlight an average of 3.6 times a year from May 1980 to April 1990. This cadence is set by the availability of the 'visible light beam splitter' (most of the time the FTS is configured for the infrared) together with the vagaries of weather and telescope scheduling. Spectral resolution is about 1,000,000; see Deming (1987) for instrument details.

You-Ran Huang has prepared a list of 18 medium strength iron lines at these wavelengths (average excitation potential = 3.56 eV, average equivalent width = 73 mÅ). Her lines all have comparable bisector curvature and are unaffected by telluric blends. We parameterize bisector curvature as the displacement in mÅ between line core and a tangent to the blue shifted maximum of the line bisector. By averaging all FTS observations for a given day (typically 6 to 8) and then averaging the bisector curvatures for the 18 select lines we obtain a *mean curvature* for that date.

3. Results

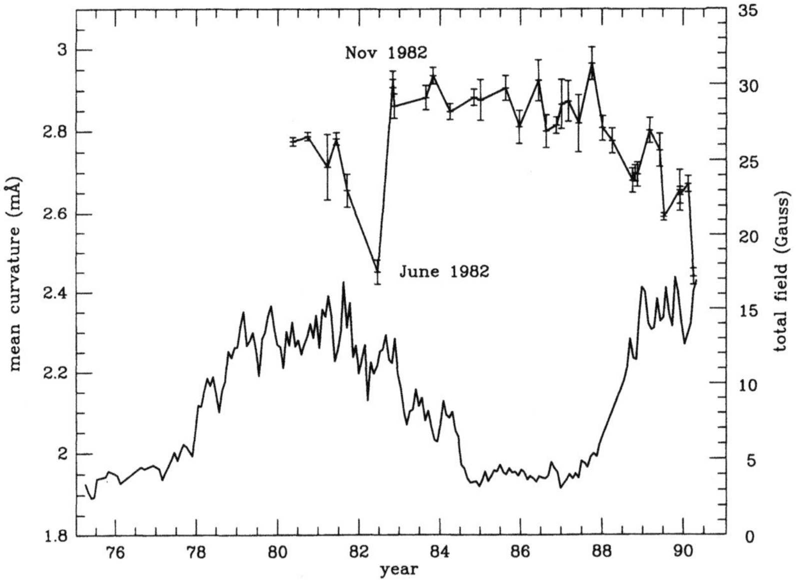

Fig. 2. Mean line asymmetry curvature as a function of date. The error bars reflect the number of FTS scans going into a single day's observation. Also plotted is the total magnetic field averaged over a Carrington rotation.

Figure 2 summarizes the archives up to the present. Of special interest is the abrupt change from June 1982 to November 1982. To demonstrate that the June 1982 bisectors were not anomalous we compare in Fig. 3 bisectors for a few specific lines between June 1982 and April 1990, epochs in which the value of the mean curvature was about the same. We see that these bisectors indeed are similar. On the other hand, if we try to match bisectors between June 1982 and November 1982 the interval of greatest change, we see that the bisectors are noticeably different, Fig. 4. This exercise lends credence to our assertion that the changes we observe are real and significant.

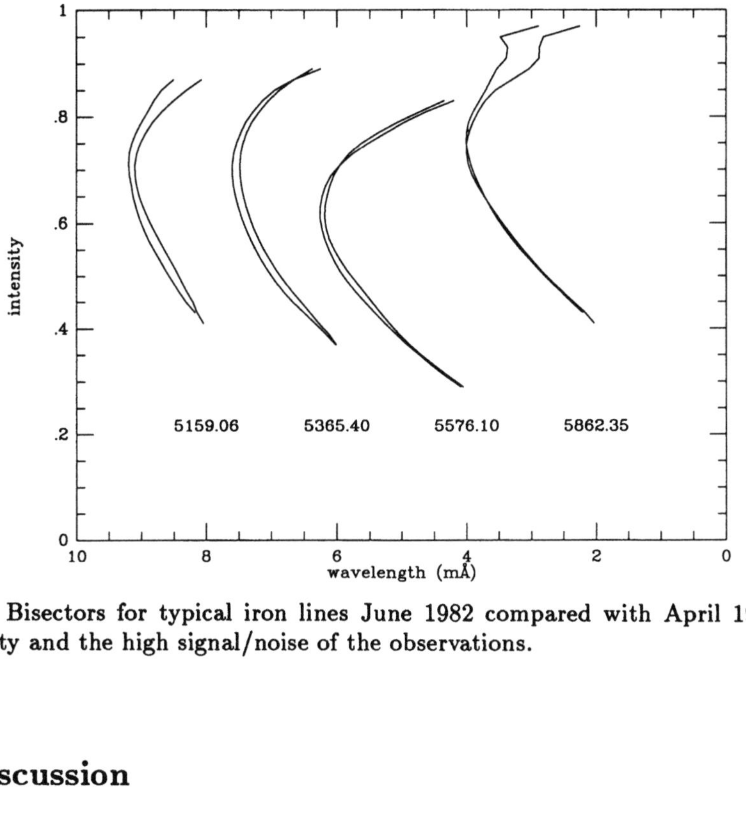

Fig. 3. Bisectors for typical iron lines June 1982 compared with April 1990, showing similarity and the high signal/noise of the observations.

4. Discussion

Included in Fig. 2 is a plot of the total magnetic field as provided by Karen Harvey. Notice that the sampling times for the two plots is disparate. Our mean curvature entries are for single days; the total field values represent averages over an entire Carrington rotation period (about 27 days synodic). Nevertheless, there is a trend toward an anti-correlation between mean curvature and total field from the time of solar minimum in 1986 through the onset of cycle 22. Our early data is also consistent with a congruent behavior during cycle 21, although our archives extend back only to 1980 and not to 1976 which was the previous minimum.

What is strange is the Jun 82 to Nov 82 discontinuity. Obviously we need to see the same process repeat following April 1990. Those observations have not yet been made. Lacking this future material we can only surmise that the process will repeat. Assuming it does, we conjecture that the cyclic process will continue as represented in Fig. 5.

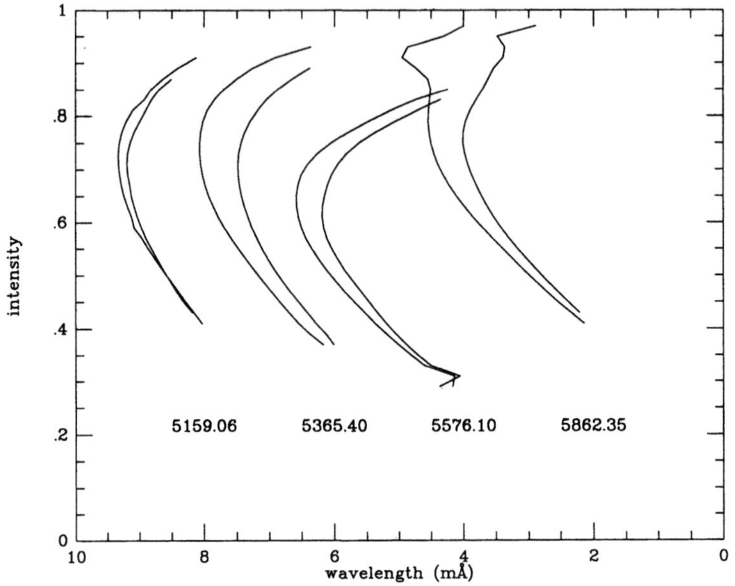

Fig. 4. How bisectors changed between June and November 1982.

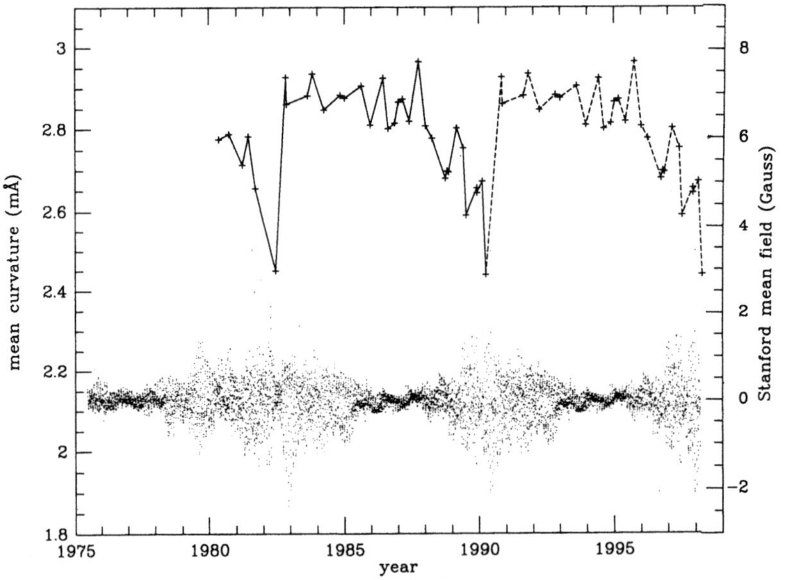

Fig. 5. Hypothetical extension of our observed line asymmetry temporal character assuming it is cyclic. Shown is the mean magnetic field as observed by the Wilcox Solar Observatory, Stanford.

We have no explanation for the Jun–Nov 82 discontinuity. Richard Muller has suggested that we may be sensing a result of the 'switching off' of ***abnormal granulation*** (Muller, 1988). Abnormal granulation is the visible fragmentation of granular structure found in magnetic regions when observed in sub-arc-sec seeing. Perhaps the creation of abnormal granulation depends in a non-linear way on total magnetic field. No doubt if our temporal sampling was more complete we would find this discontinuity replaced by a gradual transition.

In summary it appears that over the initial phase of the activity cycle line asymmetry diminishes by about 15%, and this reduction is anti-correlated with the Sun's total magnetic field. Abruptly mid cycle line asymmetry returns to near its solar minimum value. Why the relation between line asymmetry and the total field does not extend over the entire cycle is unknown. Only further observations can clarify this relationship.

References

Brandt, P.N., Solanki, S.K.: 1990, *Astron. Astrophys.* **231**, 221

Bray, R.J., Loughhead, R.E., and Durrant, C.J., 1984, *The Solar Granulation*, Cambridge Univ. Press, Cambridge

Bruning, D.H., LaBonte, B.: 1985, *Solar Phys.* **97**, 1

Cavallini, F., Ceppatelli, G., Righini, A.: 1985, *Astron. Astrophys.* **143**, 116

Deming, D., Espenak, F., Jennings, D.E., Brault, J.W., Wagner, J.: 1987, *Astrophys. J.* **316**, 771

Dravins, D., Lindegren, L., Nordlund, Å.: 1981, *Astron. Astrophys.* **96**, 345

Harvey, K.: 1990, private communication.

Immerschitt, S. and Schröter, E.-H.: 1989, *Astron. Astrophys.* **208**, 307

Kirk, J.G., Livingston, W.C.: 1968, *Solar Phys.* **3**, 510

Livingston, W.C.: 1982, *Nature* **297**, 208

Livingston, W.C.: 1987, in *The Role of Fine-Scale Magnetic Fields on the Structure of the Solar Atmosphere*, eds. Schröter, E.-H., Vazquez, M. and Wyller, A.A., Cambridge Univ. Press, p. 14

Muller, R.: 1989, in *Solar and Stellar Granulation*, eds. Rutten, R.J. and Severino, G., Kluwer Academic Publ., p. 101

The HeI λ10830 Å Line as an Indicator of the Chromospheric and Coronal Activity of the Sun

A.G. Shcherbakov and Z.A. Shcherbakova

Crimean Astrophysical Observatory,
Nauchny, 334413, Crimea, USSR

Abstract: The behaviour of the HeI λ10830 Å line profile in spectra of solar features is reviewed. The connection of line characteristics with various phenomena on the solar disc, such as plages (faculae), sunspots, jets, prominences, coronal holes, flares, fine structure and emission above the solar limb, is discussed. We present a short summary of the mechanisms for helium excitation in the solar chromosphere. Time variations of the line profile, observed in quiet and active regions, and HeI λ10830 Å absorption in the Sun as a star are discussed and compared with other indexes of solar activity. It is shown that the equivalent width of the helium line varies in an 11 year cycle and reflects the 27 day periodicity of the rotation of the Sun. It is also shown that the HeI line is a more sensitive index of activity than K-index.

1. HeI λ10830 Å line profile on the solar disc

Observations of neutral helium lines in the spectrum of the Sun are limited by the high excitation potential of the atomic levels (20 eV or higher). These levels are populated rather poorly in the solar photosphere. Therefore the absorption occurs in chromospheric regions with low density and high temperature.

This is especially true for HeI λ10830 Å, a line which is formed in the chromosphere by metastable features of its lower level. A model of the helium atom, showing the basic transitions, is presented in Fig. 1. The HeI λ10830 Å line is formed by the transition $2s^3S_1 - 2p^3P_{2,1,0}$ and is composed of three components with the ratio of *gf* values 5:3:1. The two primary components are separated only by 0.09 Å, thus being physically blended into one line at λ10830.3 Å. It is separated from its λ10829.1 Å satellite by 1.26 Å. The effective ratio of *gf*-values for the two resolvable components is 8:1. Figure 2 shows a spectral region around the HeI λ10830 line.

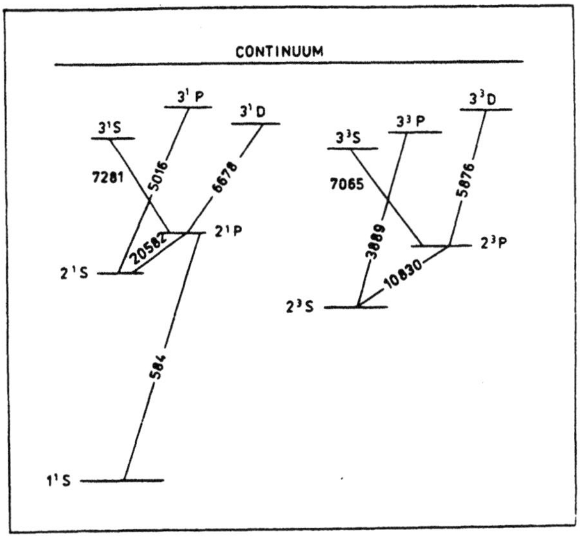

Fig. 1. Atomic levels of the HeI.

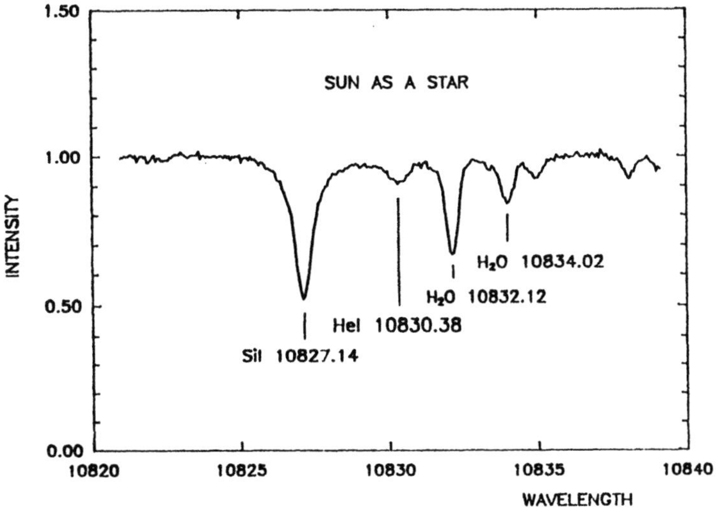

Fig. 2. The spectrum of the HeI λ10830 Å absorption region.

Babcock and Babcock (1934) were the first to observe this line as a weak diffuse absorption in the spectrum of the Sun. That the line originates in strongly inhomogeneous regions with complicated fine structure such as dark faculae, filaments and traces of chromospheric network, was discovered by D'Azambuja and D'Azambuja (1938).

1.1 Variations over the disc and fine structure

It was the equivalent width W_λ of the line that was first studied in detail (Mohler and Goldberg, 1956). They found that the centre-to-limb variation of W_λ in the equatorial direction was higher than in the polar direction. Delbouille *et al.* (1960) estimated W_λ values for several points on the solar disc. Numerous measurements of W_λ were carried out in 1961–1962 by Gulyaev (1964).

Detailed investigations of line profile in undisturbed regions by De Jager *et al.* (1966), showed that the ratio of component depths

$$r = R_0^{\mathrm{obs}}(10830)/R_0^{\mathrm{obs}}(10829) \tag{1}$$

over the whole disc is near to 4 (from 5 to 3) rather than the value $r = 8$, which corresponds to optically thin gas. On the other hand, the depth of the line is very small ($< 10\%$ in continuum units) implying a small number of absorbing atoms. This leads to a contradiction that can be solved assuming that HeI line is formed in unresolved structural inhomogeneities. In this case the observed line depth can be expressed as

$$R_0^{\mathrm{obs}}(10830) = a\, R_0(10830), \tag{2}$$

where the multiplier a is the filling factor. These authors identified such an inhomogeneities with spicules. They suggest that the HeI line occurs in fine structures (spicule type) with equal optical depths and excitation temperature, and that the medium between spicules does not contribute into the formation of line. Their data yielded $a = 0.12$. Nguen-Ngan (1970) came to the same conclusion. He postulated that the line is being formed partially in optically thin layers and partially in optically thick inhomogeneities ($\tau \sim 2$), occupying about 5% of the disc. Giovanelli and Hall (1977) showed that the line depth is twice as large at the boundaries of supergranules than in their centres. According to their measurements the line depth at the boundaries of supergranules can reach 15%.

Detailed analysis of line profile variations across the solar disc was carried out by Shcherbakova *et al.* (1983) and Shcherbakov and Shcherbakova (1983). Fig. 3 shows line profiles in the centre of the disc, at the limb and in a plage region, while Fig. 4 presents the variations of the depth and half width of the line in polar and equatorial directions. As shown in Fig. 4, in the north-south direction the line depth varies only slightly. The darkening on the limb becomes evident only at $\Theta > 70°$. From the comparison of variations of R_0 and $\Delta\lambda_{1/2}$, the main contribution to the equivalent width variations is due to the variation of half-width, but not depth (with the exception of highly disturbed active regions).

High resolution spectroheliograms by Rust and Bridgest (1975) showed even finer structure. The dark chromospheric network in HeI with an average cell size of 40" coincides in general with the bright network in Hα and CaII, but even finer structure is displayed inside the chromospheric network. At the junctions of this network darker knots are observed that are identified with dark helium points (rather stable formations with the lifetime more than 24 hours), corresponding to bright X-ray points. Harvey and Sheeley (1977) found that the most intensified He regions lie at the bases of coronal magnetic loops. The appearance of He points in

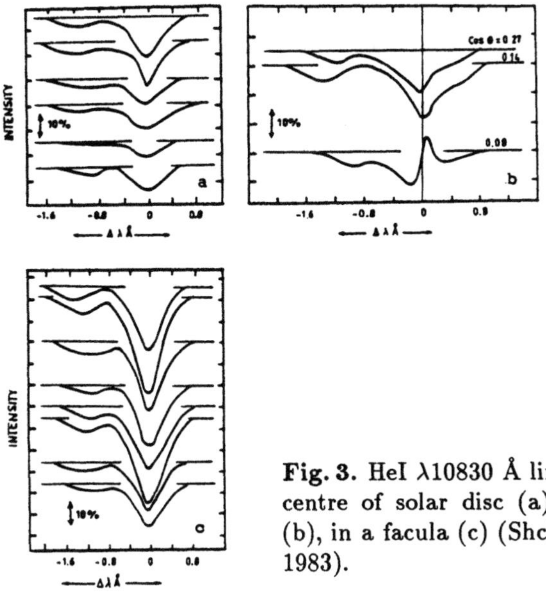

Fig. 3. HeI λ10830 Å line profiles in the centre of solar disc (a), near the limb (b), in a facula (c) (Shcherbakova *et al.*, 1983).

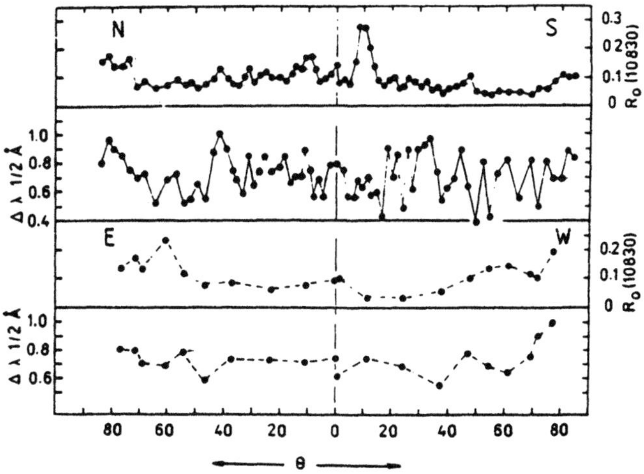

Fig. 4. Central depth and half-width variations of the HeI λ10830 Å line on the solar disc (Shcherbakov and Shcherbakova, 1983).

the network nodes is caused by the presence of intensified magnetic field in these places. Harvey (1985) showed that about 1/3 of the dark points correspond to sites of eruptive magnetic flux (ephemeral active regions), and the remainder to places where elements of the network of opposite polarities meet.

According to Rust and Bridgest (1975) absorption in the helium line has a correlation with the photospheric magnetic field configuration, though some differences are observed. Harvey and Hall (1971) also found a correlation between He line absorption and magnetic field intensity, with the greatest absorption being in places where there is a change of magnetic polarity.

1.2 Plages (Faculae)

In contrast to quiet regions, a strong absorption of the helium line is always observed in plages. Namba (1963) measured the line profiles in the spectra of 22 plage regions. He found that the average depth of the primary component of the line is 24% of the continuum intensity. Analogous results were obtained by Gulyaev (1964) and Nguen-Ngan (1970).

Shcherbakova *et al.* (1983) found that (Fig. 3) the depth and width of the line varies significantly inside plage regions. The mean values of $R_0(10830)$ for some plages does not exceed 30% while for others it exceeds 40 to 50%. The evaluation of physical conditions in plage regions was carried out by assuming plage homogeneity and stability of excitation temperature. Nguen-Ngan computed the filling factor a of absorption for plages as close to 0.5. According to Shcherbakova *et al.* (1983) the value of a was 0.3 for two observed plages, and for two other plages it was near to unity.

1.3 Sunspots

Strong helium absorption ($R = 17\%$ and 30%) was found by Fay *et al.* (1972) in the chromospheric spectrum above two sunspots, the larger sunspot having greater line depth. They suggest that line intensification occurs due to chromospheric heating by magnetohydrodynamic waves, intensified in the spot umbra by its strong magnetic field. Borodina and Papushev (1979) also note the intensification of helium line over the centre of the spot ($R = 20\%$).

At the Crimean Observatory we have investigated the line profile over the umbra and penumbra for ten sunspots (unpublished data). It was found that line strengths over the umbra of a spot and its depth changes from spot to spot, within the limits 16% – 27%. The ratio r of the two components is, on average, about 4 which is the same as for plages.

On the other hand, Lites (1986) observed that the He line was an extremely weak line over a sunspot (7.5%). After a correction for scattered light, Lites gave a value $R = 10\%$ and noted that this is close to the line depth in the undisturbed chromosphere.

1.4 Emission above the limb

The half-width of the absorption line $\lambda 10830$ Å increases toward the limb, with the centre of the line being filled by emission (Fig. 3) which changes to a full emission contour spreading wide above the limb.

According to Gulyaev (1964), the equivalent width of the line reaches a maximum value at 1400 km, with the dependence on height above the plage being similar to that of quiescent regions. Shcherbakov and Shcherbakova (1983) obtained the same result (1500 km). According to Giovanelli and Hall (1977), line emission is maximum at a height of 1700 km.

1.5 Jets and prominences

Besides plages, probably the only site on the solar disc where strong absorption in HeI λ10830 Å is observed (up to 50%) is in jets. Being a projection of intense eruptive prominences on the disc, jets display a complex asymmetrical absorption line and a broad velocity field. Individual jet knots are shifted in the line along the line of sight to 10 – 70 km/s (Shcherbakova, unpublished data). Line half-width increases with the knot velocity along the line of sight, i.e. knots with higher velocities have higher internal motions (Fig. 5).

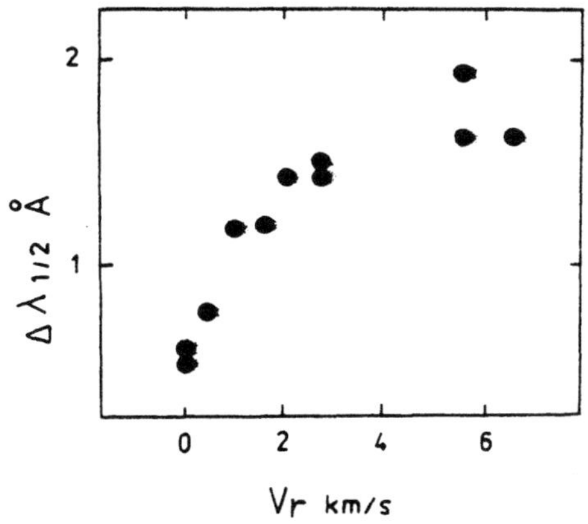

Fig. 5. The HeI λ10830 Å line parameters in a dark jet on the solar disc.

Until now only quiet prominences have been investigated. They were observed by Tandberg-Hanssen (1962), Illing *et al.*, (1975), Landman and Illing (1976), Landman (1976), Shcherbakova *et al.*, (1983), Heinzel *et al.*, (1986). It was found that the excitation temperature of the transition $2^3P - 2^3S$ in prominences corresponds to 4020 K (according to Landman) and that optically thin wings in the line cannot be explained in terms of a simple isothermal model of prominences. A temperature increase in the transition layer between the prominence and corona must be accounted for (Heinzel *et al.*, 1986).

1.6 Flares

Vaughan and Zirin (1968) showed that emission in the HeI λ10830 Å line is possible only in dense regions ($n_e > 10^{12}$ cm^{-3}), where electron collisions become a significant source of quanta at λ10830. In the solar chromosphere such conditions are realized only in regions connected with flares. Tandberg-Hanssen (1962) found that in weak flares the helium line is in absorption, and that only 2B flares show emission.

Baranovsky and Shcherbakova (1985) investigated the emission line profile for 2B and 2 - 3N flares. Line emission for these flares reached 20% of the continuum. They showed that the formation region of HeI λ10830, HeI λ10829 and HeI λ584 Å lines coincides with that of Lα–Lγ line formation and corresponds to a small optical depth $\tau \sim 0.25$ ($r = 7{:}1$)

However, helium emission is not confined to powerful flares. Rust and Bridges (1975), while studying 12 subflares, found that above developing pores bright emission points (kernels) with a size of 3–5" appear in the HeI λ10830 Å line. They came to the conclusion that dark disturbed fibrils in the He line, as well as emission points, coincide with fine structure of magnetic field. He flares develop in the regions about 5" across where there is an emerging magnetic field.

1.7 Connection with coronal holes

Solar images in the IR helium line look like a "negative copy" of the X-ray and far UV pictures. This similarity is most obvious for the negative image of the emission of the HeII λ304 Å line (Harvey and Sheeley, 1977).

While identifying solar image in HeI λ10830 line and soft X-rays, particularly during the SKYLAB mission, it became clear that by using the He line from ground-based observations, one can also observe coronal holes. Harvey *et al.* (1975) showed that the coronal holes in HeI λ10830 Å line are displayed as regions with a "faded" or weakened chromospheric network.

According to Kahler *et al.* (1983) there is a correlation between hole boundaries in HeI λ10830 line and X-rays, though they do not always coincide. Part of the boundaries may be highly diffuse. The best correspondence is observed for extended coronal holes while the correlation is rather poor for small holes in middle and low latitudes. The contrast of coronal holes in HeI line is not stable and depends upon the phase of solar activity.

1.8 Time variations

Sitnik *et al.* (1986) observed quiet and active regions on the solar disc from 1976 to 1985. They found that in the quiet chromosphere the mean values of the helium line depth R_0 during maximum solar activity are 35 – 40% higher than during the period of increasing or decreasing activity. In the chromosphere above plages the values of R_0 increase by up to 15 – 20%. Thus, during the solar maximum the "helium chromosphere" is intensified in quiet as well as in active regions on the Sun, and therefore the intensification of the HeI line in the spectrum of the Sun as a star is caused not only by the fraction of active chromosphere (number and dimension of active regions), but also by a general intensification of the helium chromospheric network.

2. HeI λ10830 Å line variations in spectrum of the Sun as a star.

In the mid 1960's techniques for observations of the "Sun as a star" were developed. For example, a variable magnetic field was detected on the Sun observed in this way. Rapid 27-day period fluctuations, connected with solar rotation, as well as long-term variations caused by solar activity, were discovered (Bumba *et al.*, 1967). Magnetic field variability is explained mainly by the irregular distribution of large-scale magnetic fields on the solar disc together with the evolution of background fields (Bumba and Howard, 1965).

Emission in the CaII K line for the whole solar disc is given by the K-index. Wilson (1968) first defined the K-index and used it for the study of stellar cycles. The correlation between CaII and various active and quiet chromospheric structure was investigated and compared with a number of other solar indexes. For more detailed analysis of the problem one may see Bumba *et al.* (1967a), Sheeley (1967), Shcherbakova (1976), White and Livingston (1978, 1981) and Keil and Worden (1984).

A number of extensive long-term programmes for the study of solar activity indexes, from the far UV to IR regions, was initiated by the National Solar Observatory in the USA. Details of the results of these programmes appear in White (1988). For sunspot cycle 21 he computed indexes such as Wolf numbers, plage-area index, radio flux at 10.7 cm, CaII K-index, HeI λ10830-index and UV fluxes in the regions 1000-4000 Å and shortward of 1000 Å.

Figure 6 shows the comparison of data obtained by White and Livingston for the K-index (a) with the equivalent width of HeI λ10830 line measured approximately monthly by Livingston (b) the same except using daily values by Harvey (c). It is seen that the K-index changes during the activity cycle by an average of 20%. However, more rapid fluctuations, caused by passage of active regions across the solar disc, are also observed. According to Lean (1987) and White (1988), the amplitude of observed variability of K-index cannot be explained entirely by the emission from chromospheric plages. Observed values of the K-index significantly exceed the total emission from all chromospheric plages. The spectrum of quiet regions at the centre of solar disc, according to White *et al.* (1987), shows only a slight variation during one cycle.

Thus, there must exist an additional source of CaII emission, contributing to the total emission of the Sun. Skumanich *et al.* (1984), Lean (1987) and White (1988) come to the conclusion that this additional source could be a so-called "active" chromospheric network appearing outside spots and plage fields. According to Leighton (1964, 1965) the "active" network consists of remnants from destroyed chromospheric plages and spots. These remnants, preserving their magnetic flux, contribute additional emission to the chromospheric network.

During the maximum of activity, the "helium chromosphere" is intensified not only in active regions on the solar disc, but also in quiet regions. Therefore the HeI λ10830 line intensification in the spectrum of the Sun as a star is caused not only by the number and sizes of the active regions, but also by an additional contribution from the "active" chromospheric network. As it is seen from Fig. 6, variations of HeI line equivalent width have a good correlation with K-index.

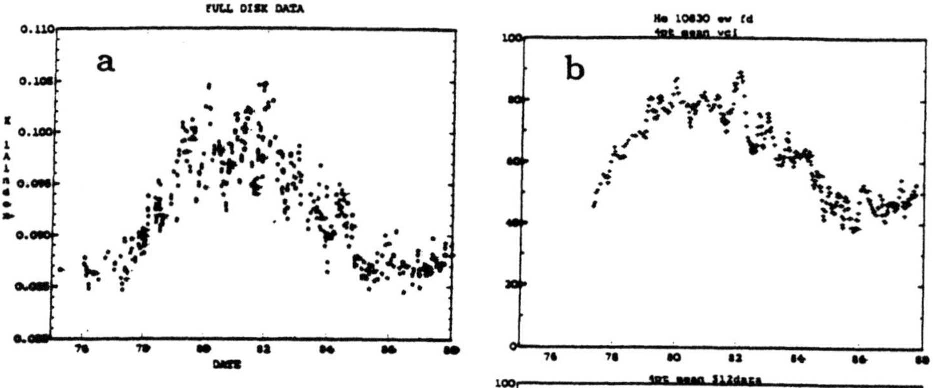

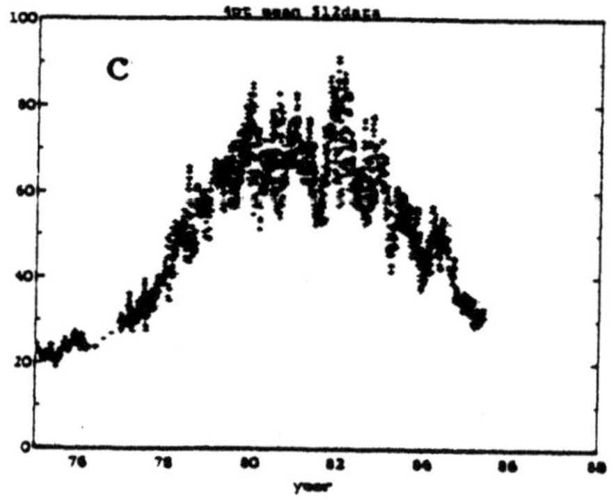

Fig. 6. Variability of K-index (a) and HeI λ10830 Å W_λ (b) Livingston's data and (c) Harvey's data within the activity cycle 21 (White, 1988).

It can see from Fig. 6 that the equivalent width of HeI line varies over the activity cycle by as much as a factor of 4, whereas the K-index shows a maximum amplitude of variability of only 29%. Therefore it follows that the HeI line is a more sensitive index of activity than the K-index. White (1988) found a satisfactory correlation between both indexes and the UV flux from the Sun. Both indexes, in addition to variability connected with activity, show the 27-day period of rotation and correlate better with the UV flux than the radio flux at 10.7 cm.

Similar results in the HeI λ10830 Å line for the Sun as a star were obtained by Shcherbakova and Shcherbakov (1987), Shcherbakova *et al.* (1989) and Shcherbakov *et al.* (1990b) (see Fig. 7). According to their data the main parameter of the line, associated with solar activity, is line depth. For the period from 1981 to 1987 the line depth varied by more than a factor of 3, whereas its half-width remained practically constant.

They confirm that the line depth shows day-to-day variations, with a dispersion that is twice as high in the period of maximum as in the period of minimum activity (see Fig. 7, also Figs. 6 b, c). This difference in amplitude is caused by the passage across the disk of inhomogeneous formations having different lifetimes and absorptions in the He-line. Close to the time of maximum activity the chromospheric network is disturbed by the birth and destruction of short-lived active

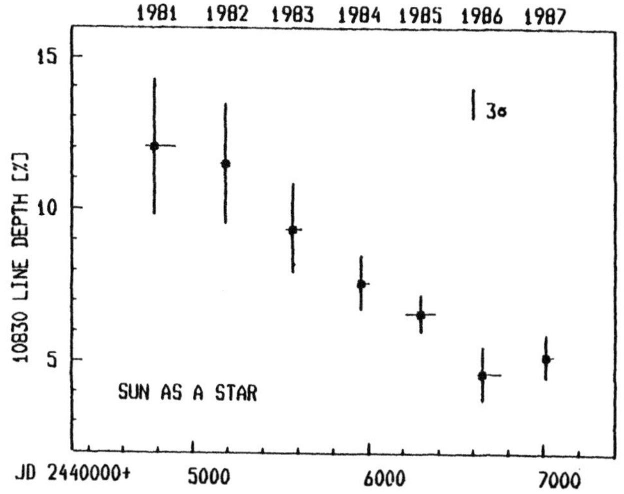

Fig. 7. Variability of the HeI $\lambda 10830$ Å line depth within the cycle 21 (Shcherbakov *et al.*, 1990).

regions having strong absorption in the HeI line and intensified in the quiet chromosphere ("active network"). Near minimum, line depth is affected by the near zero absorption in long-lived coronal holes.

According to Shcherbakov *et al.* (1990) there exists a correlation between such indexes of activity as Wolf number, radio flux at 10.7 cm and FeXIV coronal index CI (Rybansky *et al.*, 1988). The strongest correlation is observed between the line depth and coronal index ($r = 0.87 \pm 0.03$), and the weakest ($r = 0.73 \pm 0.05$) is with the Wolf number. There exists some correlation with the absolute value of the Sun's general magnetic field. The character of the correlation with the coronal index is shown in Fig. 8. As shown by Rybansky *et al.* (1988) the maximum of the coronal index does not coincide with the maximum of the Wolf number, being shifted by about 2 years. The positions of minima, however, coincide on both curves. This means that the times of increasing and decreasing of coronal index are closer to each other than the corresponding time of Wolf numbers where the time of increasing is always much shorter. This phase shift of the maximum relative to Wolf number is also observed for the HeI $\lambda 10830$ Å line (Harvey, 1984) and K CaII emission (Livingston and Wallace, 1987).

The latter circumstance is evidence in favour of a similarity of excitation mechanisms and heights of formation (upper chromosphere, transition region chromosphere-corona) for the indexes in question. The coronal index Fe XIV $\lambda 5303$ Å is determined by the forbidden transition ${}^2P_{3/2} - {}^2P_{1/2}$ and reflects the activity of the lower coronal layers. On the other hand, the absorption in the HeI $\lambda 10830$ line, which is formed by to transition $2_s^2 S_1 - 2_p^3 P^0_{2,1,0}$ is controlled by radiation shorter than $\lambda 504$ Å. The metastable level $2^3 S$ has a high excitation potential and needs electron temperatures of the order of 10000–20000 K. Line weakening

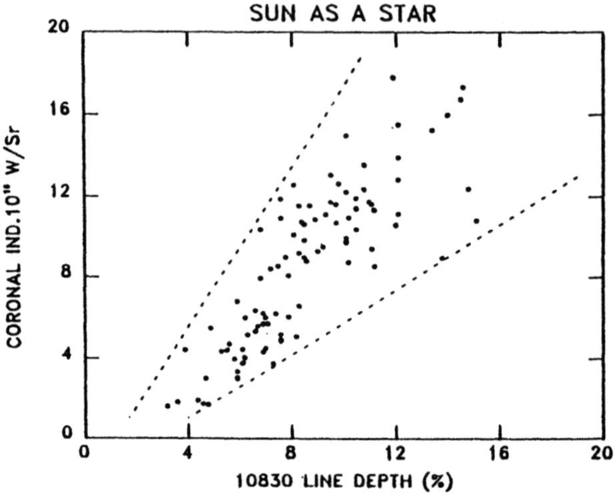

Fig. 8. The correlation between HeI λ10830 Å line depth and Fe XIV coronal index within the cycle 21 (Shcherbakov *et al.*, 1990).

is associated with the decrease of coronal UV back radiation from the corona onto the chromosphere in regions of coronal holes (Zirin, 1975, Livshits, 1975).

3. Helium excitation

Energy levels for the HeI atom are given in Fig. 1. Optical intercombination transitions between singlet and triplet levels in helium are forbidden, therefore the level 2^3S is highly metastable. The level 2^1S is also metastable, but in the conditions of the solar chromosphere it has an optical connection with the primary level through the 2^1P level ; the transition $2^1S - 2^1P$ is realized while absorbing a quantum of λ20582 Åphotospheric radiation while the transition $2^1P - 1^1S$ appears spontaneously. Because of the absence of an optical connection, the triplet and singlet systems are practically independent. As a result, helium turns out to be composed of two sorts of atoms: parahelium (IP = 24.6 eV) and orthohelium (IP = 4.7 eV), and their corresponding triplet levels appear to be overpopulated compared to singlets.

Weakening of the chromospheric HeI network in regions of coronal holes points to the connection with ionization by UV flux. The problem of the existence of helium ionization by short-wave radiation was considered firstly by Goldberg (1939) and then by Shklovsky (1951). After it became clear that triplet absorption lines appear due to the scattering of photospheric radiation (Krat and Sobolev (1960), Johnson (1960)), several authors (see Nickolskaya (1966) for spicules) developed the following mechanism of helium excitation: the neutral atom is ionized by emission with $\lambda < 504$ Å followed by the recombination and delay for atoms in the metastable level 2^3S.

On the other hand, knowing more about the nature of transition layers, it became possible to explain the appearance of regions with $T > 2 \cdot 10^4$ K at low altitudes in the chromosphere, where the excitation and ionization of helium are caused by electron collisions (Athay and Johnson, 1960, Milkey *et al.*, 1973). Milkey *et al.* propose two mechanisms: about half of the opacity in the absorption line appears in high temperature regions, where collisions populate the 2^3S level, and the other half occurs in low-temperature regions, where this level is populated by recombinations.

Gulyaev (1968) computed the concentration of HeI, HeII and HeIII atoms and the population of low levels of orthohelium for two extreme conditions: 1) the ionized emission from the corona back onto the chromosphere is absent; 2) the hard coronal emission is present. The optimal interval for orthohelium excitation appeared to be in the temperature range 20000 – 50000 K. On the other hand, the observed orthohelium concentration might be generated also in cool regions ($T < 10000$ K) by the photoionization of HeI followed by a recombination into triplet levels. Later Gulyaev (1972), from the analysis of Lyman distribution in HeI ($\lambda < 504$ Å), determined that the electron temperature in quiet helium regions is equal to 12500 K. Such a low value of T_e is a serious argument for the role of UV coronal emission in leading to helium ionization. Livshitz (1975), from the analysis of the population of the 2^3S level in flocculi, also showed the advantage of the coronal UV emission excitation mechanism of the metastable level.

On the basis of a non-LTE computation of HeI populations, and using the 10830 line depth, Pozhalova (1987) demonstrated the possibility of making a quantitative estimation of the flux of coronal emission back onto the chromosphere. The VAL-C (Vernazza *et al.*,1982) model atmosphere was used with additional coronal UV illumination flux considered as a parameter. Computations showed that the UV flux really affects the population of the HeI atom. Therefore it is reasonable to expect that the weakening of UV flux in the region of coronal holes would lead to the weakening of He λ10830 line there.

Using the above ideas Pozhalova was the first to calculate HeI line profiles in the quiet chromosphere for various values of background coronal radiation. According to her computations the helium line in quiet regions on the Sun is optically thin.

4. Conclusion

We mention that the study of spectrum of the Sun as a star can be helpful for solving a number of problems of chromospheric and coronal activity of late type stars. Investigations of time variability of the IR HeI line profile in the spectra of different surface structures of the Sun, together with HeI observations in spectrum of the Sun as a star, might lead to the integral flux calibration for the Sun itself and for other stars (solar type stars in particular). Investigations of a number of G – K and some M stars show that the helium line is observed in absorption as well as in emission. There is strong evidence of its variability with time (Vaughan and Zirin, 1968; Zirin, 1976; Zirin, 1982; Shcherbakov, 1979; Katsova and Shcherbakov, 1983; Shcherbakov *et al.*, 1990a).

References

Athay R.G., Johnson H.R.: 1960, *Astrophys. J.* **131**, 413

Babcock H.D., Babcock H.W.: 1934, *Publ. Astron. Soc. Pacific* **46**, 132

Baranovsky E.A., Shcherbakova Z.A.: 1985, *Izv. Krym. Astrofiz. Obs.* **71**, 54 (*Bull. Crimean Astrophys. Obs.* **71**, 52)

Borodina O.A., Papushev R.Zh.: 1979, *Pis'ma Astron. Zh.* **5**, 620 (*Soviet Astron. Letters* **5**, 332)

Bumba V., Howard R.: 1965, *Astrophys. J.* **141**, 1502

Bumba V., Howard R., Smith S.F.: 1967, in *"The Sun as a Magnetic Star"*, ed. C. Cameron, Mono Book, Baltimore, p. 131

Bumba V., Ruzickova-Topolova V.: 1967, *Solar Phys.* **1**, 216

D'Azambuja L., D'Azambuja M.: 1938, *Bull. Astronomique* **11**, 349

De Jager C., Namba O., Neven L.: 1966, *Bull. Astron. Inst. Netherl.* **18**, 128

Delbouille L., de Jager C., Neven L.: 1960, *Ann. d'Astrophys.* **23**, 949

Fay T.D., Willer A.A., Yun H.S.: 1972, *Solar Phys.* **23**, 58

Giovanelli R.G., Hall D.: 1977, *Solar Phys.* **52**, 211

Goldberg L.: 1939, *Astrophys. J.* **89**, 673

Gulyaev R.A.: 1964, *Astron. Zh.* **41**, 313 (*Soviet Astron.* **8**, 243)

Gulyaev R.A.: 1968, in *Solar activity*, eds. E.R. Mustel, E.I. Mogilevski, I.A. Zulin, Nauka, Moscow, p. 104

Gulyaev R.A.: 1972, *Solar Phys.* **24**, 72

Harvey K.L.: 1985, *Austr. J. Phys.* **38**, 875

Harvey J.W.: 1984, in *"Solar Irradiance Variations on Active Region Time Scales"*, NASA Conf., eds. B.J. Labonte, G.A. Chapman, H.S. Hudson, R.C. Wilson, p. 197

Harvey J., Hall D.: 1971, in *"Solar Magnetic Fields"*, IAU Symp. 43, ed. R. Howard, Reidel, Dordrecht, Holland, p. 279

Harvey J.W., Sheeley N.R.: 1977, *Solar Phys.* **54**, 343

Harvey J.W., Krieger A.S., Davis J.M., Timothy A.F., Vaiana G.S.: 1975, *Bull. American Astron. Soc.* **7**, 358

Heinzel P., Kotrch P., Sobotka M., Zloch F., Shcherbakova Z.A.: 1986, *Contrib. Astron. Obs. Skalnate Pleso* **15**, 171

Illing R.M.E., Landman D.A., Mickey D.L.: 1975, *Solar Phys.* **45**, 339

Johnson H.R.: 1960, *Ann. d'Astrophys.* **23**, 838

Kahler S.W., Davis J.M., Harvey J.W.: 1983, *Solar Phys.* **87**, 47

Katsova M.M., Shcherbakov A.G.: 1983, *Astron. Zh.* **60**, 267 (*Soviet Astron.* **27**, 153)

Keil S.L., Worden S.P.: 1984, *Astrophys. J.* **136**, 211

Krat V.A., Sobolev V.M.: 1960, *Izv. Astron. Obs. Pulkovo* **21**, 2

Landman D.A.: 1976, *Solar Phys.* **50**, 383

Landman D.A., Illing R.M.E.: 1976, *Astron. Astrophys.* **49**, 277

Lean J.L.: 1987, *Journ. Geophys. Res.* **91**, D1, 839

Leighton R.B.: 1964, *Astrophys. J.* **140**, 1547

Leighton R.B.: 1965, in *"Stellar and Solar Magnetic Fields"*, IAU Symp. No 22, ed. R. Lüst p. 158

Lites B.W.: 1986, *Astrophys. J.* **301**, 1005

Livingston W.C., Wallace L.: 1987, *Astrophys. J.* **314**, 808

Livshits M.A.: 1975, *Astron. Zh.* **52**, 970 (*Soviet Astron.* **19**, 587)

Milkey R.W., Heasley J.N., Beebe H.A.: 1973, *Astrophys. J.* **186**, 1043

Mohler O.C., Goldberg L.: 1956, *Astrophys. J.* **124**, 13

Namba O.: 1963, *Bull. Astron. Inst. Netherl.* **17**, 93
Nguen-Ngan: 1970, *Astron. Zh.* **47**, 351 (*Soviet Astron.* **14**, 283)
Nickolskaya K.I.: 1966, *Astron. Zh.* **43**, 936 (*Soviet Astron.* **10**, 751)
Pozhalova Zh.A.: 1987, *Pis'ma Astron. Zh.* **13**, 610 (*Soviet Astron. Letters* **13**, 255)
Rust D.M., Bridges C.A.: 1975, *Solar Phys.* **43**, 129
Rybansky M., Rusin V., Dzifcakova E.: 1988, *Bull. Astron. Inst. Czechosl.* **39**, 106
Shcherbakov A.G.: 1979, *Pis'ma Astron. Zh.* **5**, 542 (*Soviet Astron. Letters* **5**, 290)
Shcherbakov A.G., Shcherbakova Z.A.: 1983, *Izv. Krym. Astrofiz. Obs.* **68**, 39 (*Bull. Crimean Astrophys. Obs.* **68**, 37)
Shcherbakov A.G., Tuominen I., Jetsu L., Katsova M.M., Poutanen M.: 1990a, *Astron. Astrophys.* **235**, 205
Shcherbakov A.G., Shcherbakova Z.A., Tuominen I., Rybansky M., Rusin V.: 1990b, *Astron Astrophys.* (to be submitted)
Shcherbakova Z.A.: 1976, *Izv. Krym. Astrofiz. Obs.* **55**, 37
Shcherbakova Z.A., Shcherbakov A.G.: 1987, *Izv. Krym. Astrofiz. Obs.* **76**, 98 (*Bull. Crimean Astrophys. Obs.* **76**, 110)
Shcherbakova Z.A., Shcherbakov A.G., Shapiro B.I., Heinman A.S.: 1983, *Izv. Krym. Astrofiz. Obs.* **66**, 119 (*Bull. Crimean Astrophys. Obs.* **66**, 109)
Sheeley N.R.: 1967, *Astrophys. J.* **147**, 1106
Shklovsky I.S.: 1951, *Ann. Sternberg Astron. Inst.* **20**, 5
Sitnik G.F., Kozlova L.M., Divlikeev M.I., Porfiryeva G.A.: 1986, *Solnechnie Dannie*, No 12, 67
Skumanich A., Lean J.L., White O.R., Livingston W.C.: 1984, *Astrophys. J.* **282**, 776
Tanberg-Hanssen E.: 1962, *Ann. d'Astrophys.* **25**, 357
Vaughan A.H., Zirin H.: 1968, *Astrophys. J.* **152**, 123
Vernazza J.E., Avrett E.H., Loeser R.: 1982, *Astrophys. J. Suppl.* **45**, 635
White O.R.: 1988, in *"Solar Radiative Output Variations"*, ed. P. Foukal, Cambridge Res. and Instr. p. 88
White O.R., Livingston W.C.: 1978, *Astrophys. J.* **226**, 679
White O.R., Livingston W.C.: 1981, *Astrophys. J.* **249**, 798
White O.R., Livingston W.C., Wallace L.: 1987, *J. Geophys. Res.* **92**, D1, 823
Wilson O.C.: 1968, *Astrophys. J.* **153**, 221
Zirin H.: 1975, *Astrophys. J.* **199**, 63
Zirin H.: 1976, *Astrophys. J.* **208**, 414
Zirin H.: 1982, *Astrophys. J.* **260**, 655

The Rotational Modulation of Ca II K in the Sun

I. Sattarov and A. Hojaev

Astronomical Institute of Uzbekistan Academy of Sciences,
700052 Tashkent, USSR

The most widely used indicator of the stellar magnetic activity is the flux in the CaII K-line core (K-index) (Baliunas and Vaughan, 1985). The K-index data have also been used for measuring the rotation of stars. But using the method for the Sun gives different results (Keil and Worden, 1984; Singh and Livingston, 1987). The reason for the observed differences, besides those indicated by Singh and Livingston, may be the character of the distribution of active regions. This study is based on observations made at Tashkent Astronomical Observatory and the data published in SGD for solar cycle 21. We study the longitudional distribution of sunspots and plages. Some intervals of active longitudes (IAL) were selected and the evolution of them was studied. Active regions were found to concentrate in certain longitude intervals which are in nearly rigid rotation. Fig. 1 shows the longitudinal distribution of sunspots areas for 1983–84, as an example.

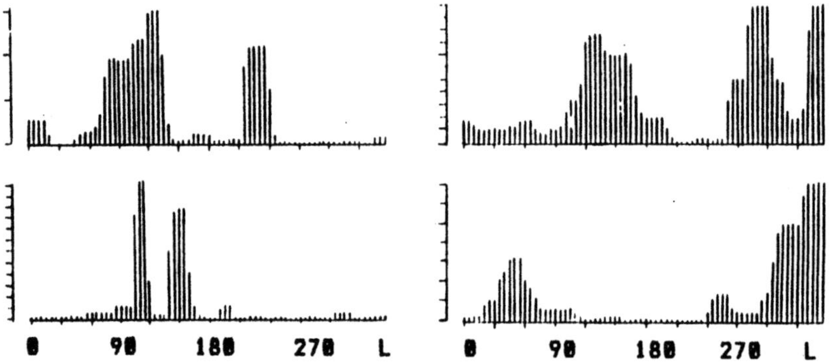

Fig. 1. Longitudinal distribution of sunspot areas for north (left) and south (right) and for 1987 (above) and 1984 (below).

The activity in the southern hemisphere was predominant in 1983 and had three IAL (in the northern hemisphere there were two). Under poor spatial resolution (the Sun as a star) the intervals merge and can form the homogeneous background. In such cases the rotational modulation of the K-index gives a poor result. In 1984 both hemispheres showed nearly equal activity, but in the north the activity was concentrated in a comparatively narrow interval of longitudes (90°–140°). Therefore the rotational modulation analysis (RMA) of the K-index from Tucson (Singh and Livingston, 1987) and Sacramento Peak (Keil and Worden, 1984) obtained in 1983 may give inexact, different results and exact, equal ones in 1984. The first groups started to form close to the eastern border of the above narrow interval and the site of formation of further new active regions is gradually shifted more and more westwards (Sattarov and Bumba, 1990). The velocity of the westward shift of the centre of activity is in the range of 40 m/sec. The rotational rate obtained from the K-index at Tucson for 1984 has an excess of 100 m/sec.

The evolution of the active regions in the narrow interval leads to the formation of large-scale magnetic fields (LMF) with a line of polarity reversal, which rotates around the pivot point anticlockwise (Fig. 2). Such a LMF rotates nearly rigidly and perhaps influences the surface rotation. So the solar rotation during summer 1982, observed earlier by Howard (1984), may be connected with the LMF on longitudes 300°–360°.

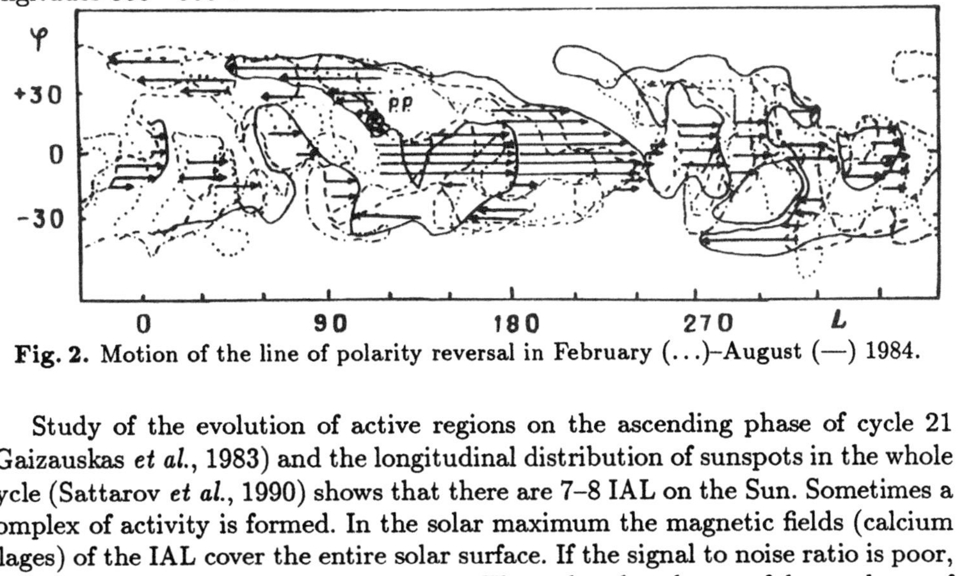

Fig. 2. Motion of the line of polarity reversal in February (...)–August (—) 1984.

Study of the evolution of active regions on the ascending phase of cycle 21 (Gaizauskas *et al.*, 1983) and the longitudinal distribution of sunspots in the whole cycle (Sattarov *et al.*, 1990) shows that there are 7–8 IAL on the Sun. Sometimes a complex of activity is formed. In the solar maximum the magnetic fields (calcium plages) of the IAL cover the entire solar surface. If the signal to noise ratio is poor, the RMA of the K-index will be inexact. The isolated and powerful complexes of activity usually appear in the ascending and particularly the descending phase of cycle. In such periods the K-index data analysis gives exact results.

References

Baliunas, S.L., Vaughan, A.H.: 1985, *Ann. Rev. Astron. Astrophys.* **23**, 379
Keil, S.L., Worden, S.P.: 1984, *Astrophys. J.* **276**, 766
Singh, J. and Livingston, W.S.: 1987, *Solar Phys.* **109**, 387
Sattarov, I., Bumba, V.: 1990, in *The Dynamic Sun*, ed. L. Dezsõ, Debrecen, p. 48
Howard, R.: 1984, *Ann. Rev. Astron. Astrophys.* **22**, 131
Gaizauskas, V., Harvey, J., Harvey, K. and Zwaan, C.: 1983, *Astrophys. J.* **265**, 1056
Sattarov, I., Mustaeva, F.G., Litvinov, O.V.: 1990, *Issledovaniya solnechnoj activnosty*, eds. Yuldashbaev, T.S., Korobova, Z.B., in press

The Prolonged Minima and Maxima of Solar Activity

V.N. Dermendijev[1], G.T. Buyukliev[1], Y.Y. Shopov[2]

[1]Department of Astronomy and NAO, 72 Lenin Blvd., 1784 Sofia, Bulgaria
[2]Faculty of Physics, University of Sofia, 5 Anton Ivanov Blvd., Sofia, Bulgaria

Abstract: We study a new indirect index of solar activity – the Intensity of Luminescence of Cave Flowstone Microzones. This index correlates directly with the solar activity. Using a time series of the index with resolution 4–5 pixels/year we study some of the statistical properties of the solar activity cycle during 28 prolonged minima and 21 prolonged maxima.

1. The new index of solar activity

Proxy data on past solar activity nowadays provide a useful means of approaching the problem of solar magnetic field generation (Dermendjiev *et al.*, 1990). For this purpose, however, we need time series of data capable of distinguishing individual solar cycles. In this work we study from this point of view 28 prolonged minima of solar activity in the past (until 22 000 B.C.) using time series of a new indirect index.

The new index was named "Microzonality of Luminescence of Cave Flowstones" (Shopov and Dermendjiev, 1990). For its detection the method of Laser Luminescence Analysis (LLMZA) was developed (Shopov, 1987). If the growth rate of the flowstone is known (e.g. from the nuclear dating of the sample), the intensity curve of the luminescence, depending on the distance from the flowstone surface, can be transformed into a time series.

In our study we use time series with resolution of 4–5 pixels per year. They make it possible to obtain information on the phase, period and epoch of the minimum, and other data about the properties of the separate 11-year cycles.

2. The prolonged minima and maxima of the LLMZA data

We had the possibility of obtaining time series of LLMZA data with time resolution about 5 pixels/year spanning 22 000 years. It contains 110 000 measurements. We use this long time series of proxy data of solar activity to seek answers to the following questions: 1) Does the 11-year cycle continue during the Maunder, Spörer, and other prolonged minima of the past solar activity? 2) Is the 11-year cycle typical of the epochs of normal activity and prolonged maxima of solar activity? and 3) Is there any correlation between duration of the minima, considered as phase shifts of a very long cycle reflecting possible global stochastic nature of solar activity, and the amplitude change of the subsequent activity?

We determined 28 prolonged minima and 21 prolonged maxima. Among them are the Modern and Medieval maxima, Maunder and Spörer minima, and the other 14 prolonged minima and maxima which have been found in ^{14}C time series. We study these peculiar intervals in the past solar activity by means of power spectral analysis.

The largest period for the Maunder minimum is 5.6 years and for the Spörer minimum 7.6 years. The power spectra for these two epochs differ insignificantly but they are rather different in comparison with the power spectra of the Medieval maximum.

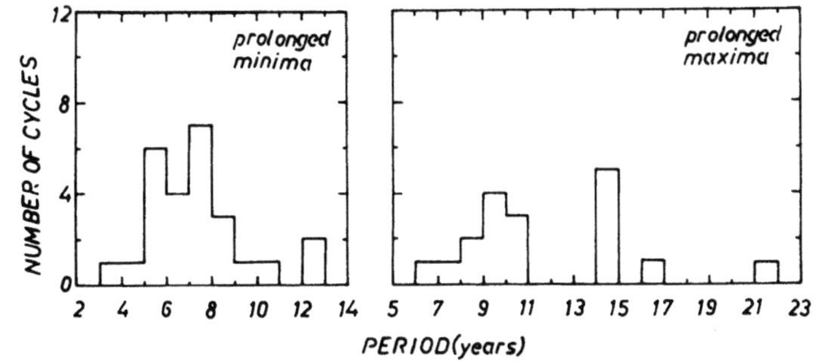

Fig. 1. Empirical distribution of the solar cycle periods for the prolonged minima and maxima.

The empirical distributions of the periods obtained for the prolonged minima and maxima determined from LLMZA data are shown on Fig. 1. As one can deduce from this figure, the solar cycle is shorter during the prolonged minima (mean value $T = 7.85$ years). The distribution of the periods during the prolonged maxima shows a special feature. When the solar cycle is longer, the preferred period is 14–15 years. It is interesting to note also that there are no periods of 11–14 years. The very good time resolution of LLMZA data also gives us the possibility of studying the correlation between the duration of the prolonged minima, considered as phase shifts of a very long cycle, and the amplitude change of the subsequent activity maxima. The results obtained from a 12000 years long time series are shown on

Fig. 2. The correlation coefficient $r = 0.422$ is positive and statistically significant. The data can be approximated by linear regressions.

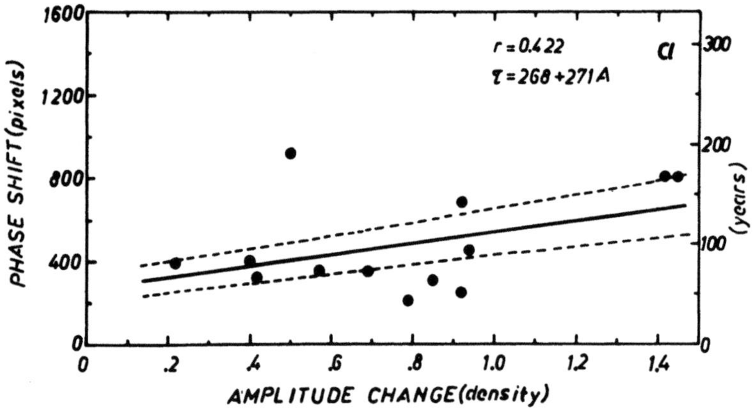

Fig. 2. Correlation diagram of the duration of the prolonged minima and amplitude change of the subsequent activity maxima for the time intervals 1985 AD – 10000 BC.

The problem of the nature of prolonged minima of solar activity is still under discussion. From a modern point of view the prolonged minima are considered as strange attractors (Ruzmaikin, 1985). For large dynamo numbers the solutions of the Lorenz system for the magnetic field are chaotic in time with episodes of very low amplitude. It will be interesting to discover if such a positive correlation exists between episodes of reduced activity and the amplitude of the envelopes of the magnetic field taken from theoretical models of nonlinear dynamos.

Acknowledgement. This work was supported by the Ministry of Culture, Science and Education under Grant No. 1162/88.

References

Dermendjiev, V.N., Shopov, Y.Y., Buyukliev, G.T.: 1990, *Solar Phys.* **128**, 217

Ruzmaikin, A.A.: 1985, *Solar Phys.* **100**, 125

Shopov, Y.Y.: 1987, in *"Proc. of Int Symp. on Complex Research of Mountain Karst"*, Tbilisi, p. 6

Shopov, Y.Y., Dermendjiev, V.N.: 1990, *Comptes rendus de l'Academie Bulg. des Sci.*, in press

The Longer Term Evolution of Magnetic Field and Mass Flow in a Decaying Active Region

Hongqi Zhang[1], Guoxiang Ai[1], Haimin Wang[2],

Harold Zirin[2], Alan Patterson[2]

[1]Beijing Astronomical Observatory, Chinese Academy of Sciences, P. R. China

[2]Big Bear Solar Observatory, California Institute of Technology, U.S.A.

During September 24–29, 1987, the longitudinal magnetic fields of a decaying active region (AR 4855) were observed almost continuously at Big Bear and Huairou Solar Observatories. The target consisted of an αp sunspot and surrounding enhanced network. We achieved a 75-hour coverage with 4 night-time gaps of six to eight hours. The images were then combined to made a continuous movie.

The BBSO Videomagnetograph (VMG) System was originally developed by Leighton and Smithson and has been improved in recent years (Zirin, 1985). This system has made it possible to study very weak photospheric magnetic structures.

The vector videomagnetograph system at Huairou Solar Observing Station of Beijing Astronomical Observatory, located on an island of the Huairou reservoir 60 km NE of Beijing, was designed by G. Ai (1987). It consists of a 35 cm vacuum telescope, a 1/8 Å birefringent filter with 3 sets of KD*P crystal modulators, a CCD camera and an Imaging Technology 151 system controlled by an AST-386 system, which transmits the data to a Vax/11-750. It works at either of two spectral lines: FeI λ5324.19 Å for photospheric magnetic and line-of-sight velocity observations and Hβ λ4861.34 Å for chromospheric observation. The field of view is about $5.7' \times 4'$ and the pixel resolution is about $0.7'' \times 0.4''$. This videomagnetograph also has high spatial and temporal resolution. With good seeing conditions, the smallest magnetic features visible are smaller than $2''$. The longitudinal component (Stokes parameter V) of the vector magnetic field in the solar photosphere was measured at 0.075 Å from the line center of FeI λ5324.19 Å and the transverse components (Stokes parameters Q and U) at the line center. The photospheric Dopplergrams were measured with the subtraction technique at $\pm$0.15 Å from the line center of FeI λ5324.19 Å and the chromospheric Dopplergrams at $\pm$0.24 Å from the center of Hβ λ4861.34 Å. After performing a 4×3 pixel ($2'' \times 2.1''$) spatial averaging and integrating 255 frames, the signal to noise ratio is 5600 corresponding to $\pm$2 G of longitudinal field, $\pm$15 G of transverse field, $\pm$2 m/s line-of-sight velocity field for the photosphere and $\pm$8 m/s for the chromosphere.

The decaying active region was located on the northern hemisphere of the Sun and the region was west of CMP. The sunspot had negative magnetic polarity. Some positive elements occurred on the west penumbra of the sunspot, which was in the opposite direction from the solar disk center. The transverse components of the sunspots magnetic field showed the spoke-like extensions from the center and connected the longitudinal fields of opposite polarity. We showed that the positive structures intruded gradually into the penumbra of the spot from September 25 to September 27 and the heliocentric angle of the target increased from about 30 to 60. The transverse components of fields in the penumbra were strong. From the distributions of the vector magnetic fields, we could infer that the longitudinal magnetic structures of opposite polarity that intrude into the sunspot were probably the projection effects of inclined magnetic force lines from a sunspot, that returned to the solar surface near to or in the penumbra, and the lines of force were almost parallel to the solar surface.

The outward flow of magnetic flux plays a very important role in the evolutionary history of a decaying sunspot. The shift and velocity pattern of the magnetic elements is shown in Fig. 1. The magnetic fields consist of the moving magnetic features (MMF) flowing outward around the sunspot. In the westward direction this flow is almost exclusively monopolar and opposite in sign to the umbral field, even though part of the moving magnetic features have mixed magnetic polarities (predominantly of positive polarity with some very weaker negative elements). However we found that some of negative magnetic fragments of the sunspot were moving outwards from the edge of the magnetic field of the sunspot at an average speed of about 0.3 km/s. The direction of these motions appeared irregular. Some of fragments formed a part of network magnetic fields of the same magnetic polarities. Some moved along the outer side of the moat and approached and cancelled with magnetic elements of opposite sign.

The decay rate of sunspot magnetic flux for the same polarity as the associated sunspot is found to be 10^{20} Mx/h on September 25 and 4×10^{19} Mx/h on September 26, but for the opposite polarity -10^{18} Mx/h on September 25 and 0.8×10^{17} Mx/h on September 26. The increase of the sunspot magnetic flux of the opposite polarity on September 25 is due to a projection effect associated with the rotation of the sunspot, even through the magnetic flux with polarity opposite to parent sunspot move out from the sunspot. The high-quality movies show these larger magnetic fragments to be eroding the sunspot edge. The fragments of magnetic fields outflowing from the sunspot probably are the main mechanism of magnetic flux loss of the sunspot.

The motions of the magnetic elements of the enhanced network also appear irregular, the average velocity of elements is about 0.3 km/s. We find that some of magnetic elements converge and merge. Some magnetic elements break and the fragments of the magnetic field move away in different directions, other elements suddenly change direction and others still move consistently in one direction. Some magnetic elements move about 3.7×10^4 km at 43 hours whereas some, of opposite polarities, collide, in which case the total flux then decreases because of cancellation.

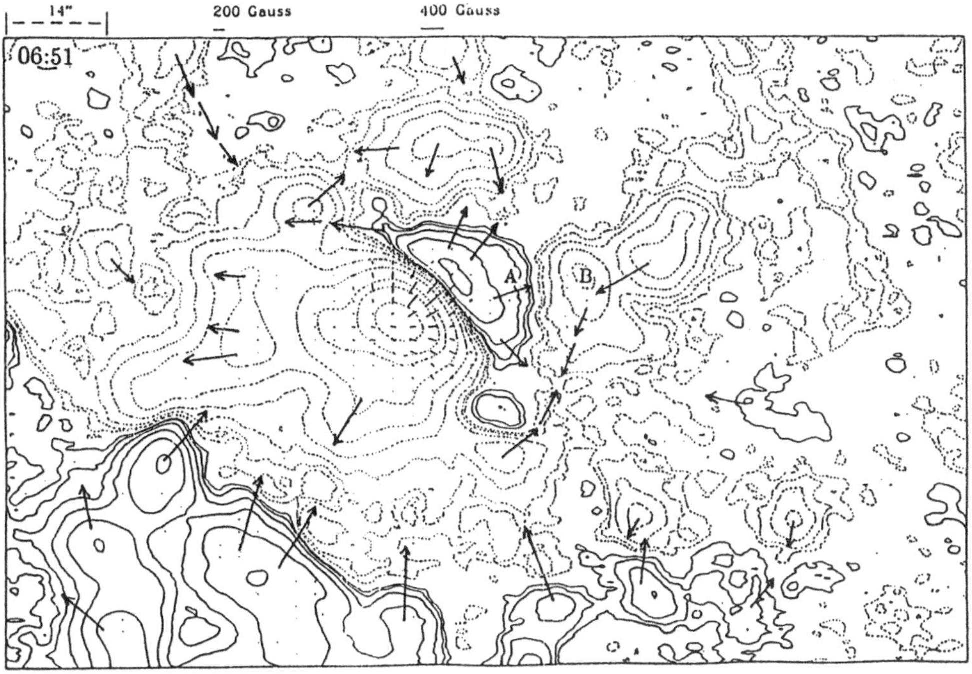

Fig. 1. The vector magnetogram of Active Region 4855 of Sept. 26, 1987. The arrows show the shift and direction of motion of the magnetic features averaged over September 26 and 27. Solid (dashed) contours correspond to positive (negative) fields of 5, 20, 40, 80, 160, 320, 640, 960 Gauss. North is at the top, west is at right.

We now present the results of attempting to measure the velocity of MMF's and velocity fields simultaneously on Sept. 26 and 27. The profiles of velocity and magnetic fields are asymmetric through the sunspot. The distributions of velocity and magnetic fields are different. By tracking individual features, we found that the speed of moving magnetic features is about 1.0 – 2.5 km/s, which is greater than photospheric Doppler measurement. On the other hand, the negative fragments of sunspot field move outward from the edge of sunspot at an average speed of about 0.3 km/s, similar to the corresponding Doppler velocity. We can infer that the mechanism driving the motion of the magnetic fragments may be different from that of discussed above for the moving magnetic features. It probably is the outward motions of the feet of the magnetic lines of force, near to the outer edge of the photospheric penumbra.

On September 26 and 27, our observed chromospheric Doppler velocities from this decaying active region mainly reflect components of the horizontal velocities. We compared the Doppler measurements of photosphere and chromosphere. The immediate result shown is that the chromospheric and photospheric velocities are generally in opposite phase, i.e. if the chromosphere has a red shift, the photosphere has a blue shift. Although the maximum of the chromospheric field has a small offset from the minimum of the photospheric field on Sept. 26, the maximum and the minimum coincide on Sept. 27. This means that the direction of the chromospheric mass motion in the superpenumbra is opposite to that of the photosphere. We find that the maximum of the chromospheric Doppler velocity fields shifts toward the penumbra from outside of the sunspot from September 26 to 27. The velocity maximum occurs to the outside of the inversion of longitudinal magnetic fields on the penumbra. Comparing the distribution of the chromospheric Doppler velocity fields with the photospheric vector magnetograms, it is difficult to infer the result that the directions of chromospheric mass motion are completely parallel to the directions of chromospheric force lines above our observed photospheric vector magnetic fields, especially near the inversion line of the photospheric longitudinal magnetic fields. It is possible that the magnetic lines of force are curved in the chromosphere or that mass flows along a complex return circuit in the sunspot atmosphere.

Acknowledgements. We are grateful to the staff at Huairou Solar Observing Station and Big Bear Solar Observatory for their support in making the observations. This research was supported by the Chinese Academy of Sciences and National Science Foundation of China and NSF of U.S.A. under the grant NSF INT-8814395.

References

Ai, G.: 1987, *Publications of Beijing Astronomical Observatory* **9**, 27
Zirin, H.: 1986, *Aust. J. Phys.* **38**, 961

The Main Component Analysis of the Longitudinal Distribution of Solar Activity

Ladislav Hejna[1], Hubertus Wöhl[2]

[1]Dept. of Computer Technique, Charles University, Prague, Czechoslovakia
[2]Kiepenheuer-Institut für Sonnenphysik, Freiburg, FRG

Abstract: In this contribution, preliminary results of the main component analysis of Bartels diagram of time series of daily values of sunspot group numbers for solar cycles 18, 19 and 20 are presented. The results obtained suggest that the most significant feature in the longitudinal distribution of sunspot activity is the existence of preferred solar hemispheres alternating with a mean period of 2.5 Bartels rotations.

The issue of the existence of preferred (active) longitudes in the solar activity distribution over heliographic longitudes is still one of the open question of present-day solar astronomy. (See, for instance, Gaizauskas, 1985). That is why it is sensible to study these phenomena using other data sets, employing classical as well as quite new methods for their analysis.

The main aim of this contribution is an attempt to study the longitudinal distribution of sunspot activity using the expansion of the time series of daily values of sunspot group numbers (for the eleven-year cycles 18, 19 and 20), arranged in the form of Bartels diagram, into the set of its proper orthogonal functions (see for instance Vertlib *et al.*, 1971).

If we interpret each row of a Bartels diagram as a concrete realization of some general longitudinal distribution (affected by the influence of evolutionary processes and so on), we may look for an expansion of this distribution into a set of its proper orthogonal functions (in the main component analysis sense) and the shape of these functions will describe single independent periods of the distribution that we are looking for. In other words, if our Bartels diagram is represented by the matrix N_{ik} ($i = 1, \ldots, m$ being the number of Bartels rotations counted from the first rotation in the solar cycle number 18; $k = 1, \ldots, 27$ being the day number in the corresponding Bartels rotation) we are interested in a development in the form $N_{ik} = a_{i\alpha} \Phi_{\alpha k}$, where $\alpha = 0, \ldots, 26$ is a summation index, and $a_{i\alpha}$ as well as $\Phi_{\alpha k}$ are both unknown before processing and where for $\Phi_{\alpha k}$ an orthogonality condition has to be satisfied. For the complete expansion we need also the least squares minimum condition and the condition of maximum information density compression also to be satisfied. It may be shown that the proper functions we are looking for must be identical to the eigenfunctions of the covariation matrix $B_{kr} = N_{ik} N_{ir}$. The corresponding eigenvalues describe the part of the total dispersion explained by the relevant proper function. In our analysis the rows of the matrix N_{ik} have been normalized to have zero mean.

The most important of the proper functions obtained, with $\lambda_\alpha / \Sigma\lambda = 0.247$, indicates that the most significant component in the longitudinal distribution of sunspot activity (explaining approximately a quarter of the total dispersion) may be recognized as corresponding to the situation when one of the solar hemispheres is preferred (its activity is above average). The boundaries between these two hemispheres are stable during the complete time interval investigated and lie at longitudes corresponding to Bartels days 9 and 23. The signs of the expansion coefficients determine which of the hemispheres is preferred in any given Bartels rotation.

The behaviour of the expansion coefficients $a_{i\alpha}$ in time seems to be at first glance rather chaotic, but a test for whether the sequence of their signs is random or not (Sachs, 1982) shows that for the first proper function the behaviour of corresponding coefficients (for level of significance $p = 0.05$) is not random. The mean length of repetitions is roughly 2.6 Bartels rotations. Periods of above-average activity on one hemisphere are therefore replaced by periods of reduced activity with just this time period.

The strongly smoothed (for whole solar cycles) values of the expansion coefficients for the most significant proper function are equal to -1.291 (for solar cycle 18), -0.732 (19) and 0.349 (20). It means that during the cycles number 18 and 19 the hemisphere corresponding to the second half of the Bartels rotation was generally preferred, whereas during the cycle number 20 we observe the opposite situation. This is in good agreement with results obtained by Balthasar and Schüssler (1983, 1984) and with their conception of solar "memory".

Generally we can conclude that these preliminary results indicate that the method of principal orthogonal components applied to data sets of daily values of selected solar activity indexes may be a very interesting and effective instrument for studying the longitudinal distribution solar activity.

References

Balthasar, H., Schüssler, M.: 1983, *Solar Phys.* **87**, 23
Balthasar, H., Schüssler, M.: 1984, *Solar Phys.* **93**, 177
Gaizauskas, V.: 1985, in *Inv. Rev. Present to Solar-Terr. Prediction Workshop*, Meudon
Sachs, L.: 1982, in *Applied Statistics*, Springer-Verlag, New-York, p. 375
Vertlib, A.B., Kopecký, M., Kuklin, G.V.: 1971, in *Issled. Geomagn. Aeron. i Fiz. Solntsa* 2, p. 194

Fe I Line Asymmetries and Shifts Caused by Pressure Broadening

V. Kršljanin, I. Vince and S. Erkapić

Astronomical Observatory, Volgina 7, 11050 Belgrade, Yugoslavia

Abstract: In order to test the validity of the usual neglect of pressure line shifts in investigations of Fe I line asymmetries and shifts, we calculated synthetic line bisectors and shifts of 96 moderate and weak photospheric lines, taking into account both hydrogen- and electron-impact shifts and neglecting the velocity fields. Absolute pressure line shifts obtained are less than or appr. equal to 50 m/s and the largest relative shifts along the bisectors are about 40 m/s. We found pronounced dependences of both shifts and bisectors on the energy of the upper level in transitions and on the transition array.

Pressure broadening produces small (Vince *et al.*, 1985b), but some cases observable (Vince and Dimitrijević, 1989) line asymmetries and shifts of NaI lines. It was shown recently (Kršljanin, 1989) that some solar Fe I lines also exhibit measurable pressure broadening effects. Estimates of Fe I lines hydrogen-impact shifts (Gomez *et al.*, 1987) appear to be insufficient to determine overall pressure shifts, since collisions with electrically charged particles can produce line shifts of the same order of magnitude as collisions with hydrogen atoms do (Vince *et al.*, 1985a,b). On the other hand, detailed theoretical calculations of Fe I lines pressure widths and particularly shifts are still missing due to complicated atomic structure of iron.

In order to estimate the importance of pressure broadening contribution to the observed asymmetries and shifts of Fe I lines we therefore decided to investigate averaged pressure broadening effects on many Fe I lines using simple theoretical approaches to hydrogen-impact (Hindmarsh *et al.*, 1967) and electron-impact shifts (Kršljanin, 1990). These approaches give good average accuracy when applied to many lines. So, such a method promise to give results comparable with previous convective motion investigations (e.g. Dravins *et al.*, 1981) performed also on large samples of Fe I lines.

We synthetisied, using a LTE code and Holweger-Müller (1974) photospheric model, profiles of 96 Fe I lines (from the list of Dravins *et al.*, 1981) in the centre of the solar disk, with hydrogen- and electron-impact shifts taken into account as the only causes of wavelength displacements in the line profiles. Lines were chosen that had observed central line depths less or equal to 0.5 and for which reliable atomic data was available (gf values at least of accuracy C according to Fuhr *et al.*, 1981). Wavelength shifts and averaged bisectors (normalised to central intensities) of the synthetic line profiles are presented in Fig. 1. The bisector shapes are mainly determined by perturbed density distribution in the line formation regions and often by opposite sings of hydrogen- and electron-impact shifts. The shifts of the line bottoms and the relative wavelength displacements along the bisectors are

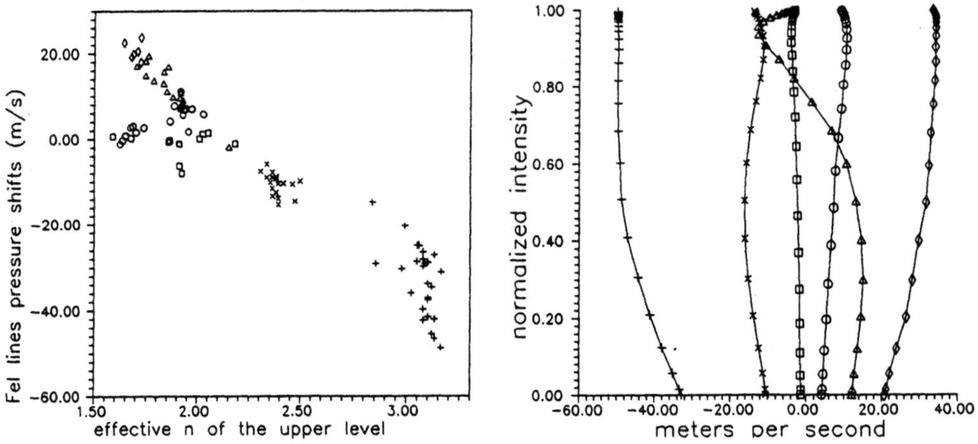

Fig. 1. Fe I lines pressure shifts (left) and averaged bisectors (right) for the transition arrays: (Δ) $3d^8 - 4p$; ($\sqcap$) $4s^2 - 4p$ with lower energy of the upper level parent terms; ($\diamond$) $4s^2 - 4p$; ($\bigcirc$) $4s - 4p$; ($\times$) $4p - 5s$; ($+$) $4p - 4d$.

very small in comparison with the observed ones (Dravins *et al.*, 1981). Another important fact which appears from our results is the pronounced dependence of the pressure shifts and line bisectors on the energy of the upper level in the transitions, i.e. on the effective principal quantum number of the upper level, and on the type of transition, i.e. on the transition array. Such a dependence, if eventually found in stellar observational data, might be an indicator of an important pressure broadening contribution to the observed line shapes. However, even in the solar case the pressure shifts should be considered for some Fe I lines (e.g. for lines from the $4p - 4d$ transition array with relatively large pressure shifts and from the $3d^8 - 4p$ transition array with relatively large bisector curvatures). Our calculations give an estimate of the value of pressure broadening effects in diagnostics of convective motions from the shifts and shapes of Fe I lines.

References

Dravins, D., Lindgren, L., Nordlund, Å.: 1981, *Astron. Astrophys.* **96**, 345

Fuhr, J.R., Martin, G.A., Wiese, W.L., Younger, S.M.: 1981, *J. Phys. Chem. Ref. Data* **10**, 305

Gomez, M.T., Marmolimo, C., Roberti, G., Severino, G.: 1987, *Solar Phys.* **112**, 227

Hindmarsh, W.R., Petford, A.D., Smith, G.: 1967, *Proc. Roy. Soc.* **A297**, 296

Holweger, H., Müller, E.A.: 1974, *Solar Phys.* **39**, 19

Kršljanin, V.: 1989, in *Solar and Stellar Granulation*, eds. R.J. Rutten and G. Severino, Kluwer, Dordrecht, p. 91

Kršljanin, V.: 1990, in *22nd EGAS*, ed. A. Wännström, Dept. Phys. Univ. Uppsala, p. 757

Vince, I., Dimitrijević, M.S.: 1989, in *Solar and Stellar Granulation*, eds. R.J. Rutten and G. Severino, Kluwer, Dordrecht, p. 93

Vince, I., Dimitrijević, M.S., Kršljanin, V.: 1985a, in *Spectral Line Shapes*, ed. F. Rostas, W. de Gruyter, Berlin, p. 649

Vince, I., Dimitrijević, M.S., Kršljanin, V.: 1985b, in *Progress in Stellar Spectral Line Formation Theory*, eds. J.E. Beckman and L. Crivellari, D. Reidel, Dordrecht, p. 373

Rotation of Large Scale Patterns on the Solar Surface as Determined from Filament and Millimeter Data

S. Pohjolainen[1], B. Vršnak[2], H. Teräsranta[1], S. Urpo[1],

R. Brajša[2], V. Ruždjak[2], S. Jurač[2], A. Schroll[3]

[1]Metsähovi Radio Research Station, 02540 Kylmälä, Finland
[2]Hvar Observatory, 58450 Hvar, Yugoslavia
[3]Sonnenobservatorium Kanzelhöhe, 9520 Sattendorf, Austria

Abstract: The rotation of large scale solar magnetic field patterns was studied using quiescent filaments and low temperature regions observed at 37 GHz as tracers.

1. Introduction

Solar filaments and low temperature regions (LTR) observed in the millimeter range are located at the inversion lines of the large scale solar magnetic fields pattern (Vršnak *et al.*, 1991). These features can be used as tracers to study the differential rotation on global scale. Filament data were obtained at the Kanzelhöhe Observatory (Austria) and 8 mm data at the Metsähovi Radio Research Station (Finland). A solar map measured at 37 GHz is presented in Fig. 1.

2. Differential rotation of filaments from the Kanzelhöhe observations

The sample of filaments which were used as tracers from the Kanzelhöhe data consisted of 124 polar crown filaments, in order to examine the differential rotation up to a latitude of 80 degrees. The positions of footpoints were measured to trace the rotation. Taking an average footpoint height to be 7000 km above the photosphere and matching the results with the measurements at low latitudes (Adams and Tang, 1977) we obtained the differential rotation law

$$\omega = 14.45 \pm 0.15 - (0.11 \pm 0.9) \sin^2 \phi - (3.69 \pm 0.9) \sin^4 \phi \qquad (1)$$

which is close to the standard law obtained by other methods. This law includes implicitly the time dependence, since the polar crown shifts towards the pole through the cycle. The differential rotation of filaments is presented in Fig. 2.

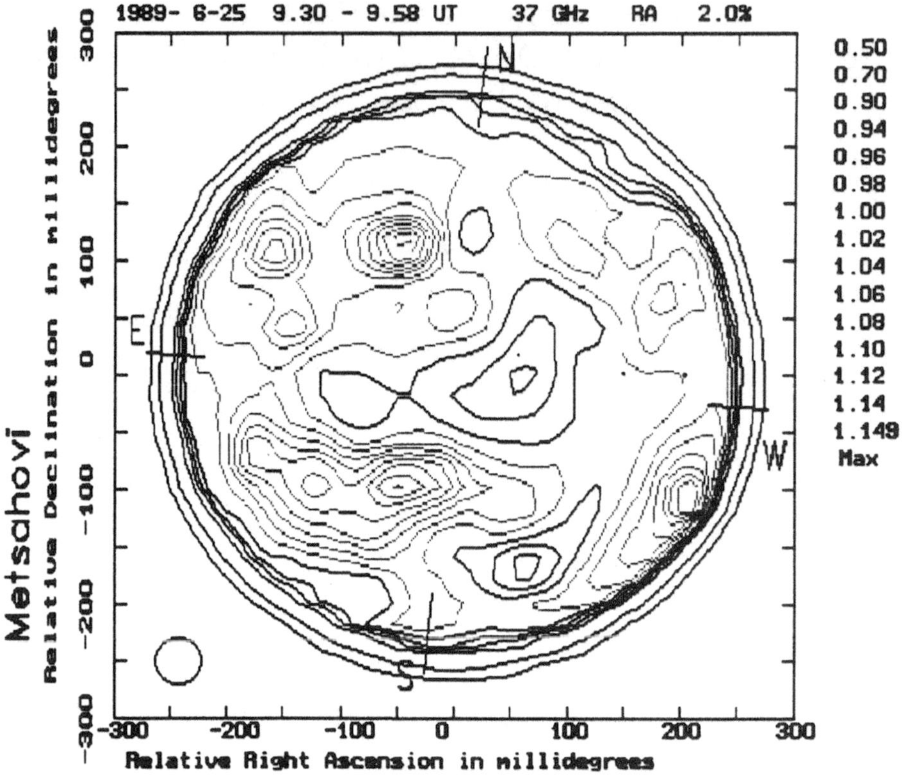

Fig. 1. Solar radio map measured at the Metsähovi Radio Research Station on June 25, 1989 at 37 GHz (λ = 8 mm). Black solid lines represent low temperature regions and grey lines active regions, compared to the quiet Sun level (estimation 7800 K at 8 mm wavelength).

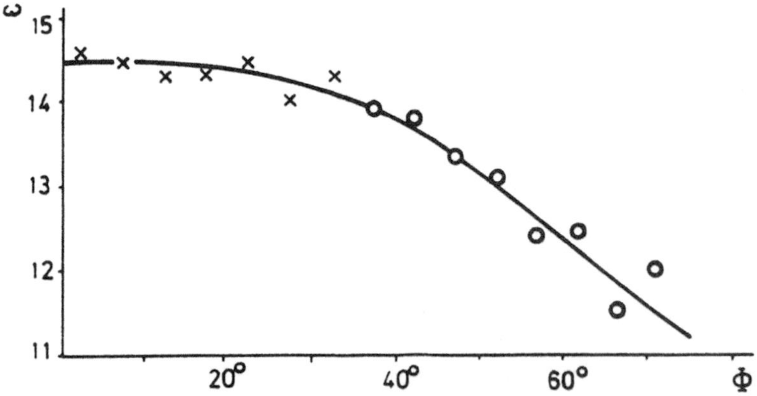

Fig. 2. Sidereal differential rotation (ω) of filaments (deg/day). Crosses represent the results by Adams and Tang (1977).

3. Differential rotation of 8 mm low temperature regions

The sample of 214 LTRs covering the years 1979–82 and 1987–89, and latitudes up to 55 degrees, were used as tracers and showed a differential rotation of the form

$$\omega = 14.46 \pm 0.15 - (2.91 \pm 0.57)\sin^2\phi. \tag{2}$$

The rotation curve is presented in Fig. 3a. This is somewhat faster at higher latitudes than in standard laws and implies an effective average height of 15000 km for the LTRs. However, when we checked the time development of the differential law we found statistically significant variations: During the maximum activity years 1979/80 and 1988/89 the rotation was almost rigid, and the effective height of LTRs would have to be 30000–40000 km to explain such behaviour in terms of projection effects. In periods 1981/82 and 1987 (close to the minimum) the rotation was following the standard differential rotation law, implying low heights of about 5000 km for LTRs. The rotation curves for the separate years are shown in Fig. 3b. The polar crown filaments also showed an increase of differential rotation at high latitudes (Brajša *et al.*, 1989), which could be explained by the same kind of time development since polar crown filaments occupy these latitudes at about solar maximum.

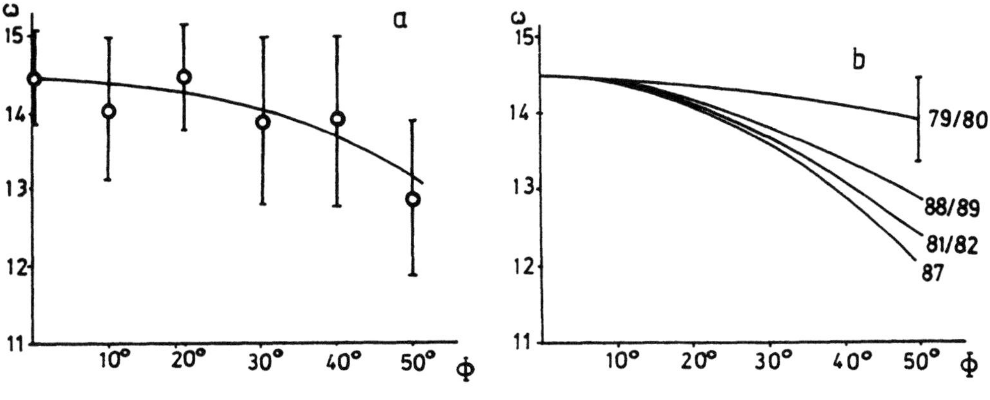

Fig. 3. Sidereal differential rotation (ω) of LTRs (deg/day). a) The complete sample with error bars representing the standard deviations, b) the time variation with error bar representing the average standard deviation at the latitude of 50 degrees.

References

Adams, W.M., Tang, F.: 1977, *Solar Phys.* **55**, 499

Brajša, R., Vršnak, B., Ruždjak, V., Schroll, A.: 1989, *Hvar Obs. Bull.* **13**, 449

Vršnak, B., Pohjolainen, S., Teräsranta, H., Urpo, S., Brajša, R., Ruždjak, V., Schroll, A., Jurač, S.: 1991, these Proceedings

Large Scale Patterns on the Solar Surface Indicated by Microwave Observations

B. Vršnak[1], S. Pohjolainen[2], H. Teräsranta[2], S. Urpo[2],

R. Brajša[1], V. Ruždjak[1], A. Schroll[3], S. Jurač[1]

[1]Hvar Observatory, 58450 Hvar, Yugoslavia
[2]Metsähovi Radio Research Station, 02540 Kylmälä, Finland
[3]Sonnenobservatorium Kanzelhöhe, 9520 Sattendorf, Austria

Abstract: A large set of observations of the Sun at 37 GHz is analysed. An association of 99% is found between the regions of brightness temperature depression and the magnetic field inversion lines. Observations indicate a possible existence of "giant cells" with duration of 1–2 years and a longitudinal extension up to 90°.

1. Introduction

Observations of the Sun in the millimetric range disclose two categories of features which are not related to active regions. These are the areas of low amplitude brightness temperature depressions (Schmahl *et al.*, 1981) and low amplitude enhancements occurring in polar regions (Urpo and Pohjolainen, 1987).

We present here a study of low temperature regions (LTRs) based on a large set of observations measured at 37 GHz at the Metsähovi Radio Research Station during the years 1979-1989. The beam size of the 13.7 m radio telescope is 2.4 arc min at 8 mm wavelength. The temperature resolution is typically better than 1% of the quiet Sun brightness temperature (estimated as 7800 K at 8 mm).

2. Results

When the positions of the temperature minima within LTRs are compared with the inversion lines of the inferred magnetic fields, Fig. 1, one finds that 99% of the LTRs are lying on the inversion lines presented in the Solar Geophysical Data (SGD) or in the Solnechnie Dannye Byulleten (SDB). Comparing the positions of LTRs with the positions of filaments (Meudon Observatory, courtesy Z. Mouradian) one finds a 60% association.

The latitudinal distribution of the LTRs discloses two belts of LTRs which embed a belt of active regions of the actual cycle, Fig. 2. A cradle in the LTR distribution associated with the forthcoming cycle is formed at latitudes of about 50°, already at the maximum of previous cycle.

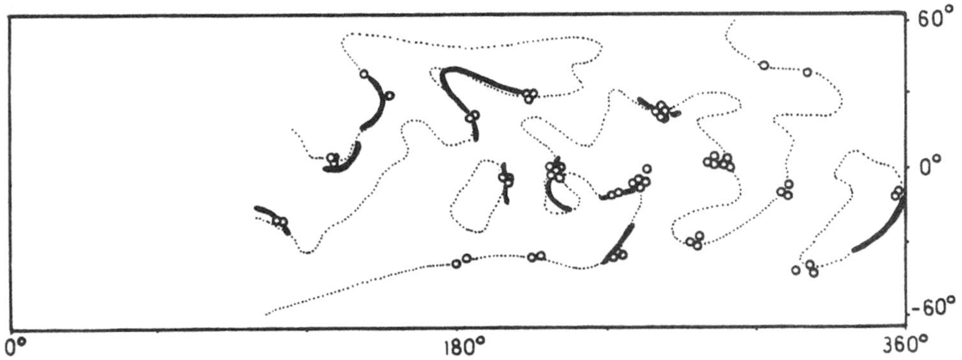

Fig. 1. Association of LTRs (circles) with the SGD magnetic inversion lines (dotted line) and filaments (bold lines).

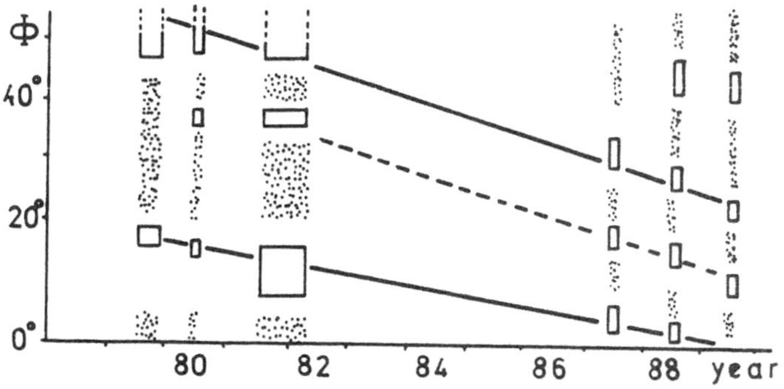

Fig. 2. Latitudinal distribution of microwave features (dotted - LTR belts; rectangles – minima in LTR distribution). Active region belts of the Cycles 21 and 22 (bold lines) are separated by a broken line.

In Fig. 3 the superposition of LTRs in the period from May 1979 to May 1980 is presented. The distribution of LTRs indicates a possible existence of large-scale cell-like patterns which are arbitrarily outlined by a thin line. Such patterns could also be found in other analysed periods. The typical longitudinal and latitudinal extensions are 60°–90° and 30°–40°, respectively. The life-time could be estimated to be 1–2 years. In order to check the reliability of these patterns we compared our results with the results based on magnetographic data (Bumba, 1987). In Fig. 3 the positions of the cell centers taken from Bumba's study are presented, and there is a very good correspondence.

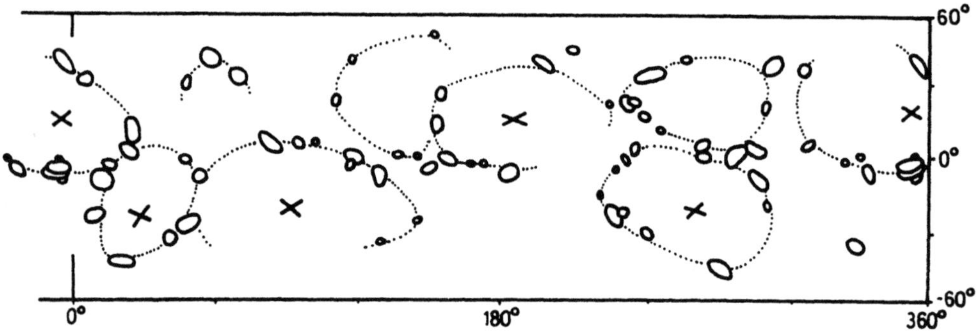

Fig. 3. Superposition of LTRs observed from May 1979 to May 1980. The "cells" are sketched by dotted lines and the centers of Bumba's cells are indicated by crosses.

3.1 Discussion and conclusions

The LTRs are directly related to the magnetic inversion lines, while the association with filaments could be accidental since filaments occupy a considerable fraction of the inversion line length. The LTRs represent a large-scale phenomenon related to the evolution of the solar cycle. Although a belt of LTRs corresponding to the polar crown filament belt could be detected on the microwave maps, it was not possible to find any poleward migration as LTRs are rarely detected at latitudes higher than 50°.

Finally, let us note that we can not claim that our results undoubtedly indicate the existence of giant cells but we would like to stress the very good correspondence with the results obtained by Bumba (1987).

References

Bumba, V.: 1987, *Bull. Astron. Inst. Czech.* **38**, 92

Schmahl, E.J., Bobrowsky, M., Kundu, M.R.: 1981, *Solar Phys.* **71**, 311

Urpo, S., Pohjolainen, S.: 1987, *Hvar Obs. Bull.* **11**, 137

The Solar Corona over the Recent Saros

Jay M. Pasachoff

Williams College–Hopkins Observatory, Williamstown, Mass. 01267, USA;
Institute for Advanced Study, Princeton, N.J. 08540, USA

Abstract: We are now at both the maximum of the solar activity cycle and at the most populated part of the saros. I discuss the solar corona over the recent saros and its changes with the solar activity cycle. We consider the scientific value of eclipse studies and how they relate to other ongoing coronal studies on the sun and other stars.

1. The saros

Similarly to the way that the beauty of total solar eclipses depends on the coincidence of the angular sizes in the sky of the sun and the moon, the saros depends on the coincidence between 223 synodic months (the period of the phases), which take 6585.32 days, and 19 eclipse years (returns of the sun through one of the nodes where the lunar orbit meets the ecliptic), which take 6585.78 days. The difference is only 0.46 day. In addition, 242 nodical months (also called draconic months, after the mythological Chinese dragon that eats the sun at eclipses), each of which marks the passage of the moon through the node, are only 0.04 day different from the 223 lunar months. Further, 239 anomalistic months (the period of variation of the moon–earth distance) are also within a fraction of a day, so not only the existence of an eclipse but also its general appearance – total or annular, long or short – repeat with this 18 year 11 1/3-day period, the saros (Pasachoff, 1991a). The current saros has a dozen eclipses in it. Thus 22 July 1990, saw the repeat after a saros of the eclipse of 11 July 1972. The 1/3–day brought the eclipse from Canada in 1972 to Finland, the Soviet Union, and the U.S. Aleutians, in 1990.

2. The value of eclipse observations

Total solar eclipses still provide the best view astronomers can get of the middle solar corona. The very lowest corona can be seen from coronagraphs usually out to less than a solar radius. The outermost corona has been visible for long periods from spacecraft, such as Solar Maximum Mission. SMM's coronagraph occulted out to 1.75 solar radii, and additional corona was lost because of diffraction rings. Thus the region between 1.1 and 2.0 solar radii is best observable at eclipses.

Further, no spaceborne coronagraph is now aloft. The corona is observable in x-rays on occasion from rockets, but the frequency with which Golub *et al.* can make x-ray solar images is similar to that of total solar eclipses – a few minutes about once a year. These x-ray images show only the lower corona, and can be mosaiced nicely with white–light eclipse images whenever a rocket flight can be made on the day of an eclipse (which I therefore suggest as desirable).

Other advantages of eclipse observations include the possibility of taking newly developed equipment into the field. Lead times for space missions now approach 10 years and costs are hundreds of millions of dollars. For less than 0.1% of this cost, one can take an eclipse expedition into the field. Even allowing for cloudy occasions, solar eclipse expeditions are very cost–effective.

Since different eclipses take place at different times in the 11–year solar activity cycle, studying the corona at different activity phases requires a concerted effort over many years. Studies of the solar corona are made all the more important by our new abilities to observe stellar coronas. Important problems like the cause of coronal heating seem best solvable by increased scrutiny of the sun.

3. Solar eclipses through the most recent saros

The 1972 eclipse, one saros ago, took place halfway down the cycle. The major e vent in the saros is the long eclipse that took place on 30 June 1973 in Africa and that will repeat on 11 July 1991. The year 1973 was a time of low sunspot number; the corona had mainly equatorial str eamers. My own experiments at that time observed the ratio of the [Fe XIII] spectral lines at 10 747 Å and 10 798 Å, whose ratio is especially density sensitive.

Weather at the 1974 eclipse in Australia was cloudy for ground-based observers, preventing measurement of the coronal electron temperature by studying the Doppler scattering of Fraunhofer lines, following calculations by Cram. We hope a sensitive CCD will allow us to make a temperature map by this method in 1991. The 1976 eclipse also occurred in Australia under patchy clouds.

For the 1977 eclipse, we brought two inertial-guidance platforms aboard a ship. We made an image in the [Fe XIII] 10 747 Å line through a hole in the clouds (Pasachoff *et al.*, 1978), but the skies were not photometric. We were thus not able to make the desired coronal density map. At the 1979 eclipse, seen through hazy skies over the U.S. and Can ada, the corona took second place to the prominences, which heralded the coming solar maximum.

The 1980 solar-maximum eclipse crossed from Africa to India. We started testing methods of coronal heating via surface Alfvén waves by searching for 1-Hz oscillations in coronal loops, and found signs of excess power in that frequency range in the Fourier transforms (Pasachoff and Landman, 1984). The isophotes showed the round, solar-maximum appearance of the corona. The 1981 corona was still of the solar-maximum type.

The 1983 eclipse was already at a time of declining solar-activity cycle, as the isophotes showed. Though the eclipse was a long one, the sky was not very clear. We again found excess power at about 1 Hz (Pasachoff and Ladd, 1987). I also collaborated in interpreting Fabry-Perot profiles in terms of the coronal velocity field (Chandrasekhar *et al.*, 1991) and how it varies over the cycle.

The eclipse of 1984 in Papua New Guinea, much shorter and with inferior weather forecasts, was superior scientifically since the sky was exceptionally clear. Our isophotal contours show that we were close to solar minimum. We have used a computer algorithm to bring out coronal streamers to about 5 solar radii, a range usually reached only with spacecraft, illustrating again how important it is to observe all th e eclipses from the ground. Since then, there was only an inaccessible 1985 eclipse and a 1988 eclipse that crossed Sumatra and the Philippines in mostly cloudy skies.

The 22 July 1990 corona was of the solar-maximum type, with round isophotes except for a major coronal hole. We can expect similar morphology for the 11 July 1991 eclipse. We will continue testing magnetohydrodynamic models of coronal heating with an improved version of our coronal-oscillation experiment (Pasachoff, 1991b). On 30 June 1992, an eclipse will start in Uruguay and cross the Atlantic. On 30 November 1994, a better eclipse will cross South America. Solar observers should take the best advantage of them.

Acknowledgments. The expeditions have been supported by NSF grants AST-7922104, CDP-7922926, PRM-8114631, RII-8304403, ATM-9005194, and USE -9050643; NSF Expedition grants to NCAR; and the National Geographic Society.

References

Chandrasekhar, T., Desai, J.N., Ashok, N.M., Pasachoff, J.M.: 1991, *Solar Phys.*, in press

Pasachoff, J.M.: 1991a, *Eclipses of the Sun* in The Reference Encyclopedia of Astronomy and Astrophysics, in press

Pasachoff, J.M.: 1991b, *Proc. Heidelberg Conf. on Mechanisms of Chrom. and Coronal Heating*, ed. by P. Ulmschneider, E. Priest, R. Rosner (Springer-Verlag), in press

Pasachoff, J.M., Landman, D.A.: 1984, *Solar Phys.* **90**, 325

Pasachoff, J.M., Ladd, E.F.: 1987, *Solar Phys.* **109**, 365; **110**, 412

Pasachoff, J.M., Sandford, II, M.T., Keller, Jr., C. F., Jr.: 1978, *Bull. Am. Astron. Soc.* **10**, 431

The History of Activity of the Sun and its Rotational Period on the ZAMS

Kazimierz Stępień

Warsaw University Observatory, Al. Ujazdowskie 4, 00-478 Warszawa, Poland

Abstract: The rotation period of the Sun after it reached ZAMS is estimated from the present rotation rate, average X-ray emission flux and average calcium emission flux. Taking into account all existing uncertainties it is concluded that this initial period was within the range 1–9 days, with the most probable value 2–3 days. Possible influence of the solar activity on evolution of life on the Earth is briefly discussed.

1. Introduction

Recent observations of cool stars possessing subphotospheric convection zones suggest that stars reaching the Zero Age Main Sequence (ZAMS) rotate rapidly velocity and activity decrease continuously in time (Skumanich, 1972; Barry, 1988; Kawaler, 1988; Stępień, 1988b; Stauffer and Soderblom, 1990). It is generally accepted that the slow down of rotation is caused by the activity-related mass loss in the presence of the dynamo generated magnetic fields. The spin down decreases the efficiency of the dynamo action, which results in the weaker magnetic activity.

The loss of angular momentum can be described by (Mestel, 1984):

$$\frac{dJ}{dt} \sim \frac{dM}{dt} R_*^2 \Omega \left(\frac{r_A}{R_*}\right)^n , \qquad (1)$$

where J is angular momentum, dM/dt rate of mass loss via a stellar wind, R_* and Ω radius and angular velocity, and r_A the Alfvénic radius of the star. The exponent n ($0 \leq n \leq 2$ depends on geometry of the stellar magnetic field. Assuming a formula for the dependence of the stellar magnetic field on distance from the star one can replace r_A with the surface magnetic field B_{surf}. The next step is expressing B_{surf} as a function of Ω. This should in principle be supplied by the dynamo theory but at present the theory is not yet enough developed. In the absence of anything better a power law has usually been adopted: $B_{surf} \sim \Omega^p$, with p assumed (Endal and Sofia, 1981; Mestel, 1984; Kawaler, 1988; Pinsonneault *et al.*,

1989). Unfortunately, the usefulness of such an approximation is very doubtful. The observations of activity and surface magnetic fields indicate that for periods shorter than about 2–3 days, a saturation effect appears (Saar and Linsky, 1986; Vilhu and Walter, 1987) – contrary to an unbound increase of B_{surf} with Ω, predicted by the power law. A method of replacing the unknown function $B_{surf}(\Omega)$ by a semi-empirical relation was suggested by Stępień (1988a,b). The method permits also the determination of a value of the geometrical exponent from the observational data.

It assumes that the heating of stellar coronae is caused by the Alfven waves generated in tubes of the strong magnetic fields at the photospheric level. Taking into account the back-reaction of the wave generation on the generating medium a new estimate of the expected flux per unit area, F_A, was calculated. This flux can be compared with the observed X-ray flux, F_x. These two are connected by a relation: $F_x = f_e f_A F_A$, where f_A is the filling factor of the magnetic tubes heating the corona and f_e the efficiency factor describing all processes connected with the transmission of waves upwards, converting their energy into X-rays, geometry of the radiating region etc. Now, let us introduce a parameter $f_x = F_x/F_A$. Assuming that f_e varies much less than f_A among the discussed stars, we see that f_x is proportional to the filling factor f_A. The exact value of the proportionality coefficient is not needed as long as we are interested only in the differential analysis.

The parameter f_x, which can also be called the (uncalibrated) filling factor, turned out to be well correlated with the directly measured surface magnetic fields for those few stars for which both quantities were known (Stępień, 1988a). This confirms the correctness of the adopted assumptions. On the other hand, f_x has the advantage that it can be determined for each star with known F_x, i.e. for a relatively large sample of stars.

We expect that the most active stars have surface magnetic fields close to their maximum (saturation) value, hence the filling factor f_x of these stars should be close to its maximum permissible value. Figure 2 in Stępień (1988a) shows indeed that such stars follow the line $f_x \approx 0.1$. The lines $f_x =$ const. have an inclination (Stępień, 1989b): $\log F_x \sim -1.43(B-V)$, or, with $\log F_{bol} \sim 4 \log T_e \sim -(B-V)$ (Böhm-Vitense, 1981):

$$\log(F_x/F_{bol}) \sim -0.43(B-V). \tag{2}$$

Because these lines were determined theoretically, the resulting values of f_x calculated for the investigated stars were called semi-empirical.

Recently Pallavicini *et al.* (1990) published X-ray data for a number of very active Ke and Me dwarfs. They found a tight correlation between L_x and L_{bol} of these stars: $\log L_x \sim 1.21 \log L_{bol}$. The dependence of $\log L_{bol}$ on $(B-V)$ is given by Schmidt-Kaler (1982). In the range of spectral types K5–M6 it is approximately linear with the inclination $\log L_{bol} \sim -2.3(B-V)$. Inserting this into the Pallavicini *et al.* relation we get:

$$\log(L_x/L_{bol}) \sim -0.48(B-V), \tag{3}$$

in an excellent agreement with the theoretical value given by (2). Using the Pallavicini *et al.* relation one can calculate purely empirical filling factors from the ratio of the observed X-ray flux to its saturated value given by this relation. Due to very similar inclinations of both, theoretical and empirical, reference levels, thus determined filling factors are essentially identical with the f_x values, apart from a constant factor.

Stępień (1988a) showed that the filling factor f_x is well correlated with the Rossby number $R = P_{rot}/\tau_c$, where P_{rot} is the rotation period and τ_c a function of the spectral type, which can be determined empirically (Stępień, 1989a), and is sometimes identified with the turnover time near the bottom of the convection zone. The resulting relation is (Stępień, 1989b): $\log f_x = -1.7 - 1.55R$.

We can now replace B_{surf} in the formula for loss of angular momentum with f_x which is a *known* function of R, hence Ω. After integrating the resulting equation one gets the relation $P_{rot}(t)$. The constants appearing in this relation (including the value of the exponent describing geometry of the coronal magnetic field) were determined by Stępień (1988b) from the observations of activity in open clusters of different age, obtained by Barry *et al.* (1987). Since that, however, several new data have appeared and it became useful to revise the calibration of $P_{rot}(t)$.

2. The initial period of rotation of the Sun

The relation $P_{rot}(t)$ was recalibrated in the same way as described by Stępień (1988b) but with the following modifications:

1. The Barry *et al.* (1987) data were reduced anew using a revised value of the basal flux given by Rutten (1987)
2. The empirical relation $\tau_c(B-V)$ determined by Stępień (1989a) was used to calculate the stellar Rossby numbers
3. The empirical relation between the excess calcium emission flux and R, found by Stępień (1989a) for field stars was adopted.

The new values of the calibrating constants turned out to be not much different from the old ones: $\beta = 3$ (same as previously) and $T_0 = (2.0 \pm 0.5) \times 10^8$, as compared to $(1.5 \pm 0.5) \times 10^8$ found previously.

Thus the calibrated $P_{rot}(t)$ relation permits the backward projection of the present rotation period of the Sun to its initial value on the ZAMS, or, more strictly speaking, to the value the Sun had after the phase of the possible rapid spin down suggested by observations of the youngest clusters (Stauffer and Soderblom, 1990). The result is:

$$P_0(\text{Sun}) = 2.0 \pm 0.5 \quad \text{days},$$

where the uncertainty comes solely from the formal errors of the used relations.

The moderate error given above looks quite satisfactory. Unfortunately, it is definitely overoptimistic. The quoted uncertainty does not take into account all the possible systematic effects which may influence $P_0(\text{Sun})$. Some of the important effects are:

1. A number of $\tau_c(B-V)$ relations have been published (see Stępień 1989a for discussion and comparison of them). Different τ_c result in different relations between activity indices and the Rossby number
2. Two different measures of the activity-related calcium emission flux have been suggested in the literature (see again Stępień 1989a for discussion). It is not yet clear which one is more correct and should be used with the Barry *et al.* (1987) observations to calibrate the $P_{rot}(t)$ relation
3. Even when one uses the excess calcium emission flux ΔF_{CaII} (which is a better, in the opinion of the present author, measure of the chromospheric activity) one has to allow for uncertainty in the basal flux subtracted from the total measured flux to get Δ_{CaII}. Its value was recently revised (Rutten, 1987) but is still known with insufficient accuracy. This influences particularly strongly the determination of ΔF_{CaII} for the least active stars, for which the total measured flux barely exceeds the basal flux.

The problem with calibration of the basal flux is well illustrated in Fig.1. Here the predicted excess calcium emission flux as a function of time for solar type stars with different initial periods is given as solid lines (from Stępień, 1989b). Filled circles correspond to the original data listed by Barry *et al.* (1987). One can notice a systematic trend between the predicted and observed fluxes, which would suggest a need for correcting the theory. Open circles correspond to the same data reduced with the new calibration. Now a trend in the opposite direction is seen. It is clear that the uncertainty of the observational data is still considerable.

Repeating the calculations of the initial period of the Sun with different relations, listed above, one gets values over a much broader range: $P_0(\mathrm{Sun}) = 1-5$ days.

The initial rotation period of the Sun can also be determined directly from the present level of the solar calcium emission (see Fig. 1). Unfortunately, the same possible systematic effects influence this method. In addition, the net calcium emission flux of the Sun is known within about 30 % (judging from the values given in literature). The bars associated with the solar symbol in Fig. 1 correspond to this uncertainty. As a result, we get even broader limits: $P_0(\mathrm{Sun}) = 0.5-9$ days.

A similar procedure can be applied to X-ray emission (see Fig. 7 in Stępień 1989b). In case of X-ray flux we have no problem with calibrating its activity-related part – we simply assume that the whole emission is due to magnetic activity (Stępień and Ulmschneider, 1989). However, again, several diverging values of the average solar X-ray flux are given in the literature. Together with the above discussed uncertainties they lead to the following estimate: $P_0(\mathrm{Sun}) = 3-8$ days.

Summing up we conclude that the rotation period of the Sun, after it had reached ZAMS was between 1 and 9 days, with the most probable value between 2 and 3 days. The rather broad range of uncertainty is caused by a low number and a poor accuracy of the presently available observations of activity of late type stars (including the Sun itself), and an insufficient understanding of relations connecting observations and physical parameters.

In spite of the all discussed uncertainties on can attempt to reproduce, at least approximately, the evolution of rotation and activity of the Sun during its main sequence life. Figure 2 shows schematically how the rotation period, excess calcium

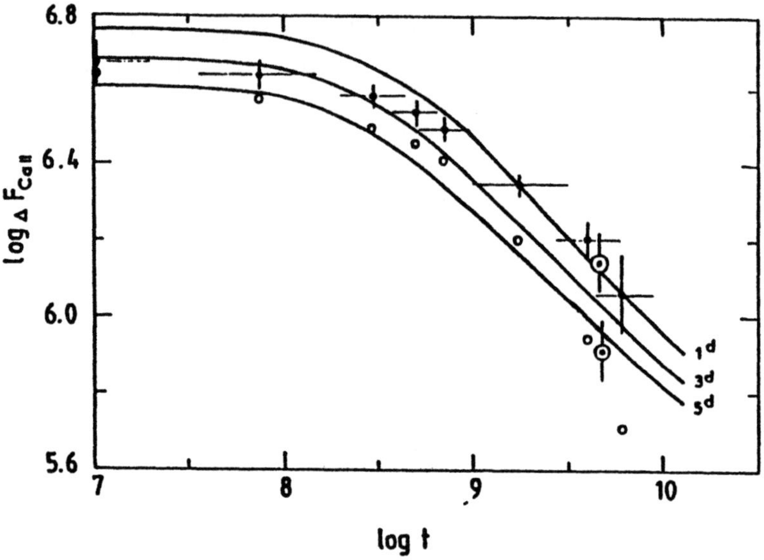

Fig. 1. The predicted calcium emission flux as a function of age for stars with different initial rotation periods (solid lines) and the observed data from Barry *et al.* (1987), reduced with the old (filled circles) and the new (open circles) basal flux.

emission flux and X-ray flux of the Sun varied with its age. The plotted curves are just illustrative – they do not reflect the discussed systematic effects. Nevertheless, we can draw some conclusions.

The Sun rotated on the ZAMS probably about 10 times faster than today. Its X-ray flux was 500–1000 times larger and the chromospheric flux 4–5 times larger. It was also surely flaring very frequently and strongly. During the first billion years the Sun was losing angular momentum quite fast. As a result its period of rotation increased to a value of about 14 days, whereas the X-ray emission decreased by a factor of about 30. After 3.5 billion years, i.e. one billion years ago, the Sun rotated only slightly faster than presently and was nearly as active as now.

When discussing the origin and development of organic life on Earth, suggestions have been made that large amounts of ionizing radiation associated with a very high activity of the young Sun could influence the evolution of life by producing mutations at a much higher rate than presently. The above picture shows that the Sun could possibly influence in this way only the earliest forms of life existing about 3.5 billion years ago. The evolution of life during the last billion years has taken place in the presence of solar activity being essentially at the present day level.

This paper was partly supported by the grant CPBP 01.20.

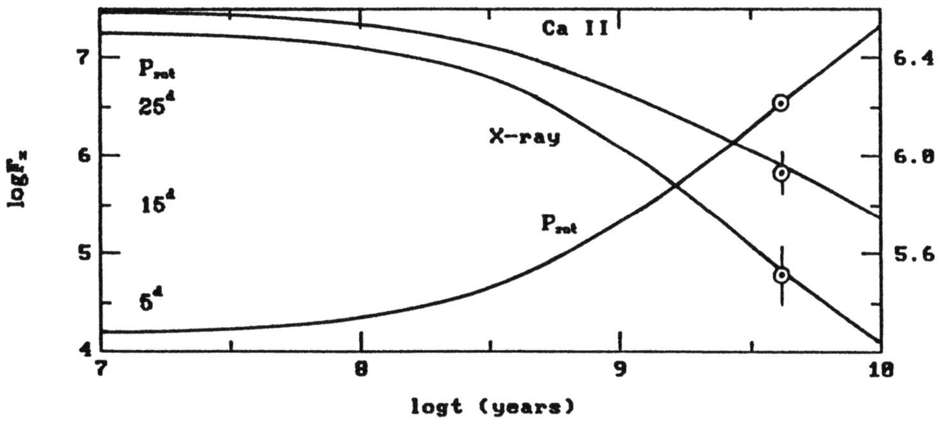

Fig. 2. The predicted evolution of rotation and activity of the Sun. Error bars correspond to uncertainties of the present levels of the solar emission.

References

Barry, D.C.: 1988, *Astrophys. J.* **334**, 436
Barry, D.C., Cromwell, R.H., Hege, E.K.: 1987, *Astrophys. J.* **315** 264
Böhm-Vitense, E.: 1981, *Ann. Rev. Astron. Astrophys.* **19**, 295
Endal, A., Sofia, S.: 1981, *Astrophys. J.* **243**, 625
Kawaler, S.D.: 1988, *Astrophys. J.* **333**, 236
Mestel, L.: 1984, in *Cool Stars, Stellar Systems, and the Sun*, ed. by S.L. Baliunas and L. Hartmann, Springer, Berlin, p. 49
Pallavicini, R., Tagliaferri, G., Stella, L.: 1990, *Astron. Astrophys.* **228**, 403
Pinsonneault, M.H., Kawaler, S.D., Sofia, S., Demarque, P.: 1989, *Astrophys. J.* **338**, 424
Rutten, R.G.M.: 1987, *Astron. Astrophys.* **177**, 131
Saar, H., Linsky, J.L.: 1986, *Adv. Space Res.* **6**, 235
Schmidt-Kaler, Th.: 1982, in *Landolt-Börnstein*, ed. by K. Schaifers and H.H. Voigt, Springer, Berlin, Vol. 2, p. 1
Skumanich, A.: 1972, *Astrophys. J.* **171**, 565
Stauffer, J.R., Soderblom, D.R.: 1990, Space Telescope Science Institute Preprint Series, No. 432
Stępień, K.: 1988a, *Astrophys. J.* **335**, 892
Stępień, K.: 1988b, *Astrophys. J.* **335**, 907
Stępień, K.: 1989a, *Astron. Astrophys.* **210**, 273
Stępień, K.: 1989b, *Acta Astr.* **39**, 209
Stępień, K., Ulmschneider, P.: 1989, *Astron. Astrophys.* **216**, 139
Vilhu, O., Walter, F.M.: 1987, *Astrophys. J.* **321**, 958

IV

Stellar activity

- Surface mapping of active stars
- Stellar cycles, starspots and plages
- Stellar magnetic field measurements
- Chromospheric and coronal activity as a dynamo indicator
- Evolution of magnetic activity, activity in pre-MS-stars, subgiants and giants

The Differential Rotation and Evolution of Spots on UX Arietis from a Sequence of Doppler Images

Steven S. Vogt and Artie P. Hatzes

University of California Observatories/ Lick Observatory, U. of California, Santa Cruz, CA 95064

Abstract: We present a sequence of three Doppler images of the spotted RS CVn star UX Arietis obtained over a 5-month interval from August 1986 through January 1987. The spot distribution was quite complex and consisted of a large stable polar spot, a spot near the equator, and several other spots at intermediate positive and negative latitudes. The time intervals between successive images were small enough that we were able to reliably track the evolution of the spot distribution, measuring accurate longitudes, latitudes, and areas of the major spots, as well as their drift rates in longitude and latitude. The longitudinal drifts of spot features at equatorial, intermediate, and high latitudes yielded an accurate measurement of differential rotation. We find that the spotted primary of UX Arietis is indeed rotating differentially and in the sense opposite to that of the Sun, *i.e.* the poles rotate faster than the equator. The equator is synchronized to the orbital angular velocity, and the angular velocity increases towards either pole. The angular velocity distribution can be expressed as $\Omega(°/day) = -55.91 + 1.09(\pm 0.09)\sin^2\phi$, where ϕ is the latitude. The amount of differential rotation, parameterized as the ratio of the difference between the equatorial and polar angular velocities to the equatorial angular velocity, is then $\alpha = -0.020(\pm 0.002)$, as compared to a value of $\alpha = +0.2$ for the Sun.

1. Introduction

For the past 9 years, we have been obtaining Doppler images of a set of relatively bright spotted RS CVn stars. Our aim is to track the emergence and evolution of spots on these stars from one year to the next so as to obtain information on the systematics, patterns, and periodicities in spot behavior as well as to provide a direct measurement of differential rotation or latitudinal shear. Such information will provide insight into the underlying dynamo processes and large scale global convective and circulation patterns in much the same way as long term monitoring of sunspots on the Sun has revealed the 11-year and 22-year sunspot cycle and 'butterfly' diagram. These systematics are important observational anchor points for theories of the solar dynamo, and, through Doppler imagery of starspots, we hope to provide similar observational anchor points for theories of the extremely powerful dynamos of the RS CVn stars.

The technique we use is called Doppler imaging (Vogt *et al.*, 1987; hereafter Paper I). Doppler imaging exploits the correspondence between wavelength position across a rotationally Doppler-broadened spectral line and spatial position across the stellar disk to derive a 2-d image of the star. Cool spots on a rotating star produce distortions in the star's spectral line profiles. If the line profiles are dominated by rotational broadening, a high degree of correlation exists between the position of a given distortion in a line and the longitude on the star of the feature which produced it. A high resolution spectrum of the line is thus, to first order, a 1-d image of the star in longitude, but completely blurred in latitude. As the star is observed from other aspects (other rotation phases) additional 1-d images are obtained. Then, if the inclination, rotation period, and other basic physical parameters of the star are known, the set of 1-d images can be combined into a 2-d image of the star.

We use Doppler imagery since the other available method of obtaining information on spot sizes and locations, modeling integrated flux light curves, is highly nonunique. There just isn't enough information in integrated flux light curves to be able to determine accurate longitudes and latitudes of multiple spots on a star, nor to extract any detail about the structure and shapes present in a complex spot distribution.

UX Arietis is one of our primary Doppler imaging candidates. It is a bright, noneclipsing RS CVn spectroscopic triple system. It consists of a spotted K0 IV primary and a G5 V secondary in a 6.44 day orbit. In the course of our Doppler imaging studies, we noticed a faint third component in the spectra which remains stationary near the γ velocity of the system and which is apparently from a star in a long-period orbit about the binary.

In this contribution, we present a sequence of three Doppler images of the spotted K0 IV primary of the UX Arietis system which we obtained in late 1986 and early 1987, and from which we are able to derive an accurate measurement of differential rotation. For this symposium, we will present only a brief overview of our findings.

2. Observations

Data were acquired using the coudé focus of the Shane 3-meter telescope at Lick Observatory and a TI 800×800 3-phase CCD. The observations were centered on 6435 Å and covered the wavelength region from about 6425 Å to 6453 Å. The spectral resolution was about 0.13 Å. A modified Bowen-Walraven image slicer was used to reformat the light from a 3-arcsecond diameter hole to a 0.67-arcsecond slit. Slicing not only minimized slit losses due to seeing but it also spread the light (perpendicular to dispersion) over a larger area of the CCD. This accomplished two things important for acquisition of the high signal-to-noise required for Doppler

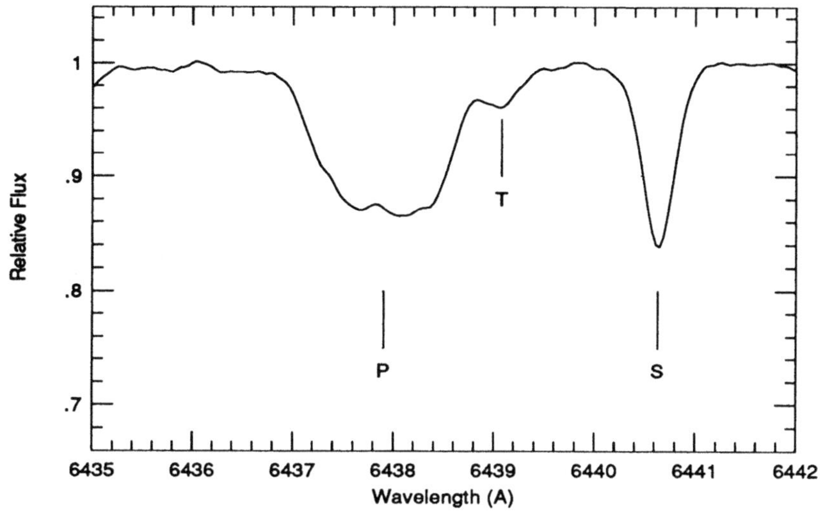

Fig. 1. Spectral region around Ca I 6439 Å in UX Arietis showing the three stellar components.

imaging: it allowed for a higher total photon count without saturation of individual CCD pixels and it removed any additional pixel-to-pixel variations that may have been present after performing standard flatfielding.

It is important not only to have data with low noise, but also to remove the contribution of atmospheric features and stellar companions which may cause systematic errors. This is somewhat complicated for UX Arietis which is actually a triple system. Figure 1 shows an observation of Ca I 6439 Å when all three components are visible. Before deriving the Doppler image care was taken to remove both secondary and tertiary components from the observed spectral line profiles.

The Ca I 6439 Å line was used for modeling the spot distribution on the stellar surface. The data set was divided into three subsets. The 'August' set consisted of 4 observations whose median date was 18 August 1986. The time span of the observations was 4 days (0.6 rotations). The resulting Doppler image thus represents a 'snapshot' of the star in that the spot distribution would have changed little during the course of the observations. However, the phase coverage is incomplete. The observed Ca I profiles as a function of phase are shown as crosses in Figure 2. The fit to the observed spectral line profiles from the derived spot distribution is shown as a solid line.

The 'November' data set consisted of 5 phases (partial phase coverage) with a median date of 18 November 1986. Virtually all the observations were taken in December with the exception of one taken at phase 0.207 in October. This time span is long enough that if the spots had moved significantly during this time then the Doppler image would show some phase smearing, but only around phase 0.2. The observed Ca I spectral line profiles as well as linefits are shown in Figure 3.

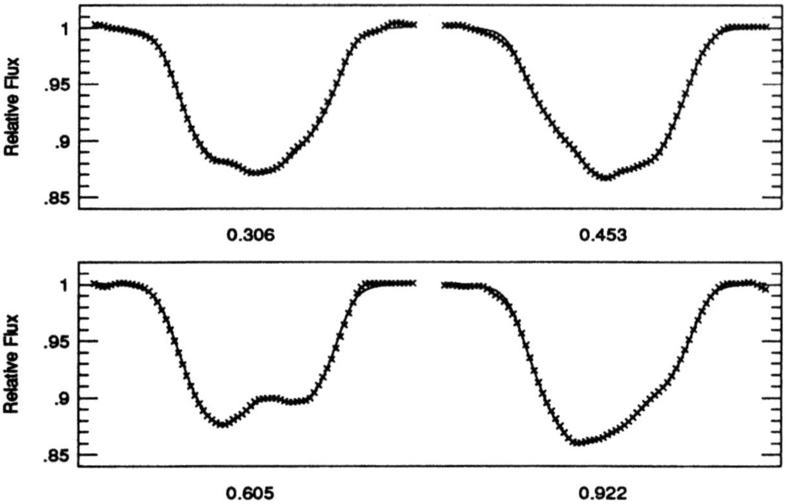

Fig. 2. The observed spectral line profiles (crosses) for Ca I as a function of phase for the August data set. The solid line represents the fits to the profiles from the derived spot distribution.

The 'January' data set represents our best data set for the observing season. The observations were all taken within an 8 day time span (1.3 rotations) centered on 16 January 1987. The phase smearing due to spot migration is thus negligible. This is the only data set for which we have complete phase coverage for the star. Again the observed line profiles are shown as crosses and the linefits as a solid line in Figure 4.

3. Modeling the spot distribution on UX Arietis

3.1. Stellar parameters

The adopted parameters used for modeling the spot distribution on UX Arietis are shown in Table 1. The parameters for the spotted primary are appropriate for a K0IV star and include the projected rotational velocity $v \sin i$, the inclination i, the temperature of the photosphere T_{phot}, the logarithm of the surface gravity $\log g$, the radial-tangential macroturbulent velocity ξ_{rt}, and the coefficient for the standard limb darkening law ϵ. The parameters for the secondary component are also listed and are appropriate for a G5V star. These were used to create a synthetic profile for the secondary which was subtracted from those observations where the primary and secondary were blended. For the third component we found that its profile could be adequately fit with a scaled version of the secondary profile. This component appeared to be stationary at 6.6 km s^{-1} blueward of the γ velocity of the system.

Specific intensity profiles for the Ca I line were generated using model atmospheres from Bell *et al.* (1976) and Kurucz's WIDTH5 program. Thirty limb angles

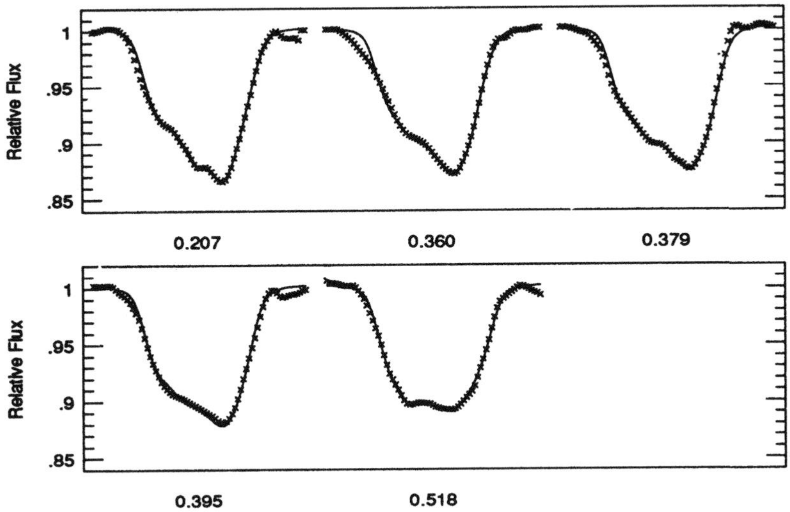

Fig. 3. The observed spectral line profiles (crosses) and linefits for the November data set.

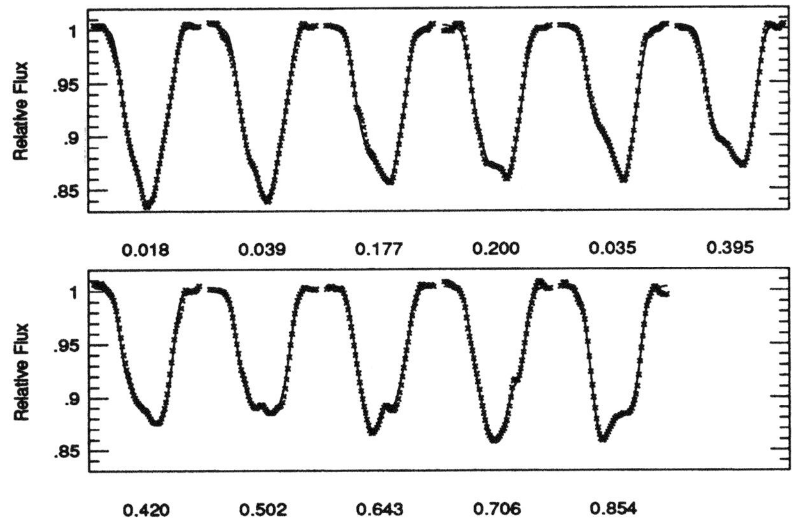

Fig. 4. The observed spectral line profiles (crosses) and linefits for the January data set.

were used to take into account the center to limb variations in the line profile. The limb darkening coefficient in Table 1 represents an approximation to the limb darkening coefficient produced by the model atmospheres. The spectral line profile for the spot was taken to be a scaled version of the photospheric profile. This assumes, naturally, that the line in the spot has the same shape as in the photosphere and thus does not change its equivalent width with temperature. This is not strictly true as the Ca I line increases in strength with decreasing temperature. We feel, however, that this is a valid representation for the spot profiles for two reasons. First, extensive testing in Paper I indicated that the final Doppler image is insensitive to changes in the line strength in the spot. This is primarily due to the fact that the spots are dark and do not contribute a significant fraction of the flux to the integrated profiles. The changes in the integrated line flux shapes therefore come primarily from changes in the continuum level and not from changes in line strength. Secondly, in our observations of the RS CVn's we find that the shape of the Fe I 6430 Å line, which has a weaker temperature dependence than the Ca I line, is always a scaled version of the Ca I line. This would not be the case if the spot contribution to the spectral line profiles were significant.

Table 1: Stellar Parameters

Parameter	Primary	Secondary
Spectral Type	K0IV	G5V
$v \sin i$	39 ±1 km s^{-1}	7.25 ±0.25 km s^{-1}
i	60°	–
T_{phot}	4750 K	5700 K
log g	3.5	4.27
ξ_{rt}	5 km s^{-1}	3.5 km s^{-1}
ϵ	0.64	0.6

3.2. The Doppler images

The Doppler images derived from the three data sets are shown in Figures 5–7. The reconstruction procedure uses the principles of maximum entropy reconstruction as outlined in Paper I. Each figure shows the star in a "mashed" polar projection. Latitude lines are drawn equally spaced at intervals of 4.5° down to a latitude of −50°. Each square represents an image pixel of 4.5° × 4.5°. A total of 2310 pixels was used in the modeling. The darkest latitude circle represents the stellar equator while medium dark latitude lines are drawn every 30°. The star rotates in a counter-clockwise manner for an observer looking up from phase 0.0.

The radial tic marks surrounding the star represent those phases at which observations were made. In Figures 5 and 6, the missing pixels represent phase 'gaps' of missing pixel information. At a given phase, pixels about ±0.2 in phase from the center make a discernible contribution to the observed line profile. Naturally, those pixels near the limb make less of a contribution due to foreshortening and ideally one should display the pixels in a manner that is weighted to their visibility. At the present time there is no convenient way for us to do this. Instead we chose only to display those pixels for which we have at least some information.

Pixels contained in the spotted region are marked with an 'X' and thus the displayed images represent a binary brightness distribution (spot or no spot). The actual Doppler images produce a smooth variation in brightness between spot and photosphere. Also the maximum entropy method tends to favor producing a

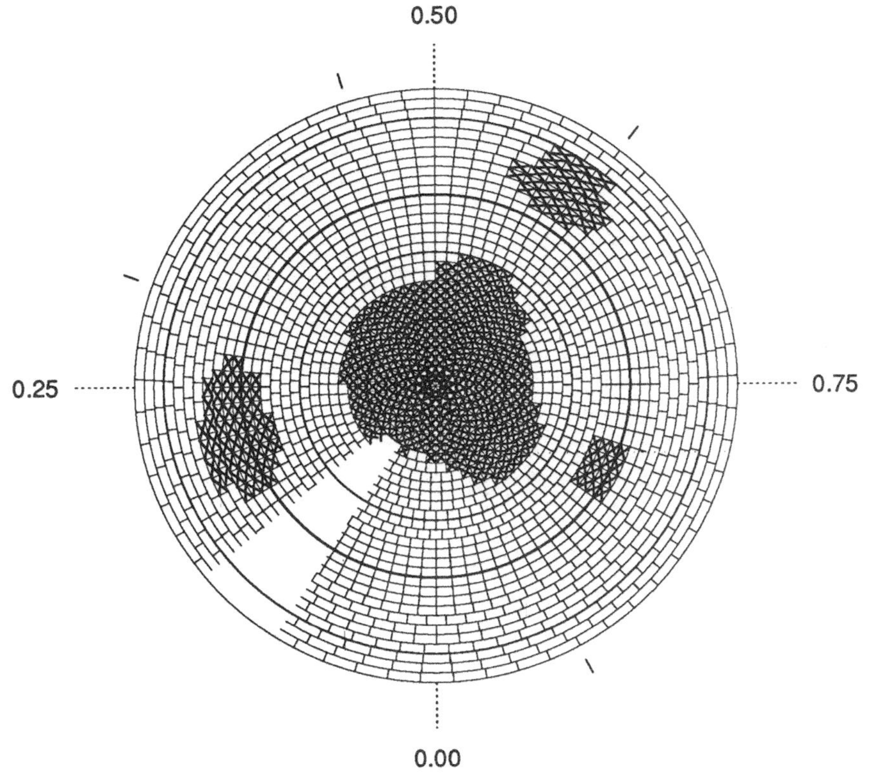

Fig. 5. Doppler Image of UX Arietis in August 1986.

large warm feature as opposed to a small cool one. We therefore must decide, in a consistent manner, below what intensity level a pixel must have in order to be in a 'spot'. This thresholding procedure was chosen as follows. A temperature of 3550 K (ΔT=1200°) was chosen for the spot, slightly warmer than the value of 3360 K found by Poe and Eaton (1985) using photometry. A threshold value is chosen such that all pixel values below this threshold are replaced by the spot temperature and all pixels above the threshold are replaced by the photospheric temperature. The integrated line profiles from this thresholded image are then compared to the observed profiles. The threshold level is then varied until a good match is made between the observed and computed profiles. This procedure essentially turns the original continuous brightness distribution into a discrete 2-level one *that is still consistent with the observed profiles.* Experience has also shown that for those Doppler images for which published photometry is available, images thresholded in this way are also consistent with the photometry.

Figure 5 shows the thresholded Doppler image for UX Arietis in August 1986. The spot distribution is characterized by a large polar spot and three low-latitude spots; one near the equator at phase 0.25, one at latitude –25° and phase 0.59 and

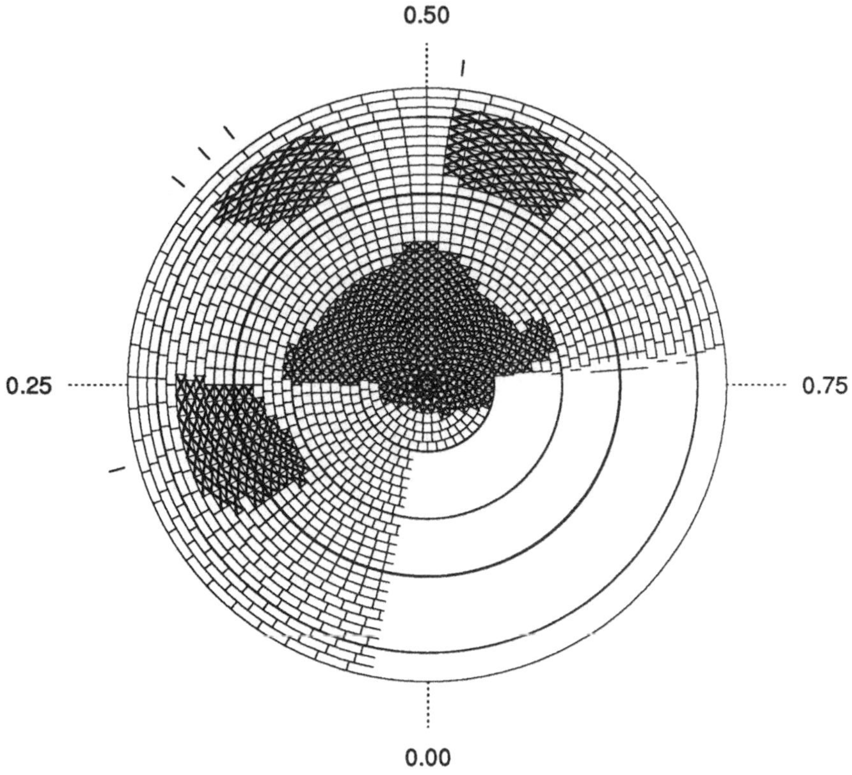

Fig. 6. Doppler Image of UX Arietis in November 1986.

another at +10° and phase 0.85. Note that the polar spot has two appendages: one at phase 0.56 and another near phase 0.88.

The spot distribution for UX Arietis derived from the November data set is shown in Figure 6. Both low-latitude spots from the August image are still present although the one below the equator near phase 0.6 has migrated counter-clockwise (to a lower phase). Note that a new low-latitude spot near phase 0.40 has also appeared. The reality of this spot is somewhat uncertain since it is at such a low latitude and it may be a 'shadow' of the polar spot due to poor phase coverage. There are also several polar appendages two of which (at phase 0.5 and 0.69) appear to have counterparts in the August image.

The Doppler image of UX Arietis in January 1987 is shown in Figure 7. Two of the low-latitude features from the November image are present in the January image. However, the low-latitude spot which was present at phase 0.40 in the November image (but not in the August image) has disappeared. This leads us to believe that this spot in the November image was real and not a shadow of the polar spot because the phase coverage for this part of the star was the same as for the November image and no such shadow spot appears in the January image. The ultimate fate of this spot is unknown at the present time due to

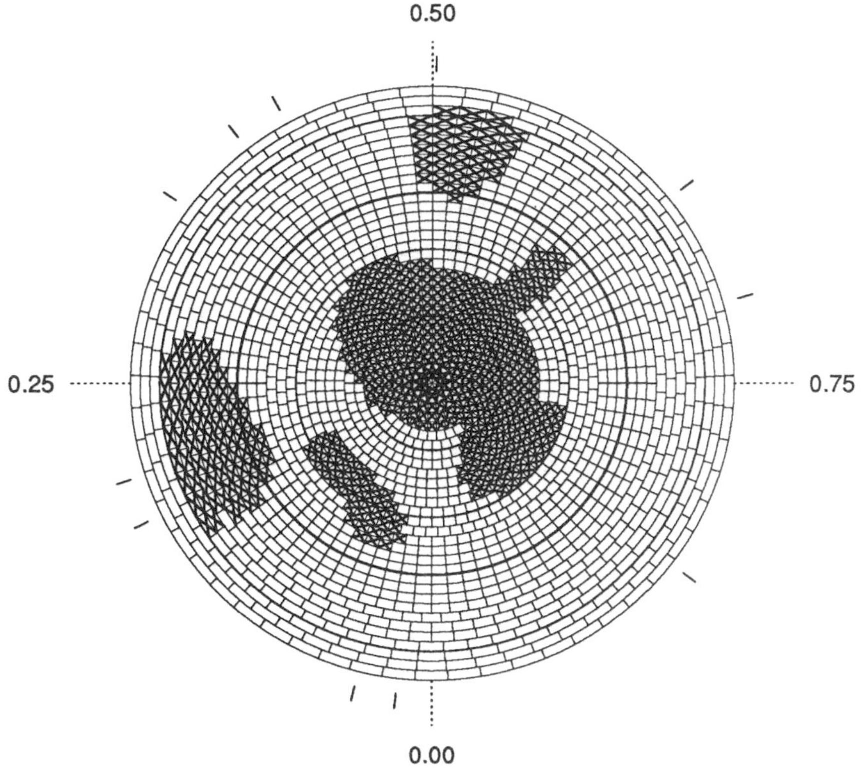

Fig. 7. Doppler Image of UX Arietis in January 1987.

the poor time resolution of the Doppler image sequence. This spot could have simply disappeared (possibly by submerging back into the subphotospheric layers) or it could have migrated in latitude and/or longitude from its original position in the November image. Most likely the spot migrated towards the south pole since a small amount of southern motion would cause the spot to drift out of view. Significant longitudinal drift seems unlikely due to the apparent absence of phase smearing for this feature in the November image as well as the fact that its disappearance in the January image would require a merger with one of the other low-latitude spots. This requires a much larger migration rate than is observed for the other features. The January image also marks the appearance of a mid-latitude spot near phase 0.1 as well as the disappearance of the +10° latitude spot from the August image. The lack of phase coverage near this latter spot in the November image prevents us from knowing the fate of this spot although we strongly suspect that it migrated polewards possibly becoming part of the polar appendage near phase 0.85 in the January image.

4. Differential rotation in UX Arietis

Although the spot distribution on UX Arietis visibly changes in the sequence of images presented in Figures 5–7, there are certain features which one can clearly follow from image to image. These include two polar appendages (at phases 0.43 and 0.63 in the January image), and two low latitude spots (near phases 0.2 and 0.6 in the August image). This enables us to measure the differential rotation rate for UX Arietis over a latitude range of –29° to +45°.

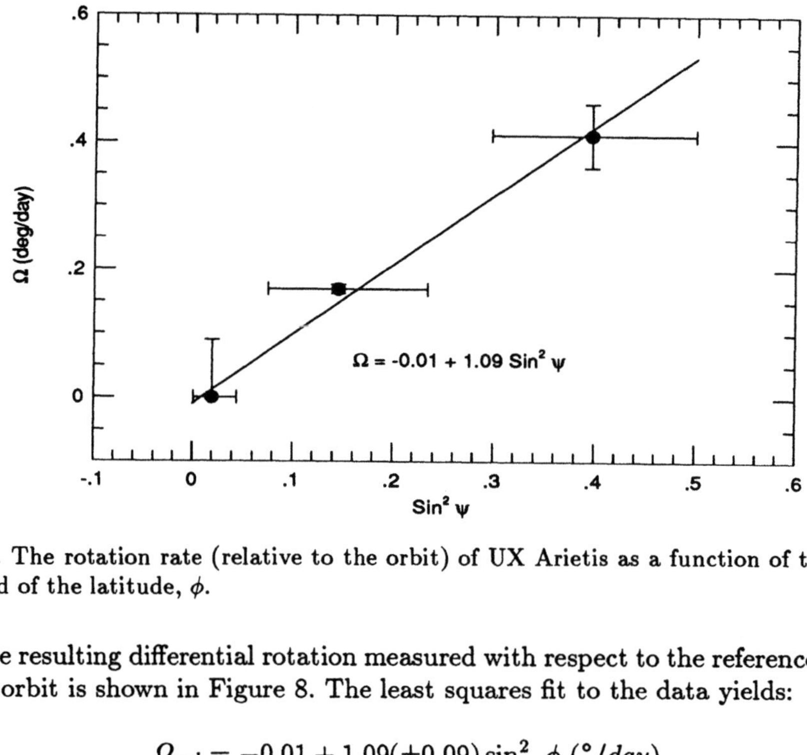

Fig. 8. The rotation rate (relative to the orbit) of UX Arietis as a function of the sine-squared of the latitude, ϕ.

The resulting differential rotation measured with respect to the reference frame of the orbit is shown in Figure 8. The least squares fit to the data yields:

$$\Omega_{rel} = -0.01 + 1.09(\pm 0.09)\sin^2\ \phi\ (^\circ/day) \tag{1}$$

The first term on the right hand side of Eqn. 1 represents the rotation rate of the equator with respect to the orbital frame. To within our uncertainties, it is zero, indicating that the equator is tightly synchronized to the orbital angular velocity. The second term gives the degree of latitudinal shear and, being positive, indicates that the angular velocity increases with increasing latitude (polar acceleration). This measured differential rotation as measured from various features in the images was consistent in that the rates derived from the polar and low-latitude features between the August and November images were consistent with those derived from the November and January images.

By comparison the solar value (relative to the solar equator) is:

$$\Omega_{rel} = -3.0\sin^2\ \phi\ (^\circ/day) \tag{2}$$

where ϕ is the stellar latitude.

The amount of differential rotation may be parameterized by α, the ratio of the difference between polar and equatorial velocities to the equatorial velocity:

$$\alpha = \frac{\Omega_{equator} - \Omega_{pole}}{\Omega_{equator}} \tag{3}$$

Solid body rotation results in $\alpha = 0$, whereas equatorial acceleration has $\alpha > 0$ and polar acceleration results in $\alpha < 0$. The differential rotation law for UX Arietis is $\alpha = -0.02 \pm 0.002$ whereas $\alpha = +0.197$ for the Sun.

5. Discussion

The most surprising result from our sequence of images is that the differential rotation in UX Arietis is small and negative, that is, the poles rotate faster than the equator. In the case of the Sun the coupling between rotation and convection may be able to explain the equatorial acceleration (Glatzmaier, 1988; Gilman and Miller, 1986), however, one must be very wary of applying the solar analogy to the RS CVn stars. For one, the structure of UX Arietis is quite different from that found in the Sun; UX Arietis is more evolved and has a much larger radius and rotation rate than the Sun. It also has a much deeper convection zone. Also, UX Arietis is in a close binary system and tidal forces certainly must play a role in controlling stellar motions, unlike for the solar case. Clearly UX Arietis may play a key role for testing theories of dynamos and differential rotation.

The synchronization of the equator of UX Arietis with the orbital period is no coincidence since UX Arietis nearly fills its Roche lobe (Huenemoerder *et. al.*, 1989) and one expects tides to play a dominant role in determining the rotation rate. Also 88% of all RS CVn systems exhibit synchronization with the orbital period. What is not known is what role tidal stresses have in suppressing (or creating) differential rotation in UX Arietis. One may naively think that tidal stresses would work to keep the rotation rate of the various latitude bands the same. However, theoretical calculations by Scharlemann (1982) suggest that tides in RS CVn binary systems should not suppress differential rotation.

Scharlemann (1982) also predicted that, when the tidal coupling torque vanishes (maximum degree of synchronism achieved), part of the star should be rotating faster than the orbit, and part more slowly. Co-rotation with the orbit should occur at some intermediate latitude (the co-rotation latitude). This concept has often been proposed to explain the observed forward and reverse migration of waves in the RS CVn photometric curves as spots migrate above or below the co-rotation latitude. However, if our differential rotation results are correct, UX Arietis does not fit this model. Rather, the co-rotation latitude is at the equator, and no part of the star rotates more slowly than the orbital velocity. Perhaps the tidal coupling torque has not yet vanished and UX Arietis is not yet in a state of maximum synchronization.

Furthermore, if, as we believe, no part of the star rotates more slowly than the orbit, then one should never see migration of the photometric 'wave' to later phases in the light curves. Rather, the wave should always migrate to earlier orbital phases. Busso *et al.* (1986) do find a dominant trend for migration to earlier phases as expected from our result, but also find times where the wave migrates to later phases. We suspect that the latter must be attributable to changes in the brightness centroid of a complex spot distribution rather than to drift of the longitude of some presumed single 'effective' spot. Such observations of migration of the wave to later phases on UX Arietis may indicate that wave migration is not a reliable indicator of differential rotation.

In spite of the very small amount of apparent differential rotation at the surface, UX Arietis shows a great deal of activity indicative of the presence of a very powerful dynamo. If differential rotation is indeed a necessary ingredient for dynamo action, then it must either be hidden below the surface of UX Arietis, or else the spot motions are not reflecting the true surface latitudinal shear (*i.e.* the spots are anchored to a deep and more uniformly rotating inner zone).

Finally, it is clear that the spot distribution on UX Arietis can change on time scales of a month. This is evident from changes in the shape of the polar spot as well as the appearance and disappearance of isolated spots. This demonstrates that one must exercise great care in using photometric light curves for deriving differential rotation rates. This also shows that to accurately derive the morphology of spots on the surface of these stars, one would like at least 3 or more Doppler images a year. This requires prodigious amounts of observing time and severely taxes the resources of most major observatories. Collaborative observing efforts between several observatories is required in order to get adequate time resolution between images as well as to get complete phase coverage.

References

Bell, R.A., Eriksson, K., Gustafsson, B., Norlund, Å.: 1976, *Astron. Astrophys. Suppl.* **23**, 37

Busso, M., Scaltriti, F., Cellino, A.: 1986, *Astron. Astrophys.* **156**, 106

Gilman, P., Miller, J.: 1986, *Astron. Astrophys. Suppl.* **61**, 585

Glatzmaier, G.: 1985, *Astrophys. J.* **291**, 300

Huenemoerder, D., Buzasi, D., Ramsey, L.: 1989, *Astron. J.*, **98**,1398

Poe, C., Eaton, J.: 1985, *Astrophys. J.* **289**, 644

Scharlemann, E.: 1982, *Astrophys. J.* **253**, 298

Vogt, S.S., Penrod, G.D., Hatzes, A.P.: 1987, *Astrophys. J.* **321**, 496

The Art of Surface Imaging

N.E. Piskunov

Astronomical Council of the USSR Academy of Sciences, Moscow, USSR
and
Observatory and Astrophysics Laboratory, University of Helsinki, Finland

Abstract: We intend to analyze the reliability of surface imaging of stars based on high resolution spectroscopy and the technique of inverse problem solution. Both astrophysical and mathematical aspects including different regularization methods are reviewed. The influence of the different factors on the resulting map is discussed and it is shown that the simultaneous use of different kinds of observational data (spectroscopy, photometry, polarimetry etc.) is very useful in providing additional constraints for the solution. The recent results in the surface imaging of Cp- and late-type stars show the way for further progress: the use of more adequate mathematical description of the stellar atmosphere and the simultaneous consideration of various surface inhomogeneities.

1. Introduction

The idea of the surface imaging of the stars was invented in 1958 by Deutsch who suggested using the Doppler shift produced by rotation in order to trace inhomogeneities on the surfaces of Ap-stars. He developed a method based on Fourier analysis (Deutsch, 1970) that in principle could provide information about the distribution of the spectral line intensity from its equivalent width variation as well as the changes in the apparent radial velocity. The subsequent modifications and improvements of the Deutsch method were made by Pyper (1969), Rice (1970), Falk and Wehlau (1974) and Mégessier (1975).

Further progress was made by Goncharskij and his collaborators (Goncharskij *et al.*, 1977; Goncharskij *et al.*, 1982; Goncharskij *et al.*, 1983) which lead to the understanding that surface imaging is a fundamentally ill-posed problem. It has a multiplicity of solutions and any expansion method will provide just one of them but not necessarily the real distribution. Instead they proposed fitting the observed line profiles with the calculated ones by changing the surface distribution. The fitting is done according to a special mathematical procedure called the inverse problem solution that provides the unique solution.

Recently the surface imaging technique has been expanded to temperature mapping of the cool stars (Vogt and Penrod, 1982) and to magnetic field mapping of Cp-stars (Piskunov, 1985; Landstreet, 1988; Donati *et al.*, 1990). Also, efforts have been made to replace the use of an approximation formula for the local line profile with the numerical solution of the transfer equation (Piskunov *et al.*, 1990; Piskunov and Wehlau, 1990), to add different kinds of observational data, and to use several spectral lines simultaneously (Hackman *et al.*, 1991). Several aspects of the surface imaging technique are discussed and illustrated by numerical examples in works by Piskunov (1985), Piskunov and Wehlau (1990), Vogt *et al.* (1987), Rice *et al.* (1989) and Semel (1989). An extensive reference list on the application of surface imaging technique to different stars is given in the paper by Khokhlova (1985).

2. Mathematical formulation

We have to establish a mathematical formulation of the problem of fitting observational data to the results of numerical simulation based on the theory of stellar atmosphere structure. Let $I(X(M), \lambda, \mu)$ be the specific intensity of the radiation with wavelength λ emitted at point M of the star surface in the direction μ (for strict definitions and notation see Mihalas, 1970). $X(M)$ represents a local characteristic of the point M that determines the specific intensity I. It can be chemical element abundance, temperature, magnetic field, etc. The monochromatic flux at wavelength λ will be given by a surface integral over the visible hemisphere:

$$F_\lambda(\phi) = \iint I(X(M), \lambda + \Delta_\lambda, \mu)\mu d\sigma, \tag{1}$$

where ϕ is the rotational phase and Δ_λ - the Doppler shift due to the projected rotation velocity at the point M. If we can find a way to calculate $I(X(M), \lambda, \mu)$ for arbitrary set of parameters, we will be able to calculate fluxes for a given set of wavelengths and phases, and to compare them to the observational data. The function $X(M)$ that gives the best fit is the required map.

The inhomogeneous structure of the stellar surface produces the variations of F_λ because of two different effects:

1) the contribution of any surface feature to the flux changes with the rotational phase because of limb darkening and the visibility of the feature;
2) the Doppler effect shifts the wavelengths of the contribution as the the projected velocity of the surface feature varies.

The difference between these two effects is that the first of them works independently of the rotation velocity of the star and can provide information about regions with small Doppler shifts. On the other hand, the second effect is much more pronounced in rapid rotators. Thus for producing a reliable image our simulation of line profiles must account correctly for both effects. To stress the importance of both effects we use the words *Surface imaging* rather than *Doppler imaging*, which was suggested by Vogt.

2.1 Surface imaging as an inverse problem solution

To compare the fluxes calculated from (1) with the observations we have to calculate the continuum level and allow for instrumental broadening. The continuum flux is expressed by:

$$F_\lambda^{\rm cont}(\phi) = \iint I^{\rm cont}(X(M), \lambda + \Delta_\lambda, \mu)\mu d\sigma, \tag{2}$$

where $I^{\text{cont}}(X(M), \lambda, \mu)$ is the continuum specific intensity. The dependence of I and I^{cont} on X can be very different. For example, if $X(M)$ represents the abundance of a rare-earth element at point M, I^{cont} will not depend on X at all. The Doppler shifts can be neglected, while calculating F^{cont}.

Now F_λ must be convoluted with the instrumental profile, usually approximated by a Gaussian:

$$\mathcal{F}_\lambda(\phi) = \frac{1}{\sqrt{\pi}\Delta\lambda_{\text{instr}}} \cdot \int_{-\infty}^{\infty} e^{-\left(\frac{y-\lambda}{\Delta\lambda_{\text{instr}}}\right)^2} F_y(\phi)dy. \tag{3}$$

$\Delta\lambda_{\text{instr}}$ is the half width of the instrumental profile. So finally we calculate the residual intensity r_λ:

$$r_\lambda(\phi) = \frac{\mathcal{F}_\lambda}{F^{\text{cont}}}, \tag{4}$$

and the surface imaging problem becomes mathematically equivalent to searching for such $X(M)$ that provides the minimum of the discrepancy function D:

$$D(X) = \sum_{\phi,\lambda} \omega_{\phi\lambda} \cdot \frac{\left(r_\lambda(\phi) - r_\lambda^{\text{obs}}(\phi)\right)^2}{N_\phi N_\lambda}, \tag{5}$$

where summing is performed over all available rotational phases and wavelengths. The weights $\omega_{\phi\lambda}$ are meant to account for the possible differences in the observational errors for each $r_\lambda^{obs}(\phi)$. If $\sigma_{\phi\lambda}$ is the observational error and σ_{min} is the minimum of all errors, the weights can be defined as $\omega_{\phi\lambda} = \sigma_{min}/\sigma_{\phi\lambda}$. The problem (5) is known as the inverse problem. The name comes from the fact that if D is formally treated as an operator over X, the solution of (5) can be obtained by constructing the inverse operator D^{-1}. But the very bad behaviour of D^{-1} means that the problem (5) is a so-called ill-posed problem (for strict definition see Tikhonov, 1963; Shore and Johnson, 1980). The major consequence of this is the multiplicity of solutions of (5). The theory to treat ill-posed problems was developed in the early 1960's, but it has only recently found its practical applications. The method is known as a regularization method.

2.2 Regularization method

The idea of the regularization is to replace the problem (5) with another one with better properties and a unique solution, which approximates in a certain known way the real $X(M)$. We shall assume now that $X(M)$ is a scalar function so that the mapping is performed just for one surface characteristic, e.g. temperature. The case of vector $X(M)$ is discussed in sections 6 and 7. The regularized problem can be expressed in the form:

$$\Phi(X) = D(X) + \Lambda \cdot R(X), \tag{6}$$

where Λ is a conventional Lagrange multiplier. $R(X)$ is the regularization functional which implies some additional constraints on the possible solution and thus makes it unique. Problem (6) can be considered as a typical conditional minimization problem. The value of Λ should be selected in such a way that if $X(M)$ is the solution of (6) then

$$D(X) = \sigma_{\mathrm{obs}}, \tag{7}$$

where σ_{obs} is the mean observational error (for the corresponding theorems see Tikhonov, 1963; Yagola, 1979; Shore and Johnson, 1980), so if the quality of the observations is better, the influence of the regularization $R(X)$ is smaller and we are getting closer to the solution of (5) realized in Nature.

Several forms of the regularization function are known. Two of them are quite general: the Tikhonov function and the entropy measure. Others have a more limited range of application and are based on additional information about the physical processes that control $X(M)$ (for examples, see Goncharskij and Stepanov, 1979).

The Tikhonov regularization function

$$R^T(X) = \iint \|\nabla X(M)\|^2 d\sigma \tag{8}$$

is a measure of a smoothness of $X(M)$ so that the resulting map provides the smoothest possible solution that is able to reproduce the observations within the observational errors. The use of (8) as the regularization function for (6) means that a strong correlation between neighbouring points is assumed.

The Entropy regularization function is used in the form

$$R^E(X) = \iint X(M) \log X(M)\, d\sigma, \tag{9}$$

and assumes no correlation between neighbouring points.

Although the Tikhonov function seems to be more appropriate for the surface imaging of the stars, both (8) and (9) can be used because, as can be seen from Fig. 1, with high quality data the rôle of the regularization function is limited to providing the unique solution and is small compared to the influence of the observations. More detailed discussion of the two methods can be found in papers by Khokhlova *et al.* (1986) and Piskunov *et al.* (1990).

2.3 How close can we get to reality?

According to the theorems mentioned above, the inverse problem solution will converge to the exact surface distribution of $X(M)$ as the quantity and the accuracy of the data increase. But while our observations are not perfect we still can get a unique solution which approximates the real surface distribution.

Also with both the Tikhonov and MEM regularizations it is difficult to say what is the upper limit for the differences between the calculated and the measured $X(M)$ all over the stellar surface. Several indirect approaches can be suggested. For example, comparison of the abundance maps, derived from the data for different

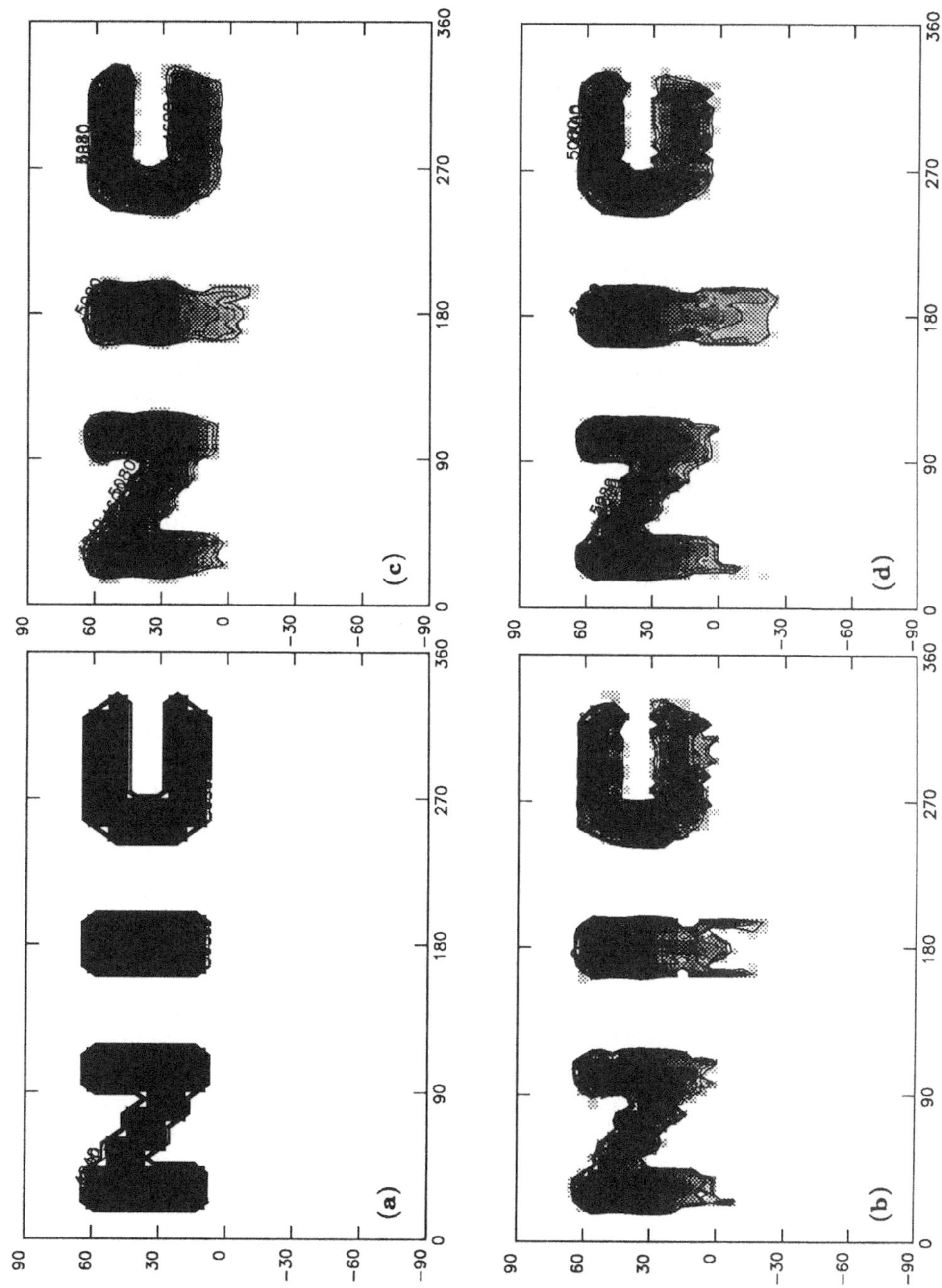

Fig. 1. A comparison of the inverse problem solutions for NIC-star **(a)** without regularization **(b)**, and using Tikhonov **(c)** and Maximum entropy **(d)** regularizations. S/N is 500. Spectral resolution 60 000. Temperature inside the letters is 1 500 K lower than on the rest of the surface. Ca I λ 6439.1 line profiles were used for the imaging.

spectral lines of a given element, might be used for Cp-stars. For late type stars the comparison of the different kinds of observations with the simulations based on the obtained maps, such as photometric light curves, can provided an estimate of the error.

3. The resulting map and the observational data parameters

The quality and the quantity of the spectral observations influence the produced image in different ways. All parameters of the observations (spectral resolution, S/N ratio, and phase coverage) are important for correct mapping. We are going to illustrate an influence of all major characteristics of the observational data using the numerical simulations of 25 profiles of Ca I λ 6439.1 line for the artificial "NIC-star" shown on Fig. 1a.

3.1 Spectral resolution

The spectral resolution should be consistent with the S/N ratio and the rotational broadening of the spectral lines. This balance also determines the surface grid size (and thus the size of the surface resolution) that it is reasonable to use for a given star (for detailed discussion see Piskunov and Wehlau, 1990). Figure 2 shows the result of NIC-star reconstruction with a reduced spectral resolution.

3.2 Signal-to-noise ratio

The signal-to-noise ratio determines how accurate the amplitude of the surface feature can be reconstructed. Numerical experiments prove that S/N = 150 is the lower limit for the observational data to be used in surface imaging. High S/N ratio is also necessary to take advantage of a high spectral resolution (if any), especially for rapidly rotating stars where the slope of the line wings is small. Figure 3 shows the map of the NIC-star obtained from the line profiles with the simulated S/N = 100.

3.3 Phase coverage

The phase distribution of the observations is supposed to cover all of the stellar surface visible from the Earth. If some longitudes are not represented in any of the line profiles, the whole solution will be quite unstable because the assumption of the continuity of the surface distribution is not anymore applicable. The computer code usually keeps the initial distribution inside a gap in longitude coverage but this is of course a purely artificial decision.

Good phase coverage is important not only for achieving a good surface resolution (smaller features will manifest themselves at many different Doppler shifts) but also for differentiating between north and south hemispheres. The surface points below and above the equator can be distinguished from each other (for intermediate values of the inclinations of the star rotation axis) only because of the difference in their times of visibility during the rotational period, and the difference in the amplitudes of the Doppler shift variations. Figure 4 illustrates the reconstruction of the NIC-star from data for phases between 0 and 0.75 only.

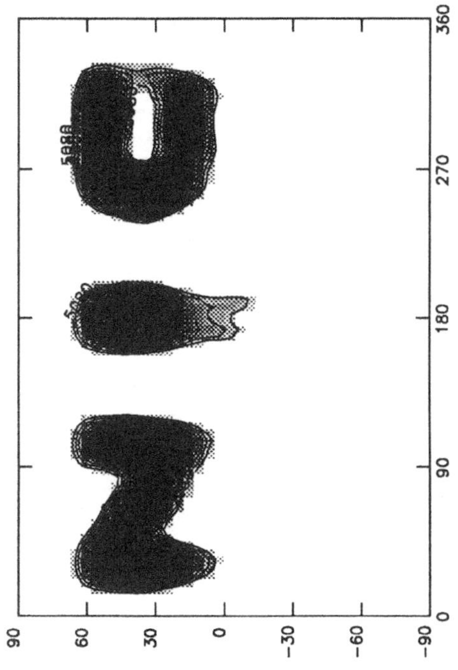

Fig. 2. The results of the NIC-star mapping using the line profiles with spectral resolution 20 000.

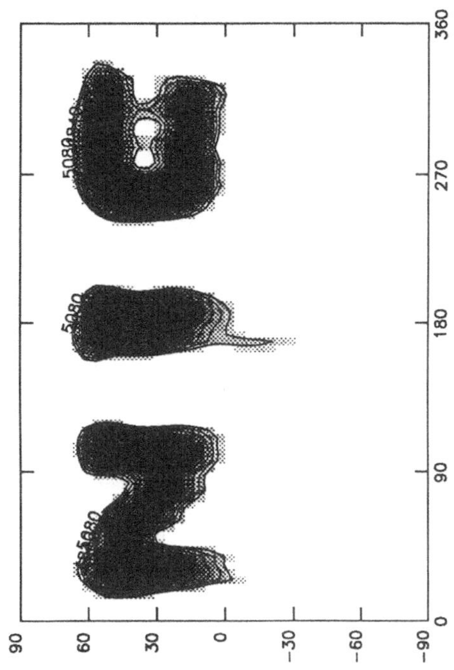

Fig. 3. The results of the NIC-star mapping using the line profiles with S/N = 100.

4. The resulting map and the stellar characteristics

Knowledge of the stellar rotation velocity and the inclination angle is necessary to start the inverse problem solution. In some cases the accuracy of those characteristics can be improved during the imaging.

4.1 Rotation velocity

Because only rapidly rotating stars are selected for surface imaging, the line profiles are dominated by the rotational broadening, so that the procedure of fitting the observations with the calculated profiles is very sensitive to the error in $V \sin i$. It is impossible to get the discrepancy function down to the observational errors if $V \sin i$ is more than 10 km/s away from its real value. Smaller errors of the order of 5 km/s may result in an artificial equatorial belt of enhanced local line strength if the assumed velocity is too low, or in the polar cap if the assumed velocity is too high. Those features are quite helpful for fine tuning of the rotation velocity value. But it is very difficult to achieve an accuracy better than $\pm$ 2 km/s (see

Vogt *et al.*, 1987). An attempt to map the NIC-star with an incorrect $V \sin i$ is illustrated in Fig. 5.

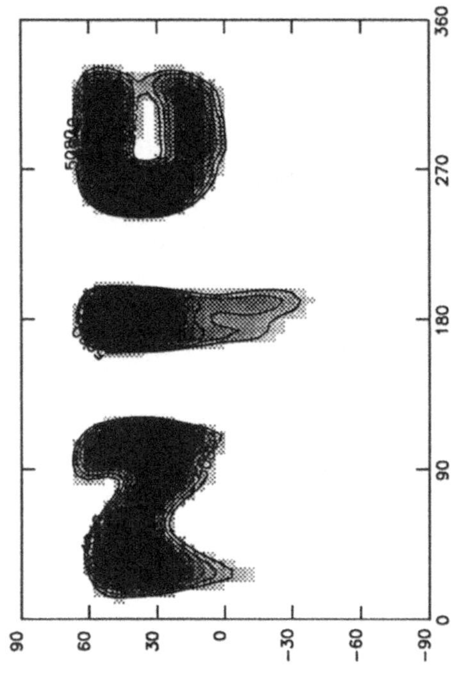

Fig. 4. The results of the NIC-star mapping using line profiles for phases between 0 and 0.75.

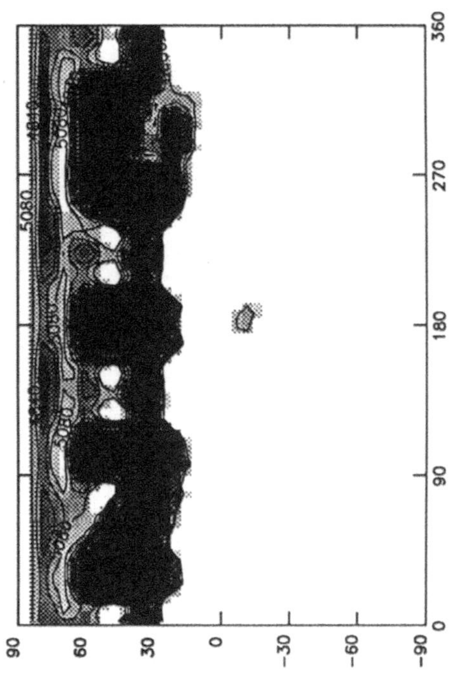

Fig. 5. The results of the NIC-star mapping using $V \sin i =$ 53 km/s instead of 55 km/s.

4.2 Inclination

The influence of the inclination angle is not so strong as that of the rotation velocity. Changing the value of i can move the surface features to a different latitudes while the shape and the amplitude of the spots remain essentially the same (see Khokhlova *et al.*, 1986). The required accuracy for i is about $\pm$ 10°.

5. The resulting map and the line profile calculation

The quality of the specific intensity calculations used for high S/N data simulation is a crucial point for the extraction of the information about the surface structure. A poor approximation for $I(X, \lambda, \mu)$ may produce some artifacts in the resulting map. Now we shall see how different methods treat the following six problems of major importance for surface imaging:

1) line parameter selection (oscillator strength and line broadening constants);

2) model atmosphere selection (numerical or simplified, e.g. Milne-Eddington);
3) accounting for line shape variation;
4) treatment of blends;
5) accuracy of the representation of the line profiles;
6) interpretation of the resulting maps.

5.1 Using the approximation formula

An approximation formula for the surface profile of the specific intensity $I(X, \lambda, \mu)$ was generally used by many researchers because it is much faster to compute and it is usually expressed in analytical form, so that the accurate calculation of gradients is much easier. Several formulae were used: Minnaert function (Goncharskij *et al.*, 1977), Unno functions (Piskunov and Khokhlova, 1983), etc.

When using an approximation formula problems 1) and 2) are replaced by the problem of the correct selection of the formula parameters. This selection can be made by comparing the approximation formula with more accurate calculations for a wide set of atmosphere conditions, although for this calculation problem 1) still exists (Khokhlova and Pavlova, 1984). Problem 3) is sometimes neglected (Vogt *et al.*, 1987) but can be accounted for if a separate approximation is used for the the line and the continuum.

The approximation formula approach has nowadays mainly historical interest because:

- it is unable to deal with high S/N observations (the approximation accuracy is usually between 1 and 2 percent);
- it is practically impossible to treat blends;
- the interpretation of the results is not straightforward because we usually get maps of approximation formula parameters and an extra step is needed to obtain physical parameter distributions over the surface.

5.2 Transfer equation solution

The explicit solution of the transfer equation reproduces the observed data in a much more satisfactory manner, accounts for continuum variation and can be easily extended to treat blends (see 5.3). Also, no additional interpretation of the results is required. Advanced numerical model atmospheres and even several different models for different surface points can be used, although the problem of obtaining a reliable model atmosphere still exists. The numerical solution of the transfer equation must be done many times for all surface points, so it is worthwhile to pre-calculate the interpolation tables for specific intensity in order to reduce the computing time. One possible approach that we are currently using (Piskunov *et al.*, 1990; Piskunov and Wehlau, 1990) requires the calculation of two tables. The first is wavelength dependent and contains $I(X, \lambda, \mu)$ values for a set of nodes over all three parameters. The second stores the continuum specific intensities and does not depend upon the wavelength. Storing tables for the values of μ used for surface integration reduces the calculation of the local profile to a 2-dimensional interpolation (if $X(M)$ is a scalar function, such as in the cases of abundance or temperature mapping).

5.3 Spectral synthesis

The use of spectral synthesis is a further development of the surface imaging technique based on the numerical solution of the radiation transfer equation. The spectral synthesis is not simply a tool for the correct treatment of spectral blends but also provides the additional constraints for the inverse problem solution. The reliability of the resulting image can be significantly improved by a simultaneous fitting of a set of spectral lines with a different sensitivity to the mapping parameter. This possibility was first noticed by Landstreet (1988) and he successfully used it for the imaging of the magnetic field on 53 Cam, selecting lines with different Landé factors. The results of temperature mapping of EI Eri, using a spectral synthesis technique for a blend of five lines, is presented by Hackman *et al.*, in these Proceedings.

6. Using additional data

Using a combination of different kinds of observational data can be very helpful in improving the reliability and the quality of the imaging. It can also provide information about star geometry (orientation of the rotation axis), magnetic field strength and distribution, etc.

6.1 Photometry

When mapping the temperature on the surface of late-type stars the photometry provides an important test for the correct regularization of the solution. Too small a Λ value will result in extra temperature spots on the surface which will distort the light curves. The transfer equation (or spectral synthesis) method of line profile simulation can be easily checked for consistency with photometric data by producing a temperature map from spectral line(s) only, calculating the emerging flux for a set of rotation phases and convoluting it with a corresponding filter functions. The results can be compared with the photometric light curves. An example of this kind of comparison for EI Eri is shown in Hackman *et al.*(these Proceedings). We want also to stress here the importance of a good knowledge of the rotational period as well as the use of *simultaneous* spectroscopy and photometry for late-type stars with strong surface activity.

6.2 Polarization

The achievements from using polarization observations for surface imaging are rather modest. For Cp-stars, only some restricted multipole expansion approximation for magnetic field have been considered (e.g. Borra and Landstreet, 1978; Glagolevskij *et al.*, 1985; Donati *et al.*, 1990).

The big challenge when using polarization observations for the surface imaging is that this kind of data is much more sensitive to the surface structure than the line profiles. According to the numerical modelling (Piskunov and Khokhlova, 1983, 1984; Semel, 1989) with the polarization data it may be possible to map polar region and to break symmetry in the equatorial belt. For late-type stars it may provide a unique opportunity to measure the field strength inside the active regions.

6.3 Several spectral regions

Additional information can also be obtained simply by using several spectral regions with different properties and fitting them simultaneously. When mapping temperature in this way one can select lines with different temperature sensitivity belonging to different ionization stages. For Cp-stars this method provides a possibility of increasing the amount of observational data or of mapping the distribution of several chemical elements (if $X(M)$ is taken in vector form).

Landstreet (1988) suggested using several spectral regions for magnetic stars in order to extract information about the surface field by fitting the intensification of lines as well as the distances between Zeeman components of lines with different Landé factors.

7. Conclusions

Surface imaging has become an important tool for studying individual stars with strong surface activity. Further improvement of the quality of the spectroscopic observations and in the understanding by the observers of the requirements of surface imaging will provide possibilities for new kinds of research, for example, studying the evolution of the surface features in late-type stars or making a statistical analysis of elements and magnetic field distribution in Cp-stars.

To meet these approaching opportunities efforts should be directed towards determining better spectral line data, and producing better atomic data, atmosphere models, etc.

The implementation of spectral synthesis techniques, together with the use of several kinds of observations, will enable us to produce the most reliable and detailed maps.

High quality polarization observations are the only way to study the magnetic field on the surfaces of rapidly rotating stars.

References

Borra, E.F., Landstreet, J.D.: 1978, *Astrophys. J.* **222**, 226

Deutsch, A.: 1958, in *Electromagnetic Phenomena in Cosmological Physics*, IAU Symp. 6, ed. B. Lehnert, Cambridge Univ. Press, Cambridge, p. 209

Deutsch, A.: 1970, *Astrophys. J.* **159**, 895

Donati, J.-F., Semel, M., del Toro Iniestia, J.C.: 1990, *Astron. Astrophys. Lett.* **233**, L17

Falk, A.F., Wehlau, W.H.: 1974, *Astrophys. J.* **192**, 409

Glagolevskij, Yu.V., Piskunov, N.E., Khokhlova, V.L.: 1985, *Sov. Astron. Lett.* **10**, 187

Goncharskij, A.V., Stepanov, V.V.: 1979, *Sov. Math. Dokl.* **20**, 414

Goncharskij, A.V., Stepanov, V.V., Khokhlova, V.L., Yagola, A.G.: 1977, *Sov. Astron. Lett.* **3**, 147

Goncharskij, A.V., Stepanov, V.V., Khokhlova, V.L., Yagola, A.G.: 1982, *Sov. Astron.* **19**, 576

Goncharskij, A.V., Ryabchikova, T.A., Stepanov, V.V., Khokhlova, V.L., Yagola, A.G.: 1983, *Sov. Astron.* **27**, 49

Khokhlova, V.L.: 1985, *Astrophys. Space Phys. Rev., Soviet Scientific Rev.*, Sect. E., ed. R. Sunyaev, **4**, p. 99
Khokhlova, V.L., Pavlova V.M: 1984, *Sov. Astron. Lett.* **10**, 158
Khokhlova, V.L., Rice, J.B., Wehlau, W.H.: 1986, *Astrophys. J.* **307**, 768
Landstreet, J.D.: 1988, *Astrophys. J.* **326**, 967
Mégessier, C.: 1975, *Astron. Astrophys.* **39**, 263
Mihalas, D.: 1970, *Stellar Atmospheres*, Freeman, San Francisco
Piskunov, N.E.: 1985, *Sov. Astron. Lett.* **11**, 18
Piskunov, N.E., Khokhlova, V.L.: 1983, *Sov. Astron. Lett.* **9**, 346
Piskunov, N.E., Khokhlova, V.L.: 1984, *Sov. Astron. Lett.* **10**, 187
Piskunov, N.E., Tuominen, I., Vilhu, O.: 1990, *Astron. Astrophys.* **230**, 363
Piskunov, N.E., Wehlau, W.H.: 1990, *Astron. Astrophys.* **233**, 497
Pyper, D.: 1969, *Astrophys. J. Suppl.* **18**, 347
Rice, J.B.: 1970, *Astron. Astrophys.* **9**, 189
Rice, J.B., Wehlau, W.H., Khokhlova, V.L.: 1989, *Astron. Astrophys.* **208**, 179
Semel, M.: 1989, *Astron. Astrophys.* **225**, 456
Shore, J.E., Johnson, R.W.: 1980, *IEEE Trans.* **IT-26**, 26
Tikhonov, A.N.: 1963, *Sov. Math. Dokl.* **4**, 1624
Vogt, S.S., Penrod, G.D.: 1982, in *Activity in Red-Dwarf Stars*, IAU Coll. 71, eds. P. Byrne, M. Rodonò, Reidel, Dordrecht, p.379
Vogt, S.S., Penrod, G.D., Hatzes, A.P.: 1987, *Astrophys. J.* **321**, 496
Yagola, A.G.: 1979, *Sov. Math. Dokl.* **20**, 275

Surface Imaging of EI Eri

T. Hackman [1], N.E. Piskunov [2], M. Poutanen [3],

K.G. Strassmeier [4], I. Tuominen [1]

[1]Observatory and Astrophysics Laboratory, University of Helsinki, Finland
[2]Astronomical Council of the USSR Academy of Sciences, Moscow, USSR
[3]Finnish Geodetic Institute, Helsinki, Finland
[4]Institute for Astronomy, University of Vienna, Austria

Abstract: We present maps of the temperature distribution on the spotted RS CVn-type binary EI Eri (HD 26337), obtained by means of the surface imaging technique described in the Colloquium (Piskunov, 1991). Images were calculated for two separate lines for two epochs of observation. For one epoch we also calculated an image using a blend of several lines. The reliability of the maps is confirmed by comparing the simultaneous photometric observations with the light curves calculated from the temperature maps.

1. Observations

The spectroscopic observations for the present surface imaging study were obtained with the McMath solar telescope at Kitt Peak National Solar Observatory during two epochs. The data for "image 1" consists of spectra for eight different phases obtained between October 22 and November 21, 1988. The data for "image 2" consists of spectra for eleven different phases taken between November 27 and December 22, 1988. Simultaneous Johnson BV photometry was available from the 0.4 m Vanderbilt University APT. More information about the observations can be found in papers by Strassmeier (1990 and 1991). For image 1 and 2 the spectra of the lines Fe I λ6411.66 Å and Ca I λ6439.073 Å were used to produce separate temperature maps. In addition a blend of five lines, V I λ6430.50, Fe I λ6430.85, Ni I λ6432.02, Fe II λ6432.65 and Si I λ6433.46, was used to derive a third temperature map for image 2. The largest contribution to the main component of the blend Fe I λ6430 Å comes from the line Fe II λ6432 Å, without which it is impossible to produce the red part of the line profile.

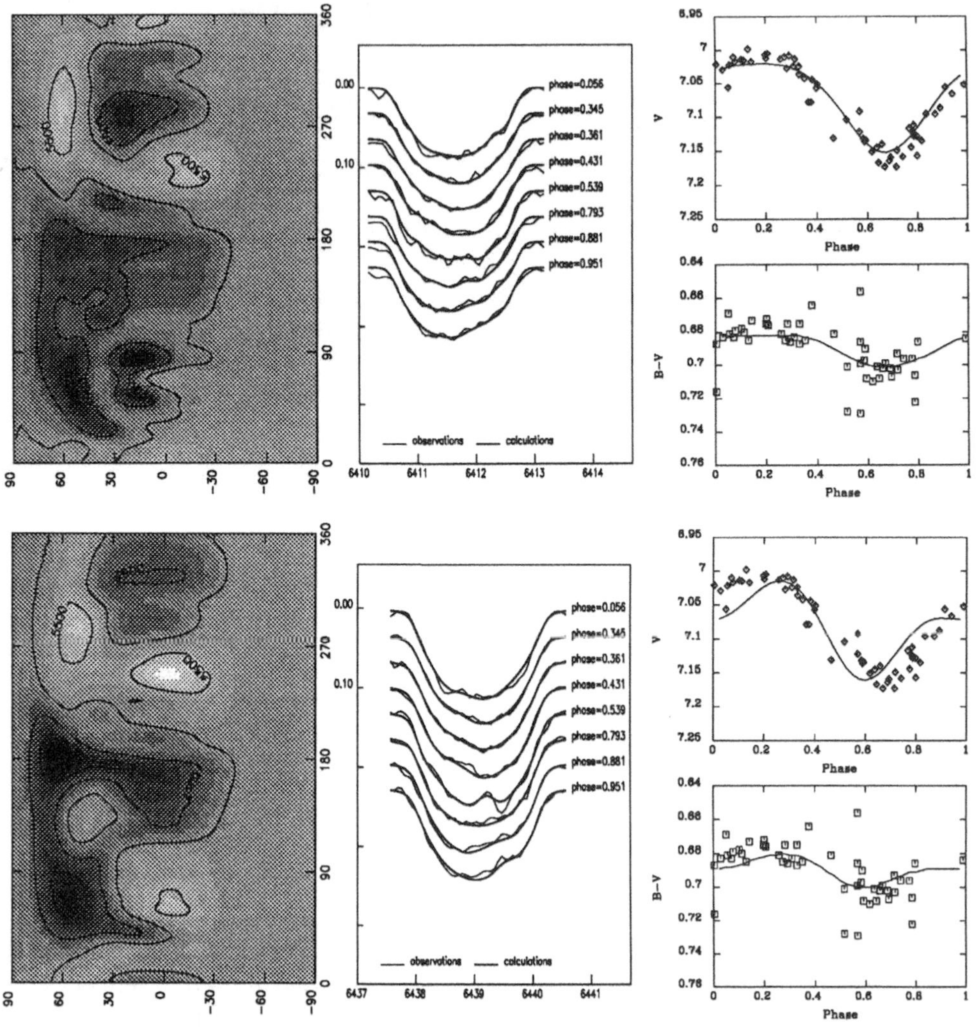

Fig. 1. Surface imaging solution of image 1 using the Fe I λ6411.66 Å line (upper row) and the Ca I λ6439.073 Å line (lower row). The solutions are presented by temperature maps in Mercator projection, comparison of observed spectra and line profile resolved by a numerical solution of the equation of radiation transfer, and comparison of the simulated V- and B-V -photometric curves (lines) and observations (squares). The grey shade scale for the temperature maps ranges from 4200 K (black) to 5700 K (white).

2. Analysis technique

We used a surface imaging technique with Tikhonov regularisation (Piskunov 1991) to produce temperature maps of the two images. The tabulated local line profiles and continuum fluxes were calculated using four atmospheric models, $T_{\text{eff}}= 4000$, 4500, 5000 and 5500 K, with $\log g = 3.75$ and solar element abundances (Bell *et al.* 1976). Local line and continuum fluxes were calculated for 40 points from the centre of the star disk to the limb. The microturbulent velocity was adjusted in such a way that the temperature of the unspotted surface was the same for the different temperature maps of the different lines and images. The damping parameter is important, effecting especially the line wings of the strong lines Fe I λ6430 Å and Ca I λ6439 Å.

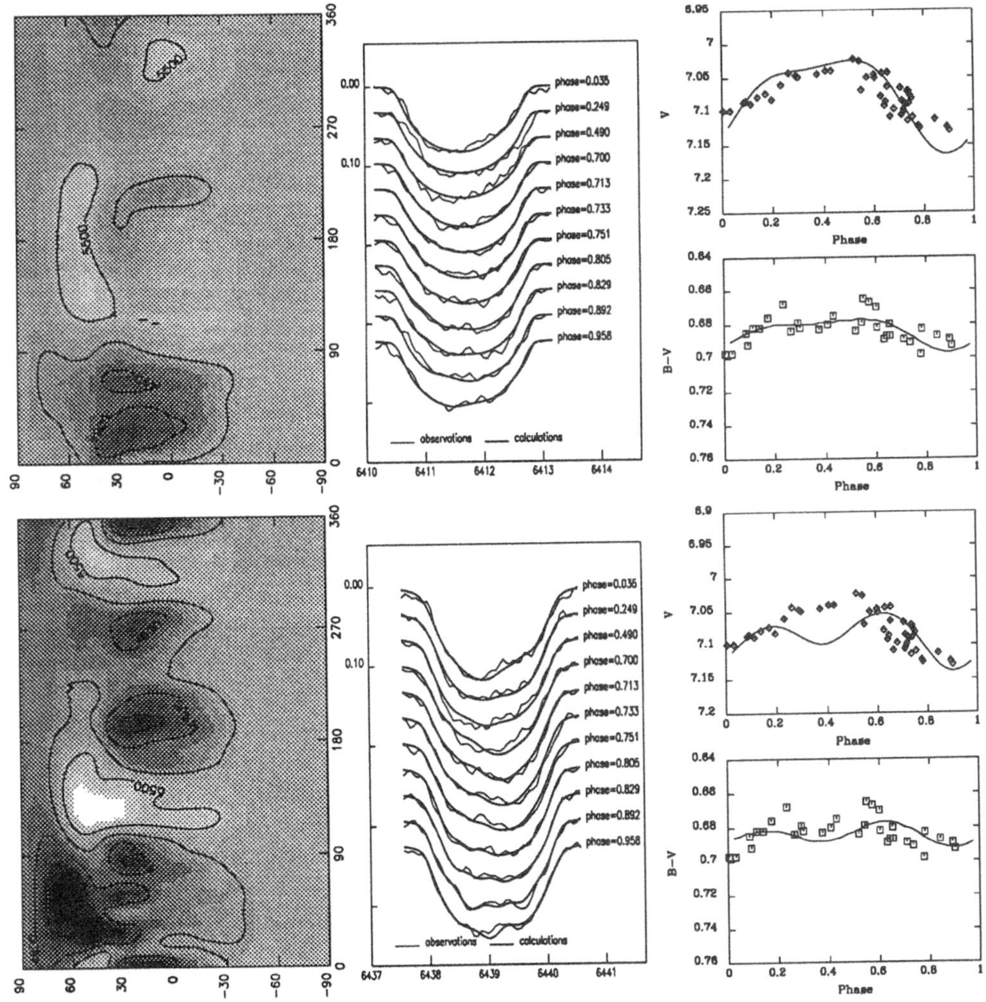

Fig. 2. Solution for image 2, using the Fe I λ6411.66 Å line (upper row) and the Ca I λ 6439.073 Å line (lower row), presented as temperature maps, line profiles and photometry. The grey-shade scale for the temperature maps is the same as in Fig. 1.

By using the transmission functions of B and V passbands and integrating the continuum flux distribution of the radiative transfer solution over the visible hemisphere in different phases, we calculated theoretical V, B, and B-V light curves for the temperature maps. We used an arbitrary zero point scaling in order to fit the curves to the observed magnitudes.

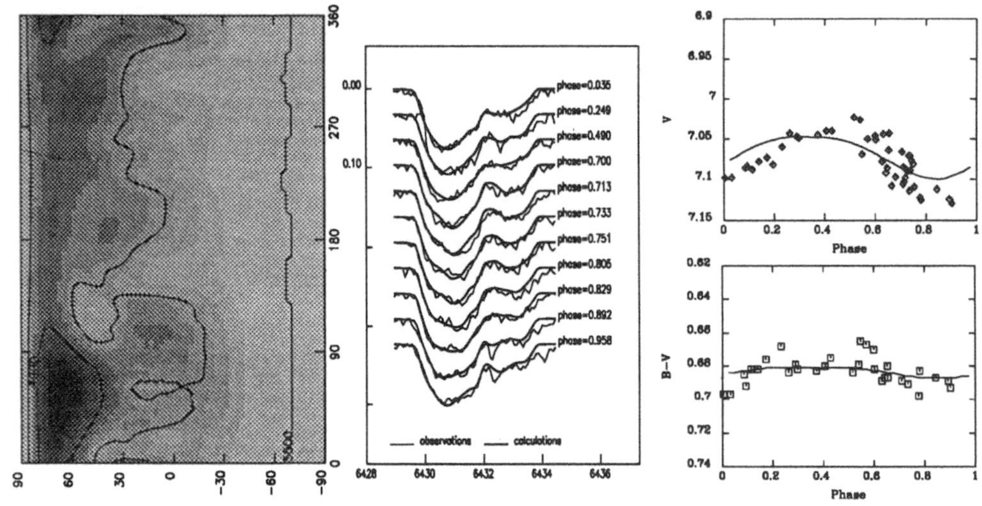

Fig. 3. Solution for image 2, using the λ6430 Å blend of lines, presented as temperature map, line profiles and photometry. The grey-shade scale for the temperature map is the same as in Fig. 1.

3. The images and conclusions

The temperature maps (Figs. 1–3) show large cool spots on both low and high latitudes. The spots are 500-1000 K cooler than the unspotted surface. The spot structure is somewhat similar to that derived for the FK Comae star HD 32918 by Piskunov *et al.* (1990). In addition, the maps show some smaller bright spots, with temperatures up to 500 K hotter than the unspotted surface. None of the temperature maps show large caps on the pole. This result differs significantly from those derived by two other groups (Vogt and Hatzes, Rice and Wehlau) using the same data set and presented by Strassmeier *et al.* (1991). The main difference between the present analysis and theirs is the direct calculation of the local line profiles in our analysis.

The overall fit of the calculated light curves to the observed photometry is quite satisfactory, giving the general variation with phase correctly. The light curves from the temperature maps produced by using the Fe I λ6411.66 Å line coincide better with the observations than those produced using the Ca I λ6439.073 Å line and the λ6430 Å blend. The cool regions have a tendency to be at higher latitudes for the strong Ca I line than for the weaker Fe I λ6411.66 Å line. Whether this results from the fact that the stronger line is formed in higher layers of the atmosphere or form inaccuracies in the line formation calculation, should be investigated. The differences in the temperature maps produced for the same image can also partly be explained by lack of data for some phases. This means that noise in spectra more easily may produce artificial features on the temperature maps. The overall pattern of the spots is, however, the same on temperature maps of the same image.

Simultaneous photometry thus provides very useful extra constraints for the inversion. This is especially important since some spectral line parameters, which significantly effect the line profile, are not well enough known for all lines.

The main difference between images 1 and 2 seems to be a shift in the longitude of the spot features. The same shift can be seen in the light curves. The difference in the light curves can be eliminated if the period is increased by about 0.01 days. This suggests that the differences can be due to the use of an inaccurate period rather than changes in the spot structure. Since one of the purposes of this work was to compare different surface imaging techniques we have used the same period (1.945 days) as the other groups analysing the same data (Strassmeier 1991, Strassmeier *et al.* 1991).

The use of the λ6430 Å blend of lines (Fig. 3) for surface imaging is especially interesting, since the five lines have different temperature behaviour. This puts extra constraints on the fitting procedure which means that noise in the spectra is not as easily interpreted as spot features. The use of a blend of lines instead of single lines, however, complicates especially the calculation of local line profiles. Even though the calculated line profile for the 6430 blend fits the observed line profile satisfactorily, the deviation from the observed profile is larger than that of the Fe I λ6411.66 Å or the Ca I λ6439.073 Å line.

Acknowledgement. The calculations were carried out on the Cray X-MP EA/432 supercomputer of the Centre for Scientific Computing in Espoo, Finland.

References

Bell, R.A., Eriksson, K., Gustafsson, B., Norlund, Å.: 1976, *Astron. Astrophys. Suppl.* **23**, 37

Piskunov, N.E.: 1991, these Proceedings

Piskunov, N.E., Tuominen, I., Vilhu, O.: 1990, *Astron. Astrophys.* **230**, 363

Strassmeier, K.G.: 1990, *Astrophys. J.* **348**, 682

Strassmeier, K.G.: 1991, in *Surface Inhomogeneities on Late-type Stars*, Armagh Observatory Bicentenary Coll. 1990, Springer-Verlag, Heidelberg, in press

Strassmeier, K.G., Rice, J.B., Wehlau, W.H., Vogt, S.S., Hatzes, A.P., Tuominen, I., Piskunov, N.E., Hackman, T., Poutanen, M.: 1991, *Astron. Astrophys.*, submitted

Tikhonov, A.N.: 1963, *Sov. Math. Dokl.* **4**, 1624

Towards Magnetic Images of Rapidly Rotating Late-type Stars

J.-F. Donati[1], M. Semel[2]

[1]DESPA, Observatoire de Paris, Section de Meudon, F-92195 Meudon Principal Cédex, France
[2]DASOP, Observatoire de Paris, Section de Meudon, F-92195 Meudon Principal Cédex, France

Abstract: We present the first magnetic detections using Zeeman-Doppler Imaging (ZDI) in the two bright rapidly rotating RS CVn systems HR 1099 and σ^2 CrB, and discuss their compatibility with various recent results on magnetic activity in cool stars.

1. Introduction

In a large number of late-type stars, activity indicators are found to be essentially similar to those observed in the sun despite the different strengths and time scales: this common phenomenology is usually attributed to the dynamo mechanism. Solar physicists have long known that the major parameter in the study of activity is the magnetic field. However, stellar magnetic geometries are very difficult to derive since the star itself is unresolved. In the last 10 years, the first direct estimates of magnetic field strengths B and filling factors f_B have been obtained for slowly rotating cool stars from Zeeman broadening measurements of unpolarized profiles. Recent studies (e.g. Saar, 1990) report that the field intensity correlates with the surface pressure, while the magnetic flux $f_B B$ correlates with both rotation rate and inverse Rossby number. However, the study of the origin and evolution of activity processes in cool stars still remains elusive, partly due to the lack of spatial information on the magnetic structures. The direct study of individual magnetic starspots should lead to a more informative solar-like way of studying activity.

A breakthrough may come from the recent developments of indirect imaging techniques. Using the correspondence between wavelength position of a rotationally Doppler-broadened spectral line and spatial position across a stellar disk, it becomes in principle possible to derive 2D images of a rapidly rotating star in terms of surface brightness inhomogeneities. ZDI (which can be roughly described as Stokes V Doppler Imaging) should similarly help unravel the spatial structure of the field.

We present here new magnetic measurements for the bright RS CVn systems HR 1099 and σ^2 CrB, using ZDI. For both of them a field is detected. In HR 1099, the rotational modulation of the Zeeman signature is briefly interpreted in terms of the distribution of magnetic field. The implications of these results for photospheric activity are discussed.

2. Deriving spatial magnetic information

Polarization in line profiles is an unambiguous signature of magnetic field. In particular, the circularly polarized light is related to the longitudinal component of the field. Because the shape of spectral lines from a rapidly spinning star is dominated by rotational broadening, a high resolution spectrum becomes essentially a 1D image of the star (Vogt, 1988): for such stars, the heterogeneous structure of photospheric magnetic fields may be resolved from the rotational modulation of the circular polarization profile $V(\Delta\lambda)$ (Semel, 1989).

Let $F(\Delta\lambda)$ denote the Doppler-resolved distribution of the longitudinal component of the stellar magnetic field and w the equivalent width of the local profile f. In the weak field approximation, the convolution of $F(\Delta\lambda)$ with the *normalized* local profile $f_N(\Delta\lambda) = f(\Delta\lambda)/w$ is proportional to the indefinite integral of $V(\Delta\lambda)$ (Donati and Semel, 1990):

$$F * f_N(\Delta\lambda) = \frac{1}{g'w}\int^{\Delta\lambda} V(u)\,du$$

in which g' denotes the "reduced" Landé factor of the transition, equal to $4.67\times10^{-13}\lambda^2 g_{eff}$ Å/G. Note that the magnetic distributions derived are in principle almost independent of the line used, so that averaging the results from many different lines may significantly increase S/N.

From the time modulation of these 1D magnetic distributions, a complete map may be recovered: the first results of a maximum entropy code applied to the recovery of synthetic magnetic structures are presented in Brown and Donati (1990). Like all other methods (e.g. Saar, 1990), ZDI detects the magnetic field in plages rather than in cool spots, due to the very low relative contribution of dark regions to stellar profiles.

From the formal expression for the photon noise level in the magnetic distribution (Donati *et al.*, 1989), we find that $S/N = 900$ per 80 mÅ pixel is needed to detect a 1000 G field over 10% of the stellar surface in a star with $v \sin i = 38$ kms^{-1}, with a confidence level of 10σ, if one uses a single $g_{eff} = 2$ line at 5500 Å with $w = 180$ mÅ. To satisfy these stringent requirements, we use the special observation and reduction procedure described in Donati *et al.* (1990).

3. Results for HR 1099 and σ^2 CrB

HR 1099 = V711 Tau (K1IV + G5V) and σ^2 CrB (F6V + G0V) are very bright RS CVn binary systems, the cool components of which show manifestations of intense activity. Both are well suited to ZDI. Three magnetically sensitive Fe I lines ($g_{eff} \simeq 2$) around 5500 Å were simultaneously used in the analysis.

The complete results of a 2 day observing campaign at the Anglo-Australian Telescope on the active primary star of HR 1099 are reported in detail in Donati *et al.* (1990). Phases 0.50 and 0.85 were observed. In the two corresponding distributions, the only feature exceeding the 5σ level is a large positive redshifted bump at phase 0.85. With the rough approximation of a radial field (of constant strength B) inside a circular region, we can try to mimic the two observed distributions and infer basic information on the field structure. For a spot latitude l greater than 10°, the amplitude of the feature generated in $F * f_N(\Delta\lambda)$ at phase 0.50

would be too high to go undetected. The magnetic field should also be lower than the equipartition "pressure-balancing" magnetic field B_{eq} (Saar, 1990), equal to 1060 G in the particular case of HR 1099, so that l must be larger than 0°.

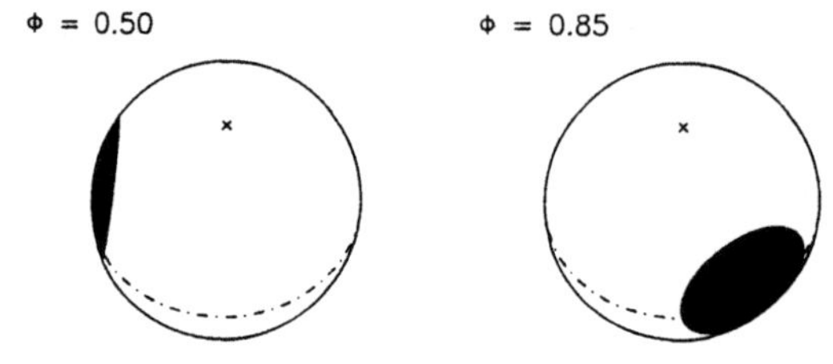

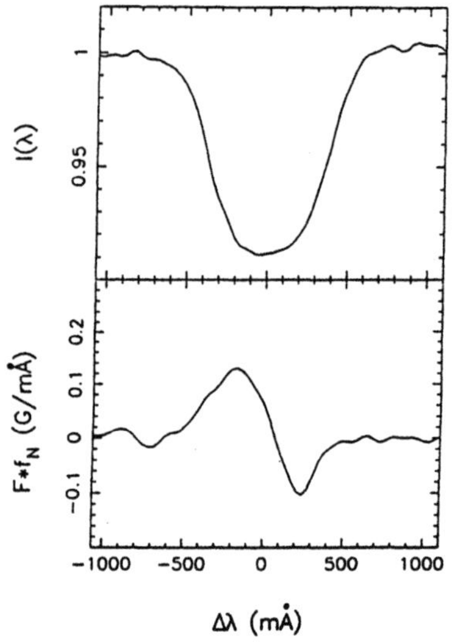

Fig. 1. a) HR 1099 in two observed phases with the detected magnetic region (up) b) Simultaneously measured intensity profile (upper curve) and magnetic distribution (lower curve) for σ^2 CrB around phase 0.25.

The final parameters are $B = 985 \pm 270$ G, $L = 86 \pm 4°$, $l = 5 \pm 5°$ and $r = 0.53 \pm 0.08$, where L and r denote the spot longitude and fractional radius. The region therefore covers $f_{ph} = 8 \pm 2\%$ of the total stellar surface [note the mistake in Donati *et al.* (1990)] and $f_B = 14 \pm 4\%$ of the visible hemisphere at phase 0.85. The geometrical configuration of the star with the detected magnetic region on its surface is shown in Fig. 1 a) for both observed phases. This region is found to be largely monopolar and may coincide with a hot region if associated with a brightness inhomogeneity (Donati *et al.*, 1990).

The very new data on σ^2 CrB were obtained in April 1990 at the Canada-France-Hawaii Telescope and the system was observed around phase 0.25. At that particular phase, a bipolar magnetic field is unambiguously detected for the first time on the G0 secondary component, and nothing is seen on the F6 primary star. The lower curve of Fig. 1 b) shows the 1D magnetic distribution of the secondary star around first quadrature. More than only one of these distributions is required to obtain spatial information on the field structure, such as we derived for HR 1099.

4. Discussion

It is now possible to derive spatial information on the magnetic structure of rapid rotators: this may be especially interesting for the study of stellar activity given the various correlations already reported between rotation rate and magnetic flux for slow rotators (Saar, 1990). The size of the monopolar magnetic region detected on the primary star of HR 1099 is similar to that of the cool spots derived with spectroscopy (Vogt, 1988). The typical horizontal size scale of the field distribution is therefore also much larger than in the sun. As a result, a large rotational modulation of the 1D magnetic flux distribution is noticed between the two observed phases. Similar huge magnetic regions may also be responsible for the high magnetic variability of the cool G8 dwarf ξ Boo A (Saar, 1990). Conversely, some other cool stars (like ϵ Eri) show little magnetic variability along the rotational phase (Saar, 1990): there the photospheric field structure may consist of many small regions (flux tubes) for which the integrated filling factors and average field modulus over the visible hemisphere are almost independent of rotational phase. Different emerging modes of the magnetic field may be associated with different horizontal size scales in the magnetic distribution and to different values of the stellar parameters (rotation rate, convective zone thickness, age...).

Extrapolating the linear correlation reported by Saar between magnetic flux $f_B B$ and rotation rate in single slowly rotating late-type stars would imply $B \simeq 4$ kG and $f_B \simeq 1$ for a star with $\Omega = 0.35$ day^{-1} like HR 1099. Our estimate of the magnetic flux $f_B B$ for this star is at least 25 times lower, with approximate values of 150 G at phase 0.85 and $\leq$50 G at phase 0.50. Part of this difference may be attributed to tiny dipoles undetected in circular polarization (bipolar structure smaller than the local line width) but they would need then to be very strong and to almost cover the entire stellar surface, which is somewhat unlikely. Our $f_B B$ value is also 20 times lower than the magnetic flux in some of Saar's most rapid rotators, so that the saturation regime that his relation probably reaches for high rotation rates does not explain everything. Saar's stellar sample contains main sequence stars almost exclusively: his relation may not be valid for subgiants and giants. In particular, other parameters than the rotation rate alone may be invoked to explain stellar activity levels in these RS CVn stars.

At the moment, very few magnetic measurements are available on rapid active rotators ($\Omega \geq 0.17$ day^{-1}) and ours is the only one in such a spotted RS CVn system. Some more data are needed for this particular stellar class to study how magnetic flux correlates with rotation at high rotation rates. Zeeman-Doppler images of late-type stars will be soon available (Brown and Donati, 1990) to provide this type of information.

References

Brown, S.F., Donati, J.-F.: 1990, in *Surface Inhomogeneities on Late-Type Stars*, Armagh Bicentenary Colloquium, in press

Donati, J.-F., Semel, M., Praderie, F.: 1989, *Astron. Astrophys.* **225**, 467

Donati, J.-F., Semel, M.: 1990, *Solar Phys.*, in press

Donati, J.-F., Semel, M., Rees, D.E., Taylor, K., Robinson, R.D.: 1990, *Astron. Astrophys.* **232**, L1

Saar, S.H.: 1990, in *Solar Photosphere: Structure, Convection and Magnetic Fields*, IAU Symp. 138, ed. J.O. Stenflo, Kluwer, Dordrecht, p. 427

Semel, M.: 1989, *Astron. Astrophys.* **225**, 456

Vogt, S.S.: 1988, in *The Impact of Very High S/N Spectroscopy on Stellar Physics*, IAU Symp. 132, eds. G. Cayrel, M. Spite, Kluwer, Dordrecht, p. 253

Chromospheric Surface Structures on EI Eridani and HD 199178

James E. Neff

Laboratory for Astronomy and Solar Physics, Code 681
NASA Goddard Space Flight Center, Greenbelt, MD 20771

1. Introduction

Several groups at this meeting are presenting maps of the spatial distribution of either brightness or effective temperature in the *photospheres* of rapidly-rotating, late-type stars. It is generally believed that structure seen in these maps traces the magnetic topology, in analogy with the Sun. We expect the structure of the outer atmospheres (i.e., chromosphere and corona) of these stars to be even more directly tied to the magnetic topology; the magnetic structure is *three-dimensional.* In order to probe the radial dimension of stellar atmospheres, we need to combine maps of the spatial distribution of emission from *chromospheres* and coronae with these detailed photospheric maps.

Along with collaborators at Armagh, Catania, Boulder, Paris, Helsinki, and Stony Brook, I have been obtaining high-dispersion ultraviolet spectra of several rapidly-rotating, late-type stars using the IUE spacecraft. I discuss results for two stars, EI Eridani and HD 199178, for which photospheric maps are presented elsewhere at this conference.

2. EI Eridani = HD 26337

We observed EI Eri with IUE continuously for one rotational cycle in September 1988, and almost continuously for most of the following rotational cycle. As the light curves clearly demonstrate (Figure 1), most of the features did not repeat from one cycle to the next (the vertical dashed line is drawn at the end of the first two days, or one rotational cycle). The top panel shows the variation of the transition region emission lines. There appear to have been several flares in the first cycle. The relative enhancement of the transition region lines during these flares is higher than for the chromospheric lines (second panel). The third panel shows the flux modulation in the Mg II k line, which doesn't appear to correlate very well with anything else. The bottom panel shows the FES light curve obtained as accurately as possible with one data point between each IUE exposure.

The Mg II k fluxes shown in Figure 1 were measured using a two-component gaussian fit (one for the stellar emission and one for the interstellar absorption) to the observed profiles. The centroids of these fits are systematically *redshifted* by $\sim$10 km s^{-1} with respect to the known photospheric velocity. Either there is an unidentified error in the IUE wavelength scale, or there is a net flow of the emitting plasma. The most direct way to test the wavelength scale would be to compare the measured Mg II k interstellar component velocity with an independent measure of

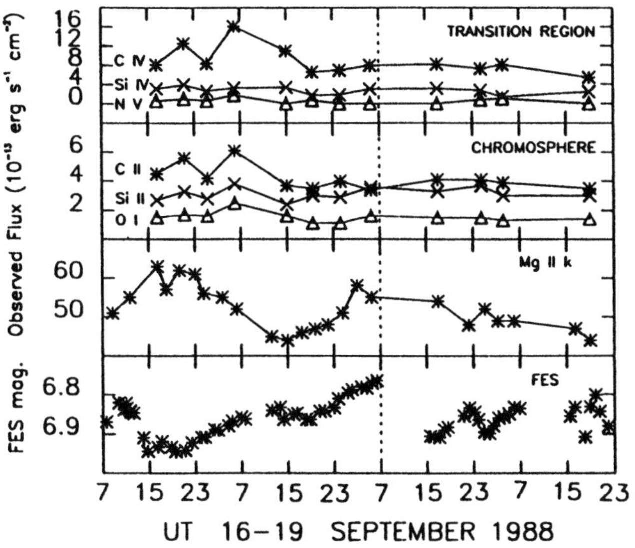

Fig. 1. The ultraviolet emission line fluxes from the transition region (top panel) and chromosphere (second panel) of EI Eri are plotted v. time for the September 1988 observing run. Also shown are the Mg II k emission line flux and the FES magnitude (roughly equivalent to B-band). The vertical dashed line is drawn at the end of the first rotational cycle (~2 days).

the interstellar flow velocity in this line of sight. The interstellar Mg II k absorption component could then serve as an *absolute* wavelength reference (it is currently used only as a relative reference, because we do not know the LISM flow velocity *a priori*). There are some noteworthy Mg II k profile variations. Because of the uncertainties introduced by the systematic redshift, we have not yet mapped the phase-dependent asymmetries into chromospheric active regions, but such work is in progress.

The chief lesson to be learned from these data is that active stars do vary on timescales comparable to the rotational cycle. Becuase only light-curve or line-profile features that **repeat** at the same phase in several cycles can be mapped into spatial structures, it is crucial that observations be obtained *continuously for at least two rotational cycles.*

3. HD 199178

We observed HD 199178 with IUE in September 1986 and September 1987. Also included in our study are all the spectra obtained at other epochs that are available in the IUE archives. Preliminary results were presented by Neff *et al.* (1988), and our final results are presented by Neff *et al.* (1990). I will highlight a few of the most significant results.

Figure 2 (left panel) shows the emission line fluxes for C IV (1550 Å), C II (1335 Å), and O I (1305 Å) as a function of rotational phase, using the ephemeris of Jetsu *et al.* (1990). Plotting all of the data together, it is evident that the highest fluxes indicate stellar flares (see Neff *et al.*, 1989), which are most enhanced for

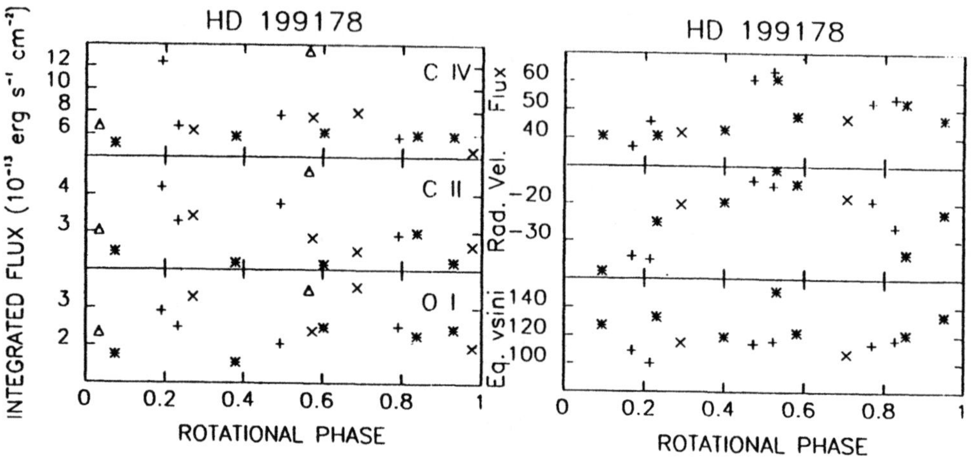

Fig. 2. (left) The integrated emission line fluxes from low-dispersion IUE spectra are plotted using the ephemeris of Jetsu *et al.* (1990). The different symbols represent different epochs. (right) The parameters of the unconstrained round of two-component fits to the Mg II k profiles.

C IV. The results from two-component gaussian fits (stellar plus interstellar) are shown in the righ-hand panel of Figure 1. Several more flares are evident here (these are discussed by Neff, 1990). The most significant result from this first round of fits, which was unconstrained except that the interstellar absorption line width is assumed to be instrumental, was a quasi-sinusoidal velocity curve with a peak-to-peak amplitude one-third of $v \sin i$. This is surprising because Huenemoerder (1986) demonstrated conclusively that HD 199178 is a single star with a radial velocity of –28 km s^{-1}. If the IUE wavelength scale is correct, then the line profile is asymmetric and variable, but *only on the red wing.* If, on the other hand, we apply a 5 to 10 km s^{-1} systematic redshift to the IUE wavelengths (as required for the EI Eri data), then the profile variations imply a large-scale brightness non-uniformity of the star. As for EI Eri, the only conclusive test would be an independent measure of the interstellar flow velocity in this line of sight.

References

Huenemoerder, D.P.: 1986, *Astron. J.*, **92**, 673

Jetsu, L., Huovelin, J., Tuominen, I., Vilhu, O., Bopp, B.W., Piirola, V.: 1990, *Astron. Astrophys.* **236**, 423

Neff, J.E.: 1990, *Mem. Astron. Soc. Italy*, special volume, ed. B.R. Pettersen, in press

Neff, J.E., Brown, A., Linsky, J.L.: 1989, in Solar and Stellar Flares, (Poster Volume), eds. B.M. Haisch, M. Rodonò, Publ. Catania Astrophys. Obs., p. 111

Neff, J.E., Vilhu, O., Walter, F.M.: 1988, in *A Decade of UV Astronomy With the IUE Satellite*, ESA SP-281, p. 291

Neff, J.E., Vilhu, O., Walter, F.M.: 1990, *Astron. J.*, submitted

Activity, Colour Anomalies and Temperature Determination in Solar-type Stars

J. Fabregat [1,2], V. Reglero [1], J. Suso [1], J.E. Armentia [3]

[1]Departamento de Matemática Aplicada y Astronomía, Universidad de Valencia, 46100 Burjassot (Valencia), Spain
[2]Physics Department, The University, Southampton SO9 5NH, U.K.
[3]Departamento de Astrofísica, Universidad Complutense de Madrid, 28040 Madrid, Spain

1. Introduction

The presence of photometric anomalies in broad band colours produced by chromospheric activity was first suggested by Campbell (1984). He defined the colour anomaly $\delta(\mathrm{B-V})_{\mathrm{V-K}}$ as the deviation from the mean relation (B−V)−(V−K) for Hyades dwarfs, and found this anomaly correlated well with several activity indicators.

Since the work of Campbell, several authors have discussed the relation between activity and colour anomalies, obtaining diverging results. This discussion is relevant because long baseline photometric colours are the most reliable temperature indicators, and the presence of such anomalies implies that no consistent temperature scales can be obtained for stars with different activity levels. These topics have been recently reviewed by Soderblom (1989).

For this work we have obtained JHKL values for 18 main sequence solar-type stars with well determined activity levels, which spread over a wider range than the Hyades stars used by Campbell, and we aim to reproduce his discussion and check his results with the same kind of data.

2. Discussion

Observations were made during three runs in December 1987, March 1988 and June 1988 in the 1.5 m Sánchez Magro Telescope at the Observatorio del Teide (Tenerife, Spain). The (V−K) indices have been obtained from our K data and the V values in Nicolet (1978). We have also adopted the (B−V) values of this reference. The colour anomaly $\delta(\mathrm{B-V})_{\mathrm{V-K}}$ has been computed in the same way as in Campbell (1984). $\delta(\mathrm{B-V})_{\mathrm{V-K}}$ is the difference between the (B−V) obtained from (V−K) by means of the Carney (1983) mean relation for the Hyades, and the observed value.

In Fig. 1 we have plotted the colour anomaly against two activity indicators related to the chromospheric emission in the CaII H and K lines. $\mathrm{R'}_{\mathrm{HK}}$ is the flux in the lines normalized to the bolometric flux and corrected for photospheric contribution, as defined by Noyes *et al.* (1984). The values used have been taken from this reference. ΔF_{CaII} is the chromospheric H and K flux density, in erg cm^{-2} s^{-1}, taken from Rutten (1987). As can be seen, no correlation between colour anomaly and CaII emission level is present.

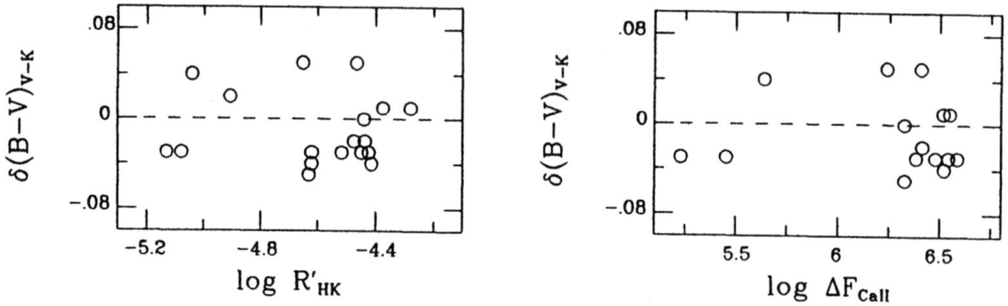

Fig. 1. The colour anomaly $\delta(B-V)_{V-K}$ against the activity indicators R'_{HK} and ΔF_{CaII}

In order to obtain further evidence of the independence of the (V−K) colour with activity, and its reliability for determining a consistent temperature scale for solar type stars, we have computed temperatures for the stars in our sample from the (V−K) values, using the calibration of Cayrel *et al.* (1985). We have also computed temperatures from the uvbyβ photometry published by Fabregat and Reglero (1990). For stars with $\beta > 2.595$ ($T > 5800$ K) we have derived the temperature from the β index, using the calibration of Saxner and Hammarbäck (1985). For the rest we have employed the calibration of Olsen (1984) in terms of the (b−y) colour. β is known to be not affected by chromospheric activity (Giampapa *et al.*, 1979; Fabregat, 1989), while (b−y) is not expected to be significantly affected. We have also considered temperatures determined by spectroscopic methods, taken from the catalogue of Cayrel de Strobel *et al.* (1985). We have computed mean differences for active and inactive stars, following the Soderblom (1985) criterion of considering as active stars those with $\log R'_{HK} > -4.5$ and inactive those with $R'_{HK} < -4.7$. The mean differences between the (V−K) scale and the others are comparable for active and inactive stars, and hence we can conclude that no activity effects are present in the temperature determination from the (V−K) colour. We have also made plots of temperature differences against the activity indicators discussed before, finding no correlation in any case. It is noticeable, however, that temperatures determined with the (V−K) calibration are systematically higher than the others by some 100 K.

We can conclude that broad band photometric colours are not affected by different levels of magnetic activity, and hence can be used as reliable temperature indicators. Our results agree with those of Soderblom (1989) and with the theoretical discussion of Spruit and Weiss (1986).

Acknowledgments. The 1.5 m Sánchez Magro Telescope is operated by the Instituto de Astrofísica de Canarias at the Teide Observatory (Tenerife, Spain). J.F. acknowledge a grant from the Spanish Ministerio de Educación y Ciencia. This work has been supported by the Spanish Comisión Interministerial de Ciencia y Tecnología (PB86-0536-C02-02).

References

Campbell, B.: 1984, *Astrophys. J.* **283**, 209
Carney, B.W.: 1983, *Astron. J.* **88**, 623
Cayrel, R., Cayrel de Strobel, G., Campbell, B.: 1984, *Astron. Astrophys.* **146**, 249
Cayrel de Strobel, G., Bentolila, C., Hauck, B. and Duquennoy, A.: 1985, *Astron. Astrophys. Suppl. Ser.* **59**, 145
Fabregat, J.: 1989, *Ph. D. Thesis*, Universidad de Valencia
Fabregat, J., Reglero, V.: 1990, *Astron. Astrophys. Suppl. Ser.* **82**, 531
Giampapa, M.S., Worden, S.P., Gilliam, L.B.: 1979, *Astrophys. J.* **229**, 1143
Nicolet, B.: 1978, *Astron. Astrophys. Suppl. Ser.* **34**, 1
Noyes, R.W., Hartmann, L.W., Baliunas, S.L., Duncan, D.K., and Vaughan, A.H.: 1984, *Astrophys. J.* **279**, 763
Olsen, E.H.: 1984, *Astron. Astrophys. Suppl. Ser.* **57**, 443
Rutten, R.G.M.: 1987, *Astron. Astrophys.* **177**, 131
Saxner, M. and Hammarbäck, G.: 1985, *Astron. Astrophys.* **151**, 372
Soderblom, D.R.: 1985, *Astron. J.* **90**, 2103
Soderblom, D.R.: 1989, *Astrophys. J.* **342**, 823
Spruit, H.C., Weiss, A.: 1986, *Astron. Astrophys.* **166**, 167

Dynamo Action in Evolved Stars

David F. Gray

Department of Astronomy University of Western Ontario
London, Ontario N6A 3K7, Canada

Abstract: Evolved stars tell us a great deal about dynamos. The granulation boundary shows us where solar-type convection begins. Since activity indicators also start at this boundary, it is a good bet that solar-type convection is an integral part of dynamo activity for all stars. The rotation boundary tells us where the magnetic fields of dynamos become effective in dissipating angular momentum, and rotation beyond the boundary tells us the limiting value needed for a dynamo to function. The observed uniqueness of rotation rates after the rotation boundary is crossed can be understood through the rotostat hypothesis. Quite apart from the reason for the unique rotation rate, its existence can be used to show that magnetic activity of giants is concentrated to the equatorial latitudes, as it is in the solar case. The coronal boundary in the H-R diagram is probably nothing more than a map of where rotation becomes too low to sustain dynamo activity.

1. The granulation boundary

The asymmetries of spectral lines betray the presence of granulation in stellar photospheres. The combination of brighter granules moving upward, producing blueward Doppler shifts, with darker material moving downward, producing redward Doppler shift, is the reason for the asymmetries. The line bisector, i.e., the locus of points through the midpoints of horizontal line segments bounded by the sides of the spectral line profile, has been used to specify the asymmetries. Solar-type line asymmetries produce bisectors sweeping redward near the continuum. A series of observations across the H-R diagram show solar-type bisectors on the cool half only. Some other (stronger) velocity fields appear to exist in the photospheres of hot stars, and the slope and curvature of their bisectors is reversed from the solar case (Gray and Nagel, 1989; Gray, 1990). The boundary between these two domains, where the asymmetry disappears, is the granulation boundary. Figure 1. shows the position of the granulation boundary. It corresponds to the onset of deep-envelope convection. It also corresponds to the onset of activity indicators, and I refer you to Gray and Toner (1986) Fig. 5.

Stars waste no time starting up activity even though from the stellar interiors point of view convection zones near the boundary are very thin. Nevertheless, the granulation structure of the velocity fields has apparently been established. Whether it is the herding of field lines to granule boundaries, or the Coriolis force twisting of flux, or something else that is the critical element here I don't know, but whatever it is, it starts up promptly at the granulation boundary. It also turns off promptly when stars evolve across this boundary from cooler to hotter temperatures, as the more luminous stars do during their blue-loop evolution.

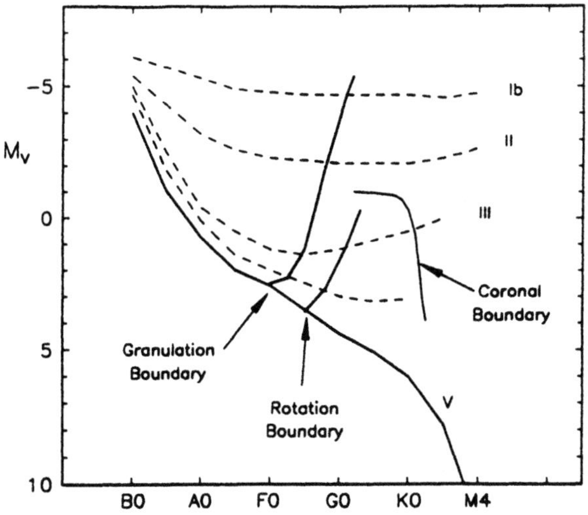

Fig. 1. The granulation boundary is clearly delineated by the line bisectors: left-leaning bisectors are seen on the left; right-leaning (solar type) on the right. Also shown here are the rotation boundary and the coronal boundary discussed below.

2. The rotation boundary

The drop of rotation along the main sequence through the early F stars has been well documented over the years. The correspondence with the depth of the convective envelope and the increase of proxy-indicators of magnetic activity were pointed out a quarter of a century ago (Wilson, 1966; Kraft, 1967). Dwarfs give us a mass sequence. Giants, by contrast, give us something close to a constant-mass time sequence. Stars of Luminosity class III have masses too small to show blue loops, so increasing spectral type corresponds to increasing time. Figure 2. shows the precipitous drop in rotation between G0 and G3 seen for single stars. A similar drop occurs for luminosity class IV (Gray and Nagar, 1985). The locus in the H-R diagram across which rotation suddenly drops is the rotation boundary shown in Fig. 1.

A dynamo-generated magnetic brake is almost certainly the cause of the sudden loss of angular momentum. Further, it must be a structural factor, as opposed to a time delay, because the rotational boundary parallels the granulation boundary even though the evolutionary times increase rapidly with lower luminosity. On the hot side of the boundary, the mean rotation is large and the distribution of rotation rates is Maxwellian. On the cool side of the boundary, the rotation is small, i.e. 6.1 km/s at G3 III, and is a single-valued function of spectral type. My pet theory for understanding this is the Rotostat Hypothesis which says that dynamo activity ceases when rotation falls below a limiting value, but angular momentum brought

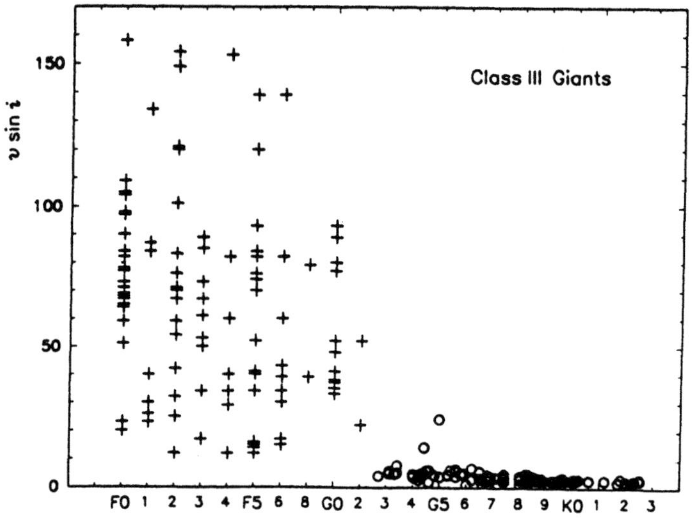

Fig. 2. The rotational break is shown for class III giants. (From Gray, 1989a.)

into the envelope as the lower boundary of the convection zone deepens with time keeps trying to spin up the star. As soon as the star rotates a bit faster than the limiting value, the dynamo comes to life, sets up its magnetic brake again, and rapidly reduces the rotation back to the limiting value. In this way rotation is a slave to the physics of the convection zone, and regulated by it.

3. Equatorial activity

We have precious little information on the latitude dependence of stellar activity, and it seems to me rather crucial information to have if we're going to understand which kinds of dynamos stars might be partial to. In the case of the sun, we naturally have detailed latitude mapping, and there we see tracers like spots localized to certain latitude bands. A glance at the "butterfly diagram" shows that sunspots prefer to stay within ±45° latitude, and in that sense they are concentrated toward the equatorial zone. It is true that the spatial resolution is so good for the sun that we can also see a narrow equatorial band where few sunspots occur. As a stellar astronomer, I ignore such fine detail.

The giants in question are those in the G3 to K2 III range where the rotation is a single-valued function of spectral type. Once we understand this fact, the mean value of $v \sin i$ at each spectral type can be converted to the actual v, under the assumption of random orientation of rotation axes, by multiplying by $4/\pi$. Then each star's $v \sin i$ yields a $\sin i$ upon being divided by the now-known v. Pole-on stars have $\sin i = 0$; equator-on unity.

Magnetic activity indicators can then be plotted as a function of orientation, as is done in Fig. 3 for the carbon emission. The carbon data is more recent and has smaller errors. But both plots show the same thing: stars seen near pole-on are weak emitters; those seen near equator-on are the strongest emitters. The curves in these diagrams show the geometrical factors, i.e., the amount of an equatorial belt that is visible as a function of axial orientation. Each curve is labeled with its latitude span, for example, 80° is a belt running from +40° to −40° latitude.

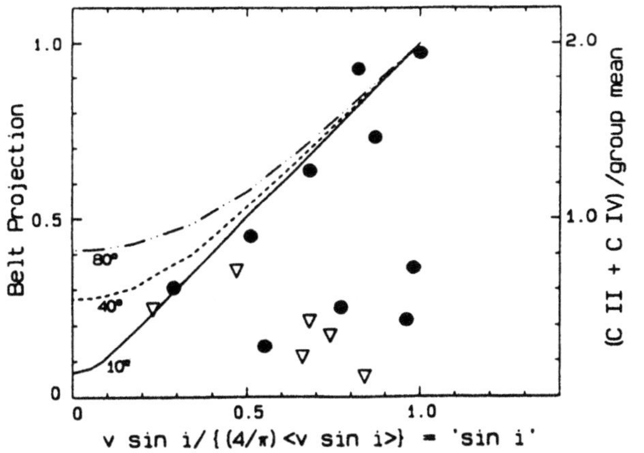

Fig. 3. Carbon emission (right ordinate) is shown as a function of axial orientation (computed sin *i*). The left ordinate shows the fraction of a photosperic equatorial belt that is visible at any orientation. Each curve is labeled with the full latitude width of the belt. (From Gray, 1990.)

Some of the points in these figures lie well below the belt curves, but that is what we would expect if the strength of activity indicators varies with rotation phase, long-term cyclic pattern, and/or rotostat flickering. Axial orientation is one more variable contributing to the range of activity-indicator strength seen in otherwise homogeneous groups of stars.

Apparently normal giant stars have a liking for the the same dynamo patterns seen on the sun.

4. The coronal boundary

Indicators of coronae such as x-rays show a domain in the H-R diagram bounded on the right and top, as shown in Fig. 1. Coronae are seen for stars "under" the boundary, but for the very luminous and the very cool stars, none of the usual coronal indicators appear (Linsky and Haisch, 1979; Ayres *et al.*, 1981; Haisch and Simon, 1982). Explanations have been given for this in terms of physical structure of the outer atmosphere (e.g., Antiochos and Noci, 1986; Böhm-Vitense, 1987), but a simpler explanation lies with the rotation itself: star outside the coronal boundary rotate too slowly to sustain dynamos. Stars of high luminosity rotate below the dynamo limit because evolutionary expansion has increased their moment of interia sufficiently to make it so across the whole domain to the right of the granulation boundary. Stars to the right of the coronal boundary continue to struggle along at the dynamo limit until they reach spectral types $\approx$ K2 III. There their evolution carries them up the asymptotic branch where the track in the H-R diagram changes from nearly horizontal to nearly vertical. This new tack involves a rapid increase in the moment of inertia, and the rotation dives below the dynamo limit, squelching the support of coronae.

Evolved stars clearly supply a wealth of information about dynamos. We need to connect it more explicitly to dynamo calculations.

To those whose thoughts may have drifted a bit, let me express it a different way:

Dynamos Keep Us Going in Circles

Dynamos, dynamos, friend or foe?
with all those equations, my head works too slow!
But of stellar activity give us a taste,
and these grand efforts won't go to waste.
Theory in one hand, data in the other,
don't go and hide, but join hands with your brother.
Bring 'em together and turn up the heat.
We'll cook up some theories that will be hard to beat.
Put the pieces in place over the years,
even though the slow pace brings us to tears.
On we struggle with truth our goal.
Cosmic detectives our chosen role.

Now magnetic fields string everything together,
but their motions and antics change like the weather.
Twist 'em and turn 'em and wiggle 'em around.
Who knows what we're doing? Is any of it sound?
Well, look at the data; look where it's led.
We can tie it together by using our head.
The weakest example of many around,
the solar dynamo we continue to hound.
While other stars show us more dramatic effects,
essential evidence one sometimes neglects.

Evolved stars are special in what they do.
So watch carefully for clue upon clue.
Across that boundary of granulation,
convection joins rotation to bring consummation.
Then rotation, once fast, is soon slowed down.
No more wild spinning or playing around!
To each spectral type the rotation's unique.
Doesn't this tell us something we seek?
You bet it does; in your heart it's felt.
Activity for giants comes round in a belt.

Do dynamos cease or do they flicker
in that rotostat mode over which we dicker?

It's all good fun; don't sit there alone.
Join in with spirit; your heart's not a stone.
Dynamos are really a tricky beast.
It's obvious from this intellectual feast.
And so on to those questions you surely will ask,
making for this day the end to my task.

D.F.G.

References

Antiochos, S.K., Noci, G.: 1986, *Astrophys. J.* **301**, 440

Ayres, T.R., Linsky, J.L., Vaiana, G.S., Golub, L., Rosner, R.: 1981, *Astrophys. J.* **250**, 293

Böhm-Vitense, E.: 1987, *Astrophys. J.* **301**, 297

Gray, D.F.: 1986, *Highlights in Astronomy*, Reidel, Dordrecht, ed. J.-P. Swings, p. 411

Gray, D.F.: 1988, *Lectures on Spectral-Line Analysis: F, G, and K Stars*, (The Publisher: Arva, Ontario)

Gray, D.F.: 1989a, *Astrophys. J.* **347**, 1021

Gray, D.F.: 1989b, *Publ. Astron. Soc. Pac.* **101**, 1126

Gray, D.F.: 1990, *Publ. Astron. Soc. Pac.*, accepted

Gray, D.F., Nagar, P.: 1985, *Astrophys. J.* **298**, 756

Gray, D.F., Nagel, T.: 1989, *Astrophys. J.* **341**, 421

Gray, D.F., Toner, C.G.: 1986, *Publ. Astron. Soc. Pac.* **98**, 499

Haisch, B.M., Simon, T.: 1982, *Astrophys. J.* **263**, 252

Kraft, R.P.: 1967, *Astrophys. J.* **150**, 551

Linsky, J.L., Haisch, B.: 1979, *Astrophys. J. Letters* **229**, L27

Wilson, O.C.: 1966, *Astrophys. J.* **144**, 695

The Boundary between the Magnetic Fields of Ap Stars and the Fields of Solar-type Stars

John D. Landstreet

Department of Astronomy, The University of Western Ontario
London, Ontario N6A 3K7, Canada

Abstract: The boundary between Ap-type magnetic fields and the magnetic fields of solar-type stars occurs near $T_e \sim 7000$K, about where deep envelope convection develops in main sequence stars. This seems natural for solar-type stars, in which the field is generated by the convection zone. However, among magnetic Ap stars the frequency of occurrence declines from about 10% of all A stars near A0 to about 1% near F0. It is not clear what produces this decline in frequency, but the convection zone is probably not responsible. In fact, it seems likely that if global fossil fields occur in main sequence F stars, such fields should be detectable even if the stars having them are not chemically peculiar.

1. Introduction

Magnetic fields are observed in two quite different forms in main sequence stars. Among lower main sequence (solar type) stars having $T_e \lesssim 6000$ K, and spectra of G0 or later, some stars have fields which seem to be analogous to the field of the sun. In these stars the fields have rather complex structure and are present in only a part of the photosphere. In contrast, a few percent of upper main sequence stars (Ap or Bp stars), with $7500 \text{ K} \lesssim T_e \lesssim 25{,}000$ K, and spectra between F0 and B1, have quite a different type of field. In these stars, the field is simple in structure, usually roughly dipolar, and appears to cover essentially all of the star.

There is a gap on the main sequence between about F0 and G0 where no magnetic fields are observed. In this paper I shall examine the relevant stellar physics and try to understand why this gap occurs.

2. Magnetic field structures on the main sequence

The differences between the fields of lower and of upper main sequence stars reflect their quite different physical origins. These are summarized in the table below.

	Solar-type stars	Ap- or Bp- type stars
Structure near surface	Isolated flux tubes embedded in non-magnetic regions, as in sun.	Fields permeate entire atmosphere. Overall structure usually roughly dipolar.
Strengt of field	1–5 kG inside flux tubes, much less elsewhere. Areal coverage in various stars ranges downward from $\sim$ 80%; about 1% for sun.	In a single star, field varies by roughly a factor of 2 over surface. In various stars, mean field ranges from 3×10^2 to 3×10^4 G.
Correlation with stellar properties	Areal coverage increases with decreasing stellar rotation period.	Field may vary slightly on average with stellar mass and/or age.
Intrinsic variability	Pattern of emerging flux at stellar surface probably changes on timescale of $\sim 10^6$ s.	Field structure seems quite stable. Observed field varies only due to stellar rotation.
Source of field	Probably a contemporary dynamo, either within the outer convection zone or at its base.	Probably a fossil field, compressed from the weak field of the interstellar medium as the star condensed.
Evolution of field	The dynamo is thought to decline as the rotation of the star is gradually braked by its stellar wind.	The field is altered by changes in stellar structure as the star evolves, by meridional circulation, and by ohmic decay of the driving currents.

Observational searches have been made for magnetic fields in the hottest solar-type stars by Landstreet (1982) and Gray (1984), among others (see also Borra *et al.* 1982). Most searches have looked for the coherent fields found in Ap stars, but Gray's survey was sensitive to the disordered fields observed by numerous other observers in cooler stars. No fields have been found in non-Ap stars earlier than G0. Quite probably the absence of fields is related to the dramatic decline of the depth of the surface convection zone with increasing T_e. This decline, which is illustrated by Clayton (1983), finds the outer convection zone reduced from a fractional depth of about 30% of the stellar radius at $T_e \sim 6000$ K to about 5% at $T_e \sim 7000$ K (assuming a mixing length to pressure scale height ratio of $\alpha \sim 1.5$). Since it is almost certainly the convection that drives the dynamo which produces fields in solar-type stars, it is reasonable to suppose that as this convection becomes less important with increasing T_e, fields should be hard to detect or even non-existent. However, it is not clear just where in this interval the hottest stars with solar-type fields should occur.

Observational searches for fields in upper main sequence stars have shown that fields are always found associated with certain types of atmospheric chemical peculiarity. Fields are found (as a function of increasing T_e) in stars with excess Sr, Cr, and Eu, with excess Si, with weak He, and at the highest T_e's, with excess He. These chemically peculiar stars, with their strong magnetic fields, make up about 10% of all stars with $T_e \sim 12{,}000 - 9500$ K (B8-A0), only $\sim 1\%$ of all late A stars ($T_e \sim 8000$ K), and 0% of stars later than F0 (Wolff, 1968). The observed fields thus end about where deep convection first begins, somewhere around $T_e \sim 7300 - 7500$ K. Is the onset of deep convection related to the low-T_e boundary of the Ap star fields? This is less obvious than the idea that weakening convection leads to an upper boundary for fields of solar-type stars.

3. Possible explanations of the low-T_e boundary of the Ap star fields

If the deepening convection is to be invoked to explain the cut-off for Ap type magnetic fields around F0, we need to look a little more closely at the interaction between surface convection and a fossil magnetic field.

The hydrostatic and thermal structure of the envelope of a late A type star ($T_e = 8000$ K, $\log g = 4.0$) is illustrated nicely in Figure 1 of Latour, Toomre and Zahn (1981). This model clearly shows the main features expected in stars near the cool boundary of the Ap stars. A surface convection region, caused by partial ionization of H and He I, occurs in a region not far below $t \sim 1$, where the pressure p is about 10^4 dynes cm^{-2}. In this region, the thermal energy density in the gas is $(3/2)nkT \sim (3/2)p$. As long as any convection is subsonic, the kinetic energy in bulk motions should be less than the thermal energy. Thus the total kinetic energy in the gas is of order $2p$. If the magnetic field has an energy density $B^2/8\pi$ much greater than $2p$, it will not be possible for the gas, which is an excellent electrical conductor, to convect, since it would have to strongly distort the field. The largest field in which convection may take place is the equipartition field, given by $B_{\rm eq}^2/8\pi \sim 2p$. For the HI convection zone, $B_{\rm eq} \sim 10^3$ G. In most Ap stars, the surface field is at least this large, and so we expect that this outer convection zone will usually be suppressed by the field. A deeper convection zone, due to partial ionization of He II (and probably linked to the upper zone by overshooting) occurs at about 10–15,000 km below the first. Here $B_{\rm eq} \sim 10^4$ G, and in all Ap stars except the very small fraction with the largest fields, convection can persist. However, in slow rotators ($v_{\rm eq} \lesssim 1 \times 10^2$kms^{-1}), which almost all Ap's are, Michaud (1982) has shown that downward diffusion of He below this second convection zone gradually reduces the He concentration to a value too small to drive further convection. Enough He diffusion to shut down this convection zone is expected to occur in about 10^6 yr. Thus, in most magnetic Ap stars, which have main sequence lifetimes of 2×10^7 – 1×10^9 yrs, no convection will occur in the outer envelope.

As one looks at lower values of T_e, the rapidly deepening zone of convective instability quickly outstrips the ability of any observed magnetic Ap star to control it. Furthermore, because the deepening of the convection zone with lower T_e is due mainly to the low ionization and high opacity of HI deeper in the envelope, this convection zone cannot be much reduced by He diffusion even if this occurs. Thus, once we reach the values of T_e for which the surface convection zone starts to deepen, convection should occur regardless of whether the star is a magnetic Ap or a normal star.

Now, what sort of effect might a deep convection zone have on a deep-seated fossil field? If we look only at the short interval in T_e in which the thickness of the convection zone, from where it is suppressed by the magnetic field to the lower boundary, is no larger than the depth of the stable level above the convection zone (which means as long as the convective zone is no more than a few thousand km thick), the effect of the convection on the observable field on the stellar atmosphere is not likely to be large. Convective eddies will be smaller than the distance to the photosphere, and although they may locally make bubbles in the otherwise uniform magnetic field lines, the field will have room enough above the convection zone to spread back to a roughly uniform surface flux distribution.

However, as soon as we go to low enough T_e that the thickness of the convective zone substantially exceeds the distance between the photosphere and its upper boundary, the size of the convective eddies should become so large that great distortion of the surface field will occur. Two possible effects seem plausible. One possibility is that the large eddies will confine the surface field to ropes or sheets, with more or less the same total flux emerging from the atmosphere that would occur in the absence of convection, but in a different structure, with larger local fields and a smaller areal coverage. Alternatively, the convection may stir the field lines threading it so vigorously that flux ropes of opposite polarity are pushed together, and reconnection occurs. One would expect that the outer loops would quickly escape from the star, and that the flux lines reconnected deep in the convection zone might stay below it. In this case, substantially less flux might appear in the photosphere than without convection. Only more detailed modelling, or perhaps appropriate observations, can choose among these (or other) possibilities.

However, this possible explanation for the cutoff of Ap star fields at low T_e does not account for the decline by a factor of order 10 in the frequency of magnetic stars, as a fraction of all main sequence stars of the same temperature, between A0 and F0. Perhaps deep convection is *not* the complete explanation of the observed cutoff.

An alternative possibility is suggested by the fact that the Henry Draper spectral type of one of the very coolest magnetic Ap stars, HD 24712 (= HR 1217), is given as F0, rather than F0p. HD 24712 has quite strong (and variable) abundance anomalies, including considerable overabundances of several rare earth elements, and a very large magnetic field, of order 5 kG at the surface. It is also near $m_v = 6$, so the Harvard objective prism plate should have had an adequate exposure for this star. Nevertheless it was missed by as keen a judge of spectral type as Miss Cannon. This suggests to me that perhaps a substantial fraction of the coolest magnetic chemically peculiar stars may not have been noticed in spectral classification, because the peculiarities are not obvious at low resolution. Since most searches for magnetic fields have concentrated on spectrally classified Ap stars, a substantial selection effect could work against finding Ap-type magnetic fields near the low-T_e boundary.

It thus seems well worthwhile to explore the magnetic fields of the A5 – G0 region again, both from a theoretical and from an observational perspective. There may well be something interesting still to be learned here.

References

Borra, E.F., Landstreet, J. D., Mestel, L.: 1982, *Ann. Rev. Astr. Ap.* **20**, 191, Table 1.

Clayton, D.D.: 1983, *Principles of Stellar Evolution and Nucleosynthesis*, (Chicago: Univ. of Chicago Press), Sec. 6-6

Gray, D.F.: 1984, *Astrophys. J.* **277**, 640

Landstreet, J.D.: 1982, *Astrophys.J.* **258**, 639

Latour, J., Toomre, J., Zahn, J.-P.: 1981, *Astrophys.J.* **248**, 1081

Michaud, G.F.: 1982, *Astrophys.J.* **258**, 349

Wolff, S.C.: 1968, *Pub. Astr. Soc. Pac.* **80**, 281

Doppler Mapping of CP-star Surfaces with Weak and Strong Magnetic Fields

V.L. Khokhlova

Astronomical Council of the USSR Academy of Sciences, Moscow, USSR

A code for Doppler mapping of the distribution of chemical elements on stellar surfaces (Goncharskij *et al.*, 1977, 1982) has been used during the past decade to make maps for several CP-stars. The code is based on Tikhonov's method of solution of so-called ill-posed inverse problems and on an analytical expression of local line profiles in the form of the Milne-Eddington solution of the radiative transfer equation. This code takes into account the atmospheric model of the star through an assumed central depth of very strong and saturated lines, and through limb-darkening coefficients. As a solution one gets local line profiles and hence local equivalent widths.

Recently some more sophisticated codes were developed by Piskunov (1985) and Vogt *et al.* (1987) which use the results of the numerical solution of the transfer equation directly for local profiles and are thus able accurately to take into account an appropriate atmospheric model. In these codes the results of solution are local abundances of an element. Only Piskunov's code provides a simultaneous solution for all Stoke's parameters (I,V,U) if observed polarization profiles are available. Up to now only in one case, namely α^2 CVn, have phase variations of I and V been observed and Doppler mapping of some elements as well as Doppler-Zeeman mapping [1] of magnetic field made (Glagolevskij *et al.*, 1985).

The original code by Goncharskij *et al.* (1982) was modified in two respects:

a) An additional block has been written to convert local equivalent widths into abundances using a theoretical curve of growth.

b) Straight introduction of input data (observed profiles) was made. This prevents loss of precision through intermediate calculations.

We note some advantages of this code as compared with more sophisticated ones in spite (or may be because) of its simplicity. First, it permits the use of lines for which $\log gf$ values are unknown and the construction of maps of local equivalent widths. It is also easy to make averages of abundance maps, obtained from different lines of the same element, and later to introduce necessary linear corrections when new improved $\log gf$ values are available, without time consuming

[1] The apt terms "Doppler imaging" and "Doppler-Zeeman imaging" were proposed for these procedures by Vogt (1984) and Donati *et al.* (1990).

recalculations. Secondly, it permits the use of less powerful computers (such as PC AT286).

During the past decade seven CP stars, over a wide range of T_e and surface magnetic field strengths, have been mapped (see Table).

Table

star	T_e K	B_e Gauss	elements	Reference
CU Vir	12000	$+700/-500$	Si	(1)
χ Ser	9500	<300	Fe, Sr	(1)
α^2 CVn	12000	$+1000/-1300$	Ti, Cr, Fe	(2)
ε UMa	9500	< 200	Si, Cr, Fe, Eu	(3)
				(4)
θ Aur	10000	$+360/-240$	Si, Ti, Cr, Fe	(5)
σ OriE	22500	> 3000	He, Si	(6)
HD 32633	12500	$+3000/-5000$	Ti, Fe	(7)
References: (1) Goncharskij *et al.*, 1983, (2) Khoklova and Pavlova, 1984 (3) Wehlau *et al.*, 1990, (4) Totochava and Khokholova, 1990 (5) Khokhlova *et al.*, 1986, (6) Khokhlova *et al.*, 1990 (7) Landstreet *et al.*, 1991				

Some useful remarks on the basis of the experience of using the code follow: as the only Stokes parameter I is used to obtain the distribution of an element, choice of stars and lines should be suitably chosen so as to avoid the distortion of results by a magnetic field (by Zeeman splitting and/or by magnetic intensification of lines).

Having chosen unblended lines for the analysis, further criteria should be taken into account. The first one is obvious and requires that rotational broadening must be larger than magnetic splitting of a line. The second requirement is that the thermal Doppler width should be larger than the magnetic Zeeman splitting over the whole surface of a star. In that case the influence of magnetic intensification is not important. This requirement is undoubtedly fulfilled for light elements such as He, C, Mg, Si, or for lines with zero or small Landé factors, or for hot stars (such as σOriE) or for stars with weak magnetic fields (such as HD124224, HD140160, θAur, ϵUMa). The star HD32633 is a limit case for this requirements and needs special consideration.

Some conclusions may be made using our experience and looking at the results obtained up to now.

1. Large chemical inhomogeneities on the surface of CP-stars correlate neither with temperature nor with magnetic field strength.
2. There is a hint that rapid rotators have a simpler surface chemical structure than slow rotators (fewer small features, larger spots).

3. It is very difficult (if not impossible) to obtain with the method of Doppler mapping a map of the major part of a stellar surface; only in the strip within 40° of the "subsolar line" is described satisfactorily. The situation is demonstrated in Fig. 1 where the results obtained with different initial approximations are shown. The solution is more or less independent of the initial approximation only within this strip. So any conclusions about the real structure of inhomogeneities outside of this strip are not very reliable.

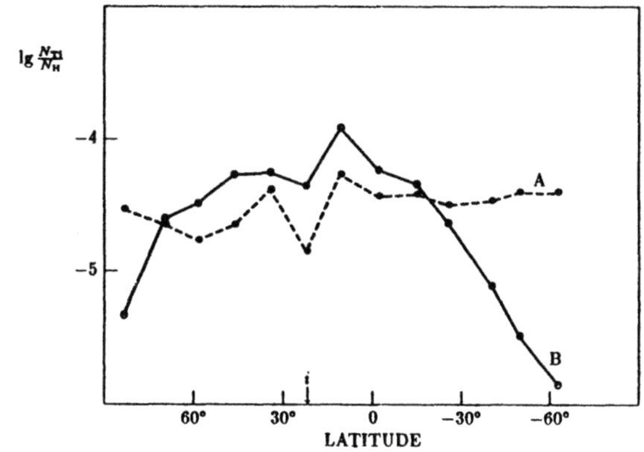

Fig. 1. Abundance of Ti along a meridian on the surface of HD 32633. A – an initial approximation of lg $\frac{N_{Ti}}{N_H} = -4.2$. B – an initial approximation of lg $\frac{N_{Ti}}{N_H} = -6.0$.

References

Donati, J.-F., Semel, M., del Toro Iniestia, J.C.: 1990, *Astron. Astrophys.* **233**, L17

Glagolevskij, Yu.V., Piskunov, N.E., Khokhlova, V.L.: 1985, *Pis'ma Astron. Zh.* **11**, 371

Goncharskij, A.V., Stepanov, V.V., Khokhlova, V.L., Yagola, A.G.: 1977, *Pis'ma Astron. Zh.* **3**, 278

Goncharskij, A.V., Stepanov, V.V., Khokhlova, V.L., Yagola, A.G.: 1982, *Astron. Zh.* **159**, 1146

Goncharskij, A.V., Ryabchikova, T.A., Stepanov, V.V., Khokhlova, V.L., Yagola A.G.: 1983, *Astron. Zh.* **160**, 83

Khokhlova, V.L., Pavlova, V.M.: 1984, *Pis'ma Astron. Zh.* **10**, 377

Khokhlova, V.L., Rice, J.B., Wehlau, W.H.: 1986, *Astrophys. J.* **307**, 168

Khokhlova, V.L., Zelvanova, E.I., Schöneich W., Bolton T.: 1990, in *"Hot chemically peculiar and magnetic stars"*, ed. G.Scholz, Potsdam, p. 85

Landstreet, J.D. *et al.*: 1991, (in preparation)

Piskunov, N.E.: 1985, *Pis'ma Astron. Zh.* **11**, 44

Totochava, A.G., Khokhlova, V.L.: 1990, *Pis'ma Astron. Zh.*, (in press)

Vogt, S.S: 1984, *Publ. Astr. Soc. Pac.* **95**, 565

Vogt, S.S.: 1987, *Astrophys. J.* **321**, 496

Wehlau, W.H., Rice, J.B., Piskunov, N.E., Khokhlova, V.L.: 1982, *Pis'ma Astron. Zh.* **8**, 30

Temperature Variations on the Surface of the Si-star CU Virginis

T.A. Ryabchikova

Astronomical Council of the USSR Academy of Sciences, Moscow, USSR

Abstract: A hot region with an increase of effective temperature of 700 K was found on the surface of Si-star CU Vir, based on the analysis of the available photometric and spectroscopic observations. This hot region is located near the phase of negative magnetic extremum and can explain both hydrogen line and infrared light variations.

The surface abundance inhomogeneity of magnetic CP stars is a well-established phenomenon. But in all investigations of chemical abundance and/or spotty structure of magnetic CP (CP2) stars the effective temperature is considered to be constant over the surface although there are some indications of regions with different temperature on the surface of CP2 stars. To explain the observed amplitudes of photometric *uvby* variations and energy distribution at different phases in the Si-star CU Vir, Weiss *et al.* (1976) suggested $\Delta\, T_{\rm eff} \sim 600\,$K, as well as $\Delta \log g \sim 0.15$, at phases of maximum and minimum light. In this paper we present arguments for temperature variations on the surface of CU Vir, based on the analysis of the available photometric and spectral data.

CU Vir = HD 124224 is the Si-star with the shortest known rotational period among CP stars ($0\overset{d}{.}52$). This star has been studied photometrically by many authors who give practically the same period of light variations. In this paper we used the ephemeris from Blanco and Catalano (1971): JD(lightmin.) = $2439995.4413 + 0.52067688E$.

The existence of a phase shift $\Delta\phi \sim 0.15$ between the visible light curves and the Hγ intensity variations (Ryabchikova, 1972) did not permit us to explain the latter by temperature variations, although Hγ and Hβ line profile variations in the case of CU Vir are indicative of a temperature rather than gravity effect (Musielok *et al.*, 1990). In order to reconcile photometric and spectroscopic observations of CU Vir we analysed the variations of different parameters available from the literature. We used photometry of CU Vir taken from White *et al.* (1980) and Pyper and Adelman (1985). Intensity variations of He I λ4026 are taken from electrophotometric measurements by Pedersen (1978) and are converted into equivalent widths using a relation proposed in his paper. Equivalent width variations of Si II λ3862 are taken from Krivosheina *et al.* (1980), those of Hβ from Musielok *et al.* (1990). β-index measurements by White *et al.* (1980) and Pyper and Adelman (1985) are converted into equivalent widths through the relation: $W(H\beta) = -19.58 + 37.857(\beta - 2.00)$ (Osawa, 1965). Hγ radial velocity measurements are taken from Abt and Snowdon (1973) and magnetic field measurements are taken from Landstreet and Borra (1977).

First of all we consider the phase correspondence between the hydrogen line variations and the colour index $[u - b]$ calculated from White *et al.* (1980) and Pyper and Adelman (1985). This index has been shown to be the most reliable temperature indicator for Si-stars (Mégessier, 1988). From Fig. 1 one can see a good phase agreement between $W(H\beta)$ and $[u - b]$. Using the Mégessier (1988) calibration we estimate two values of effective temperature: T_{eff}= 12000 K (phases 0.05–0.15) and T_{eff}= 12700 K (phases 0.6–0.7). This temperature difference agrees well with that obtained from IR observations (Catalano and Kroll, private communication). A model with a hot region near the phase 0.6 explains both the equivalent width and the radial velocity variations of hydrogen lines. To get a better reproduction of Hβ and Hγ amplitudes we must accept a small difference in gravity, $\Delta \log g \sim 0.2$, at phases 0.1 and 0.6. The resulting temperature and gravity differences are in good agreement with those of Weiss *et al.* (1976), but the position of the hot region is shifted in phase compared with theirs.

In Fig. 2 we present the visual light variations, Si II λ3862 and He I λ4026 equivalent width variations and variations of effective magnetic field. One can see that regions of different effective temperature roughly coincide with the magnetic field extrema. We performed calculations of line intensities at phases 0.1 and 0.6 using the "ABSTAR" program (Ryabchikova and Piskunov, 1988) with the corresponding Kurucz (1979) models.

We obtained the following results: 1. Near phase 0.1 CU Vir looks like a star with nearly normal helium and silicon abundance. 2. Temperature rise near phase 0.6 can fully explain the secondary maximum of He I line intensity variation, as well as a small depletion in Si II line intensity variation.

The double wave structure of visual light variations can be easily explained by joint effect of enhanced silicon opacity near phase 0.35 and of the temperature increase near phases 0.6–0.7. If the silicon and helium anomalies are caused by a diffusion process then we must conclude that this process operates near the magnetic equator.

References

Abt, H.A., Snowdon, M.S.: 1973, *Astrophys. J. Suppl.* **25**, 137
Blanco, C., Catalano, F.: 1971, *Astron. J.* **76**, 630
Catalano, F., Kroll, R.: private communication
Krivosheina, A.A., Ryabchikova, T.A., Khokhlova, V.L.: 1980, *Nautch. Inf.* **47**, 70
Kurucz, R.L.: 1979, *Astrophys. J. Suppl. Ser.* **40**, 1
Landstreet, J.D., Borra, E.F.: 1977, *Astrophys. J. (Letters)* **212**, L46
Mégessier, C.: 1988, *Astron. Astrophys. Suppl. Ser.* **72**, 551
Musielok, B., Ryabchikova, T.A., Davydova, E.S., Madej, J.: 1990, *Mitt. Karl-Schwarzschild Obs. Tautenburg*, Nr. 125, 110

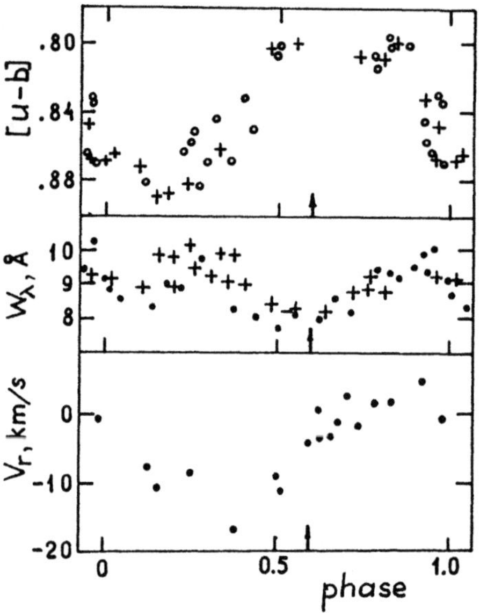

Fig. 1. Variations of index [u - b] (a), Hβ equivalent width (b), and Hγ radial velocity (c). a. open circles - data from White et al. (1980), crosses - data from Pyper and Adelman (1985). b. dots - data from Musielok *et al.* (1990). c. dots - data from Abt and Snowdon (1973). Position of the hot region is indicated by arrows.

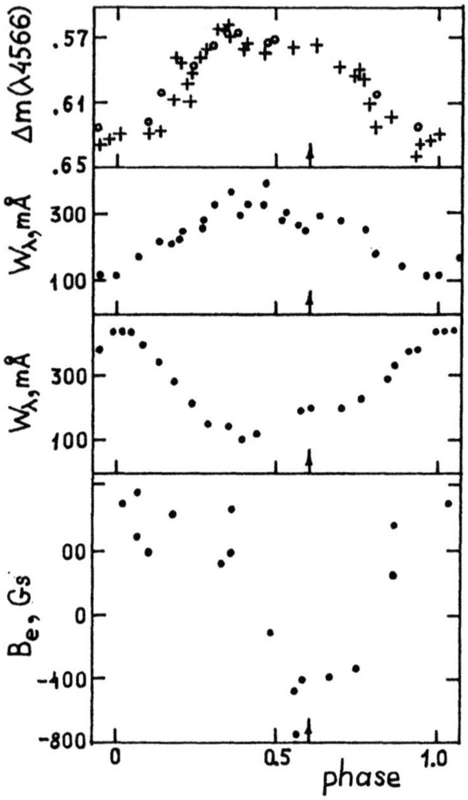

Fig. 2. Variations of visual light (a), equivalent widths of Si II λ3862 (b) and He I λ4026 (c) lines and effective magnetic field (d). For symbols on (a) see Fig. 1. The position of the hot region is indicated by the arrows.

Osawa, K.: 1965, *Ann. Tokyo Astron. Obs.* **9**, 123

Pedersen, H.: 1978, *Astron. Astrophys. Suppl. Ser.* **33**, 203

Pyper, D.M., Adelman, S.J.: 1985, *Astron. Astrophys. Suppl. Ser.* **59**, 369

Ryabchikova, T.A.: 1972, *Izv. Krymsk. Astrofiz. Obs.* **45**, 146

Ryabchikova, T.A., Piskunov, N.E.: 1988, in *Elemental Abundance Analysis*, eds. S.J. Adelman, T. Lanz, Univers. Lausanne, p. 93

White, R.E., Pyper, D.M., Adelman, S.J.: 1980, *Astron. J.* **85**, 836

Weiss, W.W., Albrecht,R., Wieder, R.: 1976, *Astron. Astrophys.* **47**, 423

Learning about Stellar Dynamos from Long–term Photometry of Starspots

Douglas S. Hall

Center of Excellence Tenessee State University Nashville, Tennessee 37203

Abstract: Spottedness, as evidenced by photometric variability in 277 late-type binary and single stars, is found to occur when the Rossby number is less than about 2/3. This holds true when the convective turnover time versus $B - V$relation of Gilliland is used for dwarfs and also for subgiants and giants if their turnover times are twice and four times longer, respectively, than for dwarfs. Differential rotation is found correlated with rotation period (rapidly rotating stars approaching solid-body rotation) and also with lobe-filling factor (the differential rotation coefficient k is 2.5 times larger for $F = 0$ than $F = 1$). Also reviewed are latitude extent of spottedness, latitude drift during a solar-type cycle, sector structure and preferential longitudes, starspot lifetimes, and the many observational manifestations of magnetic cycles.

1. Introduction

This will be a review of photometry of stars which are variable as a result of starspot activity, and how these observations can be useful in providing verification of various dynamo theories.

There is much other observational material which has been useful also (like spectroscopy of chromospheric emission, far ultraviolet emission from the transition region, x–ray emission from the corona, radio emission, flares, and direct measures of the magnetic field) but I have the impression that starspots have been somewhat neglected or, at least, not used to their full potential.

One emphasis will be on the continuous long–term behavior of starspots. Much of the existing work has provided only individual snapshots of spotted stars taken once or at random times.

My second emphasis will be to include *all* of the types of stars in which stellar dynamos may be operating. Often comparison between theory and observation will consider only one type of star, for example, only main-sequence dwarfs or only RS

CVn–type binaries. A wide net is important because it can explore wide ranges of possibly important factors:

rotation periods from less than 2 hours to longer than a year
spectral types from F to K
dwarfs to giants
strong tidal interaction (contact binaries) to weak (single stars)
shallow convection to deep convection
time scales from minutes (flares) to centuries (magnetic cycles)

Table 1 is a list of the 11 types of stars, separately defined, which show one or more aspects of chromospheric activity, *i.e.*, solar-type activity probably related in some way to dynamo action. For references to each type, see Hall (1987), Strassmeier *et al.* (1988), Hall (1989), and Hall (1990b). Note that in some types the active star is single, in some types it occurs in a binary, and some types include single stars and binaries.

2. Spottedness as a function of Rossby number

The most outstanding observational manifestation of starspots is the so–called wave. Numerous examples of waves are seen in Strassmeier and Hall (1988, figures 1 – 5) and in Strassmeier *et al.* (1989, figures 1 – 49). These are light curves of 54 chromospherically active stars, taken over a four–year interval with an automatic telescope in Arizona. The vertical axis is magnitude in V; the horizontal axis is Julian date. The brightness in the continuum varies as the star rotates because one hemisphere is much darker than the other as a result of a concentration of dark starspots distributed unevenly in stellar longitude. The amplitudes are often as large as $0.^m2$, which is about 20% and would be produced by a dark region about 20° in radius. The largest wave amplitude yet seen, $0.^m5$ in the V band, was in the late-1986 light curve of II Peg (Doyle *et al.*, 1988).

Previous attempts have been made to demonstrate a functional dependence between wave amplitude, or spottedness, and some parameter such as rotation period, spectral type, or H & K flux. Examples are Catalano *et al.* (1980, figure 4) and Strassmeier (1989, figure 4). To my knowledge, none has succeeded. The problem, I think, is that the samples studied were restricted to stars already known to be heavily spotted and did not probe extremes in parameter space.

Therefore I have searched many sources for stars of many different types which have a measured wave amplitude, even those demonstrated to vary less than 1 %, *i.e.*, a wave amplitude less than $0.^m01$. Altogether 277 such stars were found. Table 2 lists my principal sources. For binaries, in which only one of the two stars is spotted, the observed amplitude has been corrected for the dilution caused by the other star's light. For non–variable stars, the rotation period was estimated from the observed $v \sin i$ and the canonical radius for the corresponding luminosity class. One result is shown in the first three figures.

Figure 1 shows the dwarfs, *i.e*, main-sequence stars. The vertical axis is log of the rotation period. The horizontal axis is $B - V$. The open circles indicate

Table 1. Types of stars with choromospheric activity

1.	RS CVn binaries [original definition]
	A. short-period
	B. 2 days < P_{orb} < 2 weeks
	C. long-period
2.	RS CVn binaries [expanded definition]
	A. companion star = A dwarf
	B. companion star = O-B subdwarf
	C. companion star = white dwarf
3.	BY Draconis variables [single and binary stars]
4.	UV Ceti variables = flare stars [single and binary stars]
5.	solar-type single dwarfs
6.	FK Comae stars [single]
7.	other rapidly rotating G-K single giants
8.	T Tauri variables [single]
9.	W UMa binaries
10.	Algol-type binaries [the cool, contact secondary]
11.	cataclysmic variables [the cool, contact secondary]

Table 2. F–G–K stars searched for variability

1.	eclipsing F-type and G-type main-sequence binaries, [*Publ. Univ. Pennsylvania Astr. Ser.* XII]
2.	W UMa binaries, [*I.A.U. Colloq.* **29**, 287]
3.	Algol-type binaries, [*I.A.U. Colloq.* **107**, 219]
4.	T Tauri stars, [*Publ. Astron. Soc. Pac.* **98**, 1088]
5.	UBV standards, [*Astrophys. J.* **117**, 313]
6.	G and K dwarfs in the O.C. Wilson H & K emission survey, [*Astrophys. J.* **279**, 763]
7.	Lowell Observatory photometry of the Wilson survey stars, [*Lockwood and Skiff AFGL-TR-88-0221*]
8.	Arizona-Tonantzintla Catalogue (the standards), [*Comm. Lunar and Planetary Lab.* No. **63**]
9.	Catalogue of Chromospherically Active Binary Stars (plus candidates) [*Astron. Astrophys. Suppl. Ser.* **72**, 291]
10.	four years of photometry with an automatic telescope [*Astrophys. J. Suppl. Ser.* **67**, 439 - the constant stars] [*Astrophys. J. Suppl. Ser.* **67**, 453 - the single stars] [*Astrophys. J. Suppl. Ser.* **69**, 141 - the binary stars]
11.	Photometry of 50 Suspected Chromopspherically Active Variables [*Astrophys. J. Suppl. Ser.* **74**, 225]
12.	Thirteen Chromospherically Active Ellipsoidal Variables, [*Astron. J.* **100**, 554]
13.	single BY Dra variables, [*Astron. J.* **82**, 490]

constant stars, wave amplitudes less than 1 %. The plusses indicate stars with wave amplitudes of 1% or greater. The solid line is the convective turnover time taken from Gilliland (1985, figure 10) for main–sequence stars. Any star above this curve should have a Rossby number greater than unity, on the curve a Rossby number of unity, and below the curve a Rossby number less than unity. It is demonstrated very nicely, I think, that a Rossby number a little less than unity is a sharp dividing line between the non–variable and the variable stars.

Figure 2 is the same thing for the subgiants, *i.e*, luminosity class IV. The curve is the same convective turnover times for *dwarfs*, but we see now that the dividing

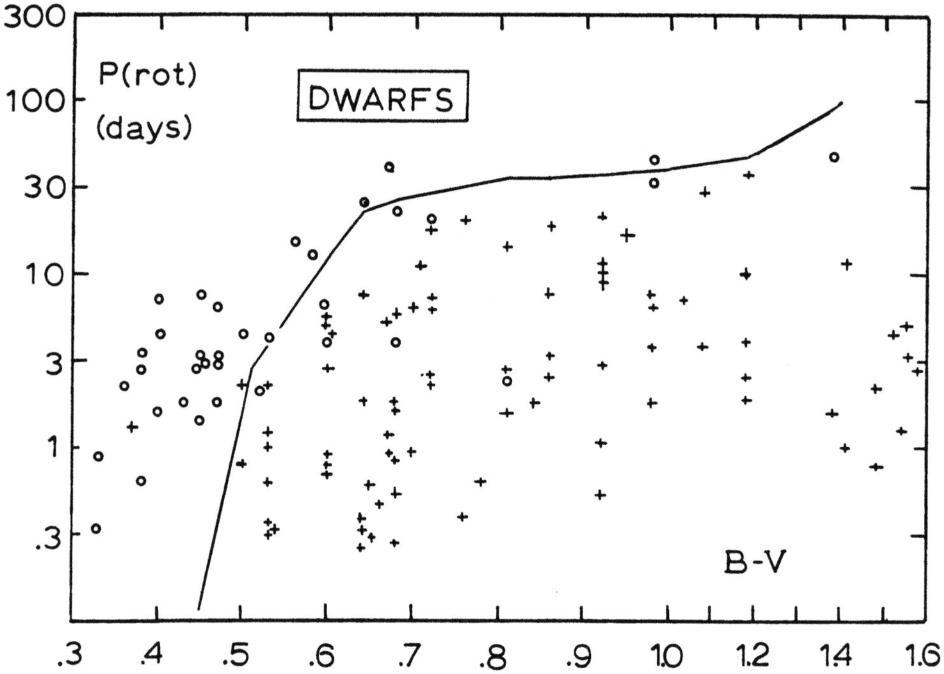

Fig. 1. P_{rot} versus $B-V$ for dwarfs. Variables (+) and non–variables (○) are distinguished. The solid curve represents Gilliland's convective turnover times.

line between non–variable and variable lies somewhat higher – about 0.3 in the log or turnover times longer by a factor 2 compared to dwarfs of the same $B-V$.

Figure 3 is the same thing for the giants, *i.e.*, luminosity class III. We see here that the dividing line between non–variable and variable lies even higher – about 0.6 in the log or turnover times longer by a factor 4 compared to dwarfs of the same $B-V$.

Gilliland's turnover times show that, as a star evolves to the base of the giant branch (this would be the subgiants), his turnover times become longer by about 0.5 in the log. So my results for the subgiants are in fair accordance with his. But as a star evolves up the giant branch his turnover times become shorter. For a typical late–type star of luminosity class III they would be shorter than the main-sequence times by about 0.6 in the log, whereas mine were *longer* by about 0.6 in the log. So my results for the giants are not in accordance with his.

For the next three figures I have computed Rossby numbers for all of my 277 stars, using Gilliland's convective turnover times for the dwarfs but longer turnover times for the subgiants and the giants. The amounts added to the log of the main-sequence turnover times were

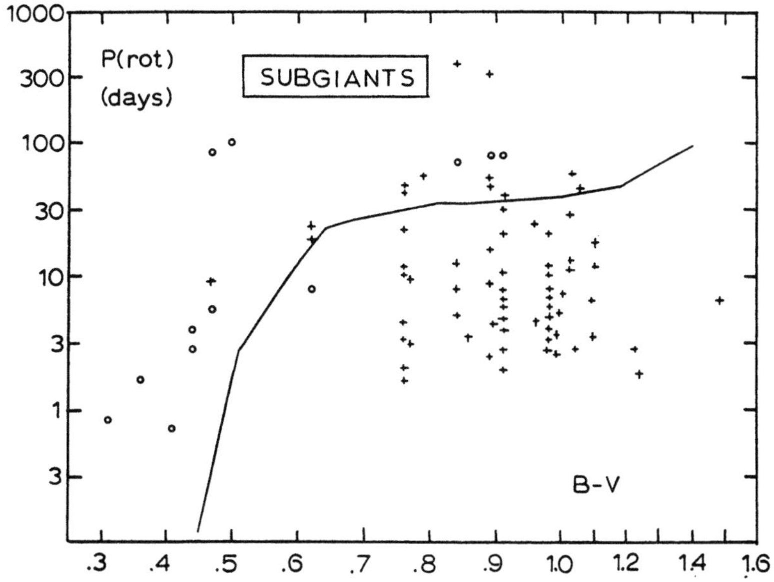

Fig. 2. Same as Figure 1, for subgiants. Note that variability occurs at rotation periods longer than the convective turnover times for dwarfs.

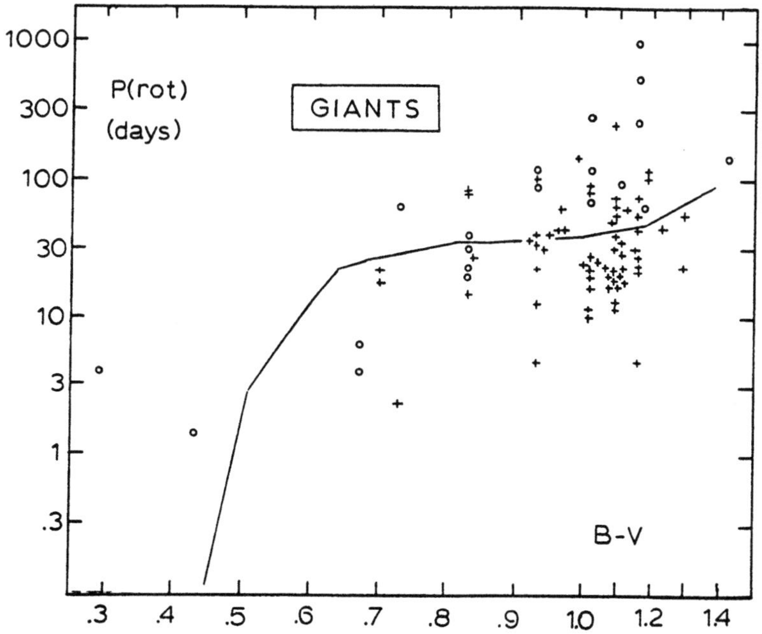

Fig. 3. Same as Figure 1, for giants. Note that variability occurs at rotation periods longer than the turnover times for dwarfs, more so than for the subgiants.

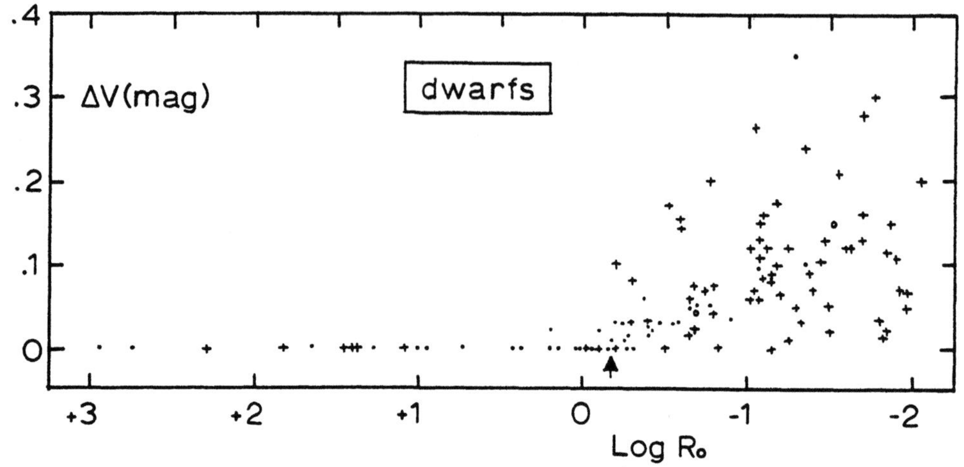

Fig. 4. Starspot wave amplitude versus log Rossby number, for dwarfs. Note the sharp onset of variability at Ro = 2/3, marked by the arrow, for both binaries (+) and single stars (•). The Sun is at log Ro ≈ 0.

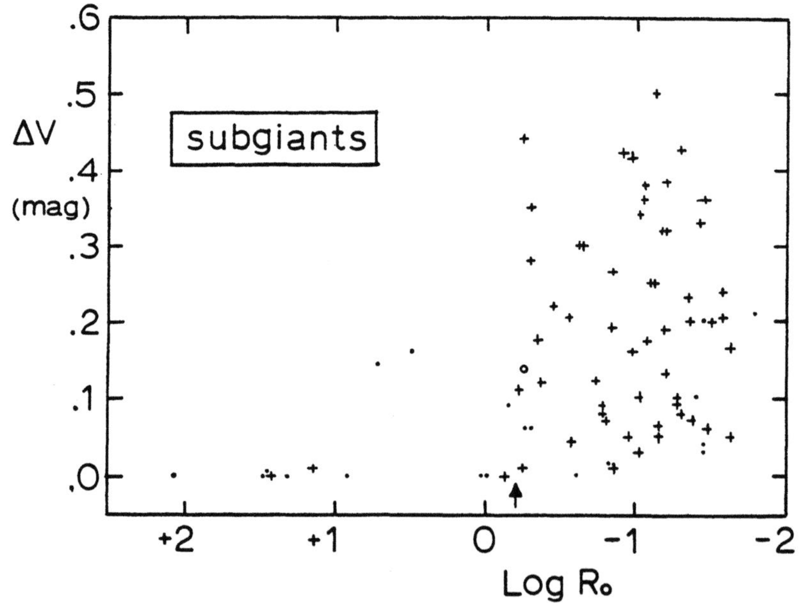

Fig. 5. Same as Figure 4, for subgiants. If HR 1362 and HD 181943 are considered anomalous, we see the same sharp onset of variability at Ro = 2/3.

Luminosity Class	*Correction*
V	0.00
V–IV	0.15
IV	0.30
IV–III	0.45
III	0.60
III-II	0.75

Figure 4 is for the dwarfs. The horizontal axis is log Rossby number and the vertical axis is wave amplitude in the V bandpass, in magnitudes. Notice the onset of measurable variability at a Rossby number somewhat less than unity, at about 2/3. Moreover, note that the onset is very sharp, within 0.1 in the log, or a factor of only 1.25. The Sun appears in this figure at a $\log \mathrm{Ro} \approx 0$; if it happened to rotate in two weeks rather than one month, it would probably vary in brightness by 2 or 3 percent rather than by only 0.1 %. There is some indication that wave amplitude increases as Rossby number decreases from 2/3 to 1/10, but that trend is not very well–defined. The plusses are the binaries, the dots are the single stars, and their distributions are indistinguishable.

Figure 5 is for the subgiants. First let me point out two stars which are probably anomalous. They are HR 1362 and HD 181943, both single G8 subgiants with rotation periods longer than a year but nevertheless quite chromospherically active in the usual ways (H and K emission, x–ray emission, etc.). Their anomalous status has been the subject of a paper by Strassmeier *et al.* (1990). If we overlook them, then we see again the onset of variability occurs at a Rossby number of about 2/3 and is very sharp, within ~ 0.1 in the log. Here, unlike the dwarfs, starspot wave amplitude rises almost immediately and does not continue to increase at smaller Rossby numbers. The amplitudes themselves are somewhat larger than for the dwarfs, up to a half magnitude. But again the distributions of the binaries (plusses) and the single stars (dots) are indistinguishable.

Figure 6 is for the giants. We see the same very sharp rise at $\mathrm{Ro} \approx 2/3$, no continuing increase as Rossby numbers get smaller, the same large amplitudes, again up to a half magnitude, and no difference between the single stars and the binaries.

The smallest Rossby numbers in my sample were about 0.01 but there is no indication that the wave amplitudes were beginning to decrease, *i.e.*, no evidence that extreme dynamo action suppresses spottedness.

The results seen here are consistent with the rotostat model advanced by Gray (1989a) to explain the abrupt decrease in rotational velocity among single giants later than spectral type G0. His model has dynamo action as the mechanism for braking the rapid rotation seen in early–type giants and maintaining slow rotation in late-type giants. (1) The dramatic increase in Gilliland's turnover times (which makes dynamo action effective) and the abrupt braking which Gray sees both occur at the same value of $B-V$. (2) If we take Gray's $\langle V \sin i \rangle = 3.02\,\mathrm{km\ s^{-1}}$ for G7 and G8 III stars, employ the factor $\pi/4$ to get $\langle V \rangle = 3.85\,\mathrm{km\ s^{-1}}$, and adopt a canonical value of $12 R_\odot$, we get a rotation period of 158 days. The turnover

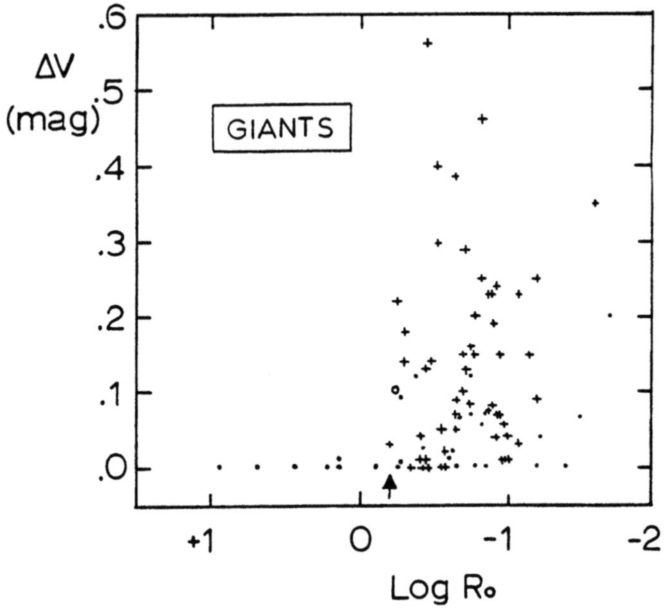

Fig. 6. Same as Figure 4, for giants. The onset of variability at Ro = 2/3 is very sharp and well-defined.

time for a G7.5 III star, based on my previously mentioned adjustment for giants, would be 152 days. The agreement is not quite so good at earlier or later types but still is not bad. At G5 III Gray's $\langle V \sin i \rangle$ leads to a rotation period shorter than my turnover time by a factor 1.5; at K0 III his rotation period is longer, again by a factor 1.5.

3. Latitude extent

We all know that sunspots are confined to approximately 30° above and below the equator. It is now established that large starspot regions can occur at much higher latitudes. In some stars the dark regions form essentially a polar cap. Thus we have an important instance where the solar–stellar analogue breaks down. A tabulation of starspot latitude determinations is given by Gray (1988, table 7-1). Good recent examples are EI Eri, described by Strassmeier (1990) and also discussed by him in a paper at this meeting (see Hackman *et al.* in these Proceedings), and UX Ari, discussed by Vogt and Hatzes in another paper at this meeting.

The *latitude* of a starspot is a relatively difficult parameter to determine, more difficult than its longitude, for example. This is true with starspot modelling of light curves and of distorted line profiles. One problem, affecting light curve modelling, is needing to know the brightness of an unspotted hemisphere. Another problem, affecting both, is needing to know the inclination of the spin axis. In a few cases, involving modelling of photometry *and* spectroscopy obtained simultaneously, the resulting spot latitudes seem to be convincing. Examples are Rodonò *et al.* (1986) and the EI Eri paper mentioned already.

On the other hand, many other spotted stars surely do have spots in their equatorial regions. An obvious example is the many eclipsing binaries with large starspot waves. Because their inclinations are very near 90°, equatorial or nearly equatorial spots are required.

A powerful tool for latitude determination is photometry of eclipsing spotted stars during the secondary eclipse, when the spotted regions are occulted by the smaller companion star. In my opinion, this technique has not been used as much as it should. The secondary eclipse of RS CVn observed at two epochs can be seen in Eaton and Hall (1977, figure 3) and in Eaton and Poe (1985). In both cases the asymmetrical shape indicates the existence of equatorial spots. This matter is elaborated further by Eaton (1990).

Recently a paper by Gray (1989b) demonstrated that active regions, if not actually spots, in giant stars occur preferentially near the equator.

There was a paper by Gray on this subject, latitude dependence, at this meeting. I think it is still true that we do not know what factors determine latitude extent. Is it spectral type or $B - V$, is it rotation speed or Rossby number, is it depth of convection zone, or is it something else

4. Differential rotation with latitude

With long-term starspot photometry it is possible to investigate differential rotation as a function of stellar latitude. The results are surprising.

A typical wave in the light curve maintains the same photometric period for a certain interval of time, which can range from months to years but seldom longer than a decade. This indicates that the dark region responsible maintains the same central longitude and a constant rotation period as long as it exists. When a spotted variable is observed continuously for many years, several different spotted regions are seen to appear, exist, and disappear, and typically they have rotation periods which are significantly different from each other. Recent examples of this behavior are V478 Lyr (Hall *et al.*, 1990c), V1817 Cyg = HR 7428 (Hall *et al.*, 1990a), and V1149 Ori = HD 37824 (Hall *et al.*, 1990b). An older example is BY Dra (Oskanyan *et al.*, 1977).

An illustration of the history of starspot rotation periods in RT Lac, covering more than 80 years, is given by Hall and Busby (1990, figure 1) and will not be duplicated here. The horizontal axis is year. The vertical axis is $O - C$, where O is observed time of light minimum (when the spot is facing Earth) and C is time computed with an arbitrarily assumed constant rotation period (in this case it is the binary's 5.074–day orbital period). In an $O-C$ curve like this, a positive slope indicates rotation slower than the orbital period, a negative slope indicates rotation faster. In this case the maximum is 0.17 % faster than orbital and the minimum is 0.25 % slower than orbital, for a full range of $\delta P/P = 0.4\%$.

I determined $\delta P/P$ in this way for 84 variable stars of different types, both binary and single. This was largely the same sample used by Hall and Henry (1990), but I have added a few and updated some of the observational material. With the sun included, the sample totals 85.

The range of spot rotation periods should be a measure of the differential rotation with stellar latitude. Differential rotation in the sun is characterized by the coefficient of a $\sin^2\theta$ term, where θ is latitude. I'll call this coefficient k and, for the sun, $k = 0.186$. Values of $\delta P/P$ for my stars can be converted to values of k if one assumes a latitude range for the starspots, as explained in detail by Hall and Busby (1990, figure 2).

Hall and Henry (1990, Figure 11) had found a relation between k and Rossby number for their sample of stars, in the sense that k decreased when the Rossby number decreased. I would like to show that the relation is just as good if one uses the rotation period instead of the Rossby number (which is defined as the ratio between the rotation period and the convective turnover time). This was the approach used earlier by Rodonò (1986, figure 6) with a much smaller sample of stars.

Linear regression gave the relation

$$\log k = -2.30(\pm 0.06) + 0.85(\pm 0.06)\log(P_{rot}), \qquad (1)$$

where P_{rot} is in days.

It has been suggested that differential rotation might be suppressed or diminished by tidal forces from a companion star. To investigate this matter I did a double regression, *i.e.,* with respect to rotation period and Roche–lobe–filling–factor, F. Because both single stars and contact binaries are in my sample, the filling factors covered the entire range $0 < F < 1$. The result was

$$\log k = -2.02(\pm 0.12) + 0.79(\pm 0.06)\log(P_{rot}) - 0.42(\pm 0.16)F \qquad (2)$$

The coefficient of the rotation term is only slightly smaller but there does seem to be a significant correlation with lobe–filling factor, in the sense that stars nearly in contact with their Roche lobes experience diminished differential rotation, about a factor 2.5 less, compared to single stars. This is in disagreement with a conclusion reached earlier by Scharlemann (1982) but it might be consistent with the newer synchronization theory of Tassoul (1987, 1988), who shows that synchronization forces are stronger than those based on the tidal theory of Zahn (1977).

When the double regression was repeated, giving F an exponent ranging from 2 to 6, the resulting coefficient ranged between -0.33 to -0.37, *i.e.,* within the uncertainty of the value found in equation (2). When the regression was repeated with the ultra–short–period SU UMa binaries omitted from the sample, the result was

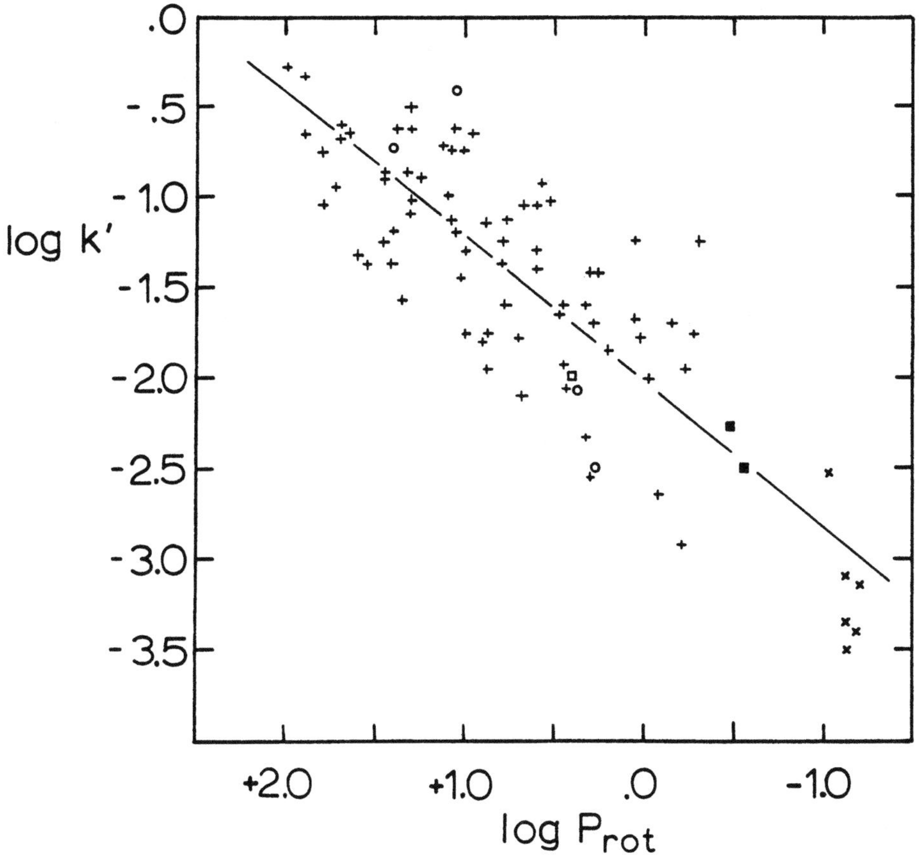

Fig. 7. Differential rotation coefficient k, corrected with equation (4), versus rotation period. The straight line thus is equation (2). Different types of stars, both single and in binaries, have different symbols, as explained in the text. Note that those with very short periods are nearly rigid rotators.

$$\log k = -1.96(\pm 0.12) + 0.71(\pm 0.07)\log(P_{rot}) - 0.38(\pm 0.16)F \quad (3)$$

Within the uncertainties, this is not significantly different from equation (2) so the observed correlation is not dependent on inclusion of the SU UMa binaries.

To display the results graphically, I have removed the lobe–filling dependence with the relation

$$\log k' = \log k + 0.4F \quad (4)$$

and then plotted $\log k'$ versus P_{rot} in Figure 7. The slope of the straight line represents the coefficient 0.79±0.06. Chromospherically active binaries of the RS CVn type or BY Dra type are plusses, the Algol–type binary U Cep is an open square, two W UMa binaries are filled squares, six SU UMa binaries are crosses, and four single stars (the spotted T Tau variable V410 Tau, the chromospherically active giant FK Com, the solar–type star ϵ Eri, and the Sun) are open circles. Note that differential rotation becomes *much* less as rotation becomes faster. For the most rapid rotators in my sample, the coefficient k is 300 times smaller than the solar value and 1000 times smaller than the largest value in my sample (for HR 7428). That is very close to rigid–body rotation.

I'm not really sure what the predictions of dynamo theory are concerning differential rotation, but I think most linear dynamo theories predict it should increase as Rossby number decreases or rotation becomes faster. That is definitely not what is seen here. On the other hand, I think non–linear dynamo theories suggest differential rotation should be suppressed by very strong dynamo action, which is what we see here. In any event, I want to emphasize this clear and dramatic observational result, which should be useful in testing predictions of dynamo theories.

Let me clarify that these results pertain only to differential rotation with latitude and say nothing directly about differential rotation with depth down into the convective zone. We can see only the outside of these stars.

5. Latitude drift during a solar cycle

We all know that sunspots form at progressively lower latitudes as the ll–year cycle progresses. Do we see the same thing in other stars

Look again at the previously mentioned spot rotation period curve for RT Lac (Hall and Busby 1990, figure 1). If RT Lac had solar–type differential rotation and solar-type latitude drift, we would see spot rotation periods becoming progressively shorter. In this curve the slopes would go gradually from positive to negative. That is not seen; they just jump back and forth.

Another example is V478 Lyr. Its spot rotation period curve during its 9–year photometric history is given by Hall *et al.* (1990c, figure 5). Again the spot rotation periods jump back and forth with no smooth trend.

Yet another example is HK Lac, which has a 14–year continuous photometric history (Olah *et al.,* 1990). In this case four long–lived spots have been followed and a fifth spot has appeared recently. Each one had a constant rotation period during its lifetime and in this case the four rotation periods were the same, again no indication of a gradual change with time.

To the best of my knowledge, the phenomenon of latitude drift in spotted stars other than the sun has not been convincingly demonstrated. Perhaps it does exist but we have not examined observations which cover a complete cycle. Or perhaps solar–type latitude drift does not exist in heavily spotted stars.

6. Sector structure and preferential longitudes

There has been some recent observational evidence that starspot formation is restricted to two regions of stellar longitude, 180° apart. See the discussion in Hall (1987, section 5.3). That would be an indication of sector structure, namely, four sectors, which are active, inactive, active, inactive.

There has also been recent observational evidence that these longitudes, if defined with respect to the line of centers in a binary system, remain stationary. That effect would be termed preferential longitudes.

One of the first examples was SV Cam. Zeilik *et al.* (1987, figure 1) showed a Mercator projection: stellar latitude and stellar longitude. Each circle represented a starspot region observed some time during the 45-year photometric history of SV Cam, with the diameter of each circle proportional to the starspot's radius. One sees a tendency for spots to occur at phases 0.25 and 0.75, *i.e.*, at the quadratures. Zeilik and collaborators have similar results for other (mostly short-period RS CVn-type) binaries, and Zeilik himself reviewed starspot activity on these systems at this meeting.

Another example is UX Ari. Hall and Strassmeier (1988, figure 1) show four light curves obtained with an automatic telescope in Arizona for four consecutive years. Notice that light minimum of the starspot wave remains stationary at phase 0.95, *i.e.*, very near one of the conjunctions. This would place the spot on the hemisphere opposite the other star. If we add earlier light curves, discussed by Wacker and Guinan (1987), we see that the spot in UX Ari has remained anchored at phase 0.95 for 7 years.

Another example may be ζ And, a chromospherically active binary known to show a relatively large ellipticity effect. In the light curve obtained between 1983 and 1986 with an automatic telescope in Arizona (Strassmeier *et al.*, 1989, figure 52) the amplitude of the ellipticity effect was about $0.^m06$ in V. At two much earlier epochs the amplitude appeared to be more than twice as large. The ellipticity effect itself, of course, should not change amplitude. Thus we may have had spots during the two earlier epochs, in both cases occurring at the two conjunctions, or we may have had spots during the later epoch, occurring at the two quadratures. Analysis by Hall (1990a) indicates the first of those two possibilities would be the more likely.

Similar results were found with V478 Lyr, as discussed by Hall *et al.* (1990c). In that case the preferential longitudes, at the moments of spot origin, were 0.0 or 0.5, *i.e.*, at the conjunctions.

A final example is HK Lac, as discussed by Olah *et al.* (1990). The four starspots have had relatively long lifetimes, about 7 years each, and each one has migrated in phase away from its point of origin. The longitude at the moment of origin, however, seems to have been the same for all four, namely, phase 0.6.

These results are quite interesting and potentially useful, because some dynamo theories can make a statement about sector structure. An important question, however, is whether it is physically reasonable that tidal forces from a companion star can influence the orientation of sectors, *i.e.*, induce preferential longitudes.

7. Starspot lifetimes

For some time photometry of starspot waves has indicated that these large spots live very long, many years, a few longer than a decade. At one time this was puzzling, because differential rotation at the solar rate would disrupt such large spots within months. People asked what mysterious force held the large spots together.

It seems there is no mysterious force. The greatly diminished differential rotation, discussed in Section IV, makes calculated disruption times correspondingly longer. Hall and Busby (1990) demonstrated that no large starspot has been observed with a lifetime longer than the calculated disruption time. The details of that demonstration are as follow.

Calculated disruption times depend mostly on 3 factors: the spot radius, the differential rotation coefficient, and the rotation period. There is a weaker dependence on the latitude of the spot. Spot radii can be estimated from the amplitude of the starspot wave (Hall and Busby 1990, figure 3). I have already discussed how one estimates the differential rotation coefficient for a spotted star. And the rotation period itself is readily available. The calculated disruption time for a given spot on a given star can be gotten from Hall and Busby (1990, figure 5 and equation 5).

Results for 40 spots on 17 different stars were shown by Hall and Busby (1990, figure 7). They found that for relatively large spots, greater than about 20° in radius, observed lifetimes (t_O) are equal to their calculated disruption times (t_C). The equality was quite close, because the rms scatter from the $t_O = t_C$ line was only 0.3 in the log, only a factor 2.

This agreement ($t_O = t_C$) cannot be the result of a triviality, because two of the three parameters which determine t_C cover enormous ranges and so does the observed lifetime itself. Rotation periods in the Hall-Busby sample ranged from 0.6 days to 385 days (600 X), k ranged from 0.0006 to 0.06 (100 X), and t_O ranged from 0.3 yrs to 30 yrs (100 X).

Hall and Busby (1990, figure 6) found a different relation for the smaller spots, smaller than about 20° in radius. For them, in general, $t_O < t_C$. Moreover, when they plotted t_O versus spot radius (but on a scale linear with radius squared, *i.e.*, spotted area), they saw that the relatively small spots have lifetimes more or less proportional to their areas. Recall that, for spots on the sun, there is also a rough area–versus–lifetime proportionality. A next step in this investigation, not done by them, would be to consider absolute area, rather than angular area, of each spot. That might decrease the scatter they found in their relation.

The remarkably close agreement between t_O and t_C (for the large spots) indicates that lifetimes of large spots are in fact limited by shear. This observation is consistent with either of two scenarios: (1) a large spot or active region originates in a deep layer which may be rigidly rotating, is magnetically disconnected after awhile, and is disrupted by differential rotation characterizing the surface, or (2) a large spot or active region is not disconnected magnetically but the deeper layers from which it originated approximately mirror the same differential rotation law which applies to the surface layers.

8. Magnetic cycles

We all know the sun has a magnetic cycle. It is also rather well established that some other stars have magnetic cycles too, probably physically analagous. There was a good review about five years ago by Baliunas and Vaughan (1985).

Table 3. Types of stars displaying magnetic cycles

1.	RS CVn binaries	8.	U Gem binaries
2.	BY Dra variables	9.	SU UMa binaries
3.	Algol-type binaries	10.	old novas
4.	W UMa binaries	11.	recurrent novas
5.	the sun	12.	low-mass x-ray binaries
6.	solar-type dwarfs	13.	RR Lyrae variables
7.	flare stars	14.	Cepheids
		15.	RV Tauri variables

Table 4. Signatures of magnetic cycles

1.	the sunspot cycle - sunspot numbers
2.	the sunspot cycle - latitude drift
3.	long-term changes in Ca II H & K emission
4.	long-term changes in mean brightness
5.	cyclic changes in orbital period
6.	frequency of flaring in flare stars
7.	variable frequency of outbursts in dwarf novas
8.	frequency of recurrent nova outbursts
9.	the high-low states of Her X-1 and HZ Her
10.	cyclic changes in the Blazhko effect in RR Lyrae stars
11.	variable magnetic fields in RR Lyrae stars
12.	variable pulsation periods in RR Lyrae stars
13.	variable pulsation periods in Cepheids
14.	variable pulsation periods in RV Tauri variables

For my contribution, I want to explain that magnetic cycles may occur on more types of stars than have been considered heretofore. In demonstrating this I want to explain also that magnetic cycles have more signatures than have been considered heretofore. Finding observational evidence of cycles on many very different types of objects should be more useful for dynamo theory than if we look narrowly only at solar-type stars.

It may turn out that I have cast my net too wide and that the cycles in some of my suggested candidates are not actually magnetic in origin. On the other hand, many of the types of objects I consider display cyclic phenomena which until now *have no other explanation.* Thus, for them, magnetic cycles represent the best, *i.e.*, the only, explanation.

Table 3 lists 15 types of objects which clearly do, probably do, or perhaps do have magnetic cycles. The RV Tauri variables have been added to the list compiled earlier by Hall (1990b) in his review of this same subject. A pertinent reference would be Percy *et al.* (1989).

Table 4 lists the different observational signatures of magnetic cycles. Hall (1990b) explained in detail how possible magnetic cycles might be expected to produce these various observational manifestations. There is not much to add except that the earlier short note by Bianchini (1988) has been followed by a much more complete discussion (Bianchini, 1990), in which he considers the situation with about half of the 15 types of objects listed in Table 3.

Hall (1990b, table 1) concluded with a summary of different types of objects which display cyclic behavior. For each he gave the median cycle length and, for those which undergo changes in a period, the typical size of those changes. Noting that the cycle lengths for all fall in the range 10 yrs to 100 yrs, he suggested that this similarity in timescale is, by itself, suggestive that a common mechanism is at work.

I conclude with an exploratory plausibility argument. Various dynamo theories have the capability of predicting a magnetic cycle. One of them is based on the time which the mean magnetic field needs to diffuse over a characteristic distance, for example, the depth of the convective zone. It predicts that cycle length should depend on the convective turnover time, the size of a convective cell, and the thickness of the convective zone. An equation was given by Hall (1990b, equation 6). If one considers four rather different types of stars (a low–mass main–sequence dwarf, the sun, a middle–G subgiant, and an early–K giant) and uses rough estimates for the three parameters mentioned above, the predicted cycle lengths cover a relatively narrow range, from about 10 years to 100 years (Hall 1990b, table 2). This is the same range found from the observations.

Work in this area needs to be continued. The observational evidence needs to be made stronger, for some of the types of objects on my list, and the comparison with theory needs to be done more rigorously than I have done. But the approach from both directions, observation and theory, promises to answer a lot of important questions, both observational and theoretical.

References

Baliunas, S.L., Vaughan, A.H.: 1985, *Ann. Rev. Astr. Astrophys.* **23**, 379
Bianchini, A.: 1988, *I.B.V.S.* **3136**
Bianchini, A.: 1990, *Astron. J.* **99**, 1941
Catalano, S., Frisina, A., Rodonò, M.: 1980, *I.A.U. Symposium* **88**, 405
Doyle, J.G., Butler, C.J., Morrison, L.V., Gibbs, P.: 1988, *Astron. Astrophys.* **192**, 275
Eaton, J.A.. 1990, in *Surface Inhomogeneities in Late-Type Stars*, eds P.B. Byrne and D.F. Mullan (Berlin: Springer-Verlag), in press
Eaton, J.A., Hall, D.S.: 1977, *Astrophys. J.* **227**, 907

Eaton, J.A., Poe, C.H.: 1985, *B.A.A.S.* **16**, 914
Gilliland, R.: 1985, *Astrophys. J.* **299**, 286
Gray, D.F.: 1988, *Lectures on Spectral Line Analysis of F, G, and K Stars*, Arva, Ontario
Gray, D.F.: 1989a, *Astrophys. J.* **347**, 1021
Gray, D.F.: 1989b, *Publ. Astron. Soc. Pac.* **101**, 695
Hall, D.S.: 1987, *Publ. Astr. Inst. Czechoslovakia* **70**, 77
Hall, D.S.: 1989, *I.A.U. Colloquium* **107**, 219.
Hall, D.S.: 1990a, *Astron. J.* **100**, 554
Hall, D.S.: 1990b, in *Active Close Binaries*, ed. C. Ibanoglu (Dordrecht: Kluwer), p. 95.
Hall, D.S., Busby, M.R.: 1990, in *Active Close Binaries*, ed. C. Ibanoglu (Dordrecht: Kluwer), p. 377.
Hall, D.S., Gessner, S.E., Lines, H.C., Lines, R.D.: 1990a, *Astron. J.* **100**, 2017
Hall, D.S., Henry, G.W.: 1990, in *Active Close Binaries*, ed. C. Ibanoglu (Dordrecht: Kluwer), p. 287.
Hall, D.S., Henry, G.W., Barksdale, W.S.: 1990b, in preparation.
Hall, D.S., Henry, G.W., Sowell, J.R.: 1990c, *Astron. J.* **99**, 396
Hall, D.S., Strassmeier, K.G.: 1988, in *Automatic Small Telescopes*, eds. D.S. Hayes and R.M. Genet (Mesa: Fairborn Press), p. 71.
Olah, K., Hall, D.S., Henry, G.W.: 1990, in preparation.
Oskanyan, V.S., Evans, D.S., Lacy, C., McMillan, R.S.: 1977, *Astrophys. J.* **214**, 430
Percy, J.R., Sasselov, D.D., Alfred, A., Scott, G.: 1989, *B.A.A.S.* **21**, 1117
Rodonò, M:. 1986, in *Cool Stars, Stellar Systems, and the Sun*, eds. M. Zeilik and D.M. Gibson (Berlin: Springer-Verlag), p. 475.
Rodonò, M. *et al.* [22 authors]: 1986, *Astron. Astrophys.* **165**, 135
Scharlemann, E.T.: 1982, *Astrophys. J.* **253**, 298
Strassmeier, K.G. 1989, in *Remote Access Small Telescopes*, eds. D.S. Hayes and R.M. Genet (Mesa: Fairborn Press), p. 223.
Strassmeier, K.G.: 1990, *Astrophys. J.* **348**, 682
Strassmeier, K.G., Hall, D.S.: 1988, *Astrophys. J. Suppl.* **67**, 453
Strassmeier, K.G., Hall, D.S., Barksdale, W.S., Jusick, A.T., Henry, G.W.: 1990, *Astrophys. J.* **350**, 367
Strassmeier, K.G., Hall, D.S., Boyd, L.J., Genet, R.M.: 1989, *Astrophys. J. Suppl.* **69**, 141
Strassmeier, K.G., Hall, D.S., Zeilik, M., Nelson, E., Eker, Z., Fekel, F.C.: 1988, *Astron. Astrophys. Suppl.* **72**, 291
Tassoul, J.P.: 1987, *Astrophys. J.* **322**, 857
Tassoul, J.P.: 1988, *Astrophys. J.* **324**, L71
Wacker, S., Guinan, E.F.: 1987, *I.B.V.S.* **3018**
Zahn, J.P.: 1977, *Astron. Astrophys.* **57**, 383
Zeilik, M., De Blasi, C., Rhodes, M., Budding, E.: 1987, in *Cool Stars, Stellar Systems, and the Sun*, eds. J.L. Linsky and R.E. Stencel (Berlin: Springer-Verlag), p. 503.

Starspot Activity on Short-period RS CVn Stars

Michael Zeilik

Institute for Astrophysics, The University of New Mexico
Albuquerque, New Mexico 87131, USA

1. Introduction

We have yet to understand the magnetic activity cycles of cool close binary systems of sunlike stars. Mutual tidal interactions, as well as magnetic ones, may result from a regime of dynamo models not yet tested, because these have been developed for single stars. To arrive at the basic physics, though, requires that we first examine the phenomenology of magnetic activity for binary systems. In particular, we would like to discover if such activity has a clearly-defined *cycle*, such as the sun exibits.

Among the proxy indicators of magnetic activity are the Ca II H and K lines. Strassmeier *et al.* (1988) used the strength of these lines as the primary criterion for the inclusion of systems in *The Catalog of Chromospherically Binary Stars.* Of the RS CVn stars in the catalog, 12 have orbital periods of one day or shorter; 9 are eclipsing systems. As part of a decade-long program, we have focussed our observations and models on eight of the short-period group (Hall, 1976): XY UMa, UV Psc, SV Cam, RT And, CG Cyg, ER Vul, BH Vir, and WY Cnc. These close systems are tidally-locked in synchronous rotation and tidally-distorted into Roche lobe configurations.

A few of these systems have archival observations that span decades, hence giving us the long time base to probe for cyclical activity. We can extend these time series with current observations with our CCD-camera at Capilla Peak Observatory and those data collected by other groups. We have completed work on data for SV Cam (Zeilik *et al.*, 1988), RT And (Zeilik *et al.*, 1989), BH Vir (Zeilik *et al.*, 1990a), and WY Cnc (Zeilik *et al.*, 1990b). Most of my comments will relate to these four systems. (We note that these stars are too faint for the current generation of APTs).

2. Analysis

Our goal is to apply the simplest possible starspot model (Budding, 1977) to parameterize photometric light curves in a way that reliably and objectively extracts information revelant to photospheric magnetic activity. Using circular, dark spotted regions (Budding and Zeilik, 1987) as the representation of the essential phenomenon, we can optimize for the longitude, latitude, radius, and temperature (relative to the photosphere) of the active area. Note we do *not* believe, in fact, that all regions of activity are circular and without structure. Rather we use this model as an adequate representation to find out information relevant to magnetic activity.

Previous criticisms of photometric modelling (cf. Rodonó, 1986) have aimed at the lack of uniqueness of the starspot solutions. That limitation is especially true when applied to *non-eclipsing* systems, as has usually been the case so far. *Eclipsing* systems present a different aspect; the orbital inclinations are known within a degree or so and the eclipses provide fiducial markers for longitude. Hence, eclipsing systems provide a much more constrained modeling problem; the drawback is that we need to separate eclipse and proximity effects to reveal the maculation (distortion) waves. We have done so by modifying Budding's 16-parameter, optimizing curve fitter (Budding, 1973; Budding and Najim, 1980) to generate the model curves for the eclipsing system. These models use Roche lobes for the distorted stars, and include both the reflection and ellipticity effects. The difference between an optimized theoretical curve and the observed one defines the maculation effect. Another curve fitter, corresponding to the case of one or two dark, circular spots, generates an optimized solution to these residuals.

Our programs (Budding and Zeilik, 1987) contain an optimizer routine that uses a variety of methods to search efficiently for a minimum in χ^2 hyperspace; the two main ones are a vector search and a parabolic interpolation (Banks, 1989). The χ^2 routine contains the particular fitting function appropriate to generate a representation of the eclipsing binary light curve or of the distortion wave from starspots (Budding, 1977). What sets our procedure apart is our *information* limit analysis, which provides an objective evaluation of the results from the optimized parameters, in the sense of the information content of the data.

As detailed in Budding and Najim (1980), our programs evaluate the *curvature Hessian* in the vicinity of the optimal solution in χ^2 hyperspace. We also calculate the *error matrix*, which is the inverse of the curvature Hessian. For mathematically-determinate solutions, the curvature Hessian will have a positive eigenvalue for each optimized parameter. We approach the information limit by increasing the number of parameters until the Hessian becomes nonpositive definite; we then know that we have exceeded the information inherent in the data.

3. Results for the short-period systems

Our main results to date, which apply to the primary (hotter) star in these short-period systems, are: (1) The primary activity signature is two active longitude belts (ALBs) at the quadrature points; one belt is more active than the other at any given time. (2) The secondary activity signature is that the active regions appear most often at high latitudes (near 45°). (3) The spotted regions cover large areas of the primary, active stars (radii are typically 10°). (4) These cool areas have temperatures about 1500 K below that of the photosphere with a variation of only a few hundred Kelvins (for those light curves for which we have simultaneous V and R or I data).

A few comments about specific systems. SV Cam: over the time from 1928 to 1984, the ALB around 270° has been more active. The mean latitude value for determinate solutions is 58°.4. The differential rotation rates within the ALBs are much smaller than the sun's. RT And: again, the ALB near 270° seems the most active (from 1920 to 1989). The mean value of determinate latitudes was 53°. The orbit does have a small eccentricity (e ~ 0.03). BH Vir: both ALBs have roughly the same activity from 1953–1986. The mean value of the determinate latitudes was 31°. The temperature contrast averaged only 1100 K. WY Cnc: The ALB at 90° seems more active in 1964–1989. The mean latitude of determinant's solutions was 15°.8, this is the only system that we have analyzed that resulted in low-latitude spots. The temperature contrast was about 1500 K.

We note that the ALB activity appears binary. When one is "on", the other is "off", at least in terms of large spotted regions. (We cannot "see" the spots if their radii is less than 5°). We have so far found no instances in which we have a clear indication of both ALBs containing activity at the same time. Hence, we may have a magnetic resonance occurring in the longitudinal distribution of the magnetic fields.

This research was supported in art by NSF Grant AST-8903174.

References

Budding, E.: 1973, *Ap. Space Sci.* **22**, 87

Budding, E.: 1977, *Ap. Space Sci.* **48**, 287

Budding, E., Najim, N.N.: 1980, *Ap. Space Sci.* **72**, 369

Budding, E., Zeilik, M.: 1987, *Astrophys. J.* **319**, 827

Nelson, E., Zeilik, M: 1990, *Astrophys. J.* **349**, 163

Rodonó, M.: 1986, in *Cool stars, Stellar Systems, and the Sun*, ed. M. Zeilik and D. Gibson (Heidelberg: Springer-Verlag), p. 470

Strassmeier, K.G., Hall, D.S., Zeilik, M., Nelson, E., Eker, Z., Fekel, F.C.: 1988, *Astrophys. Astron. Suppl.* **72**, 291

Vogt, S.S., Penrod, G.D.: 1983, *Pub. Astron. Soc. Pacific* **95**, 65

Zeilik, M. De Blasi, C., Rhodes, M., Budding, E.: 1988, *Astrophys. J.* **332**, 293

Zeilik, M., Cox, D.A., De Blasi, C., Phodes, M., Budding, E.: 1989, *Astrophys. J.* **345**, 991

Zeilik, M., Ledlow, M., Phodes, M., Arevalo, M.J., Budding, E.: 1990a, *Astrophys. J.* **345**, 352

Zeilik, M., Cox, D.A., Ledlow, M.J., Phodes, M., Heckert, P.A., Budding, E.: 1990b, *Astrophys. J.* **363**, in press

Variations in Spottedness of Photosphere and Chromosphere of the Flare Star EV Lac

R.E. Gershberg, I.V. Ilyin and N.I. Shakhovskaya

Crimean Astrophysical Observatory
Nauchny, Crimea 334413, USSR

Abstract: Quantitative analysis of the variations of EV Lac as being caused by spottedness leads to the conclusion that the total areas of starspots vary strongly – up to two or threefold – from one season to another but the starspot temperatures remain constant.

During 9 nights in September 1986 and 1987 we carried out photometrical and spectroscopical observations of the flare star EV Lac. Excluding flare occurrence time intervals, we obtained data allowing the detection of variations of stellar photosphere and chromosphere. The photometric data are shown in Fig. 1 and the spectroscopic data in Fig. 2. From these plots the following qualitative conclusions are made.

First, in 1986 periodic brightness variations occurred with an amplitude $\Delta V = 0\overset{m}{.}15$ and in 1987 the amplitude decreased to $0\overset{m}{.}07$.

Second, the phases of periodic brightness variations in these two seasons differ by about half of the period and neither of them coincides with the phase in 1980–81.

Third, in 1986 the maximum brightness of EV Lac was higher by at least $0\overset{m}{.}055$ than in 1987, and in 1986 the minimum brightness of the star was about $0\overset{m}{.}025$ lower than in 1987. This means that the least spotted hemisphere in 1986 was less spotted than the least spotted one in 1987, and the most spotted hemisphere in 1986 was more spotted than the most spotted one in 1987.

Fourth, in 1986 a clear anti-correlation between Hα emission line intensity and stellar optical brightness was observed; in 1987 such an anti-correlation disappeared. However, the Hα line intensity in 1987 was close to the highest level of 1986.

These four points lead to the conclusion that the decrease of the periodic brightness variation amplitude between 1986 and 1987 is due to more homogeneous distribution of dark spots over the stellar surface with the increase of total area of starspots. In this case an equatorial belt of spots seems much more probable than several high latitude or polar spots.

The quantitative analysis of the data obtained leads to following results.

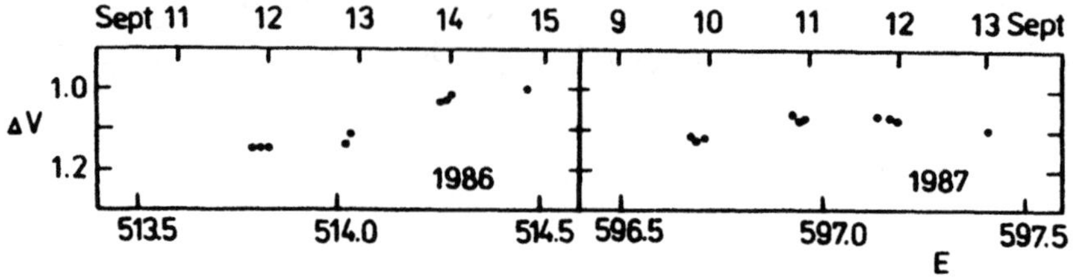

Fig. 1. Light curves of EV Lac: ΔV is a difference between stellar magnitudes of EV Lac and of the comparison star SAO 52337, E is a period number according to Rojzman's ephemeris (1984).

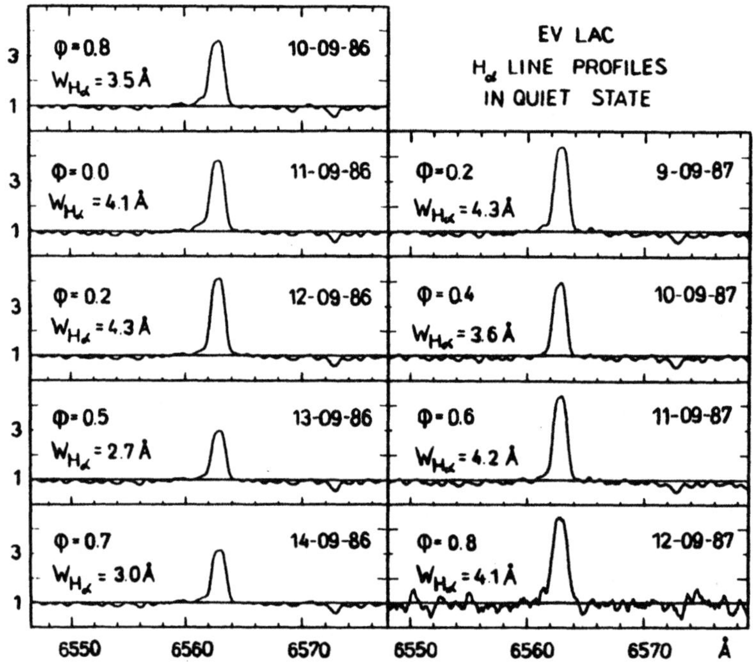

Fig. 2. Nightly averaged Hα emission line profiles in EV Lac spectra during quiet state; φ is a phase with reference to brighness minima in Fig. 1.

Brightness variations in different pass bands for the EV Lac starspots show the correlations:

$$\begin{cases} \Delta U/\Delta V = 0.82 \pm 0.12 & \text{r=0.88} \\ \Delta B/\Delta V = 1.06 \pm 0.05 & \text{r=0.98} \\ \Delta R/\Delta V = 0.60 \pm 0.13 & \text{r=0.76} \\ \Delta I/\Delta V = 0.42 \pm 0.03 & \text{r=0.96.} \end{cases} \tag{1}$$

Following Torres and Ferraz Mello (1973) and others, we model the brightness variations from starspots by the relation

$$\Delta m_\lambda = 2.5 \lg [1 - \alpha(1 - \beta_\lambda)] \,, \tag{2}$$

α being a relative spot area and $\beta_\lambda = (F_{\mathrm{spot}}/F_{\mathrm{photosphere}})_\lambda$. The ratio of specific fluxes is usually computed by the Planck formula or by energy distributions from spectra of a set of stars of different spectral classes. We assume too that starspots radiate as cooler stars. However, we take β_λ values only from photometric data. For this purpose we used Pettersen's (1976, 1980) data on stellar magnitudes and radii for K and M dwarfs and calculated values that may be called radius-unified absolute stellar magnitudes:

$$m'_\lambda = m_\lambda + (M_V - m_V) + 5 \lg(R_*/R_\odot) \,. \tag{3}$$

Then we found close correlations between variations of these stellar magnitudes along the spectral sequence:

$$\begin{cases} \Delta U'/\Delta V' = 1.23 \pm 0.04 & \mathrm{r=0.99} \\ \Delta B'/\Delta V' = 1.15 \pm 0.02 & \mathrm{r=0.99} \\ \Delta R'/\Delta V' = 0.67 \pm 0.04 & \mathrm{r=0.97} \\ \Delta I'/\Delta V' = 0.35 \pm 0.04 & \mathrm{r=0.92.} \end{cases} \tag{4}$$

According to Poe and Eaton (1985), the V and I bands are the most suitable for starspot investigations. Noting that the last line in (4) leads to $\lg \beta_I / \lg \beta_V = 0.35$, we can obtain from (1) and (2) relations defining α and β_V:

$$\begin{aligned} \beta_V &= 1 - \frac{1 - \mathrm{dex}(-0.4\Delta V)}{1 - \mathrm{dex}(-0.168\Delta V)} (1 - \beta_V^{0.35}), \\ \alpha &= \frac{1 - \mathrm{dex}(-0.4\Delta V)}{1 - \beta_V} . \end{aligned} \tag{5}$$

Applying relations (5) to our photometry, we find $\alpha = 0.25$ and $\beta_V = 0.48$ (or $\Delta T = 280$ K) for 1986 and $\alpha = 0.13$ and $\beta_V = 0.51$ for 1987; these parameters characterize differences between bright and dark hemispheres of the star in these seasons. Applying relations (5) to the Mavridis and Avgoloupis (1986) data, we find $\alpha = 0.39$ and $\beta_V = 0.42$. These parameters characterize the difference between mean spottedness of the star in 1975 and 1979. Thus, we have found large variations in total starspot areas but practically the starspot temperatures are almost constant in different seasons.

References

Mavridis, L.N., Avgoloupis, S.: 1986, *Astron. Astrophys.* **154**, 171

Pettersen, B.R.: 1976, *Institute Theor. Astrophys. Report* N 46, Blindern, Oslo

Pettersen, B.R.: 1980, *Astron. Astrophys.* **82**, 53

Poe, C.H., Eaton, J.A.: 1985, *Astrophys. J.* **289**, 644

Rojzman, G.Sh.: 1984, *Pis'ma Astron. Zh.* **10**, 279

Torres, C.A.O., Ferraz Mello, S.: 1973, *Astron. Astrophys.* **27**, 231

Surface Inhomogeneities on the W UMa star AC Bootis

Albert P. Linnell

Department of Physics and Astronomy, Michigan State University,
E. Lansing, MI 48824, USA

Abstract: New photometric data and two older data sets for AC Boo can be represented with a common set of Roche model parameters and two starspot pairs which drift only in longitude and are located on the cooler component.

The W-type W UMa star AC Bootis is well known for the variability of its light curve. A persistent light curve feature is an O'Connell effect with the maximum following primary minimum (max.I) brighter than the one preceding primary minimum (max.II). Existing light synthesis studies have not modeled this feature, and they have been hampered by lack of a cross-correlation radial velocity determination of the mass ratio.

In a collaborative effort, B. Hrivnak has obtained a cross-correlation spectroscopic mass ratio q^* ($q^*=1/q$) of 0.35 (preliminary value). E. C. Olson has observed a precise 435.5 nm light curve (1989 data). Although the O'Connell effect can be represented by a single dark spot on the advancing face of the more massive component, a second persistent feature cannot be so represented. This feature is a displacement of max.I to orbital phase 0.27. An eccentric orbit is an unacceptable possible explanation. In the absence of a persuasive argument for bright starspots, the max.I displacement must be the result of depression of the light curve egress from primary minimum by dark spot(s). To do this, starspots must be on the trailing face of the more massive component. The O'Connell effect then becomes a relatively delicate competition between spots separated in longitude by roughly 180 degrees, but differing somewhat from that value.

Experiment demonstrates that the synthesized light curve is relatively insensitive to spot latitude in the range 0–25 degrees. In an effort to minimize the number of free parameters, and by analogy to the sun, the light curve simulation adopted two spot pairs at latitudes (+20,−20 degrees) and with fixed temperature contrast from the adjacent photosphere of −1500 K. Both spots of a given pair have identical longitudes.An accurate fit to the Olson data, using the Hrivnak mass ratio, required adoption of 8 percent third light.

Having achieved a precise fit to the Olson data, it is of interest whether other data sets can be represented with the same system of parameters, varying only the spot longitudes. The data sets tested included the *V* data by Manusco *et al.* (1976) and the *UBV* data by Schieven *et al.* (1983). An accurate fit resulted in each case. The additional data sets required a small admixture of third light approximating that required by the Olson data. The difference in longitude of the spot pairs was maintained constant while changing the spot longitudes to fit the two additional data sets. It is of interest that the O'Connell effect was absent in the Manusco *et al.* data, while the Schieven et al. data show an O'Connell effect opposite in sense to the Olson data. Moreover, the Schieven et al. data show a positive light curve slope during total eclipse which the light synthesis simulation represents accurately.

The Olson data were obtained over an interval of about two months, in two separate observing sessions. Plots of the separate sessions show a slight systematic change on ingress to primary minimum, but with coincident egress. The later session ingress is slightly brighter, at all phases, than the earlier one. Based on the adopted longitude of the spot pair that produces the O'Connell effect, the light curve change indicates that the spot pair drifted forward on the star, implying a faster than synchronous rotation at the starspot latitudes. Since the light curve change persists into totality, when the less massive component is eclipsed, the observations unequivocally place the spots on the larger, more massive component. A detailed joint paper with B. Hrivnak and E. C. Olson is in preparation describing AC Boo.

References

Manusco, S., Milano, L., Russo, G.: 1978, *Astron. Astrophys.* **63**, 193

Schieven, G., Morton, J. C., McLean, B.J., Hughes, V.A.: 1983, *Astron. Astrophys. Suppl. Ser.* **52**, 463

Long-term Starspot Behavior of BY Dra and EV Lac

B.R. Pettersen[1], K. Oláh[2], and W.H. Sandmann[3]

[1]Institute of Theoretical Astrophysics, University of Oslo, P.O.Box 1029 Blindern, N-0315 Oslo 3, Norway
[2]Konkoly Observatory, Hungarian Academy of Sciences P.O.Box 67, H-1525 Budapest XII, Hungary
[3]Department of Physics, Harvey Mudd College, Claremont, CA 91711, USA

Abstract: New starspot photometry of BY Dra and EV Lac, extending over a decade is analyzed in combination with data from the literature to study the long-term behaviour of spottedness on these stars. Both stars show brightness suppression as well as brightness modulation so spots exist both near the poles and at lower latitudes. Polar spots have been continuously present on BY Dra for the last 14 years. (O–C) diagrams for lightcurve minima reveal systematic, sinusoidal variations in BY Dra with a period of 13–14 years. In EV Lac the spots have been located on the same hemisphere for the last ten years.

1. Introduction and observations

New photometric observations of the binary BY Dra and the single star EV Lac have been obtained during the last decade. We analyse them in combination with previously published data to determine long-term properties of spottedness.

The observations were made with various combinations of telescopes and photometers at different observatories: the Piszkéstető observing station of Konkoly Observatory in Hungary, McDonald Observatory, Texas, and Table Mountain Observatory, California, USA.

2. Results and discussion

BY Dra. The brightness level and range of variability for BY Dra is shown as a function of time in the upper part of Fig. 1. When rotational modulation due to starspots was discovered by Chugainov (1973) the brightness suppression and amplitude were at record values (in 1965). The large brightness suppression is presumably due to a polar spot on the primary star visible at all rotational phases, since $i \approx 30°$. The brightness modulation signifies a large spot at lower latitudes that rotates in and out of view. It could have been located on either of the two stars. When these spots decayed in area during 1967–69 the brightness level of BY

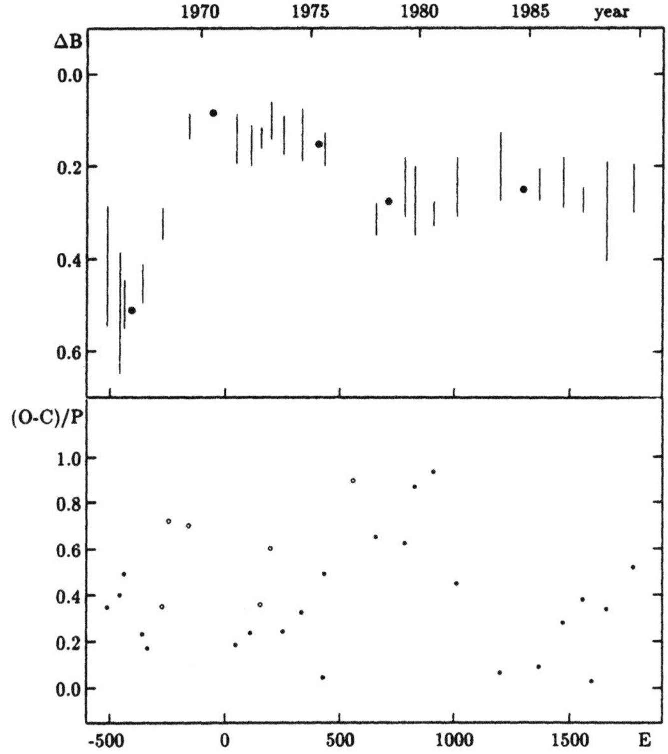

Fig. 1. Seasonal brightness changes of BY Dra from 1965 to 1989 (upper panel). The length of each bar represents the range of brightness variations. Episodes of flat lightcurves are shown as filled circles. The lower panel shows the (O–C) diagram for the same time interval, with uncertain minima identified by open circles. The sinusoidal variation of the accurate determinations (filled circles) has a period of 13–14 years.

Dra gradually increased to the immaculate level. This level seems to have been maintained for several years. The lack of Hα emission from the primary in 1974 suggests that this star had entered a state of no or very little atmospheric activity while the secondary remained active with Hα in emission. A year later the primary developed a weak emission which over the next two years grew to match that of the secondary in 1977 (Vogt, 1980). During this development a new episode of brightness suppression began, presumably due to new polar spots on the primary. 14 years later, in 1989, this state still persisted.

We have used the *average* period (P = 3.8285 days) for the entire 1965–1989 dataset to produce a linear ephemeris for the stellar rotation.

For each lightcurve we compare the *observed* time of minimum with the *computed* time and plot $\frac{(O-C)}{P}$ versus time (Figure 1, lower panel). The general trend is that $\frac{(O-C)}{P}$ varies in a sinusoidal fashion with a period of 13–14 years. The peak-to-peak amplitude is at least 5 times the typical scatter of datapoints.

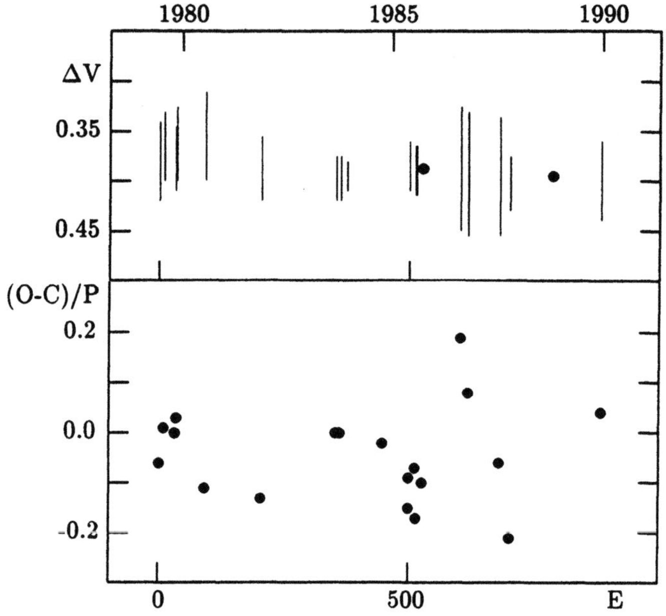

Fig. 2. The brightness changes of EV Lac from 1979 to 1990 (upper panel). Episodes of flat lightcurves are identified by filled circles. The (O–C) diagram (lower panel) demonstrates that spots were located on the same hemisphere of the star during the entire decade.

EV Lac. The brightness of EV Lac usually shows variations less than 0.1 mag. Small suppressions of the kind produced by polar spots are present during most of the past decade. Times of minima can be read off directly from the lightcurves and are used to form the (O–C) difference relative to a linear ephemeris with a period of 4.376 days (Fig. 2). The plot of $\frac{(O-C)}{P}$ versus time shows lightcurve minima to occur within 0.2 of the same phase during the entire decade. That means that spots were located preferentially on one hemisphere of EV Lac during a ten year interval.

References

Chugainov, P.F.: 1973, *Izv. Krymsk. Astrofiz. Obs.* **48**, 3
Vogt, S.S.: 1980, *Astrophys. J.* **240**, 567

The Spot Activity of FK Comae Berenices

Lauri Jetsu [1], Jaan Pelt [2], Ilkka Tuominen [1], Harold Nations [3]

[1]Observatory and Astrophysics Laboratory, University of Helsinki, Tähtitorninmäki, SF–00130 Helsinki, Finland
[2]Tartu Astrophysical Observatory, Tóravere, Estonia, USSR
[3]Franklin and Marshall College, Lancaster, Pennsylvania 17604, USA

Abstract: The active regions of FK Comae Berenices show a flip-flop behaviour, *i.e* the concentrated part of spot-activity shifts exactly to the other side of stellar surface, and then remains on the same longitude for a time interval from a few years to a decade. The activity shows excellent phase coherence with respect to these two active longitudes separated 180 degrees from each other. FK Comae may provide a physical example of a non-linear dynamo, which shows surprisingly simple observational changes in the pattern of the magnetically induced spot configurations.

1. Introduction

FK Comae (HD 117555) is an extremely rapidly rotating single late-type giant (Bopp and Stencel, 1981; McCarthy and Ramsay, 1984: G2 III; $P_{\mathrm{phot}} = 2\overset{\mathrm{d}}{.}4000$; $v \sin i = 120\,\mathrm{km/s}$; $v_{\mathrm{rad}} = -21.3\pm6.2\,\mathrm{km/s}$). It is also the prototype for the definition of the FK Comae class, which originally contained two other candidates, UZ Librae and HD 199178 proposed by Bopp and Stencel (1981). A long-term photometric study of FK Comae has now been performed using basically the same methods as those presented by Jetsu *et al.* (1990a,b) for HD 199178.

2. The long-term study

The collected UBVRI photometry contains the previously published observations combined with new observations. The total time interval of the collected observations made at six observatories is nearly a quarter of a century (from 1966 to 1990). A more detailed description of the observations and the results derived from them will be published in a separate paper. This paper concentrates on one of the main results, which has been derived from the normalized magnitudes of

FK Comae. The normalization has been done by dividing the data into individual seasons, which have an average length of about one month, since during this time the light curve of FK Comae does not change significantly. The observations were normalized inside individual seasons in every UBVRI passband with exactly the same method as used for HD 199178 by Jetsu *et al.* (1990a). The flares were naturally excluded before the normalization.

The time series analysis of the normalized magnitudes was performed by the methods developed by Pelt (1983). However, there is no time series analysis method available that can deal with phase shifts of the light curve. The result presented in this paper was derived by dividing the data into separate pieces, which surprisingly gave the same value for the photometric rotation period, but at the same time indicated the presence of a phase shift. The time interval between two consecutive phase shifts seems to vary from a few years to nearly a decade. The phase shift is the simplest possible *i.e.* half a rotation period. We call this phenomenon the "flip-flop" of the photometric minimum.

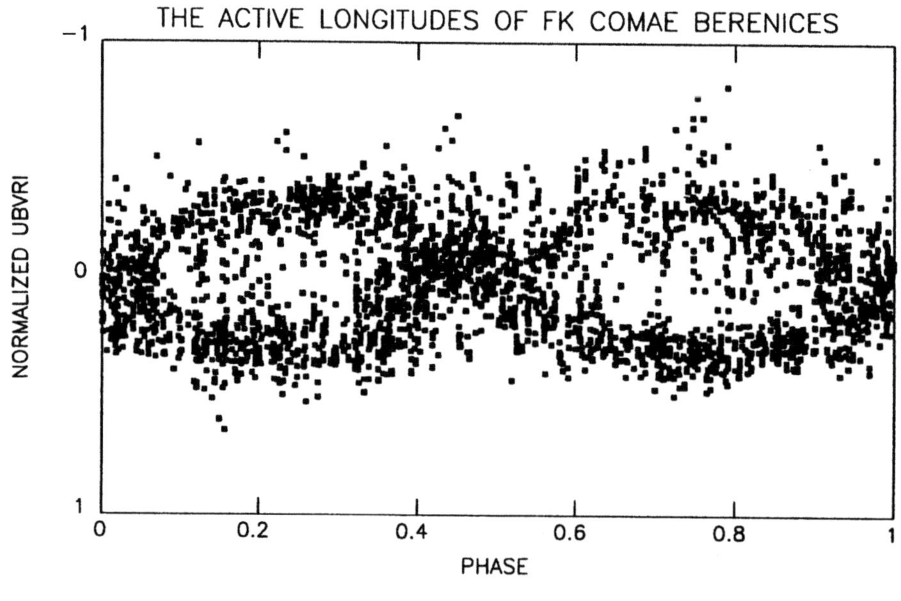

Fig. 1. The phase coherence of ϕ_{min} of all normalized UBVRI-observations from 1966 to 1990 plotted as function of phase derived from the ephemeris HJD 2439253.4375 + 2.400285 E.

3. Conclusions

Nearly a quarter of a century of photometry shows excellent phase coherence for the active longitudes of FK Comae Berenices. Inside individual observing seasons the activity has definitely been concentrated on only one of the *two active longitudes* separated by 180° from each other. The non-axisymmetric component of spot activity may suddenly shift exactly to the other side of the star after a long interval (from a few years to a decade) of phase coherence. The amount of differential rotation is very strictly limited, since the whole collected photometry can be described by one unique photometric rotation period with very small drifts of the photometric minimum. Furthermore, the "flip-flop" excludes the accretion stream model proposed by Walter and Barsi (1982) as the cause of rotational modulation of brightness of FK Comae.

The "flip-flop" phenomenon derived from the long-term photometry should give new impetus to the studies of nonlinear dynamos. It also stresses the need to develop time series analysis methods, which can deal with the phase shifts inherent in such data.

References

Bopp, B.W., Stencel, R.E.: 1981, *Astrophys. J. Letters* **247**, 131

Jetsu, L., Huovelin, J., Tuominen, I., Vilhu, O., Bopp, B.W., Piirola, P.: 1990a, *Astron. Astrophys.* **236**, 423

Jetsu, L., Huovelin, J., Tuominen, I., Bopp, B.W., Efimov, Yu.S., Linnaluoto, S., Nations, H. L., Piirola, P., Shakhovskoy, N.M., Virtanen, H.: 1990b, *Astron. Astrophys. Suppl.* **85**, 813

McCarthy, J.K., Ramsay, L.W.: 1984, *Astrophys. J.* **283**, 200

Pelt, J.: 1983 *ESA Spec. Publ.*, ESA SP-201, 37

Walter, F.M., Basri, G.S.: 1982, *Astrophys. J.* **260**, 735

On the Existence of Solar-type Activity in the Secondary Components of Cataclysmic Variables

E.S. Dmitrienko

Crimean Astrophysical Observatory, Nauchny, SU-334413, Crimea, USSR

Abstract: Cyclic variations of the parameter *O-C* (the difference between observed and calculated times of the light minimum in eclipse) have been detected for Nova 1934 DQ Her in the *UBVRI*-bands during 1982–89 with a recurrence time of about 5–6 years and amplitudes of about 2–4 min. No *O-C* oscillations for the nova-like AC Cnc have been discovered with the resolution of 45 sec in 1985–89. Both systems consist of a white dwarf and a late-type dwarf component. Some possibilities to explain their *O-C*-behaviour are discussed.

Cyclic variations of *O-C*, observed as inherent for different types of cataclysmic objects, such as polars, post novae, nova-like, dwarf novae, allow us to propose that the physical nature of this phenomenon could be uniform for these binaries. Several hypothesis are known today for the explanation of the cyclic *O-C* variations in binary systems. There are, for example, apsidal motion or the existence of a third body (Stern, 1939), eccentricity of the accretion disc around the white dwarf (WD), solar-type magnetic activity of secondaries (Bianchini, 1988; Applegate and Patterson, 1987; Warner, 1988) and swinging of WD-magnetic axis (Andronov, 1987). To choose between these hypothesis, some general properties of cataclysmic variables (CV) can be mentioned. Gas streams from the secondary form an accretion disc around a WD if the magnetic field of the latter is smaller than $10^7 - 10^8$ G. In polars there are no accretion discs. So the hypothesis of an eccentric disc as a common cause of cyclical variations of *O-C* in all CVs is not applicable. The existence of the apsidal motion can be tested by observing systems with eclipses of both components. This motion produces oscillations of *O-C* values in such a way that the primary and secondary minima are displaced with respect to each other. When a third body is present, the time-dependence of *O-C* of both minima shows oscillations only. However, in CVs the *UBVRI*-luminosity of the secondaries is so small that we can observe both minima in systems with $P > 6^d$ only (Patterson, 1984). No published light curves of these CVs are available and hence we have observed in 1985–1989 the eclipsing nova-like system AC Cnc, which has a period $7^h\ 12^m$. We also examined the time-dependence of *O-C* in Nova 1934 DQ Her in 1982–1989, with period $4^h\ 39^m$.

The *UBVRI*-light curves of DQ Her and AC Cnc, with almost complete orbital phase coverage, have been obtained in Crimean Astrophysical Observatory using the 1.25-meter reflector. The photometer-polarimeter of Helsinki University Observatory (Piirola, 1973) was used. The observational technique is described by Dmitrienko *et al.* (1985). Here we give only the results of the *O-C* observations.

The value of *O-C* of DQ Her is decreasing with the rate $\Delta(O\text{-}C)/\Delta t = 10^{-6}$ in 1982–1989 and has cyclic variations with recurrence time of about 5–6 years and amplitude about 2–4 min (Fig. 1). No *O-C* oscillations for AC Cnc have been detected with a resolution of 45 sec in 1985–1989 for either minimum. According to our observations, there is no eccentricity in accretion disc, no apsidal motions and no third body in AC Cnc. The differing behaviours of the observed parameter *O-C* in AC Cnc and DQ Her may indicate 1) that if there is cyclic magnetic activity of the secondary in CV, the effect of this activity on the value of *O-C* can be detected only in the systems with $P < 7^h$; 2) that there exists some intrinsic difference between secondaries of cataclysmic systems with periods less and greater than 7 hours.

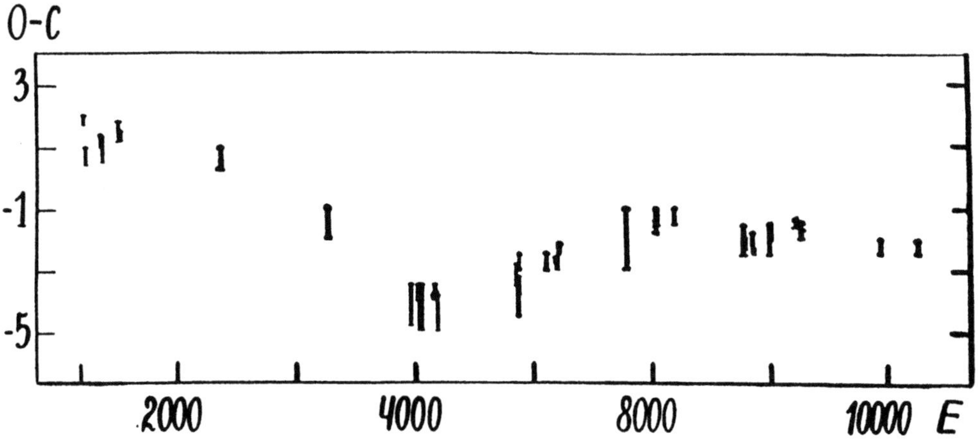

Fig. 1. The time dependence of the *O-C* for DQ Her in 1982–89. The units of *O-C* are minutes, and the units of the time the number of the double value of the orbital period. The latter is $4^h\ 39^m$.

References

Andronov, I.L.: 1987, *Astrophys. Space Sci.* **131**, 557
Applegate, J.H. and Patterson, J.: 1987, *Astrophys. J.* **332**, L99.
Bianchini, A.: 1988, *Inf. Bull. Var. Star* no 3136, 1
Dmitrienko, E.S., Efimov, Y.S., Shahovskoy, N.M.: 1985, *Astrofizika* **22**, 31
Patterson, J.: 1984, *Astrophys. J. Suppl. Ser.* **54**, 443
Piirola, V.: 1973, *Astron. Astrophys.* **27**, 383
Stern, T.E.: 1939, *Mon. Not. Royal Astron. Soc.* **99**, 451
Warner, B.: 1988, *Nature* **336**, 129

Radial Velocity Variability of K Giants

Artie P. Hatzes and William D. Cochran

McDonald Observatory, The University of Texas, Austin Texas 78712, USA

Abstract: At McDonald Observatory we have been monitoring the relative radial velocities of a sample of K giants. The technique employed uses the telluric O_2 lines near 6300 Å as a reference for measuring the stellar line shifts. We demonstrate that precisions of 10 m s^{-1} are possible with this technique. We present radial velocity data covering a 2 year time span for α Boo, α Tau, and β Gem. All of these stars show both long term variations ($\sim$ several hundred days) with a peak-to-peak amplitude of about 400 m s^{-1} and short term variations ($\sim$ few days) with a peak-to-peak amplitude of about 100 m s^{-1}. The long term variations may be due to the rotational modulation of surface active regions whereas the short term variations may be indicative of pulsations.

1. Introduction

Recent advances in techniques for measuring relative radial velocities have produced an increase in the precision of these measurements from a few tenths of km s^{-1} to a few m s^{-1}. Not surprisingly this has opened new fields of study for stellar phenomena. Recently Arcturus was shown by Smith *et al.* (1987) to be a radial velocity variable with an amplitude of several hundred m s^{-1}. This result was later corroborated by Cochran (1988). Later, Walker *et al.* (1989) found that a small sample of K giants were low-level (30–300 m s^{-1}) variables. At McDonald we have been monitoring the radial velocity of a sample of K giants. Here we present preliminary results on three K giants: α Boo (Arcturus), α Tau (Aldebaran), and β Gem (Pollux).

2. Observations

We have been employing the Griffins' telluric line technique to measure relative radial velocities (Griffin and Griffin, 1973; Cochran, 1988). This technique uses the atmospheric absorption lines of O_2 at 6300 Å as a reference for measuring stellar Doppler shifts. Data were acquired using the coudé focus of the McDonald 2.7-m telescope. An echelle grating was used in 36th order along with a Texas Instruments 800×800 CCD. This resulted in a dispersion of 0.014 Å/pixel and a resolving power of about 170,000.

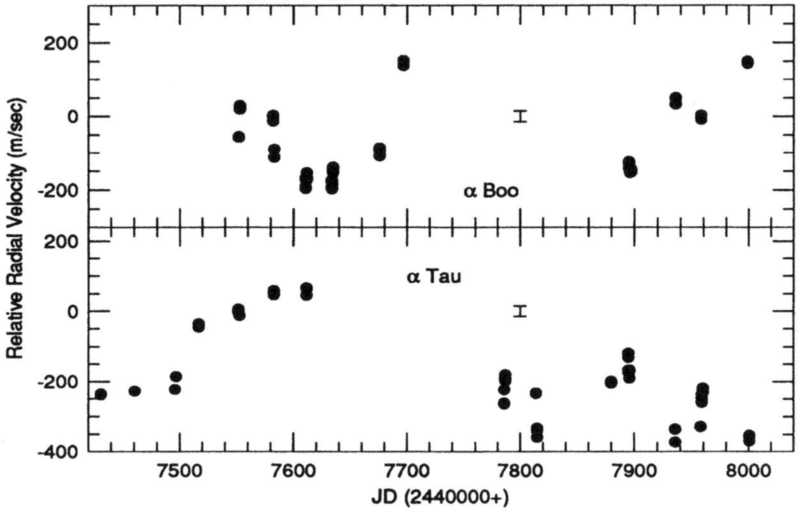

Fig. 1. Relative radial velocities for α Boo and α Tau.

3. Radial velocity measurements

The relative radial velocity measurements for α Boo, α Tau and β Gem are shown in Figs. 1 and 2. The radial velocity variations for the G8 dwarf τ Ceti are also shown in Fig. 2. The constant radial velocity of the G dwarf indicate that there are no large systematic errors in the radial velocity measurements. Typical errors for repeated measurements are about 10–15 m s^{-1} (shown as an errorbar) thus corroborating the Griffins' claim that the technique was capable of precisions near 10 m s^{-1}. All the K giants show both long term variations on timescales of hundreds of days as well as day-to-day and month-to-month variations that are larger than the estimated errors.

Arcturus exhibited long term variations in late 1988 and early 1989 (near JD 2447550) as well as in late 1989 and early 1990 (near JD 2447900). Assuming that the radial velocity curve has minima near JD 2447600 and 2447900 and maxima near JD 2447700 results in an upper limit of 250–300 days for the period of any long term variations.

Aldeberan also exhibited long term radial velocity variations in 1988/89 in the form of a steady increase of about 300 m s^{-1} within 100 days suggestive of a full period of order 200 days. This long term trend was completely absent in the measurements taken in the Fall/Winter of 1989/90.

In late 1988 and early 1989 Pollux displayed a steady increase in its relative radial velocity (amplitude $\sim$ 200 m s^{-1}). Again if this is part of harmonic variability then an upper limit of about 400 days results for the period. However, the following year such a large, long term variation in the relative radial velocity was absent from this star.

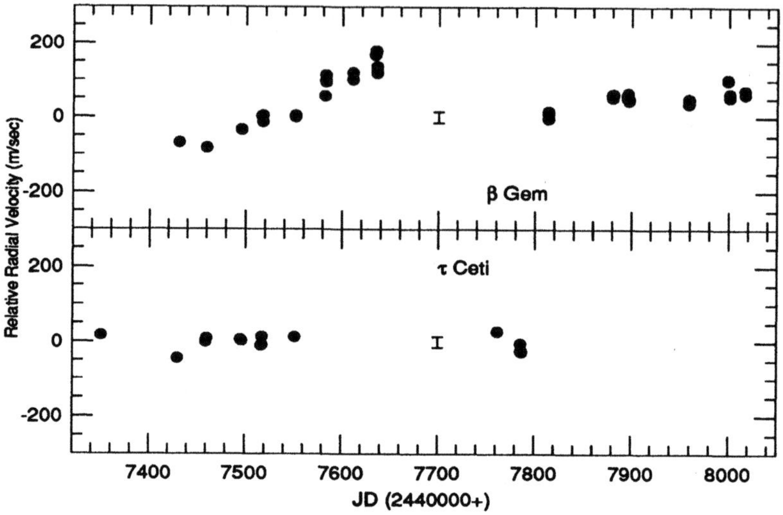

Fig. 2. Relative radial velocities for β Gem and τ Ceti (a G8 dwarf).

4. Discussion

More radial velocity data are needed before we can begin to speculate about the nature of these variations. But even with the small body of data presented here we can state with certainty that there are at least two timescales to the radial velocity variations ($\sim$ days and $\sim$ hundreds of days). It is possible that different mechanisms may be responsible for these two timescales. It is tempting to attribute the long term variations to a modulation by surface features such as active regions, plage, or spots. For instance, the measured $v \sin i$ for α Boo is about 2.7 km s^{-1}. Assuming a radius of 23 solar radii and an inclination of 90°, this yields a maximum rotation period of about 400 days, which is near the timescale of the long term radial velocity variations.

The radial velocity variations in K giants appear to be rather complicated and attributable to different pulsation modes or to a variety of mechanisms at work for producing the radial velocity variations. Clearly more data are needed before we can understand the nature of the mechanisms causing the radial velocity variations.

References

Cochran, W.D.: 1988, *Astrophys. J.* **334**, 349
Griffin, R., Griffin, R.: 1973, *Mon. Not. R. Astron. Soc.* **162**, 243
Smith, P.H., McMillan, R.S., Merline, W.J.: 1987, *Astrophys. J. Lett.* **317**, L79
Walker, G.A.H., Yang, S., Campbell, B., Irwin, A.W.: 1989, *Astrophys. J. Lett.* **343**, L21

Recent Advances in the Observation and Analysis of Stellar Magnetic Fields

Steven Saar

Harvard-Smithsonian Center for Astrophysics
60 Garden Street MS 58 Cambridge, MA 02138 USA

Abstract: There has been considerable progress recently in the study of magnetic fields on late-type stars. Advances in the theory include investigation of systematic effects, new and improved methods of analysis for unpolarized and circularly polarized lines, new models of stellar dynamos and of broadband linear polarization, and the first studies of the thermodynamic nature of stellar magnetic regions and their vertical structure. Observationally, there have been new measurements, particularly of young and active stars (including the first detection of a field on a pre-main sequence object), a new monitoring campaign, the first indications of the relative temperatures in stellar plages, and the first measurements of fields in a single stellar active region. I discuss the results in the framework of stellar activity and surface structure.

1. Introduction

Although the ink is barely dry on the most recent (last summer's) review of magnetic field measurements on cool stars (Saar, 1990a), considerable further progress has been made over the past year. Advances have been made in exploring sources of systematic error, new and improved techniques, further measurements, and in extending field measurements to new classes of objects. Partly as a result, the beginnings of a coherent picture of properties of magnetic regions on cool stars may be starting to emerge.

In this review, I will concentrate on these recent results. Integrating them with other recent work, I will update the main discoveries in the field to date, investigate trends in measurements, and summarize the clues gathered so far concerning the structure and characteristics of fields on cool stars. I also refer the reader to recent reviews by Saar (1990a), Linsky (1989), and Gray (1988).

2. Investigation of sources of systematic error

Nearly all successful methods of measuring the magnetic parameters of cool stars have followed Robinson (1980), and modeled the differential broadening of absorption lines in unpolarized light. The most simple analyses treat the effect of a magnetic field on an observed line profile, F, as the sum of two components: $F = fF_{\rm m}(B) + (1-f)\,F_{\rm q}(0)$. Here, $F_{\rm m}$ is the line profile in magnetic regions (with a mean field strength B), which are assumed to uniformly cover a fraction f of the stellar surface; $F_{\rm q}$ is the quiet region profile ($B = 0$). Lines of various magnetic sensitivities (Landé $g_{\rm eff}$) are compared to determine f and B, with low $g_{\rm eff}$ lines [where $F_{\rm q}(0) \approx F_{\rm m}(B)$] often used as an initial estimate of $F_{\rm q}$.

Of course, many simplifying assumptions must be made in the analyses due to the lack of data about the nature of the unresolved stellar surface. Two major sources of systematic error in stellar magnetic measurements are possible gradients in B with height and the unknown thermodynamic differences between $F_{\rm q}$ and $F_{\rm m}$. Up to now, studies have generally assumed that $\nabla B = 0$ and the atmospheres making up $F_{\rm m}$ and $F_{\rm q}$ are identical. Much of the disagreement surrounding the f and B values of ϵ Eri, however, can be explained by a combination of a non-zero ∇B and differences in the height of the formation of the various magnetic lines used (Grossmann-Doerth and Solanki, 1990; see also Saar, 1990a). Sun *et al.* (1986) and Saar (1988a) made some preliminary investigations of the effect of non-identical atmospheres. More recently, Basri *et al.* (1990) have studied these problems with more physically rigorous models of the Fe I 7748 Å ($g_{\rm eff}$= 1.1) and 8468 Å ($g_{\rm eff}$= 2.5) lines. They constructed theoretical profiles from combinations of the Holweger-Müller quiet solar atmosphere with various empirical plage and spot models. They find that fits to these, assuming identical atmospheres for $F_{\rm m}$ and $F_{\rm q}$, can lead to underestimates of fB by as much as 40%. Intriguingly, Basri *et al.* found they could not simultaneously fit λ7748 and λ8468 profiles computed with the plage + quiet solar atmosphere if the model assumed identical $F_{\rm m}$ and $F_{\rm q}$ atmospheres. I will discuss some implications of this in §5.

Most stellar measurements to date have used simpler atmospheric models than those of Basri *et al.* S. Solanki and I have recently been exploring the errors in the derived f and B values when realistic, self-consistent atmospheres embedded with fluxtubes are analyzed using a simple Unno (1956) type code and a Milne-Eddington atmosphere (e.g., Saar, 1988a; Bopp *et al.*, 1989). We have calculated full radiative transfer along rays passing through a two-dimensional fluxtube + quiet atmosphere model (Steiner and Pizzo, 1989), and constructed theoretical profiles by summing over all rays. For our initial tests, we computed strong, simple triplet Fe I lines with $g_{\rm eff}$ =1 and 2.5 at disk center with constant microturbulence. Since in practice, Fe/H, the oscillator strengths ($\log gf$) and the collisional broadening must usually be adjusted, we adopted the following procedure. First, the low $g_{\rm eff}$ line was fit (minimizing χ^2) with the line opacity (η_0), source function slope, and collisional broadening (Voigt α) as free parameters. These values then became initial guesses in a fit of the high $g_{\rm eff}$ line with η_0, f and B free. A final fit was then made to both lines simultaneously, with η_0, α, f, and B free. Sample

Table 1. Results of Unno-type Fits to 2-D Atmosphere Line Models with $f \approx 0.30$

quiet* atm.	mag.* atm.	λ (Å)	$\log gf$	χ_L (eV)	$B(\tau = 1)$ (G)	$B(\tau = 0.01)$ (G)	B(fit) (G)	f(fit) (%)
HSRA	NET	6000	-5.58	0	2000	970	1240	13
HSRA	NET	6000	-3.54	2	2000	970	1360	15
HSRA	NET	6000	-1.60	4	2000	970	1450	18
HSRA	NET	16000	0.07	5	2000	970	1380	17
HSRA	NET	22000	-0.70	5	2000	970	1170	15
HSRA	HSRA	6000	-5.58	0	2000	1030	1460	24
HSRA	HSRA	6000	-3.54	2	2000	1030	1510	25
HSRA	HSRA	6000	-1.60	4	2000	1030	1530	25
HSRA	HSRA	16000	0.07	5	2000	1030	1520	30
HSRA	HSRA	22000	-0.70	5	2000	1030	1350	31
K2V	K2V	6000	-5.58	0	3195	1750	2100	27
K2V	K2V	6000	-3.54	2	3195	1750	2070	33
K2V	K2V	6000	-1.60	4	3195	1750	2210	31
K2V	K2V	16000	0.07	5	3195	1750	2500	28
K2V	K2V	22000	-0.70	5	3195	1750	2240	31

*References: HSRA = Harvard–Smithsonian Ref. Atm.; NET = network model (Solanki 1986); K2V = K2 dwarf model (Basri and Marcy 1988)

results are given in Table 1, for the case of $f \approx 0.30$ at an "average" level of line formation.

A detailed interpretation of the results is not possible at present, as we have not yet calculated the "correct" B values where the lines are formed. Still, some general comments can be made. To first order, the derived B values appear to be reasonable $[B(\tau_{5000} = 0.01) < B(\text{fit}) < B(\tau_{5000} = 1)]$. Qualitatively, we obtain underestimates of f and fB similar to Basri *et al.* in the case of the solar + network model. For the same model, the magnetic flux density ($=fB$) inferred from λ6000 increases as a function of excitation potential (χ_L), consistent with a stronger contribution to the line from the (hotter) magnetic portion of the atmosphere for lines of higher χ_L (see Grossmann-Doerth and Solanki, 1990). Fits were poorer for the infrared models, where the effects of ∇B on the Zeeman sigma components were evident. We plan to extend this work in the near future, studying center-to-limb effects and disk-integrated models with various analysis methods.

3. Improved analysis methods and new measurements

As the above analyses suggest, one way to investigate the differences between $F_{\rm m}$ and $F_{\rm q}$ (which are not even well determined in the Sun; *e.g.*, Stenflo, 1989) is to analyze lines with various $\chi_{\rm L}$ and $\log gf$, thus probing different heights of formation and sensitivities to density and temperature. An approximate analysis along these lines has been completed by Mathys and Solanki (1989). They applied the Stenflo and Lindegren (1977) method (which determines f and B from a regression analysis correlating line widths and strengths with $\chi_{\rm L}$ and a modified $g_{\rm eff}$) to several cool stars. For the well-studied K2 dwarf, ϵ Eri, they find evidence that the magnetic regions are somewhat warmer than the quiet photosphere. While this result represents the first direct evidence of the thermodynamic differences between $F_{\rm m}$ and $F_{\rm q}$, one would ultimately like to test the results against magnetic measurements made using more rigorous analyses (spectrum synthesis).

Valenti (1991) has recently begun such a study, using the Hamilton Echelle at Lick to obtain over 2000Å of spectrum at moderate resolution ($\approx$ 40,000). Simultaneously fitting 25 Fe I lines in the active G8 dwarf ξ Boo A, he finds $B \approx 2000$ G and $f \approx 0.20$, generally consistent with previous results (Marcy and Basri, 1989; Saar, 1990a). At the moment, Valenti (1991) is using identical, scaled Holweger-Müller atmospheres for both F_q and F_m, and a coarse grid of parameters, but eventually analysis of the fit residuals as a function of χ_L and $\log gf$ should indicate how F_q and F_m differ.

Another method of studying the quiet/active region differences has been proposed by Sánchez Almeida and García López (1991). They focus on two lines (Fe I 5247 Å and 5250 Å) which are nearly identical (*e.g.,* $\Delta \log gf \approx 0.008$) except for $g_{\rm eff}$. Subtraction of one profile from the other should thus very nearly cancel F_q, leaving only a residual ΔF_m. The authors model ΔF_m using a Holweger-Müller atmosphere, including disk-integration, known blends, and magneto-optical (MO) effects (which may be important in some cases; Landolfi *et al.* 1989). Preliminary results for stars were somewhat confusing, however, with similar magnetic fluxes found for both λ Ser (G0V, inactive) and ξ Boo A (G8V, active). The problem may lie in the model atmosphere (*cf.* Valenti, 1991) or the disk-integration: analysis of solar plage yields reasonable results, but fits to ΔF_m in solar *flux* spectra show a spuriously large fB values (Sánchez Almeida, private comm.).

Unidentified blends may also be a source of confusion, stressing the need for accurate $\log gf$ values for *all* lines and blends in the regions to be modeled. Unfortunately, the gf data are often incomplete or of insufficient accuracy. For stars with $T_{\rm eff}$ similar to the Sun, one can circumvent the problem by deriving empirical gf values from solar spectra (*e.g.,* Valenti, 1991). This strategy fails for stars significantly cooler than the Sun, however, where many lines which are weak or absent in the Sun become important. Saar *et al.* (1990) have developed a method to obtain approximate gf values for lines in cooler active stars by carefully selecting comparison stars of similar $T_{\rm eff}$ but very low magnetic activity levels (see also Saar, 1990b). In our test study, the low activity comparison HD 32147 ($P_{rot} \approx 47$ days) proved a nearly identical match to the K5Ve BY Dra star BD +26°730 once broadened to the same $v \sin i$. The only exception in over 20 Å of spectrum was at the lone high $g_{\rm eff}$ line (Fe I 6173 Å; $g_{\rm eff}$= 2.5), indicating the presence of substantial magnetic flux. The $\log gf$ values found for HD 32147 were applied to a magnetic field analysis of BD +26°730 (using the disk-integrated model of Bopp *et al.* 1989), yielding $B = 2600$ G and $f = 0.50$. I recently recalculated the BD +26°730 spectrum including MO effects, using Calamai *et al.* (1975). Changes in the 6173 Å line amounted to $\leq$0.05%, implying the derived f and B are negligibly affected by MO effects in this case. A similar, preliminary analysis of the young K2 dwarf HD 17925 yielded $B \approx 1500$ G and $f \approx 0.35$ (Saar, 1990b).

All of the preceding analyses have difficulties measuring f and B on stars where $v \sin i \geq 10$ km s^{-1}, due to problems discerning the small magnetic broadening component in lines strongly smeared by rotation. But while it may be difficult to actually separate f and B, it is still possible to see some effects of the magnetic fields. The Zeeman effect redistributes opacity away from line center, so in optically thick lines, magnetic fields will increase the equivalent width. This so-called "magnetic intensification" effect (*e.g.*, Leroy, 1962) is being employed by Basri and Marcy (1991) to search for evidence of magnetic flux on weak-lined T Tauri stars. The basic strategy is to look for deviations from expected equivalent width (for $B = 0$). They find $fB \approx 1000$ G on Tap 35, a G8 T Tauri ($v \sin i \approx 17$ km s^{-1}), the first field measurement for this important class of stars.

White light spectra (Stokes I) have almost exclusively been used for determinations of f and B on cool stars, since the mixed polarities present on stars effectively cancel the integrated circular polarization (Stokes V) in most cases (*e.g.*, Borra *et al.*, 1984). Similarly, complex magnetic topologies reduce linear polarization (Stokes Q and U) to small levels ($\leq 10^{-3}$) while complicating its interpretation (see Huovelin and Saar, 1991). Recently, however, Semel (1989) has proposed a new kind of analysis employing Stokes V profiles which can extract magnetic field information from rapidly rotating stars. The technique, dubbed "Zeeman Doppler imaging", relies on the stars' rapid rotation to separate magnetic regions of opposite polarity *in velocity*, thereby preserving a measurable residual in the integrated signal. The method thus measures the net magnetic flux in a given velocity resolution element. Donati *et al.* (1990) have employed the technique to measure a net *unipolar* filling factor of 14% at one rotational phase of the primary (K1 IVe) star in the RS CVn system HR 1099, assuming $B \equiv 1000$ G. This implies $f \geq 0.28$ for a bipolar geometry (I use this value in §4). The shape of the I profile at the position of the V profile signature indicates that the region was bright relative to the quiet photosphere.

4. Trends in the measurements

I now take the new magnetic measurements and combine them with previous data (from Saar, 1990a) to update the trends seen between f and B and various stellar parameters. The new data are compiled in Table 2; due to the aforementioned calibration problems, the results of the line subtraction method have been omitted. I have estimated the rotational period, P_{rot}, of Tap 35 from its $v \sin i$ and radius (Walter *et al.*, 1988). Convective turnover times, τ_c, were taken from Stępień (1989) for dwarfs, Basri (1987) for RS CVns, and Gilliland (1985) for Tap 35. For latter two groups, τ_c was normalized to Stępień's $\tau_c(\odot) = 13.7$ days (note, this correction was *not* applied in Saar, 1990a). Random errors in the f and B measurements are generally ≤ 0.1 dex; poorly known *systematic* effects (see §2) dominate the uncertainties. A good part of the spread in the relationships discussed below are likely due these, plus the use of differing techniques and lines. The absolute scale of f, in particular, is uncertain. Finally, fB and $\sqrt{f}B$ are better determined than either f or B separately (Gray, 1984; Saar, 1988a; Basri *et al.*, 1990).

With these caveats in mind, I first consider relationships between B and stellar parameters. Saar and Linsky (1986) first noted that B seemed to correlate well with the gas pressure equipartition field $B_{\rm eq} \propto P_{\rm gas}^{0.5}$. Values of $B_{\rm eq}$ were computed using the formulation of Saar (1990a): $B_{\rm eq} = B_\odot \times (P_{\rm gas,*}/P_{\rm gas,\odot})^{0.5}$ kG, where $B_\odot \equiv$

Table 2. Summary of Recent Magnetic Field Measurements

Star	Sp. Type	B–V	B (kG)	f (%)	Ref.	adopted B_{eq}(kG)	P_{rot} (days)	τ_c (days)	age (Gyr)
HD 10700	G8V	0.72	...	...	1	1.7	31.9	16.9	...
HD 131156 A	G8V	0.76	2.0	20	1	1.7	6.2	19.0	0.3[a]
Tap 35	G8	0.79	$fB \approx$	1.0 kG	2	1.4	≈3.5	≈40	0.008[b]
HD 17925	K2V	0.87	1.5	35	3	2.2	6.6	20.5	0.07[c]
BD +26°730	K5Ve+	1.12	2.6	50	4	2.6	1.85	20.5	0.7[d]
HR 1099	K1IVe+	0.92	≡1.0	≥14	5	1.1	2.84	≈50	...

References: [1]Valenti (1990); [2]Basri and Marcy (1990); [3]Saar (1990b); [4]Saar *et al.* (1990) [5]Donati *et al.* (1990); ages: [a]UM Stream; [b]Walter *et al.* (1988); [c]Saar (1990b); [d]Hyades

1500 G and P_{gas} is taken at some reference level in the photosphere (*e.g.*, $\tau_{5000} = 1$). I have taken $f_{\odot} = 0.02$. For Tap 35 and HR 1099, I used the main sequence B_{eq} for the appropriate spectral type and scaled this by a factor $(g_*/g_{MS})^{0.3}$ (where g is the surface gravity), since $P_{gas} \propto g^{0.6}$ (Gray, 1976). Figure 1 shows that in general, $B \leq B_{eq}$ (with the *possible* exception of some of the RS CVn data), consistent with Saar (1990a) and theoretical estimates (Galloway and Weiss, 1981; Zwaan and Cram, 1989). There is no correlation between B and f or $\sqrt{f}$, as would be expected if errors in separating f and B dominated the results.

Dynamo theories and stellar activity diagnostics strongly imply that some combination of magnetic field-related parameter(s) should be a function of the stellar rotation rate. The field strength, however, does not correlate strongly with either normalized angular velocity, Ω ($\equiv P_{rot}^{-1}$; Fig. 1) or inverse Rossby number, $\tau_c \Omega$. (The apparent increase in B for $\log \Omega > -0.7$ and $\log \tau_c \Omega > 0.5$ is likely a selection effect caused by the lack of M dwarfs detected for small Ω and the inability to measure small B values when $v \sin i$ is large).

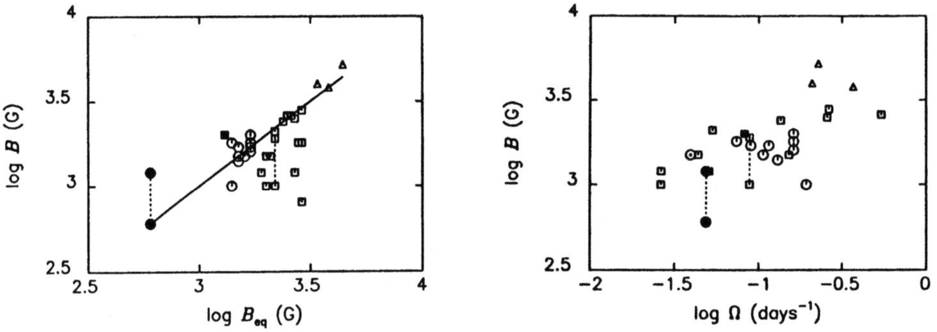

Fig. 1. B vs. $B_{eq} \propto P_{gas}{}^{0.5}$ (left) and B vs. Ω (right). Circles, squares and triangles represent G, K, and M stars; open symbols are dwarfs and filled are lower gravity stars; the Sun is denoted by $\odot$. Different measurements of the same object are connected by dotted lines and the solid line depicts $B = B_{eq}$. See text for details.

The magnetic flux density, fB, on the other hand, shows significant correlations with Ω both and $\tau_c \Omega$ (Figs. 2 and 3). Least-squares power law fits [$fB \propto \Omega^{\alpha}$ or $(\tau_c \Omega)^{\alpha}$] yield essentially linear relationships, although modified fits (*e.g.*, giving the Sun greater weight, or ignoring some K stars with more uncertain fB values; see Marcy and Basri, 1989) can increase α to 1.5 or 2.0. The correlation with $\tau_c \Omega$ yields somewhat smaller residuals. Three of the new measurements (BD +26°730, HR 1099 and Tap 35) plus AD Leo combine to give indicate a possible "saturated" fB level for log $\tau_c \Omega > 0.75$ (note that $fB_{\mathrm{HR1099}} \leq 1.0$ kG if its $B \leq B_{\mathrm{eq}}$; Donati *et al.*, 1990). Further data for rapid rotators are needed to confirm the existence of this saturation level, which has been predicted by a recent dynamo model (MacGregor and Brenner, 1989). Montesinos *et al.* (1991) have had some success matching measured fB values for G and K dwarfs by using an improved Durney and Robinson (1982) dynamo model.

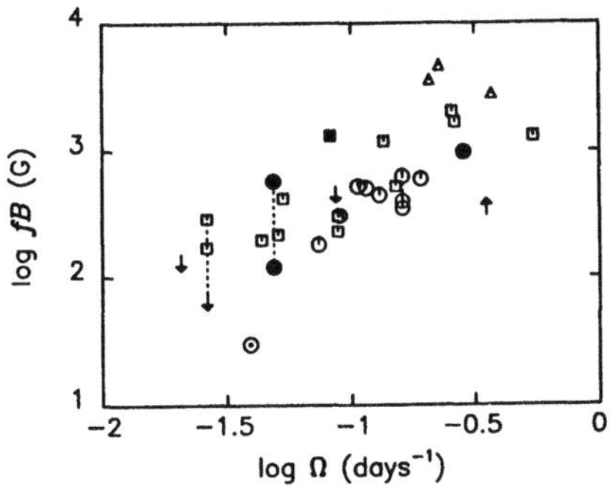

Fig. 2. fB vs. Ω (symbols as in Fig. 1; see text for details).

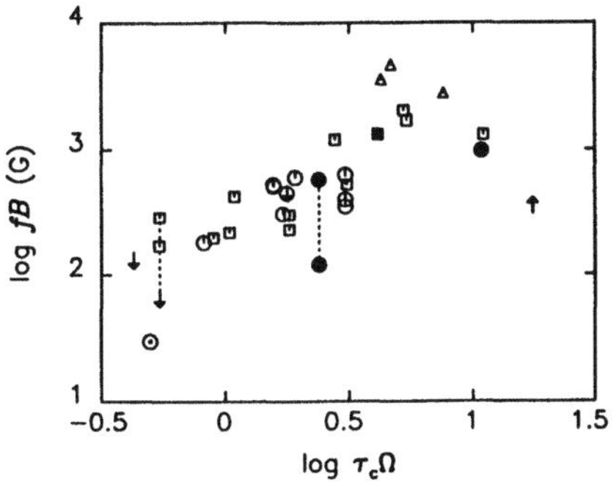

Fig. 3. fB vs. $\tau_c \Omega$ (symbols as in Fig. 1; see text for details).

Since B is apparently not a function of rotation, the correlations between rotation and fB are thus the result of a connection between rotation and f. This inference is borne out by the magnetic data (Fig. 4; where I have assumed $B_{\mathrm{Tap35}} = B_{\mathrm{eq}}$, so that $f_{Tap35} \approx 0.70$). The relationship between f and $\tau_c \Omega$ is significantly tighter than for f and Ω. A saturated level is clearly seen for log $\tau_c \Omega > 0.5$ to 0.75, roughly consistent with the $\tau_c \Omega$ at which magnetic activity "saturates" (*e.g.*, Vilhu, 1984). For log $\tau_c \Omega < 0.75$, $f \propto (\tau_c \Omega)^{\alpha}$ with $0.8 \leq \alpha \leq 1.8$ (depending on *e.g.*, how strongly the Sun is weighted). An important implication of these results is that rotation governs magnetic activity by affecting the *area coverage* of active regions and not their field strengths.

Two of the new detections (Tap 35 and HD 17925) significantly extend the range of stellar ages (t) over which f and B have been measured. The well known decline in magnetic activity with time (*e.g.*, Hartmann and Noyes, 1987) is clearly reflected in fB (Fig. 5, updating Fig. 2 from Saar, 1990b). An exponential decay law provides the best fit: fB (G) $= 840\ e^{-0.82t(\mathrm{Gyr})}$. The decline of activity with age is thus a direct result of a decline in f with stellar age. The spread of activity at a given age may then partly be due to the range of B at a fixed t.

It is natural to also look for relationships between fB and the activity diagnostics (*e.g.*, chromospheric and coronal emission) which long have been used as surrogates for the magnetic flux itself. Figure 6 (from Saar, 1990b) shows the result of an analysis of the magnetic, X-ray (F_X), and residual Ca II (ΔF_{CaII})

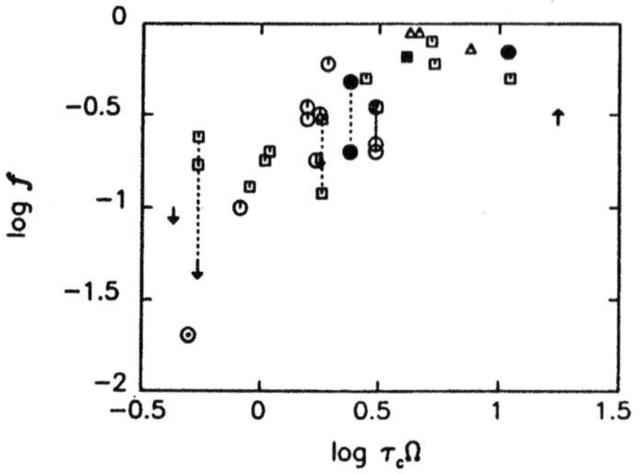

Fig. 4. f vs. $\tau_c\Omega$ (symbols as in Figure 1; see text for details).

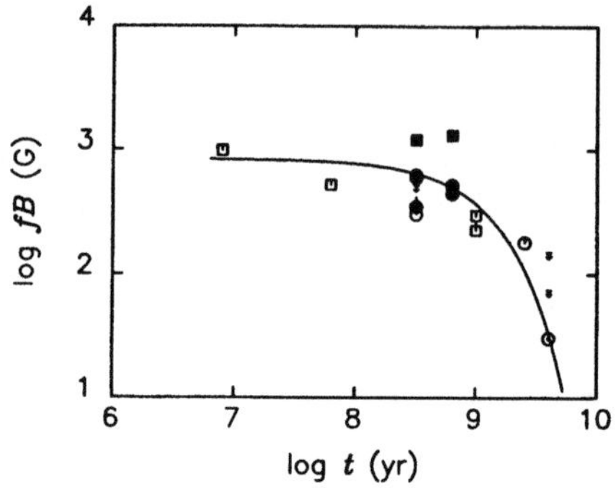

Fig. 5. fB vs. age t. Here, filled symbols represent cluster members, open symbols have ages based on Li I. An exponential fit is shown (solid; see text).

flux densities. Clear correlations are seen, which, combined with studies of the transition region (TR) C IV line (Saar, 1988b; Schrijver, 1990a) appear to define a sequence (chromosphere-TR-corona) of power laws (Schrijver, 1990b): $\Delta F_{CaII} \propto (fB)^{0.5}, F_{CIV} \propto (fB)^{0.7}$, and $F_X \propto (fB)^{1.0}$. (Including the new data for Tap 35 and HR 1099 has little effect on the derived power laws). These relations are consistent with flux-flux relations for stars *and* for fB-flux relations in spatially averaged solar regions (*e.g.*, Schrijver *et al.* 1989). Some implications for atmospheric heating are discussed in Saar (1990b) and Schrijver (1990a,b).

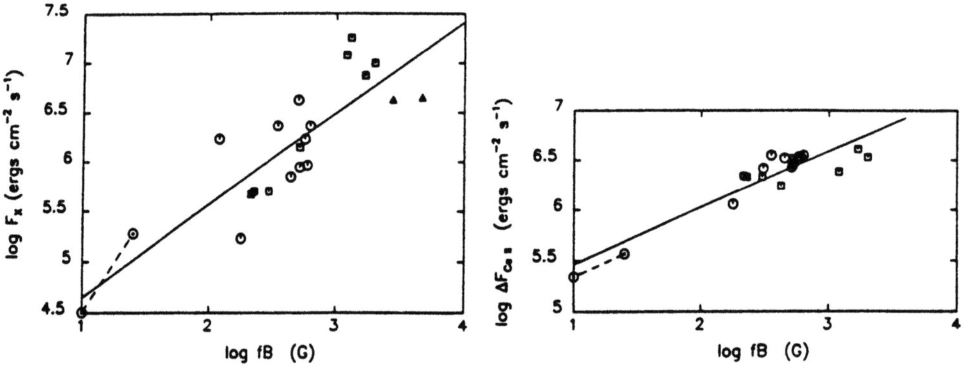

Fig. 6. fB vs. F_X (left) and ΔF_{CaII} (right). Symbols as in Fig. 1, except dwarfs and low gravity stars are not distinguished. The best fit power laws, $\Delta F_{CaII} \propto (fB)^{0.56}$ and $F_X \propto (fB)^{0.92}$, are shown (solid). From Saar (1990b).

5. Clues to the nature of stellar magnetic regions

The BD +26°730 field measurement (§3) was made during an extensive observing campaign which studied the star over several rotations in 1988 (Saar *et al.*, 1990). No variability (outside of flares) in activity or fB was detected over three weeks. This was not unexpected; the star's low inclination ($i \leq 20°$) implies that any changes in activity would almost certainly be due to *evolution* of active regions (AR) rather than rotational modulation. However, we also found no change in the level of Hα or UV line emission between our data and observations in early 1981 – despite the fact that BD +26°730 became significantly fainter (more spotted) during this period as it approached starspot cycle minimum (Hartmann *et al.*, 1981). While a lack of correlation between activity and spot coverage has been noted before in some cases (*e.g.* Butler *et al.*, 1987), BD +26°730 is the first case where f, and B and an *absolute* starspot filling factor are also available ($f_{\rm spot} \approx 0.20$; Saar and Neff, 1990).

Since starspots contribute little to Zeeman line profiles in the optical (*e.g.*, Saar, 1988a), the fields detected on cool stars must reside in "brighter" AR – the stellar analog of plage and network. This "brighter" component probably *always* dominates the total f in stars. BD +26°730 is thus nearly covered with AR, with $f \approx f_{\rm plage} = 0.50$ and $f_{\rm spot} \approx 0.20$. Perhaps because of this "saturated" level of AR (note the star's position in Fig. 4), BD +26°730 can increase $f_{\rm spot}$ during its magnetic cycle only at the expense of $f_{\rm plage}$, so that the total f and associated activity levels remain constant. The creation of more magnetic flux then only leads to the coalescence of more plage into spots (having higher B; Mullan, 1984), with no net change in observed activity. Note that with "bright" $f_{\rm plage}$ converting into $f_{\rm spot}$ and vice versa, spot cycle photometric variations are increased over the normal assumption of quiet $\Leftrightarrow$ spot, helping to explain the large magnitude changes for some stars. Also, since $i \approx 20°$, the large AR coverage must be near the poles, confirming the existence of high latitude AR on active stars (*cf.* Vogt, 1988; Hackman *et al.*, 1991). AR on stars with large f also appear to be more highly organized than on the Sun – significant unipolar concentrations of flux suggest large coherent regions (Kemp *et al.*, 1987; Donati *et al.*, 1990).

But what does it mean to have $f \approx 50\%$ or more? Clearly, the stellar surface must be considerably different from the Sun if we are to take f at face value. Solanki and Steiner (1990) suggest that the solar chromospheric magnetic canopy may be formed closer to (or in) the photosphere in active cool stars, partly explaining the large f values. The largest f values are indeed seen among K and M dwarfs, where the lowering of the canopy is predicted to be largest. These same stars are often flare stars as well, consistent with the greater probability of energetic magnetic reconnection when f is large (Saar *et al.*, 1987).

Alternately, atmospheric differences between $F_{\rm q}$ and $F_{\rm m}$ not included in the in measurements may be causing overestimates of f. Stellar plage *is* apparently warmer than the surrounding photosphere (Mathys and Solanki, 1989; Simon *et al.*, 1985; Radick *et al.*, 1988) but it is unclear just how much warmer, or how this effect impacts f determinations. Detailed comparison of quiet and active star spectra by

Basri *et al.* (1989) indicate the situation is complex. Many lines show little change between stars of very different activity levels, suggesting the differences between F_q and F_m are not large, yet some strong lines show large differences. Recall, too, that Basri *et al.* (1990) were unable to fit theoretical profiles, computed with a (hot) Solanki (1986) network model for F_m, with models having identical F_q and F_m atmospheres (see §2). The typical ease of fitting *stellar* data with the same $F_q = F_m$ models (Marcy and Basri, 1989) then suggests that AR on stars differ less from F_q than their solar counterparts. Consistent with this idea, the smaller temperature differences seen in solar plage relative to network (*e.g.*, Keller *et al.* 1990) imply that as f increases, relative F_q–F_m differences decrease. These ideas suggest that current f measurements may not need significant revision.

Another possibility, however, is that since the simple two component picture of stellar surfaces used is incomplete, proper revision of the models would reduce f (see also Marcy *et al.*, 1991). An important part of fluxtube models is a dark, non-magnetic downflowing region immediately outside (*e.g.*, Grossmann-Doerth *et al.*, 1988). Inclusion of such a component in stellar models may be useful in a number of ways. First, the dark circumtubular (CT) area could balance a hot F_m, leading to little net change in most lines or overall spectral type-color relations (Basri *et al.*, 1989; 1990). The hot F_m atmosphere might then imply a significant reduction in measured f values (due to underestimated continuum and overestimated line strengths in the $F_q = F_m$ analyses).

Cool stellar CT regions may also have an observational counterparts (*cf.* Brandt and Solanki 1990). While their restricted $g_{\rm eff}$ range and simplified magnetic analysis may have prevented detection of fB in surrounding AR (Saar, 1990a; Saar *et al.*, 1988), the "starpatch" model of Toner and Gray (1988) for the active dwarf ξ Boo A shows strong similarities to CT downflows (cool temperatures, enhanced velocities and closely associated Ca II emission; see also Toner and LaBonte, 1990). Convection is suppressed in magnetic regions on the Sun (*e.g.*, Livingston, 1982). It seems possible that on very active stars, convection is restricted over such a large area that it must be enhanced in the remaining quiet/CT regions. Such enhancements could explain greater macroturbulence values (Gray, 1984) and the larger line bisector spans and variability (Bruning and Saar, 1990; Saar and Bruning, 1990) seen in active stars. Some fraction of the "spot" coverage on stars may in fact be due to the CT medium as well (thanks to G. Marcy for suggesting this).

Finally, it is not clear at the moment how low gravity stars (especially the RS CVns) fit in to the general scheme laid out in §4 (see Donati *et al.*, 1990, and these Proceedings). They may have $B > B_{\rm eq}$ and/or lower fB than dwarfs at the same activity levels. Schrijver (1990b) has noted "over-active" stars among this class and suggests the need for additional parameters to fully describe their activity.

Many issues raised in this section remain unresolved and more questions beg to be answered. Clearly, further observations and model explorations are needed to extract more realistic f and B measurements and ultimately unravel the complex role of magnetic fields in the physics of cool stars. Significant progress has been made, however, and I think we may look forward to 1991 with hope for still more exciting revelations.

This work has been supported by a Smithsonian Institution fellowship, NASA grant NAGW-221, and NSF grant INT-8900202. I grateful to NOAO and ESO for their generous allocation of telescope time. I would also like to offer special thanks to G. Basri, J.-F. Donati, G. Marcy, B. Montesinos, J. Sánchez Almeida, C. Schrijver, S. Solanki, C. Toner, and J. Valenti, for very helpful discussions and for kindly sharing results with me prior to publication.

References

Basri, G.S.: 1987, *Astrophys. J.* **316**, 377

Basri, G.S., Marcy, G. W.: 1991, these Proceedings

Basri, G.S., Marcy, G. W., Valenti, J.: 1990, *Astrophys. J.* **360**, 650

Basri, G.S., Wilcots, E., Stout, N.: 1989, *Publ. Astron. Soc. Pas.* **101**, 528

Bopp, B.W., Saar, S. H., Ambruster, C., Feldman, P., Dempsey, R., Allen, M., Barden, S.P.: 1989, *Astrophys. J.* **339**, 1059

Borra, E.F., Edwards, G., Mayor, M.: 1984, *Astrophys. J.* **284**, 211

Bruning, D.H., Saar, S.H.: 1990, in *Cool Stars, Stellar Systems, and the Sun* ed. G. Wallerstein, Provo, ASP press, p. 165

Butler, C.J. *et al.* : 1987, *Astron. Astrophys.* **174**, 139

Calamai, G., Landi Degl'Innocenti, E., Landi Degl'Innocenti, M.: 1975, *Astron. Astrophys.* **45**, 297

Donati, J.-F., Semel, M., Rees, D., Taylor, K., Robinson, R.: 1990, *Astron. Astrophys.* **232**, L1

Durney, B.R., Robinson, R.D.: 1982, *Astrophys. J.* **253**, 290

Galloway, D.J., Weiss, N.O.: 1981, *Astrophys. J.* **243**, 945

Gilliland, R.L.: 1985, *Astrophys. J.* **299**, 286

Gray, D.F.: 1976, *Observation and Analysis of Stellar Photospheres*, Wiley, New York

Gray, D.F.: 1984, *Astrophys. J.* **281**, 719

Gray, D.F.: 1988, *Lectures on Spectral-line Analysis: F, G, and K Stars*, Arva

Grossmann-Doerth, U., Schüssler, M., Solanki, S.K.: 1988, *Astron. Astrophys.* **206**, L37

Grossmann-Doerth, U., Solanki, S.K.: 1990, *Astron. Astrophys.* , in press

Hackman, T., Piskunov, N.E., Poutanen, M., Strassmeier, K.G., Tuominen, I.: 1991, these Proceedings

Hartmann, L., Bopp, B.W., Dussalt, M., Noah, P.V., Klimke, A.: 1981, *Astrophys. J.* **249**, 662

Hartmann, L. W., Noyes, R.W.: 1987, *Ann. Rev. Astron. Astrophys.* **25**, 271

Huovelin, J., Saar, S.H.: 1991, these Proceedings

Keller, C.U., Solanki, S.K., Steiner, O., Stenflo, J.O.: 1990, *Astron. Astrophys.* **233**, 583

Kemp, J.C. *et al.* : 1987, *Astrophys. J. Letters* **317**, 29

Leroy, J.L.: 1962, *Ann. d'Astrophys.* **25**, 127

Landolfi, M., Landi Degl'Innocenti, M., Landi Degl'Innocenti, E.: 1989, *Astron. Astrophys.* **216**, 113

Linsky, J.L.: 1989, *Solar Phys.* **121**, 187

Livingston, W.C.: 1982, *Nature* **297**, 208

MacGregor, K.B., Brenner, M.: 1989, *Bull. A. A. S.* **21**, 1078

Marcy, G.W., Basri, G.S.: 1989, *Astrophys. J.* **345**, 480

Marcy, G.W., Basri, G.S., Valenti, J.A.: 1991, these Proceedings

Mathys, G., Solanki, S.K.: 1989, *Astron. Astrophys.* **208**, 189
Montesinos, B., Fernández-Villacañas, J.L., Jordan, C.: 1991, these Proceedings
Mullan, D.J.: 1984, *Astrophys. J.* **279**, 746
Radick, R.R., Lockwood, G.W., Skiff, B.A., Baliunas, S.L.: 1988, *Bull. A.A.S.* **20**, 995
Robinson, R.D.: 1980, *Astrophys. J.* **239**, 961
Saar, S.H.: 1988a, *Astrophys. J.* **324**, 441
Saar, S.H.: 1988b, in *Hot Thin Plasmas in Astrophysics*, ed. R. Pallavicini, Kluwer, Dordrecht, p. 139
Saar, S.H.: 1990a, in *IAU Symp. 138, The Solar Photosphere: Structure, Convection, Magnetic Fields*, ed. J. O. Stenflo, Kluwer, Dordrecht, p. 427
Saar, S.H.: 1990b, in *Mechanisms of Chromospheric and Coronal Heating*, in press
Saar, S.H., Linsky, J.L.: 1986, *Advances in Space Physics* **6**, No. 8, 235
Saar, S.H., Bruning, D.H. : 1990, in *Cool Stars, Stellar Systems, and the Sun* ed. G. Wallerstein, Provo, ASP press, p. 168
Saar, S.H., Neff, J.E.: 1990, in *Cool Stars, Stellar Systems, and the Sun* ed. G. Wallerstein, PASP Conf. Series, Provo, p. 171
Saar, S.H., Linsky, J.L., Giampapa, M.S.: 1987, in *Observational Astrophysics With High Precision Data*, eds. L. Delbouille, A. Monfils, U. de Liège, Liège, p. 103
Saar, S.H., Huovelin, J., Giampapa, M.S., Linsky, J.L., Jordan, C.: 1988, in *Activity in Cool Star Envelopes*, eds. O.Havnes *et al.*, Kluwer, Dordrecht, p. 45
Saar, S.H., Golub, L., Bopp, B.W., Herbst, W., Huovelin, J.: 1990, in *Evolution in Astrophysics*, ed. E. Rolfe, ESA SP-310, p. 431
Sánchez Almeida, J., García López, J.: 1991, these Proceedings
Schrijver, C.J.: 1990a, *Astron. Astrophys.* **234**, 315
Schrijver, C.J.: 1990b, in *Mechanisms of Chromospheric and Coronal Heating*, in press
Schrijver, C., Coté, J., Zwaan, C., Saar, S.H.: 1989, *Astrophys. J.* **337**, 964
Semel, M.: 1989, *Astron. Astrophys.* **225**, 456
Simon, T., Herbig, G., Boesgaard, A.M.: 1985, *Astrophys. J.* **293**, 551
Solanki, S.K.: 1986, *Astron. Astrophys.* **168**, 311
Solanki, S.K., Steiner, O.: 1990, *Astron. Astrophys.* **234**, 519
Stenflo, J.O.: 1989, *Astron. Astrophys. Rev.* **1**, 3
Stenflo, J.O., Lindegren, L.: 1977, *Astron. Astrophys.* **59**, 367
Steiner, O., Pizzo, V.: 1989, *Astron. Astrophys.* **211**, 447
Stępień, K.: 1989, *Astron. Astrophys.* **210**, 273
Sun, W.-H., Giampapa, M.S., Worden, S.P.: 1987, *Astrophys. J.* **312**, 930
Toner, C.G., Gray, D.F.: 1988, *Astrophys. J.* **334**, 1008
Toner, C.G., LaBonte, B.: 1990, *Astropys. J.*, in press.
Unno, W.: 1956, *Pub. Ast. Soc. Jap.* **8**, 108.
Valenti, J.A.: 1991, these Proceedings
Vilhu, O.: 1984, *Astron. Astrophys.* **133**, 117
Vogt, S.S.: 1988, in *IAU Symp. 132*, eds. G. Cayrel, M. Spite, Kluwer, Dordrecht, p. 253.
Walter, F.M. *et al.*: 1988, *Astron. J.* **96**, 297
Zwaan, C., Cram, L.E.: 1989, in *FGK Stars and T Tauri Stars*, eds. L.E. Cram, L.V. Kuhi, NASA SP-502, p. 215

Limits on the Magnetic Flux of a Pre-Main Sequence Star

G. Basri [1] and G.W. Marcy [2]

[1]Astronomy Department, University of California, Berkeley
[2]Dept. of Physics and Astronomy, San Francisco State University

Abstract: We attempt to detect a magnetic field on the weak T Tauri star, TAP35, via the enhanced equivalent widths of Zeeman-broadened absorption lines. We synthesize 25 Fe I lines, having a range of Zeeman sensitivities, using an LTE Stokes line-transfer calculation. The oscillator strengths of all lines are empirically determined *a priori* using the same line-transfer code applied to the spectrum of the magnetically quiet star, τ Ceti. The Fe abundance of TAP35 was established by synthesizing lines that are *insensitive* to Zeeman splitting. *We find that the equivalent widths, W_{eq}, of Zeeman-sensitive lines in TAP35 are systematically enhanced relative to the Zeeman-insensitive lines, consistent with the presence of widespread, kilogauss fields.* The excess W_{eq} can be explained by a product of field strength and surface filling factor (Bf) of 1 kiloGauss. A strong upper limit can be placed on the product of those two quantities, $Bf < 2\,\mathrm{kG}$. This measurement bears on the physics of T Tauri coronae, chromospheres, dynamos, and accretion-disk boundary layers.

1. Introduction

Pre-main sequence, solar-mass stars exhibit many characteristics suggestive of surface magnetic fields. Spectroscopically, T Tauri stars (TTS) exhibit a wide variety of emission lines such as Hα, Ca II H and K and Fe II at optical wavelengths, as well as lines of Mg II, Si I, C IV, Si IV, and He II in the UV, all reminiscent of the spectrum of active magnetic regions on the sun (cf. Herbig, 1970; Imhoff and Giampapa, 1982). These lines form above the photosphere in gas at 10^4 to 10^5 K that is heated by some non-radiative process, perhaps related to magnetic fields.

Several efforts failed to reproduce theoretically the width and strength of these lines, notably broad Hα, Ca II and Mg II, in the context of a "deep chromosphere" (Cram, 1979; Calvet *et al.*, 1984), thus casting doubt on magnetic fields as the energy source for the heating. Doubt also arises owing to the lack of correlation between Hα emission and rotation in TTS as would be expected if a dynamo generates the magnetic fields (Hartmann *et al.*, 1986). Clearly, lines of Hα, Ca II, and Mg II, when broad ($\sim 500\,\mathrm{km\,s^{-1}}$), originate in a more extended circumstellar region.

However, in the weak-emission T Tauri Stars (WTTS) the emission lines are narrow ($< 50\,\mathrm{km\,s^{-1}}$) and deep chromosphere models are successful in reproducing the observed line fluxes and continua. The WTTS UV line fluxes are typically

10–50 times solar levels, suggesting a corresponding enhancement in the magnetic fields on WTTS relative to the Sun, on which only about 2% of the surface is covered by kiloGauss fields. Thus one might expect TTS to contain kiloGauss magnetic fields covering a large fraction of their surfaces. Other indicators of surface magnetic fields on TTS include their X-ray properties (Walter, 1987), optical photometric variations, suggestive of starspots (e.g. Vrba *et al.*, 1989), and "flares" (Cohen and Beiging, 1986). Magnetic fields on TTS have been invoked to explain local phenomena such as Alfvén waves (Hartmann *et al.*, 1982), angular momentum loss (Frank Shu, private comm.) and boundary-layer turbulence (Basri, 1990).

Despite the many signatures of widespread surface magnetic fields on TTS, fields have not been detected. Spectropolarimetric efforts have been unsuccessful, perhaps because the fields have opposite polarities locally on the surface, thereby suppressing the net longitudinal field (Brown and Landstreet, 1981; Johnstone and Penston, 1986, 1987). One cannot detect Zeeman broadening of photospheric lines because of the dominance of rotational broadening ($v \sin i \sim 20\,\mathrm{km\ s^{-1}}$), though this approach may work if a bright, pole-on TTS could be found.

Here we implement a Zeeman technique that makes use of the enhancement of the equivalent widths of absorption lines owing to the splitting of the line-absorption coefficient. The technique involves observing absorption lines of different Zeeman sensitivities in the WTTS, TAP35, and in the comparison star, τ Ceti, in an effort to detect systematic enhancement of the Zeeman-sensitive lines.

2. Observations

Spectra were obtained with the Lick Observatory coude echelle spectrometer ("Hamilton") and CCD which has a resolution of 40,000 and has a wavelength coverage from 5000 Å to 8600 Å. This setting includes some lines used in past Zeeman work, notably, λ8468 and λ6173, as well as several lines having low Landé g values. The target TTS, TAP35 (=NTTS 042417+1744, V=11, Walter *et al.*, 1988) was chosen because it has both weak Hα and no IR excess. Thus its absorption lines are not contaminated by emission from a stellar wind or circumstellar disk. In choosing TAP35, we have avoided selection criteria that would bias against the presence of magnetic fields. Its rotation velocity ($v \sin i = 17$ km s^{-1}) falls near the average for TTS (Hartmann *et al.*, 1986) as does its X-ray luminosity (Walter *et al.*, 1988).

Four exposures, totaling 3.6 hours, were obtained on the 3-m telescope for TAP35, yielding a final S/N ratio in the summed spectrum of ~100:1, slightly higher in the near IR than in the yellow region. An exposure of the chromospherically quiet star, τ Ceti (G8V) was obtained with the same setting. The data were reduced in a standard way: the raw CCD images were divided by normalized flat-field exposures of an incandescent lamp, corrected for scattered light by subtracting the signal between orders, mashed into 54 individual spectral orders, and wavelength-calibrated using exposures of a Th-Ar lamp.

We searched the entire echelle format for Fe I absorption lines amenable to precise measurement of equivalent width, rejecting lines that were either too weak or blended. This task was complicated by the rotational broadening of TAP35 which both promotes blending and depresses the pseudo-continuum in regions of numerous weak lines. The final selection contains 25 Fe I lines, ranging in excitation potential from 1.0 to 4.5 eV, and in landé value from 0.0 to 2.5. We measured the equivalent width of each line in both TAP35 and τ Ceti by the following procedure. For TAP35, the continuum level was determined by eye on both sides of the profile, and a straight line was drawn between the two, under which the area of the line was computed by numerical integration. We preserved plots showing the regions used for all continuum placements so that identical placement could be employed for τ Ceti, thereby minimizing human bias in the measurements for the two stars. (We found that any "automatic" continuum determination occasionally erred owing to the large rotational broadening of TAP35.) Before carrying out the measurement of equivalent widths for τ Ceti, we convolved its spectrum with the rotational broadening function (Gray, 1974) with $v \sin i = 17\,\mathrm{km\ s^{-1}}$ so that the continuum placement and weak blends affected the measurements in the same way as TAP35. Indeed, the resulting broadened spectrum of τ Ceti was remarkably similar to that of TAP35, though the Fe lines were clearly weaker in τ Ceti, an old disk star.

3. Zeeman analysis

3.1 Basic approach

The basic idea in this Zeeman analysis is to detect an enhancement of the equivalent widths of Zeeman-sensitive lines over those expected if no magnetic field existed. We model the line-transfer of all 25 lines self-consistently to predict their equivalent widths as a function of the surface magnetic field strength, B. The line-transfer technique has been described previously in detail (Basri and Marcy, 1988). Briefly, we calculate the LTE line transfer in all Stokes parameters, thus treating the Zeeman-induced polarization across each line profile. We construct model atmospheres for both TAP35 and τ Ceti by scaling $\mathrm{T}(\tau)$ of the solar atmosphere (Holweger and Müller, 1974) to the respective effective temperatures (see below) of the two stars and recomputing the densities and ionization equilibria to satisfy self-consistently hydrostatic equilibrium and the Saha equation (Gray, 1976).

3.2 Oscillator strengths

In principle, the proper procedure is to adopt laboratory gf values (Fuhr *et al.*, 1988; Blackwell *et al.*, 1982) and to construct a model stellar atmosphere which minimizes the differences between observed and calculated equivalent widths. Instead, we construct a model atmosphere for τ Ceti, scaled from the Holweger-Müller solar model, and calculate empirical gf values by forcing agreement between calculated and observed equivalent widths.

To do this, we adopt atmospheric parameters for τ Ceti from the fine analysis by Arribas and Crivellari (1989). However, we find that using their value of Fe abundance and lab gf values yields calculated values of W_{eq} that are systematically too low by ~20%. A similar effect in the same direction was found by Arribas and Crivellari when they used the Holweger-Müller solar model. So, we have derived a new Fe abundance for τ Ceti which, when using lab gf values and scaled Holweger-Müller atmosphere, yields calculated equivalent widths that agree on average with those observed. The final Fe abundance for τ Ceti is: $\log \mathrm{Fe/H} = -4.81$ (solar case, $\log \mathrm{Fe/H} = -4.5$). We do not recompute the atmosphere itself for τ Ceti using this lower abundance since it makes little difference in predicted equivalent widths (Arribas and Crivellari, 1989). Final gf values were then determined by trying a range of values for each line and determining the value that gave exact agreement between calculated and observed equivalent width for τ Ceti. We note that these empirical gf values were found to be correlated with line strength, probably indicating an error in the Holweger-Müller atmosphere model. (Thanks are due to J. Valenti for pointing this out.)

3.3 Zeeman enhancement in TAP35

We computed the line transfer in TAP35 using an atmosphere and Fe abundance determined from *the 10 lines having lowest sensitivity to the Zeeman effect.* We computed the equivalent widths of these lines for a grid of cases spanning T_{eff} and Fe/H, and determined (by interpolation) the values that yielded the best fit to the observed equivalent widths. The resulting values were: $T_{\mathrm{eff}} = 5250 \pm 100\,\mathrm{K}$ and $\log \mathrm{Fe/H} = -4.21$. Thus, TAP35 is metal r ich, [Fe/H]=+0.29, and has an effective spectral type of G8. We note that the determination of T_{eff} was greatly aided by the following diagnostic. We demanded that the residuals between observed and calculated W_{eq} be uncorrelated with excitation, χ_{ex}, since any such correlation implies a systematic error in photospheric temperature.

We now consider all 25 lines and compute the equivalent widths for TAP35 for a range of assumed magnetic field strengths. Surprisingly, the Zeeman enhancement of equivalent width does not monotonically increase with Landé factor, g_{eff}, or with $\lambda^2 g_{\mathrm{eff}}$. The reason for this is that a complicated saturation effect takes place when multiple Zeeman components overlap, as is the case for $B = 0\,\mathrm{G}$. As one considers increasing field strengths, the Zeeman components begin to separate, diminishing the saturation. The rate of growth (with increasing B) of equivalent width is greatest when the magnetic field is strong enough for the components to become partially resolved. The critical issue is the separation of components having the same polarization. The result is that Zeeman W_{eq}–enhancement depends not only on Landé factor but also on the distribution of many π and σ Zeeman components over the profile thereby modulating the polarization saturation. Ironically, the simple Zeeman triplet, so useful for circular–polarization and Zeeman–broadening studies, experiences no further enhancement in W_{eq} once the field strength is such that the components are fully separated.

Because of these effects, we develop a theoretical sensitivity index for Zeeman equivalent-width enhancement, defined as the ratio of the equivalent width for B=1000 G to that for B=0 G. This index depends (weakly) on the atmosphere under consideration, so we determine this index using the TAP35 atmosphere. Lines of highest sensitivity have equivalent-width ratios of 1.15–1.20 (i.e., $\lambda\lambda$ 6173,

7491, 8468), thus indicating that increases of 15% to 20% will occur as the field is increased from 0 to 1 kG. This demonstrates the ease with which strong, kG fields can be detected by this technique.

For TAP35, we plot in Fig. 1 the ratio of observed to theoretical ($B = 0$) equivalent widths versus the above Zeeman sensitivity index. The scatter in $W_{eq}(obs)/W_{eq}(B = 0)$ is about 10% (as seen in the lower left corner) and is due to three sources of error: measurement of equivalent widths, use of a scaled solar atmosphere for a star that undoubtedly has a perturbed photosphere, and our inadequate determination of gf values. These sources of error are all unrelated to the Zeeman sensitivity of the individual lines. A weak trend is apparent in the plot such that the lines of highest Zeeman sensitivity indeed have observed W_{eq} systematically larger than predicted for $B = 0$. The correlation coefficient, 0.55, implies a probability of less than 1% that the data are uncorrelated. We therefore conclude that TAP35 has a magnetic field exhibited by the enhancement in W_{eq} of the most Zeeman-sensitive lines.

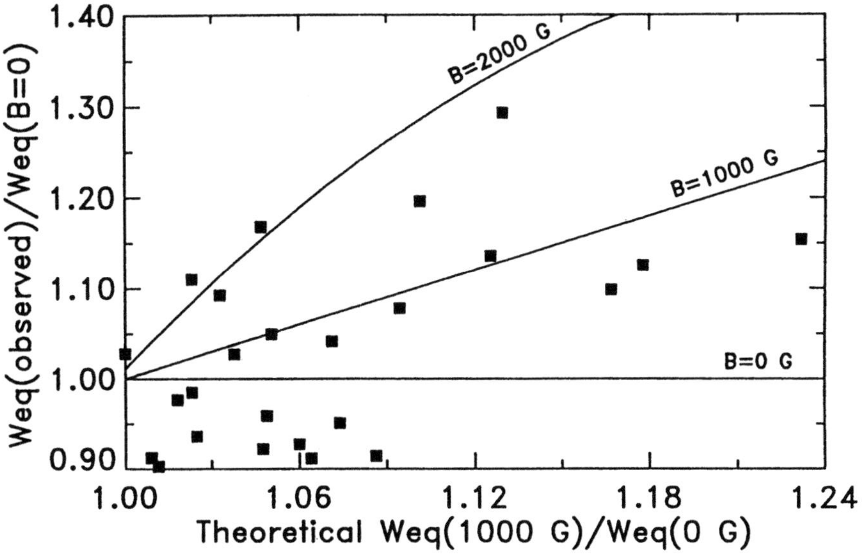

Fig. 1. Ratio of observed W_{eq} to that predicted for $B = 0$ vs. sensitivity to Zeeman-enhancement, for 25 lines in the T Tauri Star, TAP35. Lines show expected (theoretical) locus of points for $B = 0$, 1000 and 2000 G.

We now calculate the locus of points expected on Fig. 1 for three different magnetic field strengths, 0, 1000 , and 2000 G. This is done by synthesizing each line and assuming that the *entire surface* is covered by a specific field strength. The resulting theoretical loci are shown on Fig. 1. The seven most Zeeman-sensitive absorption lines (located in the right half of the figure) all lie above the locus line representing $B = 0$, supporting the suggestion of magnetically-induced W_{eq} enhancement. These seven points, however, all sit below the locus representing $B = 2000$ G, thus suggesting that the field is less than 2000 G, assuming a filling factor of 100%. Preliminary tests show that the quantity best revealed by this technique of Zeeman-analysis is the product of strength and filling factor, Bf. Thus, Fig. 1 suggests that 0.5 kG $< Bf <$ 2 kG.

4. Conclusions

The enhancement of equivalent widths of Zeeman-sensitive lines over insensitive lines serves as a diagnostic of magnetic fields, especially for stars having large rotation rates or uncertain atmospheres. Application to the weak T Tauri star, TAP35, reveals such Zeeman enhancement, and is consistent with an unsigned surface flux, $Bf \approx 1\,\mathrm{kG}$. Future refinements will include improved gf values and semi-empirical model photospheres for T Tauri stars.

We would like to acknowledge J.A. Valenti for his assistance. We acknowledge partial support for this research from NSF grants AST-8919634 (to GM) and AST-8911596 (to GB).

References

Arribas, S., Crivellari, L.: 1989, *Astron. Astrophys.* **210**, 211

Basri, G.: 1990, in *High Resolution Spectroscopy in Astrophysics*, a special issue of Memorie S. A. Ital., ed. R. Pallavicini

Basri, G., Marcy, G.W.: 1988, *Astrophys. J.* **330**, 274

Blackwell, D.E., Petford, A.D., Simmons, G.J.: 1982, *Mon. Not. R. Astron. Soc.* **210**, 595

Brown, D., Landstreet: 1981, *Astrophys. J.* **246**, 899

Cram, L.: 1979, *Astrophys. J.* **234**, 949

Calvet, N. *et al.*: 1984, *Astrophys. J.* **277**, 725

Cohen, M., Beiging, J.H.: 1986, *Astron. J.* **92**, 1396

Finkenzeller, U., Basri, G.: 1987, *Astrophys. J.* **318**, 823

Fuhr, J.R., Martin, G.A., Wiese, W.L.: 1988, *J. Phys. and Chem. Ref. Data* **17**, No. 4

Gray, D.F.: 1976, in *The Observation and Analysis of Stellar Photospheres*, Wiley

Hartmann, L., Hewitt, R., Stahler, S., Mathieu, R.D.: 1986, *Astrophys. J.* **309**, 275

Herbig, G.H.: 1970, *Mem. Roy. Soc. Sci. Liege* **19**, 13

Holweger, H.: 1967, *Z. Astrophys.* **65**, 365

Holweger, H., Müller, E.A.: 1974, *Solar Phys* **39**, 19

Imhoff, C., Giampapa, M.: 1982, in *Advances in Ultraviolet Astronomy: four Years of IUE Res.*, ed. Y. Kondo, J.M. Mead, R.D. Chapman (NASA CP 2338), p. 456

Johnstone, R.M., Penston, M.V.: 1986, *Mon. Not. Roy. Ast. Soc.* **219**, 927

Johnstone, R.M., Penston, M.V.: 1987, *Mon. Not. Roy. Ast. Soc.* **227**, 797

Kurucz, R.L., Furelid, I., Brault, J., Testerman, L.: 1984, *Solar Flux Atlas from 269 to 1300* nm, National Solar Observatory Atlas No. 1

Marcy, G.W., Basri, G.: 1989, *Astrophys. J.* **345**, 480

Saar, S.: 1988, *Astrophys. J.* **324**, 441

Shu, F.: 1990, private communication

Vrba, F.J., Rydgren, A.E., Chugainov, P.F., Shakhovskaya, N.I., Weaver, W.B.: 1989, *Astron. J.* **97**, 483

Walter, F.M.: 1987, *Pub. Ast. Soc. Pacific* **99**, 31

Walter, F.M.: 1988, Astron. J. **96**, 297

Stellar Zeeman Analyses: Effects of Multi-component Atmospheres

G.W. Marcy, G. Basri, J.A. Valenti

Astronomy Department, Univ. of California, Berkeley, CA 94720, USA

Abstract: We simulate the radiative transfer of Zeeman-broadened lines in an active star containing both quiet and flux-tube regions. A traditional one-atmosphere Zeeman analysis of these synthetic lines yields magnetic field measurements systematically inaccurate by up to 40%. However, two atmospheric components alone cannot reproduce the observed line strengths, luminosities or photometric constancy of active stars.

1. Introduction

Previous magnetic analyses of cool dwarfs (e.g., Saar, 1988; Marcy and Basri, 1989) involve fitting observed Zeeman-broadened line profiles with synthetic profiles computed by assuming that the stellar atmosphere has only one component, thus ignoring the different depth structures of flux-tubes, spots, and quiet regions (Mathys and Solanki, 1989).

To estimate the errors in the deduced magnetic field strength, B, and surface filling factor, f, caused by this oversimplification, we construct model stellar atmospheres consisting of two components, quiet and flux-tube, based on solar models. We then synthesize, in LTE, Zeeman-sensitive and insensitive profiles (λ8468 and λ7748) from this two-component atmosphere, and "analyze" them for Zeeman broadening using only a one-component quiet atmosphere.

2. Model stellar flux-tube atmospheres

We considered three models of the depth structure of magnetic flux tubes on active G2 stars, each based on solar models:

Tube A:	HSRA evacuated by 50%, $T(h)$ preserved, and Solanki (1986) solar flux tube used for $\log \tau_{5000} < -2$
Tube B:	Walton (1987) solar active region model (P1)
Starspot:	Maltby *et al.* (1986), Umbral Sunspot

For the three model atmospheres and the quiet HSRA, we calculated intensity profiles for two Fe I lines λ8468 (Landé $g = 2.5$, $\chi_{\rm ex} = 2.2$ eV) and for λ7748 (Landé $g = 1.1$, $\chi_{\rm ex} = 2.9$ eV). The gf values were taken from Blackwell *et al.* (1982). Line transfer is done by numerically solving the equation of transfer for each Stokes parameter in the model atmosphere, thus accounting for the Zeeman effect (Basri and Marcy, 1988). The important result is that the intensity profile shapes and equivalent widths that emerge from each atmosphere are significantly different, as shown in Figure 1.

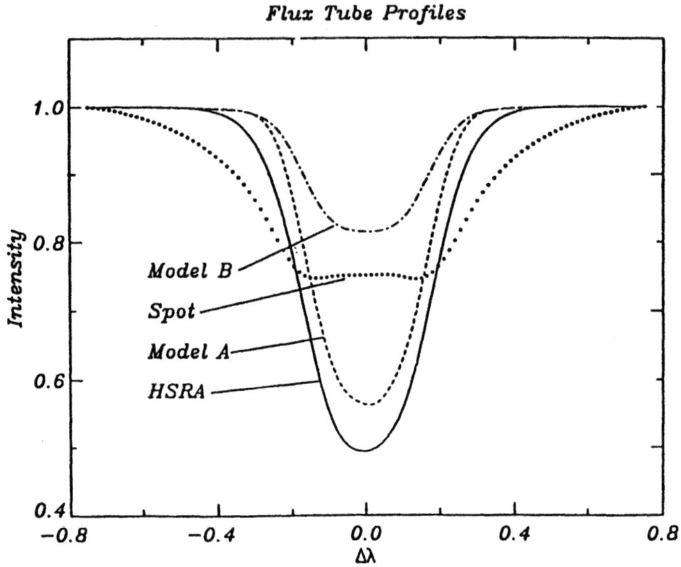

Fig. 1. Theoretical profiles for λ8468 from four model atmospheres assuming $B = 0\,G$ (but $B = 2000\,G$ for spot).

3. Two-component G-dwarf atmospheres

For each theoretical two-component atmosphere, we construct disk-integrated profiles for λ8468 and λ7748 by adding the calculated flux profiles (weighted by continuum intensity) from the flux-tube and quiet regions in the proportion 30%:70%. Thus, the three atmospheres are:

1. 30% Tube A (1250 G) and 70% HSRA
2. 30% Tube B (1250 G) and 70% HSRA
3. 30% Starspot (2000 G) and 70% HSRA

The resulting composite profiles for λ8468 and λ7748 represent profiles that would be observed if magnetic regions on active G2 stars had atmospheres similar to solar flux tubes. However, in practice Zeeman broadening is determined without knowledge of flux-tube atmospheres; it is usually assumed that the atmosphere is quiet everywhere. Accordingly, we now determine B and f that would be derived from our "observations" of a two-component atmosphere if only a one-component HSRA atmosphere were used in the analysis.

4. One-component analysis

Procedure:
1.Use HSRA ($B = 0$) to synthesize λ7748 (Zeeman insensitive). Fit to "observed" two-component λ7748. Free parameters are [Fe/H] and macroturbulent velocity.
2.Using HSRA, [Fe/H], and $V_{\rm mac}$ from step 1, synthesize λ8468. Fit to "observed" two-component λ8468. Free parameters are B and f.
3.Repeat step 1, but start with B found in step 2. Iterate.

The above procedure gives the values of B and f that would be obtained using a single quiet atmosphere to synthesize Zeeman-broadened profiles. Results are given in Table 1. The above procedure was repeated, but the cores of λ8468 were linearly scaled before fitting with B and f, simulating the scaling procedure used by some investigators. Results are in the third section of Table 1.

Comments:
1.$\sim$ 50% errors in B and f are made in a one-component analysis.
2.Model B (pure tube model) forces low Fe/H, due to line weakening.
3.Model A (evacuated plage) is milder: one-component analysis yields smaller errors.
4.Scaling cores slightly improves the performance of one-component analysis.
5.Errors in B and f are anti-correlated: Bf has smaller fractional error.

Table 1. Parameters required to achieve best fit to observations using a one-component atmosphere

Input 2–Comp.			1–Comp. Fit			Scaled 1–Comp.			
Model	B	f	B	f	$\sigma_{\rm fit}$	B	f	$\sigma_{\rm fit}$	[Fe/H]
Tube A	1250	30%	1500	27%	0.77%	1250	36%	0.31%	−0.04
Tube B	1250	30%	1900	16%	0.48%	1500	20%	0.22%	−0.38
Starspot	2000	30%	2500	12%	0.39%	1000	9%	0.39%	0.00

5. Two-component K-dwarf atmospheres

Crude two-component atmospheres have also been constructed for a hypothetical K2 dwarf, by scaling the solar flux-tube atmospheres. A Zeeman analysis of the calculated two-component profiles was done using only a quiet K2 dwarf atmosphere. Results are similar to those for G2 dwarf: We find significant errors in B and f, ($\sim$50%) that are anti-correlated. The K-dwarf tube B model (30% tube B + 70% quiet) again yields very weak profiles, and in this case was not be fit by profiles from a quiet K2 atmosphere.

Table 2. Parameters of best-fit Zeeman profile using quiet K atmosphere only

Input 2–Comp.			1–Comp. Fit			Scaled 1–Comp.			
Model	B	f	B	f	$\sigma_{\rm fit}$	B	f	$\sigma_{\rm fit}$	[Fe/H]
Tube A	1250	30%	1000	47%	0.48%	1000	53%	0.49%	0.00
Tube B	1250	30%	....			No fit attempted			....
Starspot	2000	30%	2500	9%	0.52%	2000	12%	0.41%	−0.06

6. Conclusions

1. Systematic errors in B and f occur when line profiles from two-component stellar atmospheres are Zeeman-analyzed using one atmosphere.
2. Errors in B and f are anti-correlated; error in Bf is less than 20%
3. Filling factors tend to be underestimated using a one-component atmosphere.
4. Two components are not enough to explain active stars.
   a. Including flux-tubes in a quiet atmosphere yields lines weaker than observed.
   b. Including flux-tubes also predicts active stars *bluer* than inactive stars – not seen.
   c. Luminosity would not be preserved, since tubes are twice as bright .
   d. If flux-tubes are non-uniformly placed over surface, expect optical brightness variations of $\sim 1/2$ mag – not seen; observed mod. ~ 0.02 mag
5. Active stars must have a third, cool atmospheric component that is spatially associated with flux-tubes to resolve issue 4.

References

Basri, G., Marcy, G.W.: 1988, *Astrophys. J.* **330**, 274

Blackwell, D.E., Petford, A.D., Simmons, G.J.: 1982, *Mon. Not. R. Astron. Soc.* **210**, 595

Maltby, P., Avrett, E.H., Carlsson, M., Kjeldseth-Moe, O., Kurucz, R.L., Loeser, R.: 1986, *Astrophys. J.* **306**, 284

Marcy, G.W., Basri, G.: 1989, *Astrophys. J.* **345**, 480

Mathys, G., Solanki, S.K.: 1989, *Astron. Astrophys.* **208**, 189

Saar, S.: 1988, *Astrophys. J.* **324**, 441

Solanki, S.K.: 1986, *Astron. Astrophys.* **168**, 311

Solanki, S.K., 1987, in *The Role of Fine-Scale Magnetic fields on the Structure of the Solar Atmosphere*, ed. E.-H. Schroter, M. Vazquez, and A.A. Wyller (Cambridge University Press), p. 67

Walton, S.R., 1987, *Astrophys. J.* **312**, 909

Multi–line Zeeman Analysis

Jeff A. Valenti

Astronomy Department, Univ. of California, Berkeley, CA 94720, USA

Abstract: I present preliminary results of a multi–line Zeeman analysis of the active star ξ Boo A and the inactive star τ Cet. High S/N echelle spectra are compared with LTE model line profiles for 25 Fe I lines to deduce surface magnetic fields.

1. Introduction

Magnetic field strengths on cool stars are typically measured by searching for Zeeman broadening of a single magnetically sensitive spectral line (Robinson, 1980; Marcy, 1984). The accuracy of such measurements is limited not by the precision of the data, but by uncertainties in analysis. Numerical radiative transfer is required to properly interpret line shapes (Basri and Marcy, 1988), so accurate model atmospheres are imperative. Previous analyses have usually been based on single spectral lines (Saar, 1988; Marcy and Basri, 1989) and hence are subject to uncertainty. By analyzing many lines formed at a variety of atmospheric depths, the accuracy of inferred model atmospheres and resultant magnetic field determinations can be ascertained. Multi–line analysis also permits the study of multi–component atmospheres and vertical field gradients.

2. Data

Multi–line studies of magnetic fields require high S/N, high resolution spectra covering a large wavelength range. Using the Hamilton echelle spectrograph at Lick Observatory, I obtained spectra with S/N of 500/1, resolution of 40,000, and wavelength coverage of 2380 Å in the range 4929–8917 Å. The spectra were corrected for scattered light, fringing, and instrumental profile effects, and the correction procedures were tested by comparing Hamilton day sky spectra with the NSO solar flux atlas (Kurucz *et al.*, 1984). In the lines the disagreement exceeded the noise by 0.7%, due to a poorly determined instrumental profile.

Since the Hamilton has substantially poorer resolution than is typically used in magnetic field work, it is necessary to establish the Hamilton's ability to detect magnetic fields. Solar plage provide a good test case, as they are believed to have magnetic properties similar to those of active cool dwarfs as a whole. I obtained solar intensity spectra of quiet and plage regions, which should have weak and strong magnetic fields respectively. The low Landé g, magnetically insensitive 7583.8 Å line profiles for the two regions agree to 0.3%; whereas, the high Landé g, magnetically sensitive 8468.4 Å line profiles differ systematically by as much as 4%. The plage profile is broader as expected. Thus echelle spectrographs such as the Hamilton are good instruments with which to study magnetic fields, as they sample many spectral lines while maintaining the required accuracy.

3. Models

I modeled each Stokes component in a fully depth dependent scaled solar atmosphere (Holweger and Müller, 1974) under the assumption of local thermodynamic equilibrium. Line profiles based on laboratory oscillator strengths (Fuhr *et al.*, 1988) disagreed with the NSO solar flux atlas by as much as 2%, so I used the atlas to determine empirical oscillator strengths. These empirical oscillator strengths are close to Holweger's (1967), but diverge systematically from laboratory values, disagreeing by almost an order of magnitude for the strongest lines, indicating a potential problem with the Holweger–Müller atmosphere. The collisional broadening term C_6 also had to be enhanced by a factor of six above its nominal value. Stars other than the Sun were modeled by two components with identical temperature structures, one with no magnetic field and the other with a possibly nonzero, uniform, vertical magnetic field.

I generated synthetic line profiles for 25 Fe I lines present in spectra of the inactive star τ Cet and the active star ξ Boo A, for a coarse grid of iron abundances, magnetic field strength, $|B|$, surface covering factors, f, macroturbulent velocities, and $v \sin i$. The models represented in the grid were ranked according to how well they agree with observations of *all 25 lines considered simultaneously.* For τ Cet the zero field case is best; whereas, for ξ Boo A a 2 kilogauss field covering 20% of the star is best. Other models with similar unsigned magnetic fluxes, $|B|f$, are nearly as good. The best fitting model inferred from each individual line sometimes differs significantly from the best global model, demonstrating that atmospheric parameters and therefore magnetic fields should be based on the analysis of many spectral lines.

Residuals of the best model fits to observed line profiles have an average magnitude of 1.4% for τ Cet and 1.5% for ξ Boo A, which is large compared to the expected photon noise of 0.2%. These large residuals are primarily due to use of a coarse parameter grid of limited extent, an inaccurate temperature structure, and a poorly determined instrumental profile. The results are also affected to a lesser extent by small blends, unaccounted for magneto-optical effects, and a crude treatment of macroturbulence. Despite these shortcomings, the results of this analysis are similar to those of other investigators. More study is needed to determine whether this is indicative of the robustness of multi–line analyses or simply coincidence.

4. Conclusion

A preliminary multi–line Zeeman analysis of the active star ξ Boo A gives a magnetic field strength of 2 kilogauss covering 20% of the star's surface; whereas, analysis of the inactive star τ Cet gives a null field result. Although more study is required, it seems that the use of many lines compensated for the various shortcomings of this analysis. Once these shortcoming are overcome, multi–line analyses will be a useful method for determining magnetic fields along with other properties of active stellar atmospheres.

References

Basri, G., Marcy, G.W.: 1988, *Astrophys. J.* **330**, 274
Fuhr, J.R., Martin, G.A., Wiese, W.L.: 1988, *J. Phys. and Chem. Ref. Data* **17**, No. 4
Holweger, H.: 1967, *Z. Astrophys.* **65**, 365
Holweger, H., Müller, E.A.: 1974, *Solar Phys* **39**, 19
Kurucz, R.L., Furelid, I., Brault, J., Testerman, L.: 1984, Solar Flux Atlas from 269 to 1300 nm, National Solar Observatory Atlas No. 1
Marcy, G.W.: 1984, *Astrophys. J.* **276**, 286
Marcy, G.W., Basri, G.: 1989, *Astrophys. J.* **345**, 480
Robinson, R.D.: 1980, *Astrophys. J.* **239**, 961
Saar, S.: 1988, *Astrophys. J.* **324**, 441

A New Technique to Measure Magnetic Field Strength in Active Stars

J. Sánchez Almeida, R. J. García López

Instituto de Astrofísica de Canarias, 38200 La Laguna, Tenerife, Spain

Abstract: We explore the use of the flux difference between the lines Fe I 5250 Å and Fe I 5247 Å as a technique to measure photospheric magnetic fields in late-type stars. The technique developed takes into account all the LTE physics of the problem, assuming a radial and homogeneous magnetic field distribution over the stellar surface. Some test calculations, in order to prove the feasibility of the method, are shown.

1. Introduction

Despite the advances achieved since the pioneer work of Robinson (1980), techniques to measure the magnetic field strength in stars still require improvement. The current methods resort to oversimplifications (see Hartmann, 1987) which can be summarized as: (i) An incomplete treatment of the radiative transfer in a magnetic medium which does not take into account line saturation or magneto-optical effects. (ii) An idealized magnetic model atmosphere, both in its thermodynamical properties and the distribution of magnetic zones over the stellar surface. In all cases magnetic and non-magnetic zones are assumed to be thermodynamically identical while in most cases magnetic zones are placed in a single position on the disk. This work presents a new technique which attempts to overcome some of these difficulties.

2. The method

Assume that we can observe two lines which are identical in every respect except for their Zeeman splittings. The difference between their profiles (flux difference versus wavelength) does not depend on the non-magnetic regions of the star (it is not even affected by a magnetically quiet unresolved companion of our problem star). This implies that if the magnetic field strength is derived by modelling the difference of the profiles, no hypothesis is needed about the regions of the star which do not contain magnetic fields.

We have developed a computer code which fits the difference between FeI 5250Å and FeI 5247Å by means of a non-linear least squares algorithm. It uses the Milne-Eddington solution of the radiative transfer equations integrated over the stellar disk (see Landolfi *et al.*, 1989). As we discuss in the next paragraph, these lines fulfil the requirements of being very nearly identical. The solutions of the radiative transfer equations take into account all the LTE physics of the problem (including magneto-optical effects), although they treat the thermodynamical properties of the atmosphere in a very simplified manner. Rotation is considered by direct integration of the specific intensity over the stellar disk, taking into account the local Doppler shift. Finally, the magnetic field is assumed to be uniformly spread over the star and radial. This magnetic field distribution (previously considered by other authors; see e.g. Basri and Marcy, 1988) resembles that of the Sun.

FeI 5250Å and FeI 5247Å have been chosen as the pair of *identical* lines, in the absence of a magnetic field, required by our method (Stenflo, 1973). Figure 1 shows the difference of specific intensity of these lines synthesized (LTE) in quiet and active solar regions. The synthesis takes into account the intrinsic difference between the lines. Note that the effect produced by the magnetic field is more pronounced in the core than in the wings. This fact is important because blends, which are sources of systematic errors, affect mainly the wings (Saar, 1988).

3. Observational and numerical tests

As a very preliminary observational test of our technique, we have applied the fitting procedure to an intensity (Stokes I) profile obtained over a facula close to the disk centre. The spectrum is described in detail by Stenflo *et al.* (1984), but it basically consists of a high signal-to-noise ratio, high spectral resolution, Stokes I and V profiles of a region partly covered by magnetic fields (10–15%). The Stokes V profile (circular polarization versus wavelength) allows an independent determination of the field strength of the region (Sánchez Almeida *et al.*, 1988), giving 1250 G. Figure 2 contains the observed data and the fitted subtraction $I_{FeI5250} - I_{FeI5247}$, which corresponds to a field strength of 1330 G.

A numerical test for the method has also been performed. We have simulated the difference of fluxes $F_{FeI5250}$ - $F_{FeI5247}$ for a star in which the line strengths are similar to those observed in the Sun, but which rotates with a velocity ($v \sin i$) of 5 km/s, has a macroturbulence of 7 km/s FWHM, and has a magnetic field of strength 2000 G covering 50% of its surface. Random noise has been added to the synthetic profiles to yield a signal-to-noise ratio 300 in the continuum. The values obtained from the fit for these parameters are within 20% of the *correct* values.

We infer that the method we propose seems to be a useful technique for measuring magnetic fields (and filling factors). It does not assume the same model

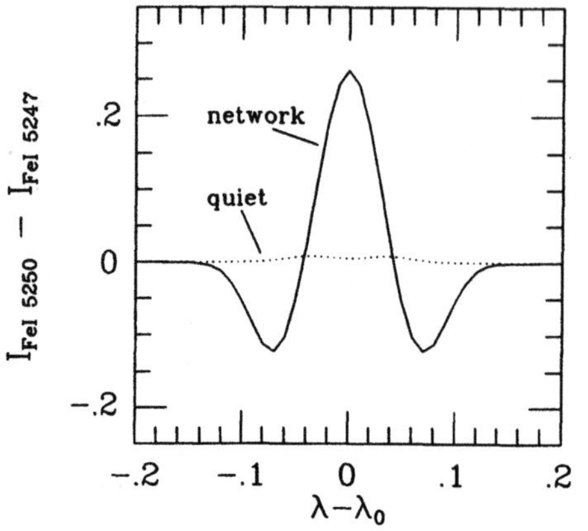

Fig. 1. Difference of specific intensity of the lines FeI 5250Å and FeI 5247Å synthesized in a quiet solar region without magnetic field, and a active region with a constant longitudinal magnetic field of 1300 G. Wavelengths (referred to the laboratory wavelengths of the lines, λ_0) are in Å.

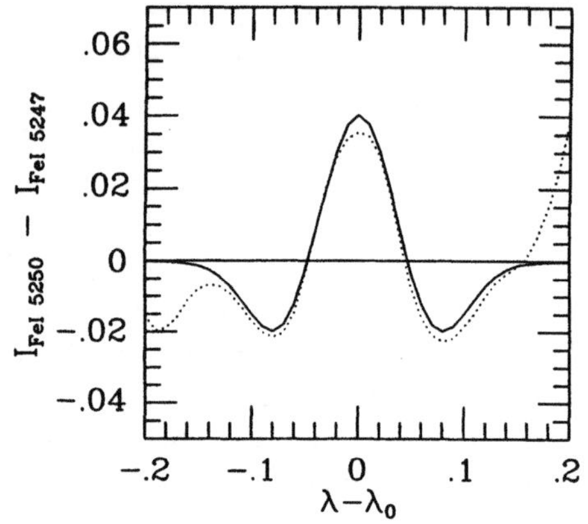

Fig. 2. The dotted line is the difference of specific intensity $I_{FeI5250}$–$I_{FeI5247}$ observed over a solar facula close to the disk centre. The solid line corresponds to the best fit provided by our method (without integration over the solar disk).

atmosphere for the parts of the star with and without magnetic fields and it includes complete polarized radiative transfer calculations.

References

Basri, G., Marcy, G.W.: 1988, *Astrophys. J.* **330**, 274

Hartmann, L.W.: 1987, in *Cool Stars, Stellar Systems and the Sun*, eds. Linsky and Stencel, Springer Verlag, p. 1

Landolfi, M., Landi Degl'Innocenti, M., Landi Degl'Innocenti, E.: 1989, *Astron. Astrophys.* **216**, 113

Robinson, R.D.: 1980, *Astrophys. J.* **239**, 961

Saar, S.H.: 1988, *Astrophys. J.* **324**, 441

Sánchez Almieda, J., Solanki, S.K., Collados, M., del Toro Iniesta, J.: 1988, *Astron. Astrophys.* **196**, 266

Stenflo, J.O.: 1973, *Solar Phys.* **32**, 41

Stenflo, J.0, Harvey, J.W., Brault, J.W., Solanki, S.K.: 1984, *Astron. Astrophys.* **131**, 333

Magnetic Flux Determination in Late-Type Dwarfs

P. Rípodas [1], J. Sánchez Almeida [1], R. J. García López [1],

M. Collados [1,2]

[1]Instituto de Astrofísica de Canarias
[2]Departamento de Astrofísica, Universidad de La Laguna
E-38200 La Laguna, Tenerife, Spain

Abstract: We present a very preliminary and simplified analysis designed to measure photospheric magnetic fields in late-type stars, using the FeI 5247.06 Å and 5250.22 Å lines. We show how the use of the equivalent widths of the lines and differences in their depth can give a rapid estimation of the magnetic flux.

1. Introduction

It is thought that rotationally-induced dynamo mechanisms are responsible for the chromospheric activity observed in late-type stars. One of the reasons for measuring photospheric magnetic fields in stars is to provide an observational basis to test this hypothesis. Confrontation of predictions of photospheric magnetic fluxes produced by dynamo models with observations should then set bounds to the basic parameters of the theory (see Montesinos *et al.*, 1990).

Apart from the uncertainties of the current methods of determining photospheric magnetic fields, the small size of the set of measurements (see Saar, 1990) still restricts the usefulness of this confrontation. We have derived a simple equation that relates the magnetic flux and the equivalent widths and depths of the FeI 5247.06 Å and 5250.22 Å lines. The determination of these parameters from a spectrum is straightforward, so that the relationship may help to enlarge the set of magnetic field strength measurements.

2. Determination of the magnetic flux

A very simple approach for the determination of the magnetic field strength can be expressed as follows:

Assuming a two-component model, the observed flux of a star, normalized to its adjacent continuum, $F(\lambda)$, can be written, in terms of the normalized fluxes of the magnetic and non-magnetic areas, $F_{m}(\lambda)$ and $F_{nm}(\lambda)$, as

$$F(\lambda) = f_{eff} F_{m}(\lambda) + (1 - f_{eff}) F_{nm}(\lambda). \tag{1}$$

λ is the wavelength and f_{eff} represents an *effective* filling factor (fraction of the surface covered by magnetic regions) given by the expression

$$f_{eff} = \frac{fr}{1 + (r-1)f}, \tag{2}$$

where r is the continuum contrast between magnetic and non-magnetic areas, and f the actual filling factor.

By observing two lines, very similar in their thermodynamic parameters and with different Landé factors, such as FeI 5247 Å and 5250 Å, recentering them and evaluating their difference, one has

$$\Delta F(\lambda) - \Delta F_{nm}(\lambda) \approx f_{eff} \Delta F_{m}(\lambda). \tag{3}$$

This means that, if the observed difference in residual flux between the two lines ($\Delta F(\lambda)$) is larger than the expected value in the absence of a magnetic field ($\Delta F_{nm}(\lambda)$), that difference can be directly attributed to the presence of a magnetic field on the stellar surface.

In order to estimate the magnetic flux, one can make use of eq. (3), and evaluate it at $\lambda = 0$ (the center of the lines used). Here we present a very simple analysis, which will be extended in the future to more realistic cases. Thus, assuming that the thermodynamic properties of the magnetic and quiet regions are the same (in which case, $f = f_{eff}$, among other implications), that the field is longitudinal and weak (Zeeman shift much smaller than the typical width of the lines) and gaussian line profiles (which is valid if the lines are not very strong), one has, for each line:

$$F_{m}(0) = 1 - \frac{W}{\sqrt{2\pi}\sigma} \exp[-\frac{\Delta\lambda^2}{2\sigma^2}] \approx 1 - \frac{W}{\sqrt{2\pi}\sigma}[1 - \frac{\Delta\lambda^2}{2\sigma^2}], \tag{4}$$

where $\Delta\lambda$ represents the Zeeman splitting, W the equivalent width of the lines (both lines are assumed to have the same W, which is true up to an accuracy of 1-2%), and σ is related to the half-width of the lines. By evaluating the difference of FeI 5247 Å and 5250 Å, using eq. (4), one has

$$\Delta F_{m}(0) = \frac{5k^2 \lambda_0^4 B^2}{2\sqrt{2\pi}\sigma^3} W, \tag{5}$$

where $k = 4.67 \cdot 10^{-10}$ mÅ Gauss^{-1} Å^{-2}. λ_0 is the central wavelength of one of the lines. Moreover, one can write for the residual flux of FeI 5247 Å (less sensitive to the magnetic field than FeI 5250 Å),

$$F_{\rm m}^{47}(0) \approx 1 - \frac{W}{\sqrt{2\pi}\sigma}[1 - \frac{\Delta\lambda^2}{2\sigma^2}] \approx 1 - \frac{W}{\sqrt{2\pi}\sigma} = F_{\rm nm}^{47}(0), \tag{6}$$

Combining eqs. (1), (5) and (6), one has an expression for the magnetic flux,

$$\sqrt{f}B = CW\frac{[\Delta F(0) - \Delta F_{\rm nm}(0)]^{1/2}}{[1 - F^{47}(0)]^{3/2}}, \tag{7}$$

where C is a constant whose value is $(\sqrt{5\pi}k\lambda_0^2)^{-1} = 19.6$ Gauss mÅ^{-1}, $\Delta F(0)$ is the observed difference between the residual fluxes of the two lines, and $F^{47}(0)$ is the observed residual flux of FeI 5247 Å (although FeI 5250 Å could also be used because of the small filling factors). All these parameters can be directly determined from observed spectra. On the other hand, $\Delta F_{\rm nm}(0)$ gives account of the intrinsic difference between the two lines, in the absence of magnetic field, and can be calculated using a radiative transfer code.

Although this is a very preliminary analysis, many of the hypotheses can be eliminated with numerical calculations. Different field topologies, integration over the stellar disk, and more realistic spectral line profiles (obtained by integration of the transfer equation) can be introduced to test the validity of eq. (7), or, instead, derive a numerical relationship between the observed parameters which might allow the calculation of the magnetic flux.

As part of an observational programme to measure stellar magnetic fields, we have applied this simple treatment to three late-type dwarfs: ξ Boo A (G8V), λ Ser (G0V), and β Com (G1V), and have detected magnetic signals in the first two (Rípodas *et al.*, 1990).

References

Montesinos, B., Fernández Villacañas, J.L., Jordan, C.: 1991, these Proceedings

Rípodas, P., Sánchez Almeida, J., García López, R.J., Collados, M.: 1990, in *Surface Inhomogeneities on Late-Type Stars*, Armagh, N. Ireland.

Saar, S.H.: 1990, in *The Solar Photosphere*, ed. J. O. Stenflo, I.A.U. Symp. 132, Kluwer Academic Publishers, Dordrecht, p. 427

Linear Polarization and Magnetic Fields in Cool Stars

J. Huovelin [1], S. H. Saar [2]

[1]Observatory and Astrophysics Laboratory, University of Helsinki, Tähtitorninmäki, SF-00130 Helsinki, Finland
[2]Harvard–Smithsonian Center for Astrophysics, Mail Stop 58, 60 Garden Street, Cambridge, MA 02138, USA

Abstract: Observations and theoretical models of broadband linear polarization (BLP) in cool stars are discussed. Two effects, 1) differential saturation in magnetically sensitive absorption lines and, 2) Rayleigh and Thomson scattering are considered as the possible causes of the BLP. Results of theoretical models, and comparisons with observations are summarized.

1. Theoretical models

We have studied the expected rotational variations, wavelength dependence, and the absolute scale of *broadband* linear polarization for two mechanisms, *1)* the magnetic intensification (MI), which is based on the cumulative effect of the differential saturation in the perpendicularly polarized Zeeman (π and σ) components of spectral lines, and *2)* scattering, either due to molecules (Rayleigh scattering), or electrons (Thomson scattering), in an optically thin medium (scattering optical thickness $\tau_s \leq 0.1$). The details of the models will be published elsewhere (Huovelin and Saar, 1991; Saar and Huovelin, 1991).

In both cases, the amount and distribution of surface inhomogeneities (in the magnetic field for MI, and surface brightness and τ_s for scattering) are very important factors in determining the total, surface integrated polarization (since polarization is a direction sensitive vector quantity). Therefore, unambiguous estimates on the expected levels of polarization cannot be made without independent knowledge on the distribution of the polarization inducers. In our models we have used Occam's razor and restricted the discussion to one region on the stellar surface. The advantage of this choice becomes clear considering more complex distributions, for which the total polarization can be derived by summing the polarization *vectors* for all individual regions.

The polarization from scattering is directly related to the inhomogeneities in the optical thickness and the surface brightness. In the optically thin case, the linear polarization is actually proportional to the product $\tau_s \times I$ in a surface element. We have made estimates of the polarization assuming homogeneous brightness, and varying optical thickness. At present, we lack estimates of the optical depth variations due to inhomogeneities in stellar atmospheres. Therefore, we used the τ_s's from the published homogeneous chromospheric models, and assumed that $\delta\tau_s \approx \tau_s$. The models predict a very small contribution from Thomson scattering in single F–K dwarfs, but in the extended envelopes of giants, supergiants, and close binaries, the effect may be significant. Rayleigh scattering from molecules and atoms also increases with decreasing gravity and temperature but, like Thomson scattering, is probably undetectable in cool dwarfs.

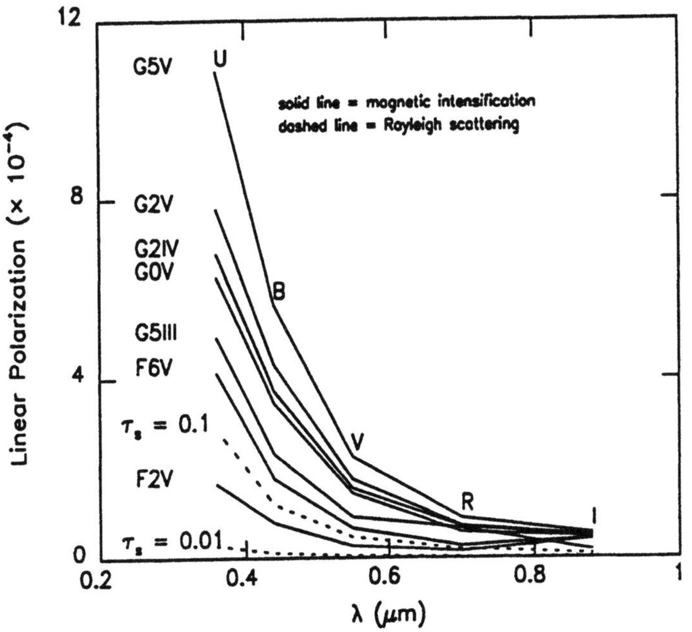

Fig. 1. Broadband linear polarization (maximum expected values with one circular region of size 24%) in Johnson UBVRI passbands. The respective upper limits for Rayleigh scattering (size 18%) are presented for comparison. $\tau_s = 0.1$ corresponds roughly to values expected in early K giants, and $\tau_s = 0.01$ in K dwarfs.

The models predict larger contribution from the differential saturation in dwarf stars, and the BLP could therefore be a direct indicator of magnetic activity in those stars. The time variations and the wavelength dependence are modelled for different spectral types for comparison with observations. The derived levels of BLP are low ($\leq 0.1\%$ in F–G dwarfs), and therefore difficult to measure. Fig.1 shows the theoretically expected broadband polarization for various spectral types in the Johnson UBVRI passbands.

2. Comparison with observations

The present observational material does not contradict the theoretical predictions. The observations show signs of stellar polarimetric variability, but the amplitudes are small, and in most cases it is impossible to firmly distinguish the details (see e.g. Huovelin *et al.*, 1989). Nevertheless, with simultaneous monitoring of other activity indicators (e.g. Ca–emission, photometry; Huovelin *et al.*, 1987, 1988), the polarimetry may be used as an additional constraint.

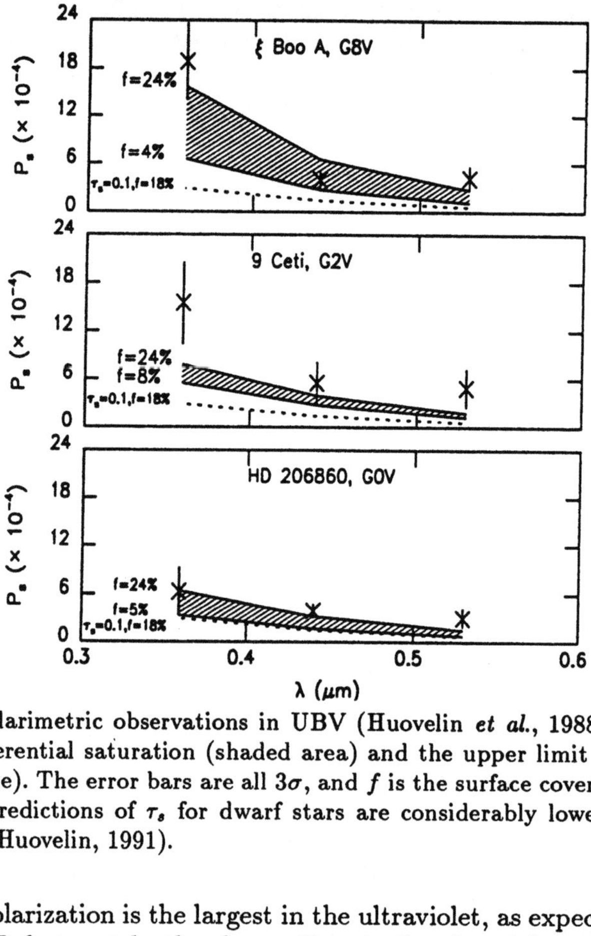

Fig. 2. Polarimetric observations in UBV (Huovelin *et al.*, 1988) compared with models for differential saturation (shaded area) and the upper limit for Rayleigh scattering (dashed line). The error bars are all 3σ, and f is the surface coverage of the region. Note, that our predictions of τ_s for dwarf stars are considerably lower than the upper limit (Saar and Huovelin, 1991).

The polarization is the largest in the ultraviolet, as expected on the theoretical grounds. Unfortunately, the observations in the ultraviolet are less accurate than in the longer wavelengths. A straightforward comparison of the BLP between different passbands may yield slopes for $P(\lambda)$ that are too steep, since the increase of random errors tends to increase also the average polarization, perhaps more than linearly with the errors. Fig. 2 gives a few examples of our comparisons between the observed and the theoretically expected polarizations. In Fig. 2 we also show indicative estimates of the surface coverages of magnetic fields, assuming that the polarization is due to the differential saturation. The complete tables of these estimates are given in Saar and Huovelin (1991).

3. Discussion

The polarimetric models predict specific variations with time and with wavelength, which set a great challenge to the observations. In principle, the measurements of the time variations could yield very detailed information on the inhomogeneities and magnetic fields on the stellar surface, and the source of the polarization could be unveiled by studying the variations with wavelength.

However, the accuracy (a few times 10^{-5}) required by quantitative tests of the theoretical models is not easy to achieve with current polarimeters. In addition to good photon statistics (i.e., bright stars and a big telescope), the observations have to be carried out with extreme care to avoid other random errors which can easily grow to a level higher than the expected polarization. The problem is most serious in the ultraviolet, where the photon statistics are lowest (cool stars are not very bright in the UV). Also, the atmospheric dispersion may cause substantial decentering of the UV stellar image in the diaphragm, which is not seen by visual inspection. The polarization caused by the edge of a metallic diaphragm, for example, may also be a serious problem in instruments where this effect has not been accounted for in the observing and reduction procedures. Finally, the differences in the wavelength dependence are largely smoothed out in the broad-band measurements. Narrower passbands would be better in this sense, but have a decreased photon statistics, which limits the observations to very bright stars and/or big telescopes.

References

Huovelin, J., Piirola, V., Vilhu, O., Efimov, Yu., Shakhovskoy, N.M.: 1987, *Astron. Astrophys.* **176**, 83

Huovelin, J., Saar, S., Tuominen, I.: 1988, *Astrophys. J.* **329**, 882

Huovelin, J., Linnaluoto, S., Tuominen, I., Virtanen, H.: 1989, *Astron. Astrophys. Suppl. Ser* **78**, 129

Huovelin, J., Saar, S.: 1991, *Astrophys. J.*, in press

Saar, S., Huovelin, J.: 1991, *Astrophys. J.*, submitted

Long-term Polarimetric Activity of the Cool Supergiant μ Cephei

J. Arsenijević and S. Jankov

Astronomical Observatory, Beograd, Yugoslavia

Abstract: The intrinsic polarisation parameters P, θ, Q and U of the star μ Cep, in the V spectral region during the period 1979–1989, are presented. The data were obtained from the Belgrade polarimetric observational program. The polarization percentage of the star exhibits change is of about 2.5% during this interval and the position angles change by about 90 degrees. These very large long-term changes of polarization parameters are discussed briefly.

1. Introduction and observations

The cool bright supergiant μ Cep (HD 206936) has been studied polarimetrically for a long period of time by many authors. The first measurement was made by Hiltner (1951). Grigoryan (1959) discovered the variability of its polarization. The scattering by grains in a circumstellar envelope was adopted as a mechanism to produce polarization. The empirical schematic model proposed by Polyakova (1984), for example, was used to explain the intrinsic polarization and brightness variation of μ Cep by the rotation modulation and intrinsic variation of two active antipodal spot regions. The polarization was supposed to be a product of starlight scattered by dust streaming from active regions. This problem is far from solution.

Polarimetric observations presented in this paper were carried out with the 65 cm Zeiss refractor and the Belgrade stellar polarimeter (Kubičela *et al.*, 1976) in the V spectral region during the period 1979–1989. The polarimeter was modified in 1979 to digitalize and record the data. The angular velocity of the analyser was one turn per minute. In most cases "one measurement" (one point in the Figs. 1–3) can represent up to 8-polarimetric phase-averaged 1-minute sine-wave signals. The typical standard deviation of the observed Stokes parameters for one 8-minute measurement of this star is not higher than 0.07%. During the whole interval of observation the instrumental system was carefully checked by measuring 11 polarized standard stars and 17 stars known to have zero polarization. For the interstellar polarization parameters in the direction of μ Cep the estimated values of Polyakova (1974) ($Q = -0.1\%$, $U = 1.5\%$) were used as the most reliable among those published. The individual values of the intrinsic polarization percentage P, the position angle θ and the corresponding Stokes parameters Q and U in the period 1979–1989 are presented in Figs. 1–3.

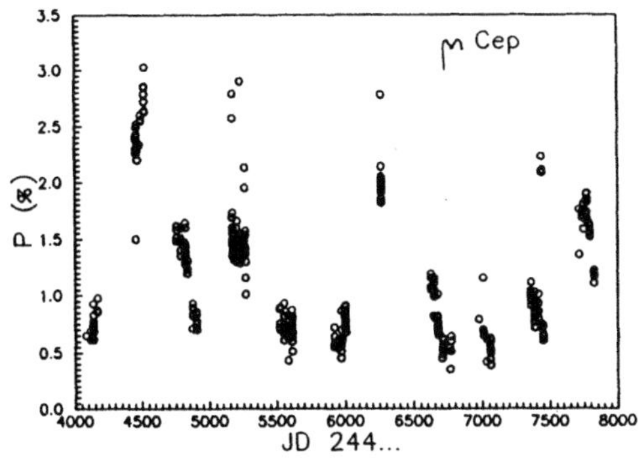

Fig. 1. The intrinsic polarization percentage P.

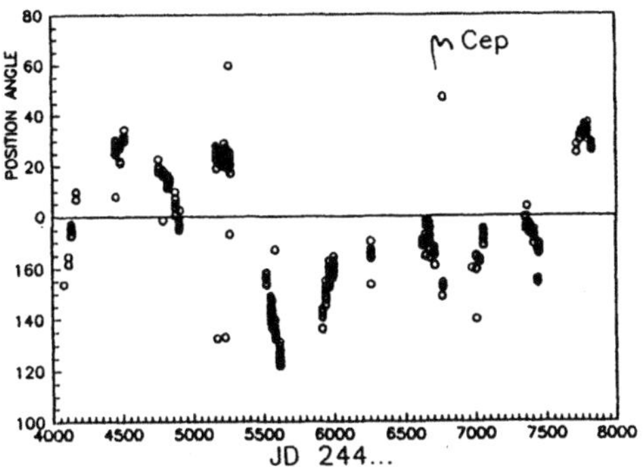

Fig. 2. The intrinsic polarization angle θ.

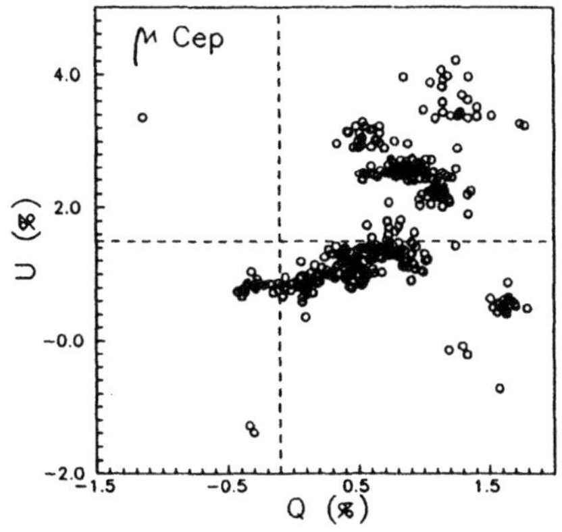

Fig. 3. The stokes parameters Q and U. The interstellar polarization is given by the intersection of the dashed lines representing also the intrinsic polarization coordinate system.

The observed and intrinsic Stokes parameters are not presented separately because the relative position of these parameters do not change after extracting the interstellar component. So, only the origin of the coordinate system changes to the values $Q = -0.1\%$, $U = 1.5\%$. In Fig. 3 the new coordinate axes (for the intrinsic polarization) are shown by the dashed lines. From Figs. 1–3 it is obvious that the polarization of μ Cep exhibit large changes. The intrinsic polarization percentage changes between values of about 0.35% and 3%. The variation in the position angle are of the order of 90 degrees.

2. Discussion

In the last several years the behaviour of the star has been extraordinary: high optical polarization during 1980 (Arsenijević *et al.*, 1980) and a possible eruptive event on August 3, 1981 (Arsenijević, 1985) were observed. In addition, some successful attempts to see the envelope structure were made. The first image of the circumstellar envelope in the Na I 589 nm emission line was obtained on 1–2 September 1984 by Mauron *et al.* (1986) and, at last, observations of dust envelope polarization were made on June 1985 and June–July 1986 by Borgne and Mauron (1989).

We shall mention now only some interesting facts seen at first sight. There is an agreement between the polarization position angle of the envelope itself measured by Borgne and Mauron (1989), covering the very disk of the star, and the intrinsic polarization position angle presented here for the corresponding period of time in 1985 and 1986. Namely, for the region of high envelope polarization (in V spectral region) in Borgne and Mauron, Table 2, the position angle of polarization is between 164 and 169 degrees while the intrinsic polarization position angle measured by us in the same period of time is almost in the same interval.

As global polarimetric parameters include the whole envelope, particularly the parts nearer to the star which are more efficient in producing polarized light, as well as the outer parts of the envelope having the angles of polarization mentioned, we can conclude that the geometry of the inner part of the star envelope was, at least during the mentioned period, similar to that of the outer one. The image of the circumstellar envelope in the Na I emission line was observed (Mauron *et al.*, 1986) in the interval of time when the envelope activity of the star, according to polarimetric observations, was small. Hence, one can perhaps conclude that the polarizing contribution of the gas component to the intrinsic polarization is also small.

Further discussion of some other physical parameters or processes (magnitude, flaring, spectral changes etc.) and their correlations with polarization parameters during the whole interval of observations will be presented later.

References

Arsenijević, J.: 1985, *Astron. Astrophys.* **145**, 430
Arsenijević, J., Kubičela, A., Vince, I.: 1980, *Inf. Bull. Var. Stars* No 1859
Borgne Le, J.F., Mauron, N.: 1989, *Astron. Astrophys.* **210**, 198
Grigoryan, K.A.: 1959, *Soobshch. Byurakan. Obs.* **27**, 55
Hiltner,W.A.: 1951, *Astrophys. J.* **114**, 241
Kubičela, A., Arsenijević, J., Vince, I.: 1976, *Publ. Dept. Astron. Beograd* **6**, 25
Mauron, N., Cailloux, M., Tilloles, P., Le Fevre, O.: 1986, *Astron. Astrophys. Letters* **165**, L9
Polyakova, T.A.: 1974, *Astrofizika* **10**, 1
Polyakova, T.A.: 1984, *Astrofizika* **21**, 125

Polarimetric Observations of Active Cool Binaries

R.H. Koch [1], N.M. Elias [1], B.D. Holenstein [1], V. Piirola [2],

F. Scaltriti [3], and G.V. Coyne [4]

[1]University of Pennsylvania, Philadelphia, U.S.A.
[2]Observatory and Astrophysics Laboratory, Helsinki, Finland
[3]Osservatorio Astronomico di Torino, Pino Torinese, Italia
[4]Vatican Observatory, Città del Vaticano

1. Introduction

For a sample of RS CVn binaries Busso and collaborators (Busso *et al.*, 1987, 1988; Busso and Scaltriti, 1990) compared the observed energy distribution from the UV to the near-IR (and to IRAS bandpasses for some objects) with those calculated from the known radii and spectral types of the stellar components. About 25% of the sample gave evidence of an IR excess up to 0.7 mag, generally starting from the I-band. The excess is significantly larger than observational error. These same authors ascribe the excess to a thin dust shell surrounding the binary. Using the "dirty silicates" model by Jones and Merrill (1976) they find for the shell a temperature range of 1500 K to 2000 K, an optical depth of about 0.01 at 10 micrometers in the shell, and a distance of the shell from the stars to be typically about 100 times the radius of the larger star.

Multi-band polarimetry represents a powerful tool for investigating circumstellar scattering envelopes. However, it has not been commonly applied to RS CVn binaries. Published data indicate that the intrinsic polarization is small, although Pfeiffer (1979) and Liu and Tan (1987) have found values in excess of 0.3% for certain binaries at unpredictable times. Piirola (in Weiler *et al.*, 1978) observed two RS CVn's (UX Ari and HR 1099) and found very small values of linear polarization ($p < 0.02\%$) in white light.

Both general arguments (e.g. Popper and Ulrich, 1977) concerning the evolutionary states of RS CVn's and observations of individual binaries indicate mass loss by winds and the presence of circumstellar and circumbinary envelopes of gas and dust in the systems. Since non-spherical concentration of neutral H, plasmas, and dust can all give rise to significant levels of polarization, we started a collaboration in 1989 to monitor selected RS CVn's polarimetrically from the U-band into the near-IR.

2. The observational campaign and instrumentation

Multiband polarimetry was performed in 1989-1990 at four sites: (1) at Kitt Peak the 90-inch telescope of Steward Observatory was used together with the 'two-holer' polarimeter of Gary Schmidt; (2) at Pennsylvania the 0.72-m Cassegrain reflector was used together with a photoelastic polarimeter described by Koch *et al.* (1985); (3) at La Palma the observations were made with the 2.56-m Nordic telescope by using the simultaneous five colour (UBVRI) polarimeter built in Turku Observatory (Piirola, 1973; Korhonen *et al.*, 1984); (4) at Torino the 1.05-m astrometric reflector fed the 5-channel (UBVRI) simultaneous polarimeter recently constructed at the Institute according to the original design by Piirola (1973, 1988). Some details of this polarimeter can be found in Scaltriti *et al.* (1989).

3. Results and discussion

In general, it may be noticed that the linear polarization values are typically small. This is consistent with the published results noted above. In order to perform a more comprehensive discussion, we divide our sample results into three sub-groups according to the amount of phase coverage of the Keplerian cycles.
(a) Single-phase measures. In B the measured polarizations for Zeta And and WY Cnc are consistent with zero (±0.02%). The B, V and R measures for RU Cnc and WY Cnc appear significant at the 0.05% level. The polarization spectrum of RU Cnc is flat within the errors of measurement. (b) Two phases sampled. Each of the stars CQ Aur, RZ Cnc, VV Mon, and AR Mon shows significant polarization (0.1–0.2%). The polarization wavelength dependence for these systems indicates that within 3-sigma, they are consistent with the wavelength dependence of interstellar polarization. (c) At least 5 phases sampled. For GK Hya there is marginal evidence of p-variability in the R and I bands. Also for AR Psc there are hints of p-variability in the B, R, and I bandpasses but these would have to be confirmed by more observations.

The best phase coverage is that for UX Ari in the B,V, and R bands and for II Peg in the V-band. The linear polarization may be phase locked. It is tempting to try to interpret this pattern as being due to a scattering envelope asymmetrically distributed with respect to the low-inclination orbital plane as in the model of Brown *et al.* (1978). The mean wavelength dependence of the linear polarization compiled from the Torino data suggests that electron scattering is not the cause of the intrinsic polarization, since the amount of polarization decreases toward longer wavelengths. A similar tendency is seen also for GK Hya, UV Psc, and AR Psc (see Fig. 1), indicating small dust particles as a possible source of the intrinsic polarization.

With respect to the main objective of the observational program our flat spectrum (interstellar type) sample contains two pairs with an IR-excess (RU Cnc, VV Mon), two pairs without the excess (CQ Aur and RZ Cnc) and one with uncertain excess (AR Mon). The binaries showing indications of polarization by circumstellar dust (Fig. 1) contain two pairs with an IR-excess (GK Hya, II Peg),

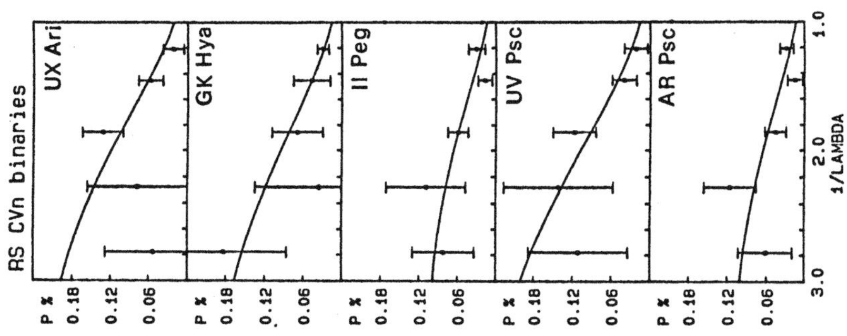

Fig. 1. The wavelength dependence of the mean polarization for five RS CVn systems. There is a tendency of smaller polarization toward longer wavelengths.

two without (UX Ari, AR Psc) and one with uncertain excess (UV Psc). In addition, WY Cnc, which has the IR excess but was observed only in two passbands, shows marginal indication of larger polarization towards longer wavelengths ($p(B) = 0.027 \pm 0.021\%$; $p(R) = 0.060 \pm 0.017\%$).

The polarization of some of these binaries may be variable in the long term and we plan to continue monitoring some of these systems. In particular, II Peg will undoubtedly repay continuous linear polarimetric monitoring, a conclusion hardly surprising in view of its activity. Moreover, the suggestive linear results for UX Ari and GK Hya indicate that detailed monitoring of the polarization with phase and wavelength is necessary.

References

Brown, J.C., McLean, I.S., and Emslie, A.G.: 1978, *Astron. Astrophys.* **68**, 413

Busso, M., Scaltriti, F., Persi, P., Robberto, M., and Silvestro, G.: 1987, *Astron. Astrophys.* **183**, 83

Busso, M., Scaltriti, F., Persi, P., Ferrari-Toniolo, M., and Origlia, L.: 1988, *Mon. Not. R. astr. Soc.* **234**, 445

Busso, M., and Scaltriti, F.: 1990, in preparation.

Jones, T.W., and Merrill, K.M.: 1976, *Astrophys. J.* **209**, 509.

Koch, R.H., Hrivnak, B.J., Bradstreet, D.H., Blitzstein, W., Pfeiffer, R.J., and Perry, P.M.: 1985, *Astrophys. J.* **288**, 731

Korhonen, T., Piirola, V., and Reiz, A.: 1984, *The ESO Messenger*, December 1984

Liu Xuefu and Tan Huisong: 1987, *Acta Astr. Sinica* **28**, 139.

Pfeiffer, R.J.: 1979, *Astrophys. J.* **232**, 181.

Piirola, V.: 1973, *Astron. Astrophys.* **27**, 383.

Piirola, V.: 1988, in *Polarized Radiation of Circumstellar Origin*, eds. G.V. Coyne *et al.*, Univ. of Arizona Press, p. 735.

Popper, D.M., and Ulrich, R.K.: 1977, *Astrophys. J. Letters* **212**, L131.

Scaltriti, F., Piirola, V., Cellino, A., Anderlucci, E., Corcione, L., Massone, G., Racioppi, F.: 1989, *Mem. S.A.It.* **60**, 243.

Weiler, E.J., Owen, F.N., Bopp, B.W., Schmitz, M., Hall, D.S., Fraquelli, D.A., Piirola. V., Ryle, M., and Gibson, D.M.: 1978, *Astrophys. J.* **225**, 919.

Magnetic Fields and Filling Factors in Late-Type Stars: Predictions from Dynamo Theory

B. Montesinos, J.L. Fernández-Villacañas, C. Jordan

Department of Theoretical Physics, University of Oxford, 1 Keble Road, Oxford OX1 3NP, England, UK

Abstract: In this paper we examine the link between observed magnetic fields and filling factors in main-sequence stars of types G and K, and the results obtained for these parameters from a simple dynamo model. We explore how the predicted magnetic fluxes, fB, for a selected sample of stars, vary with rotation, adopting two theoretical approaches to estimate the filling factor, and considering different expressions for the variation of the stellar angular velocity, ω, with depth.

1. Introduction

In the last decade, the detection of magnetic fields in late-type stars has opened a new window for a deeper understanding of all the phenomena broadly known under the name of *stellar activity*. The techniques for estimating the magnetic field strengths and the fraction of stellar surface covered by strong fields – the so-called *filling factor* – have been greatly improved since the first positive detections by Robinson (1980) up to the more sophisticated methods applied by Saar (1986, 1990) and Marcy and Basri (1989). A state-of-the-art review of the topic can be found in the work by Saar (1990, see also this volume).

On the other hand, there exists a wide range of theoretical approaches attempting to explain how the magnetic fields are generated in main-sequence stars, how they rise from the deep layers of the star to the surface and how they interact with the gas in the stellar atmosphere. The interaction between differential rotation and convection in the sub-photospheric convection zone –*dynamo action*– is so far the most plausible mechanism to explain the origin of the observed magnetic fields in the Sun and in late-type stars (see the review by Moss, 1986). The fields are supposed to rise by mechanisms combining buoyancy, Coriolis and Lorentz forces, reaching the photosphere and depositing their energy in the corona through MHD waves or by reconnection processes.

We find that there is a large gap between the observational work, which is providing a great deal of material to be analysed, and the *purely* theoretical work which creates complex models from the mathematical point of view, but which are often unable to predict what should be observed if the models were applied to a real star.

This contribution aims to help to fill that gap. Since there are many parameters of the stellar interior that are unknown we have carried out exploratory work, by taking a simple kinematic dynamo model and applying it to a sample of main-sequence late-type G and K stars which have the most reliable detections of magnetic fields and filling factors. Some extreme approaches have been chosen both for the dependence of rotation with depth, $\omega(r)$, and for the estimate of the filling factor. The model applied is summarised in Section 2 and the results are given in Section 3.

2. The model

Although a detailed description of all the characteristics of the model applied will be published elsewhere (Montesinos *et al.*, 1991) we summarize here its main points.

The available information consists of magnetic field strengths B, and filling factors, f. Our goal is to find the theoretical approach which gives an accurate prediction of both parameters, or some other related quantity, and their dependence on stellar rotation. We have chosen to study the magnetic flux fB, and we have decided to apply our formalism only to G and K stars, since F and M stars have very shallow and very deep convective zones, respectively, for which our formulation may not be appropriate.

2.1 The convection zone magnetic field

The field is assumed to be generated at the bottom of the convection zone by dynamo action. The generation region has been located one pressure scale height above the bottom of the convection zone and we assume that the depth of that region is also one scale height. The estimate of the convection zone magnetic field is described in the papers by Durney and Robinson (1982), Montesinos *et al.* (1987), and Montesinos *et al.* (1991), where further details can be found.

Two approaches have been taken to describe the variation of the stellar angular velocity with depth. These are based on helioseismological data, but some of the constants, such as the equatorial rotation velocity, are scaled to apply to each particular star. The first law, hereafter denoted as [1] is based on the approximation found by Durney (1985) and has the property that the shear in the angular velocity $d\omega/dr$ *is proportional* to the surface stellar angular velocity; the second one, [2], has been taken following a simplification of some results by Brown *et al.* (1989) and, in contrast to [1], the shear in ω *does not* depend on the angular velocity itself.

2.2 The filling factor

Starting from a toroidal configuration of tubes at the bottom of the convective zone and assuming that they rise up to the photosphere by buoyancy, it is possible to estimate the fraction of the stellar surface covered by magnetic tubes. Two approximations have been considered for the estimate of f. The first one, outlined by Durney and Robinson (1982), takes into account the diffusion time for the magnetic fields as they reach the photosphere. The expression for f is in this case

$$f \simeq \frac{3\pi^2}{16}\frac{\varphi}{l}\left(\frac{R_g}{R_*}\right)^2 \frac{B_{cz}}{B}\frac{u}{v_{ph}} L_{ph} \tag{1}$$

The second approach, which has been developed by us, only takes into account the number of flux tubes that at a given moment are present on the stellar photosphere. Whereas the first approximation can be considered as a 'time-average' estimate, our approximation can be called the 'snapshot' or 'static' approach. The expression for f is

$$f \simeq \frac{\pi^2}{16}\frac{\varphi}{l}\left(\frac{R_g}{R_*}\right)^2 \frac{B_{cz}}{B} L \tag{2}$$

In both expressions R_g is the radius of the generation zone, R_* is the stellar radius, B_{cz} and B are the convection zone and the photospheric magnetic fields, u is the rate of rise of a tube in the generation zone, L and L_{ph} are the pressure scale heights at the bottom of the convection zone and at the photosphere, respectively, and v_{ph} is the convective velocity near the photosphere. There are some unknowns in equations (1) and (2) which have to be fixed by analogy with the Sun, namely φ, the angle around the stellar equator where the bunch of magnetic tubes is supposed to be concentrated, as in the solar case, and l, the length of an eruptive section of magnetic tube when it starts rising from the bottom of the convection zone. Thus φ is taken as $\sim 2\pi/5$ and l is normalised to reproduce the solar filling factor or that of other chosen star. The photospheric field B is computed through the equipartition theorem which seems to be a good approximation to the observed values (Saar, 1990).

3. Results

Figure 1 gives the results of the calculations. The two upper graphs, (a) and (b), show the values for fB versus rotation period computed using the 'time-average' estimate for f and the two lower graphs, (c) and (d), give those computed using the 'snapshot' approach. The results in (a) and (c) have been computed using law [1] for ω, whereas those in (b) and (d) were estimated using law [2]. Open circles are observational results and solid circles are our theoretical predictions. The two open circles near the bottom of each graph and joined by a vertical bar represent the Sun. Theoretical predictions and observational results for each star lie in the same vertical line.

Although limits on space do not allow us to discuss in detail the results of the different approaches, we draw attention to several interesting results. In graph (a)

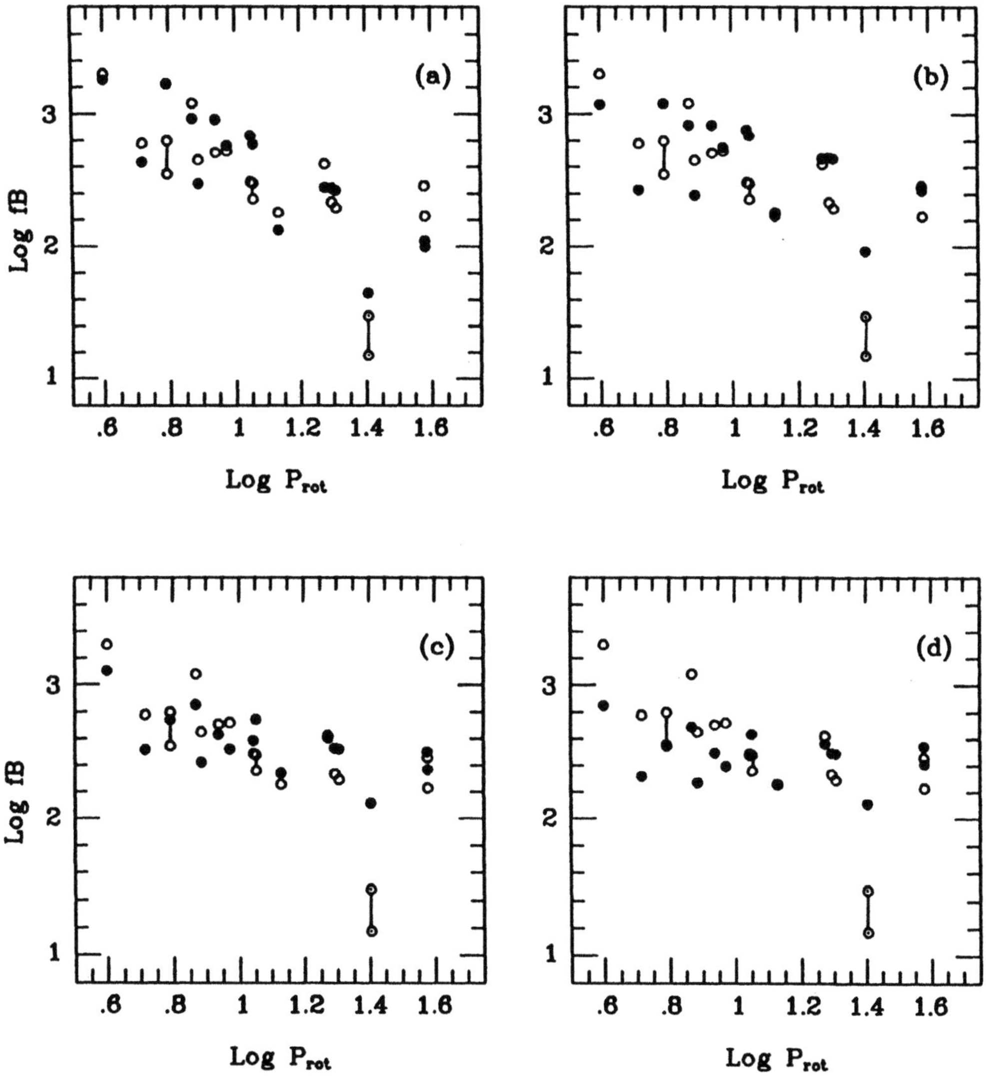

Fig. 1. Predicted values of fB (solid circles) and observational results (Saar, 1990) (open circles) versus rotation period for a sample of G and K main-sequence stars. See text for details.

both the solar and stellar values of fB are reproduced with reasonable accuracy. The agreement between the stellar values of fB observed and those computed is also good in (b) but the estimate for the Sun is a factor ~ 4 higher than the observed value. In graphs (c) and (d) the spread of the results is similar to the spread in the observational data but again the results for the solar fB-value are higher than the observed one. The different spread in the results between the top graphs and the bottom graphs is due to the dependence of v_{ph} and L_{ph} on spectral type. As far as the behaviour of the relationship $fB \propto P_{rot}^{-\alpha}$ is concerned, the slope of the observed values of fB with rotation depends on whether or not the Sun and the two slowest rotators are included in the correlation. The 'time-average' approximation for the filling factor – graphs (a) and (b) – leads to larger values of α than the 'snapshot' approach, and for a given method of computing f, larger slopes are found from adopting an explicit dependence of $d\omega/dr$ on ω, *i.e.* using approximation [1]. These points will be discussed in detail in Montesinos *et al.* (1991).

As we can see in Figure 1, the extreme approximations taken both for the estimate of the internal angular velocity and the filling factor open a range of possibilities that must be explored further. More determinations of f and B for other active stars, including both rapid and slow rotators, are necessary to enlarge the sample of data available. This will allow the behaviour of these parameters with rotation to be found more accurately. With improved data constraints on the possible dependence of the internal angular velocity and on the mechanism by which the fields originate and rise up to the photosphere could be found.

References

Brown, T.M., Christensen-Dalsgaard, J., Dziembowski, W.A., Goode, P., Gough, D.O., Morrow, C.A.: 1989, *Astrophys. J.* **343**, 526

Durney, B.R.: 1985, *Astrophys. J.* **297**, 787

Durney, B.R., Robinson, R.D.: 1982, *Astrophys. J.* **253**, 290

Marcy, G.W., Basri, G.S.: 1989, *Astrophys. J.* **345**, 480

Montesinos, B., Fernández-Figueroa, M.J., Castro, E.: 1987, *Mon. Not. R. astr. Soc.* **229**, 627

Montesinos, B., Fernández-Villacañas, J.L., Jordan, C.: 1991, in preparation

Moss, D.: 1986, *Physics Reports*, Vol. **140**, No. 1

Robinson, R.D.: 1980, *Astrophys. J.* **239**, 961

Saar, S.H.: 1986, *Astrophys. J.* **324**, 441

Saar, S.H.: 1990, In *Solar Photosphere: Structure, Convection and Magnetic Fields, IAU Symp. No. 138*, ed. J.O. Stenflo, Kluwer, Dordretch, p. 427

Activity of Relatively Close Binaries

Carolus J. Schrijver [1], Cornelis Zwaan [2]

[1]Space Science Department of ESA, ESTEC, Mail Box 299, 2200 AG Noordwijk, The Netherlands
[2]Astronomical Institute, University of Utrecht, Mail Box 80 000, 3508 TA Utrecht, The Netherlands

Abstract: Whereas the rotation–activity relationship for single cool stars can be described in terms of fundamental stellar properties, the activity in relatively close binaries is enhanced with respect to that of single stars by an amount which depends strongly on the properties of the companion. We consider mechanisms which could cause an excess in activity of binaries, and conclude that it is likely that the presence of a companion affects the interior structure of the stars in such a way that either the efficiency of the dynamo or of the atmospheric heating is enhanced.

1. Introduction

Magnetic activity, apparently ubiquitous in stars with convective envelopes immediately beneath the surface, is intimately connected with stellar rotation. Those cool stars which are either single or a component of wide binary systems define colour–dependent relationships between activity and rotation rate (e.g. Walter and Bowyer, 1981; Rutten, 1984; Schrijver *et al.*, 1984, Pallavicini *et al.*, 1985). Figure 1, however, shows that some stars are much brighter in chromospheric and coronal emissions than would be expected from the rotation–activity relationships defined by single stars (e.g. Rutten, 1987). Basri *et al.* (1985) had also noted that subgiants in synchronized binary systems were more active than single stars with the same rotation period. The "overactivity" in soft X–rays can be as large as an order of magnitude for rotation periods below about 30 days, and even larger at longer periods. These "overactive" stars are generally components of relatively close binary systems, or are otherwise unusual stars such as pre–main sequence stars, contact binaries or merged binaries. We perform a search for the parameter(s) describing the overactivity.

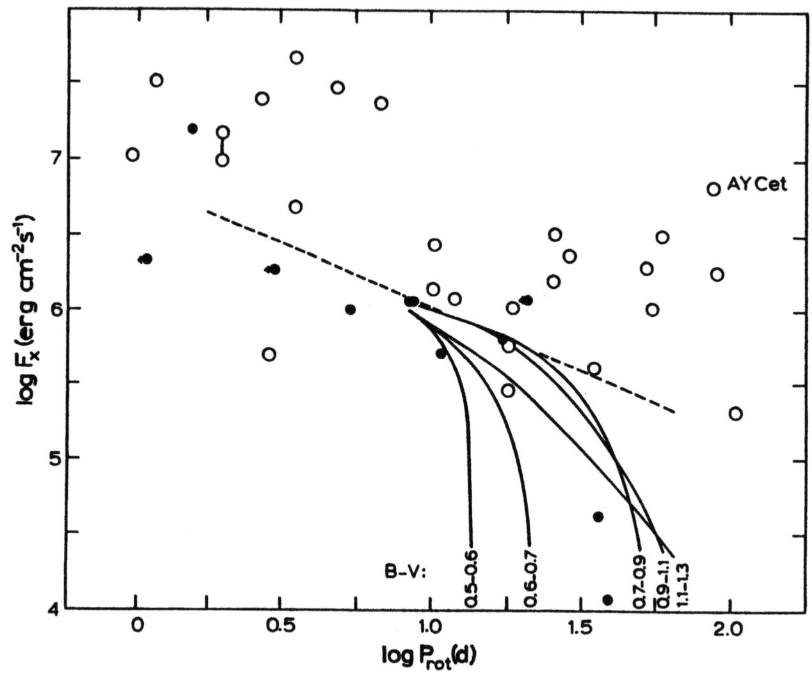

Fig. 1. Rotation–activity diagram: plotted are the surface flux density in soft X-rays, F_X, versus rotation period P_{rot}. The small dots show the location of single stars. The relatively close binaries are shown as circles. The curves show the colour-dependent rotation–activity relationships derived by Rutten (1987) using a sample of about 300 single stars and wide binaries. The dashed line shows the approximation of Eq. (2).

2. Selection of data

We selected stars from the lists of Schrijver *et al.* (1984) and of Strassmeier *et al.* (1990) for which sufficient information was available to derive the basic parameters of at least the primary (such as rotation rate, mass, radius, effective temperature, etc.), of the orbit (period, separation, inclination, and eccentricity), and of the activity level with at least the soft X-ray flux known. Our data set includes 21 binary systems for which the properties of the individual components and of the orbits are reasonably well known, and four additional systems with less certain parameters (see Schrijver and Zwaan, 1990, and Fig. 2). Eleven single stars are used for a detailed comparison, while the mean rotation–activity relations derived by Rutten (1987), based on a large sample of stars, are used as a reference. Activity is measured by coronal soft X-ray fluxes, F_X, observed with the the *HEAO–2 EINSTEIN* satellite, UV fluxes observed with the *International Ultraviolet Explorer*, and Ca II H+K fluxes observed with the Mt. Wilson HK spectrophotometer (see Schrijver and Zwaan, 1990, for more details).

Optical data show that if the two binary components are not of comparable size, the bulk of the atmospheric radiative losses appears to come from the cool primary (unless the warmer secondary is a rapidly spinning star like in Capella).

Hence the fluxes are usually assigned to the primary, except in cases where evidence exists that both components are active (see Schrijver and Zwaan, 1990).

3. Rotation and activity

A striking feature of the the radiative losses from the outer atmospheres of cool stars is that they are highly correlated among themselves, defining simple power–law relationships (e.g. Schrijver, 1987). Moreover, the basic stellar parameters do not enter into these relationships, provided one uses the radiative losses in excess of the so–called basal flux (which is probably unrelated with magnetic activity). Because of this property, Oranje *et al.* (1982), Schrijver (1983), Zwaan (1983), Oranje (1986) and Basri (1987) have suggested the existence of a single "activity parameter." This activity parameter is a function of several basic stellar parameters with the rotation rate playing a dominant role. Given the activity parameter, all radiative losses from chromosphere, transition region, and corona can be calculated without further knowledge of the star in question. The relatively close binaries generally follow the trend defined by the much larger sample of single stars and wide binaries. Although this suggests comparable differential emission measure distributions for single stars and for binaries throughout most of the atmospheres, we caution that it is not clear whether the coronal atmospheric structure is the same because too little is known about the coronal structure from the available broad–band measurements. Because of the interdependence of the diagnostics of activity, we concentrate on the soft X–ray fluxes in this analysis of overactivity.

A principal component analysis performed on the logarithmic values of the soft X–ray flux, F_X, the orbital period, P_{orb}, the binary separation, d, the radii of the hot and cool components, R_h and R_c, and the corresponding masses, M_h and M_c, suggests that the data span a two–dimensional subspace (covering 87% of the variance). We subsequently derived a two–parameter description of the activity level for the binaries in our sample:

$$F_X = 6\,10^7 \frac{R_h}{P_{orb}^{1.3}}, \qquad (1)$$

F_X in erg cm^{-2} s^{-1}, R_h in solar radii, P_{orb} in days. Eq. (1) describes the data with a 1σ deviation of a factor of 1.8 (Fig. 2), which is quite acceptable given the uncertainties in the stellar and orbital parameters and the uncertainties in and the expected variability of the soft X–ray fluxes. It is clear from Eq. (1) that some property of the *companion* is indispensible in this relationship. We cannot distinguish whether the rotation or orbital period should be used, although the fit is somewhat better if P_{orb} is used.

The expected emission for a single star of a given period with $P_{rot} \leq 30$days (Rutten, 1984) can be approximated by:

$$F_{X,0} = 7\,10^6 P_{rot}^{-0.8}. \qquad (2)$$

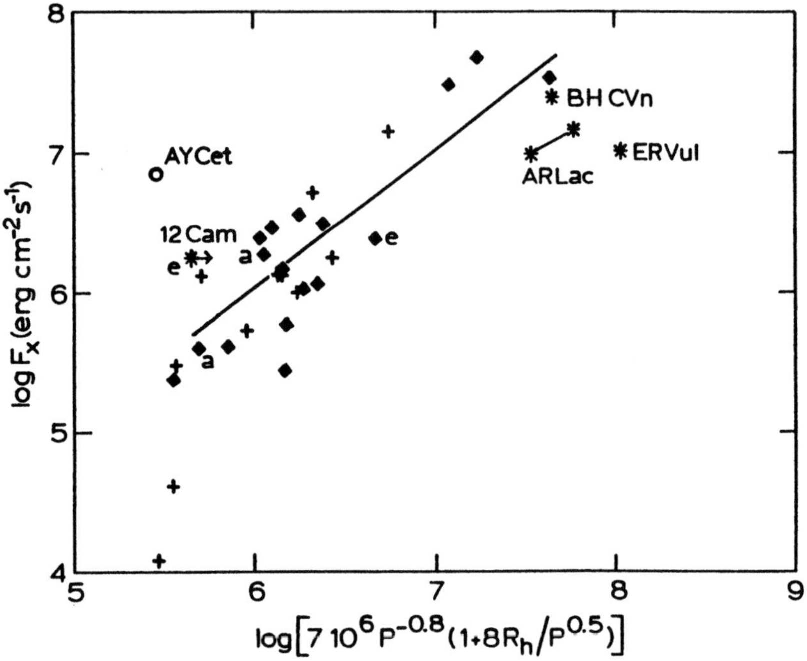

Fig. 2. Best-fit power-law parameterization of the rotation-activity relationship for single stars and for close binaries. The soft X-ray flux density is plotted vs. the best-fit value computed using Eq. (4). Symbols: ◇ close binaries, ∗ short-period systems or DA companion, + single stars.

Single stars and binaries can be combined in the approximate relationship (Fig. 2):

$$F_X = 7.10^6 \, P_{orb}^{-0.8} \, (1 + 8 * \frac{R_h}{\sqrt{P}}) = F_{X,0} \, (1 + 8 * \frac{R_h}{\sqrt{P}}) \tag{3}$$

(units as in Eq. (1)). In this relationship the activity in close binaries is expressed as a true "overactivity", i.e. an enhancement compared to the expected level of activity in a single star of the same rotation period.

Although Eqs. (1) and (3) use the radius of the companion to describe overactivity, we caution that there are strong dependences between the fundamental parameters of the stars in our sample, which could transform the relationship into a seemingly different representation of the same basic dependence. For the binary components, for instance, we find (M, d, and radii R in solar units, periods P in days): $P_{orb} = 0.12 * d^{1.34}$, (because of Kepler's law; correlation coefficient $C = 0.983$ for logarithmic values) $d = 5.5 * R_c^{0.93}$ ($C = 0.74$), $M_c = 0.55 * R_c^{0.56}$ ($C = 0.92$) and $M_h = 1.1 * R_h^{0.33}$ ($C = 0.79$).

If the secondary star can influence the activity of the primary, the primary may also affect the secondary. In close binaries with components with comparable masses and radii this appears to be the case. For binaries in which the secondaries are much smaller than the primaries, the enhanced activity of the secondary may not be detectable, or the activity enhancement may not occur in the same way.

There is some evidence that the activity (including the overactivity) in a few binary systems (ER Vul, BH CVn and AR Lac, shown by asterisks in the upper-lefthand corner of Fig. 2) is somewhat reduced compared to that in systems with longer periods. This "saturation" is a well-known property of W UMa systems (Vilhu and Rucinski, 1983, but see Zeilik, these Proceedings). These systems have therefore been excluded from the fitting procedure described above.

We have shown that the level of activity in relatively close binaries is determined not only by the rotation or revolution period, but also by basic properties of the companion. In the interpretation of the data it not clear whether the soft X-ray flux should be interpreted as an additional activity (implicit in the representation chosen for Eq. (1)), more or less unconnected with the ordinary magnetic activity, or as an enhancement of the expected level of magnetic activity given the rotation rate of the primary (as in Eq. (3)).

Schrijver and Zwaan (1990) consider several possible mechanisms which could cause overactivity in close binaries. They conclude that it is unlikely that the energy required for the overactivity is generated near the hotter companion, and then transported to the cool primary, thus eliminating accretion processes occurring near the hotter component, and magnetic coupling of the stars by the interstellar magnetic field. Instead, the presence of a companion may affect the interior structure of the cool primary in such a way that the efficiency of the dynamo (possibly through changes in the differential rotation) or of the atmospheric heating (by a modification of some characteristic of the top of the convection zone) is enhanced.

References

Basri, G.: 1987, *Astrophys. J.* **316**, 377

Basri, G., Laurent, R., Walter, F.M.: 1985, *Astrophys. J.* **298**, 761

Oranje, B.J., Zwaan, C., Middelkoop, F.: 1982, *Astron. Astrophys.* **110**, 30

Oranje, B.J.: 1986, *Astron. Astrophys.* **154**, 185

Pallavicini, R., Golub, L., Rosner, R., Vaiana, G.S., in *Second Cambridge Workshop on Cool Stars, Stellar Systems, and the Sun*, eds. M.S. Giampapa and L. Golub, SAO Spec. Rep. 392, Vol. II, p. 77

Vilhu, O., Rucinski, S.M.: 1983, *Astron. Astrophys.* **127**, 5

Rutten, R.G.M.: 1987, *Astron. Astrophys.* **177**, 131

Rutten, R.G.M., Schrijver, C.J., Lemmens, A.F.P., Zwaan, C.:1990, in prep. for Astron. Astrophys.

Schrijver, C.J.: 1983, *Astron. Astrophys.* **127**, 289

Schrijver, C.J.: 1987, *Astron. Astrophys.* **172**, 111

Schrijver, C.J., Mewe, R., Walter, F.M.: 1984, *Astron. Astrophys.* **138**, 258

Schrijver, C.J., Zwaan, C.: 1990, submitted to *Astron. Astrophys.*

Strassmeier, K.G., Hall, D.S., Boyd, L.J., Genet, R.M.: 1989, *Astrophys. J. Suppl.* **69**, 141

Vilhu, O., Rucinski, S.M.: 1983, *Astron. Astrophys.* **127**, 5

Walter, F.M., Bowyer, S.: 1981, *Astrophys. J.* **245**, 671

Zwaan, C.: 1983, in *Solar and Stellar Magnetic Fields: Origins and Coronal Effects*, ed. J.O. Stenflo, Reidel, p. 85

Two Kinds of Activity in Late-type Stars

M.M. Katsova

Sternberg State Astronomical Institute,
Moscow State University, 119899 Moscow V-234, USSR

Several years ago we proposed a method for the analysis of X-ray observations of late-type stars. It allowed the determination in a uniform manner of coronal base electron densities for more than 40 late-type stars, in terms of a one-temperature consideration of homogeneous spherically symmetric coronae (Katsova *et al.*, 1987). Fig. 1 shows the results as a function of spectral type. Comparison of our results with values for different kinds of solar regions shows that physical characteristics of F and G star coronae correspond to densities less than those in active regions on the Sun. Values for the active K–M0 stars are comparable with those of dense steady condensations found directly above large sunspots.

On this basis, activity can be explained as an increase in that part of the stellar surface that is occupied by strong local magnetic fields. This is illustrated in the table where we compare magnetic field measurements by Saar and Linsky (1988) with our calculations.

Table

Star	Sp.class	H,G	f	$n_0 \cdot 10^8\ \mathrm{cm}^{-3}$
Sun	dG2	1500	~0.02	4
40 Eri A	dK1	2800±700	0.07	2.2
HD 17925	dK2	1500	0.35	3.7
PZ Mon	dK2e	2200±200	0.80±0.20	5.6
61 Cyg B	dK7	1500	<0.1	11
EQ Vir	dK5e	2500±300	0.80±0.15	35
AD Leo	dM4.5e	3800±260	0.73±0.06	54
AU Mic	dM2.5e	4000	0.9	69

When we go from G to M stars, the filling factor f can reach 0.8. Apparently, we are dealing here with the enhancement of solar-like activity processes.

Vilhu (1987) showed that the X-ray flux increases when we consider later-type stars. The X-ray emission for stars with values of $F_x/F_{bol} < 10^{-4}$ is connected with emission of coronal gas with $T \approx 2$–3 MK. For stars with $F_x/F_{bol} \geq 10^{-4}$–10^{-3} and the same spectral type, the X-ray luminosity begins to be determined by high-temperature gas with $T \approx 20$ MK. Apparently, this conclusion is valid both for the K-stars that are the active components in RS CVn-binaries, and for the M-dwarf stars.

This is illustrated in Fig. 2, where the emission measure EM of low (T_1) and high (T_2) temperature gas is plotted for different stars, using EXOSAT data by Schmitt *et al.* (1987). For comparison, emission measures are given also for a solar active region and for the maximum of a solar flare.

We note that the slope of the lines that connect two-temperature component emission measures, differs from those in a solar active region for the case of the RS CVn-binary σ^2 CrB. Dwarf stars with the highest level of surface activity show similar behaviour. In these stars the hot gas component begins to dominate over the low-temperature component of the coronal plasma.

What causes the appearance of such hot gas in stellar coronae? Observations of RS CVn-type binary systems suggest the existence of interaction processes involving large stellar magnetic loops located on both stars. This may give rise to the development of flare-like process magnetic field reconnection. During this process the large scale loops are filled by hot plasma with T=20 MK. Such flare loops have apparently been observed in the eclipsing binary AR Lac.

Two immediate questions remain for future research: Why are such flare loops long-lived? How are they enchanced during prolonged flares on these stars?

In stars of spectral class M, the red dwarf stars, we are faced with a different situation. These stars have the same ratio $F_x/F_{bol} \approx 10^{-3}$ as the K-stars. The X-ray emission of hot plasma occurs continually. Their radii and luminosities are small, but nevertheless there is a high efficiency of X-ray generation. In contrast to the K components of the RS CVn-systems, large hot loops apparently are absent from the coronae of active red dwarf stars.

In the past few years microflaring has been discussed as a possible mechanism of heating red dwarf star coronae. Now additional arguments can be given in favour of this viewpoint. Firstly, variability of the Balmer line emission can be traced in spectroscopic observations with high temporal resolution, carried out on the 6-m telescope (Pustilnik, 1990). Secondly, the observations of the Balmer decrement in quiescent red dwarf stars, provided by Pettersen and Hawley (1989), show that stars with high X-ray luminosities have larger fluxes in the Hα-line, in contrast to stars without X-ray emission. Comparison of the Balmer decrement observations with theoretical results for different values of temperature, density and optical depth at the Lα-line centre (Katsova, 1990) shows that physical conditions in the emitting source are close to those found in the plasma of small flares.

Thus, we have certain evidence in favour of the existence of two types of stellar activity: i) a quasistationary process of enhancement of solar-like activity with increasing filling factor due to large-scale structure; ii) a nonstationary process of development of small-scale activity such as microflaring.

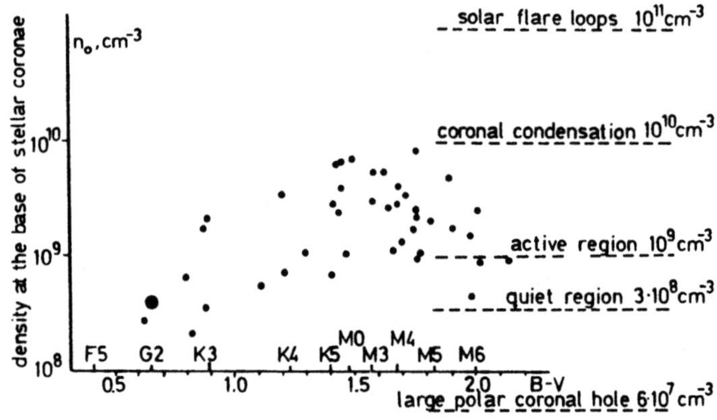

Fig. 1.

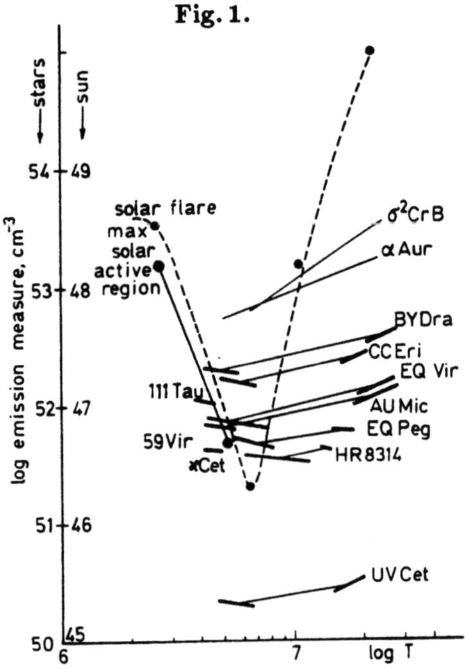

Fig. 2.

References

Katsova, M.M.: 1990, *Sov. Astron.*, in press

Katsova, M.M., Badalyan, O.G., Livshits, M.A.: 1987, *Sov. Astron.* **31**, 652

Pettersen, B.R., Hawley, S.L.: 1989, *Astron. Astrophys.* **217**, 187

Pustilnik, L.A.: 1990, private communication

Saar, S.H., Linsky, J.L.: 1986, in *Fourth Cambridge Workshop on Cool Stars, Stellar Systems and the Sun*, Santa Fe, New Mexico, p. 93

Schmitt, J.H.M.M., Pallavicini, R., Monsignori-Fossi, B.C., Harnden, F.R., Jr.: 1987 *Astron. Astrophys.* **179**, 193

Vilhu, O.: 1987, in *Fifth Cambridge Workshop on Cool Stars, Stellar Systems and the Sun*, Boulder, Colorado, p. 104

Ca II H and K and Hα, and Li Abundances in the Pleiades Late K Main-Sequence

R. J. García López [1], R. Rebolo [1], J. E. Beckman [1],

and A. Magazzù [1,2]

[1]Instituto de Astrofísica de Canarias, E-38200 La Laguna, Tenerife, Spain
[2]Osservatorio Astrofisico di Catania, Città Universitaria, I-95125 Catania, Italy

Abstract: We have observed seven main sequence stars in the Pleiades, with $B-V$ between 0.98 and 1.41 (5100–3900 K), and with a wide range of rotational velocities, in Ca II H and K, Hα and have derived Li abundances. Our results, combined with literature data, indicate that the most chromospherically active stars are rapid rotators, and that at a given effective temperature the Li-rich stars show the highest chromospheric activity. A different surface coverage of active regions could influence the dichotomy observed in the Li abundance distribution.

1. Introduction and observations

The Pleiades is a nearby, young ($6-7\times10^7$ yrs), open cluster. Differences of a few 10^7 years have been proposed between the age derived from the MS-turnoff and the age estimated by pre-MS modelling of the low mass stars (Herbig, 1962; Stauffer, 1980; Duncan and Jones, 1983; Stauffer *et al.*, 1984), implying a spread of times of star formation. Recently, however, Mazzei and Pigatto (1989) have obtained an age of 1.5×10^8 yrs, by comparing the cluster HR diagram with isochrones derived from overshooting evolutionary models, thus reducing the importance of a star formation that is dispersed in time.

In this cluster the K and M stars are particularly useful for studying the common age question. Plots of $v\sin i$ (Stauffer and Hartmann, 1987) show a trend to lower values at later spectral type, and an apparently bimodal distribution, with a gap for $20 < v\sin i < 50$ km s^{-1}. Butler *et al.* (1987) found in 4 rapid rotators out of 11 K Pleiades stars Li to be more abundant by an order of magnitude than in 4 slow rotators, suggesting that this difference in Li is due to an age difference between the two groups.

We observed 7 Pleiades dwarfs with $B-V$ between 0.98 and 1.41, on 17th and 19th Nov. 1989, with the ISIS spectrograph at the Cassegrain focus of the 4.2 m William Herschel telescope on La Palma. ISIS permits simultaneous blue and red observations. In the red, using a CCD, we could observe both Hα (λ6563 Å) and Li I λ6708 Å at a dispersion of 0.37 Å pixel^{-1} and a resolution $\lambda/\Delta\lambda \sim 1 \times 10^4$. In the blue we observed Ca II H and K ($\lambda\lambda$3933, 3969 Å) using an IPCS detector at a dispersion of 0.11 Å pixel^{-1}, with a resolution $\lambda/\Delta\lambda \sim 1 \times 10^4$. A standard data reduction was performed using the IRAF software package.

2. Chromospheric activity and Li abundances

We computed the absolute flux in Ca II H and K emission using the method of Linsky *et al.* (1979), which calibrates the continuum near H and K against $V-R$. In Figure 1 we show Ca II K flux (F(K)) versus effective temperature ($T_{\rm eff}$), including also the results of Blanco *et al.* (1974). The diameters of the symbols are proportional to the size of $v \sin i$. We see an upper envelope in the flux, which falls towards later spectral type, similar to the relation found in field stars (Kelch *et al.*, 1979; Strassmeier *et al.*, 1990). For each spectral type the most active stars tend to be the rapid rotators, as expected for dynamo-induced activity. Also, comparing the mean flux in Ca II K for a given spectral type with those from the literature for older systems, we find a clear fall-off in chromospheric activity with age, comparable to that found in G dwarfs by Barry (1988). In fact, the mean value of this activity can be used as a criterion for cluster membership (Kraft and Greenstein, 1969).

We have found that the equivalent width of Hα shows a clear trend to higher values with later spectral type, as already presented by Stauffer and Hartmann (1987). For Hα, just as for Ca II K, the rapid rotators of a given spectral type show more activity.

In Fig. 2 we show the abundances of Li against $T_{\rm eff}$, found using our equivalent widths and those from Butler *et al.* (1987), (for details see García López *et al.*, 1990). There is clearly a large scatter at a given spectral type, reaching an order of magnitude at certain $T_{\rm eff}$ values. The Li dichotomy between rapid and slow rotators found by Butler *et al.* (1987) seems to be present, but confined to stars hotter than 4500 K. Two of our stars in this range have low $v \sin i$ but Li abundances comparable with those of the rapid rotators. They could be rapid rotators whose axes are oriented near to the line of sight, a hypothesis supported by their Ca II K fluxes which are comparable with those from the rapid rotators (Fig. 1).

The abundance scatter could be real, or, in part, an observational effect due to chromospheric activity. The Li-rich stars are generally the most chromospherically active stars. The largest flux variations in Ca II K detected in Fig. 1 for a given spectral type are in the range 40–70 %. The differences in surface coverage by active regions between the "Li-rich" and "Li-poor" stars, giving rise to the detected differences in chromospheric activity, could affect the derived values of the Li abundances. Therefore the scatter in Li abundances may not imply a sequential

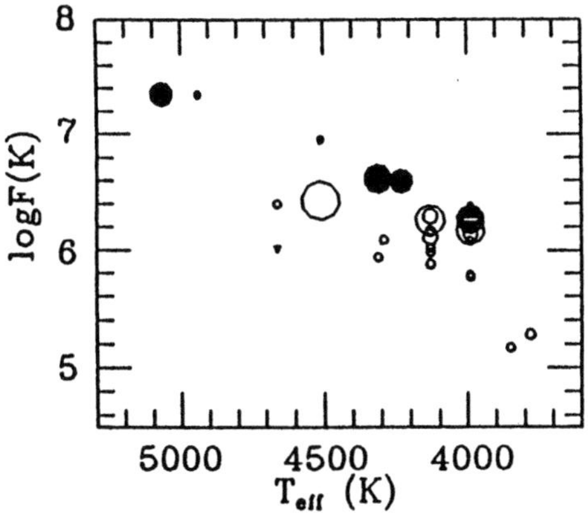

Fig. 1. Ca II K flux against T_{eff} for K and M dwarfs in the Pleiades. The diameters of the symbols are proportional to $v \sin i$. Open symbols are values given by Blanco *et al.* (1974); filled symbols correspond to this work.

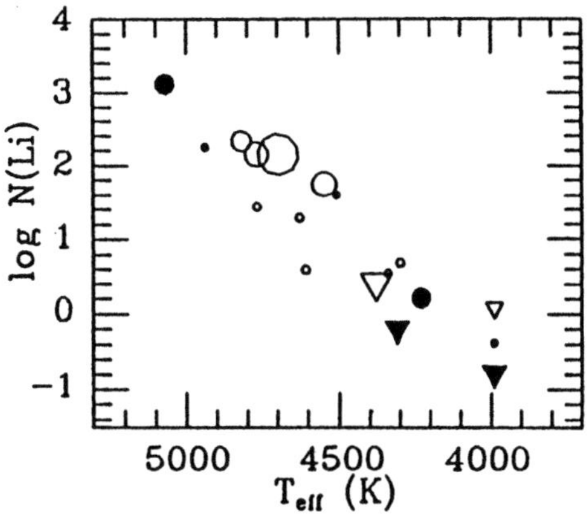

Fig. 2. Li abundances v. T_{eff} for low mass Pleiades stars. Open symbols: our measurements; filled symbols: equivalent widths given by Butler *et al.* (1987); their sizes are proportional to $v \sin i$.

formation in the cluster. For stars cooler than 4500 K there is no apparent relation between $v \sin i$ and logN(Li). In these very cool stars the differences between the photospheric and spot temperatures are less, and the resulting effect on the Li abundance correspondingly reduced. Further observations are required to test these effects in the hotter stars.

References

Barry, D. C.: 1988, *Astrophys. J.* **334**, 436

Blanco, C., Catalano, S., Marilli, E., Rodonò, M.: 1974, *Astron. Astrophys.* **33**, 257

Butler, R. P., Cohen, R. D., Duncan, D. K., Marcy, G. W.: 1987, *Astrophys. J. Letters* **319**, L19

García López, R. J., Rebolo, R., Magazzù, A., Beckman, J. E.: 1990, in IAU Symp. 145 *Evolution of Stars: The Photospheric Abundance Connection*, in press

Kelch, W. L., Linsky, J. L., Worden, S. P.: 1979, *Astrophys. J.* **229**, 700

Kraft, R. P., Greenstein, J. L.: 1969, in *Low Luminosity Stars*, ed. S. S. Kumar, Gordon & Breach, London, p. 65.

Linsky, J. L., Worden, S. P., McClintock, W., Robertson, R. M.: 1979, *Astrophys. J.* **41**, 47

Mazzei, P., Pigatto, L.: 1989, *Astron. Astrophys.* **213**, L1

Stauffer, J. R.: 1980, *Astrophys. J.* **280**, 189

Stauffer, J. R., Hartmann, L.: 1987, *Astrophys. J.* **318**, 337

Stauffer, J. R., Hartmann, L., Soderblom, D., Burnham, J. N.: 1984, *Astrophys. J.* **280**, 202

Strassmeier, K. G., Fekel, F. C., Bopp, B. W., Dempsey, R. C., Henry, W. H.: 1990, *Astrophys. J. Supp. Ser.* **72**, 191

Is ξ Boo AB a Young or an Evolved System? A Model Atmosphere Analysis

I.S. Savanov

Crimean Astrophysical Observatory, Nauchny, 334413 Crimea, USSR

Abstract: We perform a model atmosphere analysis of ξ Boo A, and make estimates for the secondary component. The abundances of 25 elements for ξ Boo A are obtained with the parameters: $T_{\rm eff}$=(5300 ± 100) K, log g= 4.1 ± 0.2, ξ_t = 0.5 km/s. The atmosphere of ξ Boo A is deficient in Y and Zr, but overabundant in Ba and rare earth elements. We confirm the high Li abundance in ξ Boo A ($\log A(\mathrm{Li}) = 2.1$), while Li is strongly underabundant in the secondary relative to the Sun. From comparison with evolutionary calculations we find that ξ Boo A has mass 0.95 $M_\odot$, radius 1.4 $R_\odot$; we also determine a rotational velocity $v \sin i = 8$ km/s. We conclude that, in spite of its relatively high rotation rate and high chromospheric activity, it is already evolved from the main-sequence.

The components of the ξ Boo AB (HD 131156) have been the objects of many investigations since 1965 when Herbig noted that the Li abundance in the atmosphere of the primary is one of the highest among the G type stars. This star has strong and variable Ca II and ultraviolet emission line fluxes (Baliunas *et al.*, 1985; Noyes *et al.*, 1984). Measurements of magnetic field on ξ Boo A have indicated the possibility of its variability (e.g. Marcy, 1984). From observations of Ca II K emission and linear polarization, Saar *et al.* (1988) constructed a map of the stellar active regions.

We performed our model atmosphere analysis on the basis of high dispersion observational data published by Boyarchuk and Eglitis (1983) and additionally obtained high S/N spectra with the CCD-camera in the coudé focus of the 2.6 m Shajn reflector in Crimean Astrophysical Observatory. Models of atmospheres by Bell *et al.* (1975) were used in the analysis. The estimates of the effective temperature and gravity were made on the basis of the $T_{\rm eff} - \log g$ diagram (Fig. 1). The statistical dependence $\log T_{\rm eff} = 3.908 - 0.234(B - V)$ was applied according to Noyes *et al.* (1984). The microturbulence was determined from the lines of Fe I.

Abundances of 25 elements were obtained with the following parameters for the A component: $T_{\rm eff} = (5300 \pm 100)$ K, $\log g = 4.1 \pm 0.2$ and $\xi_t = (0.5 \pm 0.5)$ km/s. Figure 2 shows the results of our calculations. Besides Li, Ba and heavier elements also show overabundance. The mean value of the overabundance of the rare earth elements is equal to 0.77 dex. Y and Zr, on the contrary, are underabundant.

According to the measurements of the equivalent width of the Li I λ6707 Å blend from the CCD spectrogram obtained on June 27, 1986, the Li abundance is 2.1. Our value of the ratio [Li/Ca] is 1.05 and is in agreement with the estimate

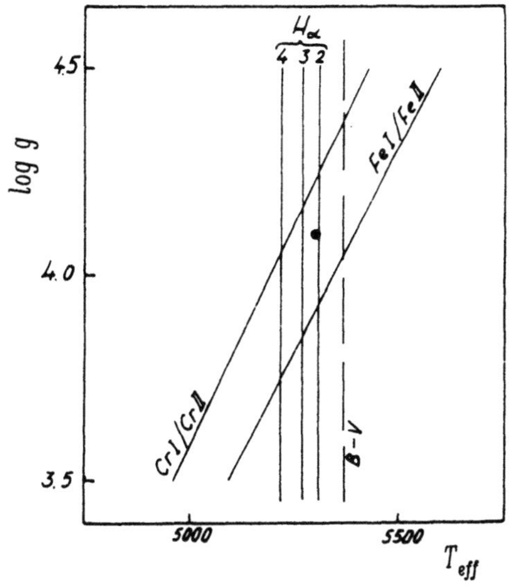

Fig. 1. Log $T_{\rm eff}$ – log g diagram. The circle refers to the adopted model.

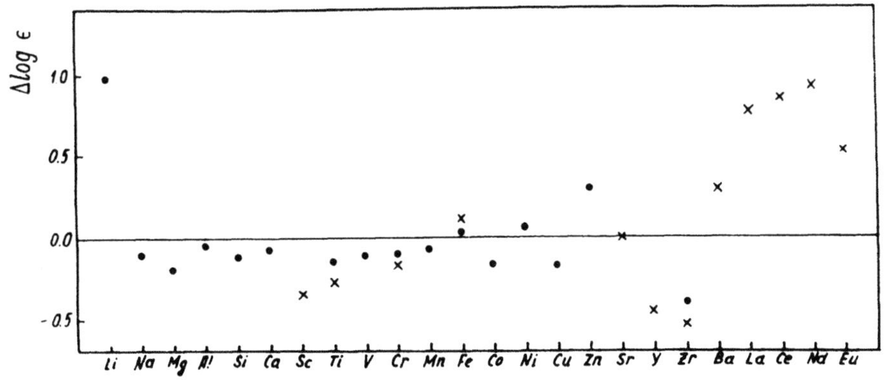

Fig. 2. Comparison between the chemical composition of the ξ Boo A and solar atmosphere. Open circles represent abundances obtained from lines of neutral elements and crosses represent the results from the ions.

by Herbig (1965) who obtained [Li/Ca]=1.39. We could not find the lithium line in the spectra of the secondary. We conclude that equivalent width of the lithium line in the spectrum of ξ Boo B is less than 4 mÅ. Assuming for the component B the atmospheric parameters $T_{\rm eff}$ = 4500 K, log g = 4.5 and ξ_t = 0.5 km/s, we found that difference in the Li abundance in the atmospheres of the components is more than 2.5 dex.

Using the evolutionary calculations by Mengel *et al.* (1979) we found the mass, radius and luminosity of ξ Boo A: $M/M_\odot = 0.95$, $R/R_\odot = 1.4$ and $L/L_\odot = 1.5$ (Fig. 3). The precision of the mass estimate is connected with the uncertainties of the $\log g$ determination. Our spectroscopic mass is slightly larger than that given by the equation $\log(M/M_\odot) = 0.38 - 0.42(B - V)$ (Noyes *et al.*, 1974). According to this equation the masses of the components are 0.91 and 0.61 $M_\odot$. These values are in agreement with the total mass of the system (1.58$\pm$ 0.19) $M_\odot$ (Wielen, 1962).

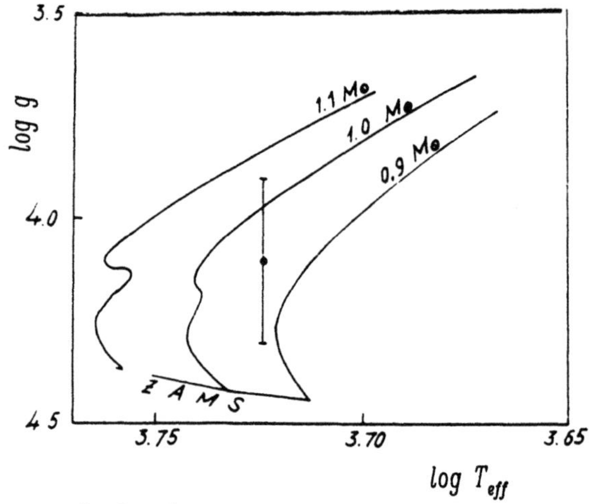

Fig. 3. Evolutionary tracks for the stars with 0.9, 1.0 and 1.1 $M_\odot$. The circle refers to the position of ξ Boo A.

From the comparison of line profile calculations of Ca I λ6717 Å with that observed in the spectrum of ξ Boo A we obtain $v \sin i = (8 \pm 2)$ km/s. Assuming $R = 1.4\,R_\odot$ and that the period of the variations in equivalent widths and spectral line asymmetries is $P = 6.43$ d (Toner and Gray, 1988) we conclude that the inclination of the rotational axis is about 45° and is close to the inclination of the orbit of the whole system, which according to Wielen (1962) is 40°.

On the basis of our results we have to conclude that ξ Boo A has already evolved from the main sequence and is not a young main sequence star. This result seems not to be in accordance with the high chromospheric activity and relatively high rotation rate of the star.

References

Baliunas, S.L. *et al.*: 1985, *Astrophys. J.* **294**, 310

Bell, R.A., Eriksson, K., Gustafsson, B., Nordlund, Å.: 1976, *Astron. Astrophys. Suppl.* **23**, 37

Boyarchuk, A.A., Eglitis, I.: 1983, *Bull. Crimean Astrophys. Obs.* **67**, 13

Herbig, G.H.: 1965, *Astrophys. J.* **141**, 588

Marcy, G.W.: 1984, *Astrophys. J.* **275**, 286

Mengel, J.D.: 1979, *Astrophys. J. Suppl. Ser.* **40**, 733

Noyes, R.W. *et al.*: 1984, *Astrophys. J.* **279**, 763

Saar, S.H., Huovelin, J., Tuominen, I.: 1988, *Astrophys. J.* **329**, 882

Toner, C.G., Gray, D.F.: 1988, *Astrophys. J.* **334**, 1008

Wielen, R.: 1962, *Astron. J.* **67**, 599

Spectroscopic Analysis of HR 1099

I.S. Savanov [1], I. Tuominen [2]

[1]Crimean Astrophysical Observatory, Nauchny, 334413, Crimea, USSR
[2]Observatory and Astrophysics Laboratory, University of Helsinki, Tähtitorninmäki, 00130 Helsinki, Finland

Abstract: We have calculated synthetic spectra for HR 1099 for the wavelength region λ6425–6445 Å, which includes the CaI λ6439 Å and FeI λ6430 Å lines, often used for the surface imaging of rapidly rotating late-type stars. The best agreement between the theoretical and observed spectra can be achieved with solar abundances in the both components of the system, except for Ca in the primary. Theoretical calculations give the possibility of estimating the influence of microturbulence, damping and blending on Ca I and Fe I line profiles.

HR 1099 (V711 Tau) is one of the most investigated RS CVn type stars. Our spectroscopic analysis is based on the spectra of HR 1099 that can be found in the literature (Fekel, 1983; Vogt, 1983). Observations were analysed using model atmospheres from a grid of Bell *et al.* (1975) with additional models obtained from K. Eriksson for the temperature range 5500–4000 K and gravities 3.00–4.50. Following the results by Fekel (1983), we used models with the following parameters: $T_{\rm eff} = 4500$ K, $\log g = 3.75$ for the primary and $T_{\rm eff} = 5500$ K, $\log g = 4.5$ for the secondary. The values of $v \sin i$ were adopted from Vogt and Penrod (1982): 38 km/s for the primary and 13 km/s for the secondary.

A composite spectrum of the double system at arbitrary phase was synthesized by the methods described by Lyubimkov and Samedov (1987) - see Fig. 1. The radii of the components and the radial velocity curve were published by Fekel (1983). From the results of our calculations for the spectral region λ6425–6445 Å (Fig. 2) we can conclude that elements such as Fe and Ca, the lines of which are best represented in this region, have abundances close to solar in the secondary component. The best fit with the profiles of the chromospherically active component can be obtained with solar abundances of Fe, but the abundance of Ca must be reduced by 0.5 dex. The remaining discrepancy between the observed and theoretical profiles may be connected with the inaccuracy of atmospheric parameters, atomic data for the spectral lines and with the existence of spotted areas on the surface of the star. In the present report we use the standard method for the rotational broadening of the lines (e.g. Gray, 1976, p. 394), without a calculation of the local line profile explicitely on the visible disc for different distances from

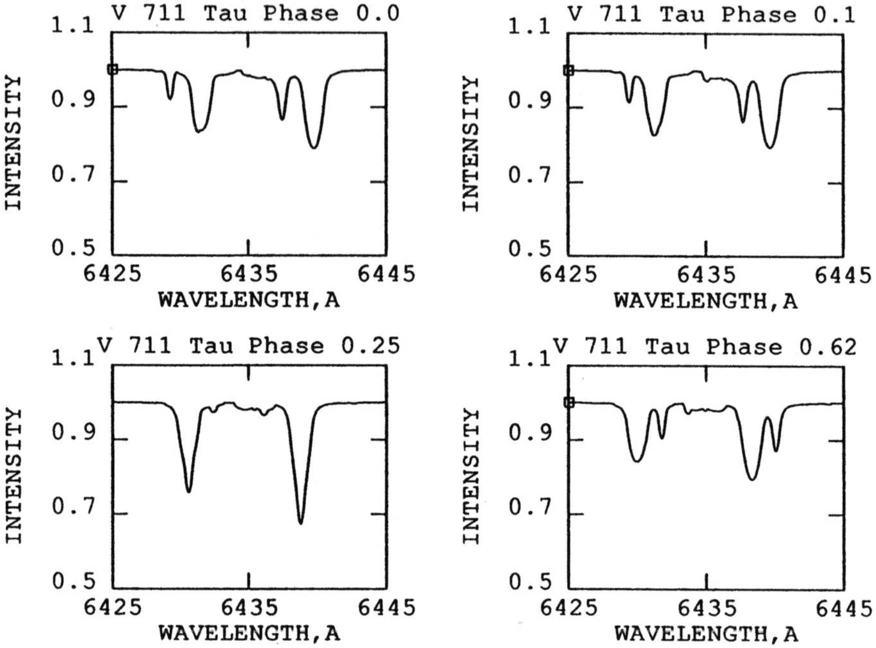

Fig. 1. Examples of the theoretical composite spectra of the HR 1099 system for four orbital phases.

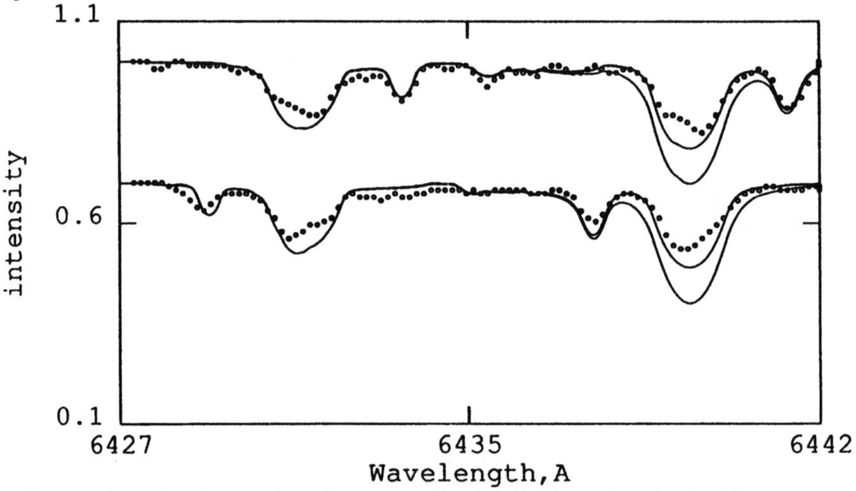

Fig. 2. Comparison between the observed (Fekel, 1983) and calculated spectrum for two phases: 0.565 (top) and 0.919 (bottom). Synthetical spectra are produced for two abundances of Ca in the atmosphere of the primary: log A(Ca) = 5.90 (upper curve) and log A(Ca) = 6.40 (lower curve). Solar abundances are assumed for other elements in the atmospheres of both components.

the disc centre. This approximation, which should be avoided in the mapping procedures of surface inhomogeneities (see e.g. Piskunov, and Hackman *et al.*, these Proceedings), influences the form of the line profile in the line centre.

Figure 3 provides the possibility of estimating the influence of microturbulent velocity ξ_t and changes in damping constant γ on the Ca I and Fe I lines. The total damping constant in our calculations is assumed to be the sum of radiative, Stark and van der Waals broadening (e.g. Gray, 1976, p. 250), the last used with a scaling factor of 1.5. The behaviour of the line profiles is different because of the different strengths of the lines and stronger saturation of Ca I line. While damping changes the whole profile, the microturbulent velocity does not influence the wings. For both Ca I and Fe I lines, the best fit can be obtained with the classical values of damping and ξ_t = 1.0 km/s. It can also be concluded that changes in the parameters γ and ξ_t can have similar effects to the changes in abundances.

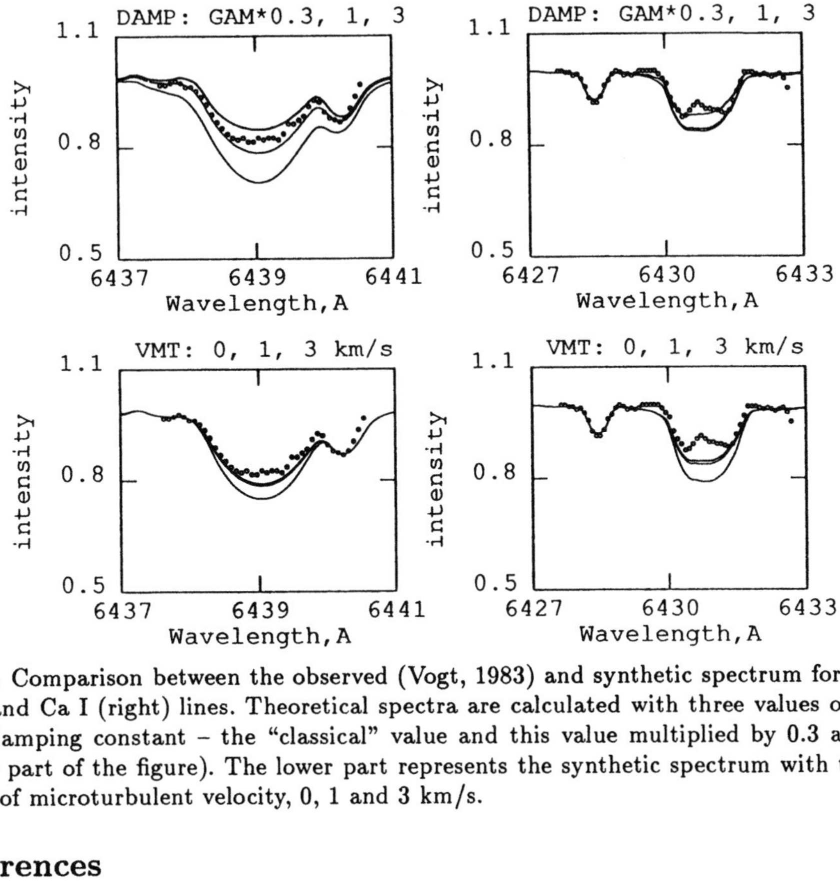

Fig. 3. Comparison between the observed (Vogt, 1983) and synthetic spectrum for Fe I (left) and Ca I (right) lines. Theoretical spectra are calculated with three values of the total damping constant – the "classical" value and this value multiplied by 0.3 and 3 (upper part of the figure). The lower part represents the synthetic spectrum with three values of microturbulent velocity, 0, 1 and 3 km/s.

References

Bell, R.A., Eriksson, K., Gustafsson, B., Nordlund Å.: 1976, *Astron. Astrophys. Suppl.* **23**, 37

Fekel, F.C.: 1983, *Astrophys. J.* **268**, 274

Gray, D.F.: 1976, in *The observations and analysis of stellar photospheres*, Wiley, New York

Lyubimkov, L.S., Samedov, Z.A.: 1987, *Bull. Crimean Astrophys. Obs.* **77**, 109

Vogt, S.S., Penrod, G.D.: 1983, *Publ. A. S. P.* **95**, 565

Recent Advances in our Understanding of Chromospheric and Coronal Heating Mechanisms

Jeffrey L. Linsky[1]

Joint Institute for Laboratory Astrophysics [JILA]
University of Colorado & National Institute of Standards and Technology
Boulder, CO 80309-0440, USA

Abstract: I will summarize some of the principal scientific results presented at the Conference on "Mechanisms of Chromospheric and Coronal Heating" held on 5–8 June 1990 in Heidelberg, Germany. The Conference included invited and contributed papers on observations that point to specific heating mechanisms and theoretical papers on the heating mechanisms themselves. There were many opportunities for useful interaction between proponents of these two approaches to understanding stellar chromospheres and coronae. I will concentrate on what is being learned from the empirical side, and then summarize the heating mechanisms discussed and for which types of stars they may be applicable.

1. Introduction

The organizers of this IAU Colloquium have asked me to summarize some of the principal scientific results presented at the Conference on "Mechanisms of Chromospheric and Coronal Heating" held on 5–8 June 1990 in Heidelberg, Germany. This topic is very germane to the present meeting, since the conversion of mechanical and magnetic energy into heat is the poorly understood intermediate stage that couples the generation of magnetic fields deep inside a star to the observable electromagnetic radiation that we use to study and characterize stellar magnetic activity. While certainly pertinent to the present IAU Colloquium, the mechanisms by which mechanical and magnetic energy may be converted into heat are both numerous and often still poorly understood. I encourage the reader, therefore, to delve deeply into the many excellent review and contributed papers that will appear in the Proceedings of the Heidelberg Conference to be published as *Mechanisms of Chromospheric and Coronal Heating*. In the meantime, here is my personal summary of some of the highlights of the meeting in a logical progression from the empirical to the theoretical.

---

[1]Staff Member, Quantum Physics Division, National Institute of Standards and Technology

2. Observations that point to specific heating mechanisms

2.1 Chromospheres

2.1.1 CO cooling and thermal bifurcation

Ayres and Testerman (1981) first called attention to the low brightness temperatures observed in the cores of the CO fundamental (4.7 μm) and first overtone (2.35 μm) vibration-rotation bands at the solar limb. Subsequent work by Ayres *et al.* (1986) and a careful non-LTE study of the formation of these molecular lines by Ayres and Wiedemann (1989) left little doubt that the low brightness temperatures in the line cores ($T_b \leq 3800$ K) demonstrate the presence of very cool plasma in the solar chromosphere, a region that was presumed to be a layer with a steeply rising temperature-height distribution. This unexpected result led to a new paradigm of a thermally bifurcated chromosphere consisting of (i) a hot component heated perhaps by magneto-acoustic waves and cooled by H^- and ionized metals, and (ii) a cool component having minimal nonradiative heating, cooled by the CO vibration-rotation bands and presumably a very large number of metallic lines.

Two-component models for the solar atmosphere have now been computed by Ayres, Muchmore, Kneer, and others. Anderson (1989) confirmed that in the absence of nonradiative heating, a radiative equilibrium model for the solar atmosphere will show a monotonic decrease in temperature with height, when a very large number of metallic lines are treated in non-LTE self-consistently. This model is similar to the cool component models computed by Ayres. Also, Anderson and Athay (1989) showed that the addition of a reasonable amount of nonradiative heating to the radiative equilibrium model can reproduce the temperature-height structure of the Maltby *et al.* (1986) Model C′ quite accurately.

Despite this agreement between two-component semiempirical and self-consistent non-LTE model atmospheres, Ayres (1991) mentioned a number of unanswered questions. One is whether the dominant cooling agent in the cool component is CO, as originally thought, or large numbers of iron and other metallic lines. A second question is whether the solar chromosphere is actually thermally unstable.

Ayres pointed out that one sees 200–300 K spatial variations in T_b for the strong CO lines in the 4.7μm band when a 5" slit is laid across the solar surface. These spatial variations are presumably caused by transient heating events, the Ca II flashes, that have time scales of roughly 200 s. By comparison, the time scale for cooling by CO at chromospheric densities, $t_{cool}(\mathrm{CO}) \sim 1000$ s, and the time scale for CO lines to become optically thick, $t_\tau \sim 200$ s, are independent of the stellar gravity. Since $t_{cool}(\mathrm{CO})$ is much longer than the transient heating time scale, the response of the solar atmosphere to heating is a rapid rise in temperature, which is observed as the large spatial variations in T_b for the CO lines. The picture described by Ayres has been criticized by Athay and Dere (1990) who concluded from HRTS data that the chromospheric temperature rise is present over ~90% of the solar surface.

Muchmore *et al.* (1987) and others have argued that cooling by CO, SiO, and perhaps other molecules is thermally unstable in stars across much of the cool half of the H–R diagram and that these stars, like the Sun, should show thermally bifurcated chromospheres. Cuntz and Muchmore (1989), for example, computed the propagation of acoustic waves in the atmosphere of Arcturus (K1 III). They found that for weak shocks the atmosphere is cool with CO and SiO dominating the cooling, whereas strong shocks lead to a hot chromosphere in which CO and SiO are not present. Ayres and Wiedemann (1989) have proposed a new two-component semi-empirical model for Arcturus. Johnson (1991) discussed the recent model of the C star TX Pyx computed by Luttermoser *et al.* (1989) in which the cool component is indicated by Mg I and CO and the hot chromosphere is indicated by the Mg II resonance lines in emission. Thus thermal bifurcation appears to be present even in stars with effective temperatures, gravities, and chemical compositions much different from the Sun.

2.1.2 Chromospheric bright points

The solar 3-minute chromospheric oscillation seen in time-resolved spectra of the Ca II resonance lines (e.g. Cram and Damé, 1983) has been explained as an upward-propagating excitation that leads to intense heating in bright points located in the chromospheric network. Rutten (1991) provided a more detailed explanation. He showed that the intensity fluctuations (seen as K_{2v} brightenings) are formed deeper in the chromosphere than the velocity changes (seen as wavelength shifts in the K_3 absorption feature). This behavior is consistent with standing waves below the temperature minimum (T_{min}) and running waves above T_{min}. He suggested that the bright points can be understood as shocks that occur when falling matter (the result of upward propagating waves having insufficient energy to escape the atmosphere) hit the next upward propagating wave. Detailed radiative-hydrodynamic calculations in which a moving piston at the bottom of the atmosphere generates a wave train are needed to understand the data better.

An important question is whether the energy in the cell grains (also called bright points) is sufficient to heat the solar chromosphere. Rutten (1991) said that the answer is yes for the quiet solar chromosphere. Sivaraman (1991) presented new observations in which he found that the bright points all correspond to magnetic network elements (consistent with the earlier work of Sivaraman and Livingston, 1982), but the reverse is not true. He noted that bright points are born where old bright points die, indicating a persistent magnetic geometry. Their lifetimes are 100–200 s and mean magnetic fluxes are 10–20 G. Thus very localized MHD shocks are likely to be major heating mechanisms for the quiet solar chromosphere.

2.1.3 Chromospheric activity in late–type giants and supergiants

Pasquini *et al.* (1991) presented results of their study of 65 stars later than spectral type F8 with luminosity classes ranging from IV to Ib. A comparison of the Ca II K line surface fluxes of these stars with stellar masses deduced from evolutionary tracks shows that the more massive stars (initial masses 5–12 $M_\odot$) have significantly higher Ca II surface fluxes, and thus higher chromospheric heating rates, than the lower mass giants with $T_{eff} < 4500$ K.

Their result has important implications for the rate of chromospheric heating in evolved stars. They argue that massive stars start on the main sequence as rapid rotators with no convective zones, but rapid evolution to the right in the H–R diagram leads to deep convective zones with minimal loss in angular momentum. Thus dynamo generation of magnetic fields is important for the massive evolved stars and their chromospheres are heated by magnetic processes. By contrast, the lower mass stars had convective zones and lost much of their angular momentum through magnetic winds while on the main sequence and during their slow evolution through the giant branch. Thus their stellar dynamo activity is weak and the chromospheres are heated by nonmagnetic processes at basal flux levels. This picture is not unexpected (e.g. Simon and Drake, 1989), but the Pasquini *et al.* data set provides the first clear empirical support. For the stars slightly more massive than the Sun, Simon and Drake (1989) find a rapid decline in ultraviolet emission lines near spectral types G0 IV and G0 III which they interpret as due to a transformation from acoustic heating in the early F stars to magnetic heating processes in the cooler stars.

Cuntz and Stencel (1991) called into question this simplistic picture for the least active cool giants like Arcturus. They mentioned the well-known observational result that these stars show variable He I 10830 Å absorption or emission that appears to require either a large amount of plasma at $T_e \sim 4 \times 10^4$ K or photoionization by x-rays. There is no evidence as yet in support of either explanation. Recent calculations by Cuntz and Luttermoser (1990) indicate that a stochastic distribution of acoustic wave periods leads to the overtaking and merging of waves. This can occasionally produce very strong shocks with $T_e \geq 4 \times 10^4$ K in the postshock region and absorption in the He I line which vary with time. This may explain the transient He I features within the context of acoustic heating. For these stars the mass loss could be produced by pulsations or acoustic waves with wavelengths similar to a stellar radius (e.g. Cuntz, 1990).

Dupree (1991) noted that episodic heating may explain the formation of the He II 1640 Å line in the Sun and late-type stars. She also suggested that pulsations may be related to the nonmagnetic heating mechanism in the M supergiants like α Ori and in giants near the tip of the red giant branch.

2.2 Transition regions

A standard procedure in the analysis of observed line fluxes from solar and stellar transition regions ($6.0 \leq logT_e \leq 4.3$) is to determine an emission measure–temperature distribution, $\mathrm{EM}(T_e) = \int_{\Delta T_e} n_e^2 dT_e$ (e.g. Jordan and Linsky, 1987). This can be done with no assumptions concerning the geometry of the emitting plasma. Jordan (1991) called attention to the very similiar ***shapes*** of the $\mathrm{EM}(T_e)$ distributions for dwarf and active giant stars deduced from IUE spectra for emission lines formed within the range $5.3 \leq logT_e \leq 4.3$. She explained this shape similarity as reflecting the radiative properties of the plasma, $\mathrm{EM}(T_e) \sim 1/P_{rad}(T_e)$, where $P_{rad}(T_e)$ is the total power radiated in all spectral features, in the absence of significant local nonradiative heating. Thus the shape of $\mathrm{EM}(T_e)$ provides information on only the redistribution of energy within the transition region and not the local heating. Cally (1991) also argued that local heating is not needed in the solar transition region, at least below 10^5 K, because the observed EM(T_e) distribution can be explained by the balance of turbulent conductivity heat transport with radiative losses.

Feldman (1983), Habbal (1991) and others have used Skylab far-ultraviolet spectra to argue that most of the solar emission measure in the temperature range $5.3 \leq log T_e \leq 4.5$ originates in cool loops that are magnetically isolated from the chromosphere and corona, although a small portion of the emission originates in a "classical" transition region at the base of hot magnetic loops. Habbal (1991) also noted that the greatest fractional variability occurs in the O IV line formed at 1–2 $\times 10^5$ K, primarily in bright points located over the chromospheric network. Thus the inference of heating rates and details of the heating mechanism(s) for the solar transition region are complicated by the complex geometry.

2.3 Coronae

2.3.1 Are stellar coronae heated by acoustic waves?

Until the *Einstein* observations of late-type stars became available, the most popular theories for the heating of solar and stellar coronae assumed the dissipation of purely acoustic or magneto-acoustic waves. In his review talk, Rosner (1991) summarized the main arguments against the purely acoustic wave heating theory. One is that the deduced x-ray surface fluxes for late-type stars correlate well with stellar rotation period and rotational velocity but poorly with T_{eff} and gravity, the two parameters that determine the properties of the convective zone where acoustic waves are generated. A second argument is that the x-ray emission for the Sun is brightest where the magnetic field is strongest, and the highest spatial resolution solar images in coronal emission lines reveal bright coronal loops that presumably trace closed magnetic field lines (Golub, 1991). Jordan (1991) provided a third argument – measured nonthermal line widths for lines formed in the chromosphere and transition region indicate a mean wave energy of 5×10^5 erg cm^{-2} s^{-1}, which is insufficient to heat the lower corona of the Sun either by acoustic or by magnetic wave modes. Calculations by Stępień and Ulmschneider (1989) and by Hammer and Ulmschneider (1990) confirm that pure acoustic waves can heat the atmospheres of late-type stars to coronal temperatures, but these coronae would have very low base pressures and be unobservable.

2.3.2 The coronal dividing line in the H–R diagram

Using the early x-ray observations from the *Einstein* satellite, Ayres *et al.* (1981) first called attention to a rather sharp dividing line in the H–R diagram separating the G giants, which were typically detected as x-ray sources and thus have hot coronae, from the later K and M giants which were not detected as x-ray sources and thus have little or no plasma at coronal temperatures. In fact, the upper limit on the x-ray surface flux from the nearby K1 III star Arcturus is 1000 times smaller than is observed from the quiet Sun (Ayres *et al.*, 1982). Zwaan (1991) noted that this picture has not changed in the past 9 years (cf. Maggio *et al.*, 1990; Haisch *et al.*, 1990), but ROSAT observations of the noncoronal stars are awaited to test it. Zwaan's (1991) explanation for the absence of hot plasma in the cooler giants is the same evolutionary argument proposed by Pasquini *et al.* (1991) to explain the weak chromospheric emission from these stars. Thus magnetic heating processes should be very weak for the cool giants, and acoustic waves can heat only an unobservably small amount of material to coronal temperatures.

2.3.3 Overactive close binaries

The dependence of many activity indicators, such as the x-ray surface flux, on rotational period or Rossby number has been confirmed by many authors and is one of the major arguments for the magnetic nature of the heating processes in rapidly rotating stars. In his review talk, Zwaan (1991) called attention to an anomaly in this behavior; namely, the G and K subgiant components in close binary systems are "overactive" in the sense that they have x-ray surface fluxes much larger than single stars with the same rotational periods. The close binaries are rapid rotators because tidal interactions produce synchronism between the orbital and rotational periods. His explanation for "overactivity" is that the tidal forces also produce a different differential rotation structure than is present naturally in a single star, and this altered angular momentum distribution leads to enhanced dynamo generation of magnetic fields. Rodonò (1986), for example, summarizes data on RS CVn systems that shows that the differential rotation with latitude in these stars is typically 1–3 orders of magnitude smaller than for the Sun.

2.3.4 Why are the coronae of active stars so hot?

Einstein Solid State Spectrometer observations of RS CVn systems (Swank *et al.*, 1981) and the dMe star Wolf 630 AB (Swank and Johnson, 1982) indicate that a substantial portion or in some cases most of the coronal emission measure at times outside of identifiable flares is at very high temperatures ($7.4 \leq logT_e \leq 7.8$). During flares, the observed x-ray emission indicates even hotter plasma; for example, EXOSAT observations of the 29 September 1983 flare on the RS CVn system σ^2 CrB (van den Oord *et al.*, 1988) indicate $T_e = 9.5 \times 10^7$ K. Rosner (1991) asked why these active stars have coronal plasma temperatures more than an order of magnitude hotter than is observed in solar active region loop and even hotter than is observed during the thermal phases of most solar flares. He did not answer his question, but he did note that the solar x-ray luminosity is near the bottom of the observed stellar L_x distribution.

In my view it is important to know that those stars with very hot plasma temperatures generally are observed to be luminous microwave sources, produced by gyrosynchrotron emission from a time varying distribution of relativistic electrons. Thus the coronal heating process is probably closely related to the process that accelerates electrons to relativistic energies. The most likely candidate is some type of magnetic reconnection process that creates hot thermal electrons at the same time as the relativistic electrons or as a thermalization byproduct.

2.3.5 Solar coronal oscillations

Another clue concerning coronal heating processes comes from Pasachoff's (1991) observations of enhanced power in the 5303 Å line near 1 Hz during the 1980 solar eclipse he observed in India and at 0.2–3 Hz during the June 1983 eclipse observed in Indonesia. He suggested that the observed oscillatory power refers to the reflection of MHD waves at the mirroring points in coronal magnetic loops. The observed timescales are consistent with either transverse or torsional body waves in these loops, as discussed by Hollweg (1991) and Berger (1991), or the LRC circuit model discussed by Cram (1991) at the meeting.

2.4 Flux–flux relations and basal and saturated heating

A critical step in determining the heating processes at work in stellar chromospheres and coronae is an accurate measurement of the total radiative loss from the plasma within a specified temperature range. This is difficult since many spectral lines and continua contribute to the total radiative loss. Linsky (1991) reviewed some of the spectral diagnostic problems encountered in deriving differential emission measures and methods for including all major emitters, especially Fe II in the chromosphere, in determining the total power radiated from a plasma. He proposed two scaling laws by which one may simply determine the total power radiated by the chromosphere (from the Mg II h and k lines) and by the higher temperature layers (from the C IV 1550 Å feature).

2.4.1 Basal fluxes

The heating of stellar chromospheres appears to consist of two components – an active component for which the heating rate depends on stellar rotation, and a basal component that depends on T_{eff} and perhaps only weakly on gravity. In his review of this topic, Schrijver (1987a) showed that the basal radiative loss rates can be measured as lower limits to the surface fluxes in the Ca II and Mg II resonance lines and the Si II 1812 Å multiplet in flux-color diagrams as functions of stellar B–V color. Zwaan (1991) extended this work by showing that the sum of the minimum radiative fluxes in these lines can be used to infer the total radiative loss rate from a basal chromosphere and thus the basal heating rate, $F_{basal} \sim T_{eff}^{\gamma}$.

This empirical heating rate may be compared with the theoretical rates for heating by pure acoustic waves computed by Bohn (1984), although Musielak (1991) mentioned that stellar acoustic fluxes are still poorly known. One should understand that (magneto-)acoustic energy fluxes are computed for the top of a stellar convective zone and thus cannot be simply compared with observed chromospheric energy losses, since most of the wave energy is dissipated by radiation in the photosphere. Thus acoustic waves are the most viable heating mechanism for the chromospheres of the least active stars, including the early F dwarfs, slowly rotating G–M dwarfs, cool giants which are not Miras, M supergiants, and the centers of solar supergranulation cells, which have chromospheric line surface fluxes close to the basal values. Judge and Stencel (1991) recently concluded that the

Table 1. A summary of proposed heating mechanisms and where they may be important

Proposed Heating Mechanism	Where It May Be Important
(1) Short period acoustic waves • Stochastic heating (**Cuntz**) • Basal heating (**Ulmschneider**) • Multipole sources (**Musielak**)	• Weak **basal** heating for late–type stars. • Dominant heating for cool giants, M supergiants. • $P \sim P_A/10$ to $P_A/5 \Rightarrow$ shock heating in chromospheres but not coronae.
(2) 3– to 5–minute oscillations with large phase shifts (pure acoustic modes)	• Standing waves in solar photosphere do not $\Rightarrow$ shocks in solar chromosphere. • Heating in cool giants (i.e. Miras) may $\Rightarrow$ mass loss.
(3) Strong acoustic shocks strengthened by radiative amplification (radiative instabilities)	• May heat density inhomogeneities in OB star winds.
(4) Resistive current dissipation following magnetic field reconnection	• Coronae of rapidly rotating late–type stars. • Heating rate depends on v_{rot}.
(5) MHD wave dissipation processes in general (**Musielak**)	• Coronae of rapidly rotating cool stars. • Heating rate depends on v_{rot}. • Generated by turbulent and convective motions in convection zones. • Energy transmitted by motions of footpoints $\Rightarrow$ heating of coronal magnetic structures. • Energy input also from magnetic fluxtube emergence.
(5a) MHD fast mode waves (**Stein**)	• May heat coronal loops when waves produced locally. • Generated by mode–mode coupling.
(5b) Transverse and torsional Alfvén body waves (**Hollweg, Berger**)	• Propagate readily through the corona. • Can heat coronal loops by resonances, phase mixing, mode coupling, turbulence, and Landau damping.
(5c) Slow mode MHD waves and longitudinal tube waves (magneto-acoustic type waves) (**Stein**)	• Propagate from convective zone to chromosphere. • Radiative damping in photosphere and chromosphere. • Heat the low and middle chromosphere by dissipation of shocks. • Easily generated by mode coupling from other waves.
(5d) Alfvénic surface waves (**Roberts**)	• Heat coronal loops by resonant absorption in thin sheaths.
(5e) Shear Alfvén waves	• Generated by phase mixing, mode–mode coupling, and turbulent heating. • May heat coronal holes. • Heating $\Rightarrow$ thermal energy to drive solar wind. • Adds momentum to high speed solar streamers.
(6) Anomalous current dissipation and and magnetic reconnection (**Priest, Spicer, Jardine**)	• May heat coronal loops. In vortex structures? • Intermittent heating mechanisms. • Heating in thin sheets or layers.
(7) Resonant LRC circuit (**Cram**)	• Useful formalism.
(8) Microflares and nanoflares (**Parker**)	• Same as current dissipation. • May heat coronal loops quasi-statically.
(9) Large scale flows	• Flows in spicules. • Initiated by other heating mechanisms.

basal flux limits deduced by Schrijver (1987b) can be extrapolated to giant stars with very low effective temperatures.

2.4.2 Saturated fluxes

At the opposite extreme, there appears to be maximum values for the observed surface fluxes in the chromospheric resonance lines of Ca II and Mg II, transition region lines like C IV 1550 Å and the coronal soft x-ray flux. Vilhu (1987) and Vilhu and Walter (1987) showed that these saturated fluxes occur in the most rapidly rotating stars of a given color. The saturated surface fluxes could be interpreted as those corresponding to the maximum rate of steady-state heating possible given the available mechanical energy in the convective motions and complete surface coverage by equipartition magnetic fields (about 1500 G in the solar photosphere). Schrijver (1987a,b) noted that for solar plages the Mg II surface fluxes are a factor of 10–30 below the saturated values, while typical magnetic flux densities are about a factor of 15 below the equipartition values, implying that strong magnetic fields only fill about 10% of the surface area. On the other hand, the heating rates for plages in RS CVn systems are consistent with saturated heating (Linsky 1991) both in the chromosphere and in the higher temperature layers.

Schrijver (1991) called attention to empirical scaling laws of the form, $F_{diagnostic} = |B|^{\alpha}$, which relate the solar and stellar surface fluxes in a specific spectral diagnostic to the magnetic flux density either for small regions on the solar surface or a mean value for stars with $|B| = fB$, where f is the magnetic filling factor and B the magnetic field strength determined using Zeeman broadening techniques (e.g. Saar, 1987). The value of α increases from 0.6 for the Ca II K line and Mg II, to 0.7 for C IV, and to 1.0 for coronal x-ray emission. Thus the chromospheric diagnostics do not count magnetic flux tubes. Schrijver (1991) argues that the change in α with temperature indicates that the heating mechanisms for the solar chromosphere and corona are probably different.

2.4.3 Explanations for flux–flux relations

Using the early observations from IUE and *Einstein*, Ayres, Marstad and Linsky (1981) showed that the normalized fluxes of diagnostics formed at different temperatures can be fit by power laws in which the index becomes larger as the temperature difference at which the diagnostics are formed increases. Zwaan (1991) reviewed more recent work that shows that tighter power law relations follow when basal fluxes are subtracted from the surface fluxes of chromospheric lines, i.e. $F_{diagnostic} \sim (F_{chrom} - F_{basal})^{\beta}$. The quantity $\beta(T_e)$, which increases with temperature, can be viewed as an activity parameter. Zwaan did not propose an explanation for the temperature dependence of $\beta(T_e)$, but Jordan (1991) argued that local heating is unimportant in the lower transition region ($T_e \leq 2 \times 10^5$ K). Therefore, the shape of $EM(T_e) \sim P_{rad}^{-1}(T_e)$ and the flux–flux plots provide information only on how energy is redistributed in the lower transition region and not on the local heating mechanism.

3. Chromospheric and coronal heating mechanisms

The remainder of the meeting was devoted to presentations of the many different mechanisms that may heat stellar chromospheres and coronae. I do not have the space to summarize these presentations in detail, but I include in Table 1 a concise summary of the different mechanisms discussed by the various speakers (their names are in **bold** type) and in what types of stars they may be important. A more complete description of these mechanisms may be found in the recent review paper by Narain and Ulmschneider (1990).

Acknowledgements. This work is supported by NASA grants NAG6–82 to the University of Colorado and S–56460–D to the National Institute of Standards and Technology. I would like to thank Dr. M. Cuntz for his comments on the manuscript.

References

Anderson, L.S.: 1989, *Astrophys. J.* **339**, 558.
Anderson, L.S., Athay, R.G.: 1989, *Astrophys. J.* **346**, 1010.
Athay, R.G., Dere, K.P.: 1990, *Astrophys. J.* **358**, 710.
Ayres, T.R.: 1991, see Ulmschneider *et al.*
Ayres, T.R., Simon, T., Linsky, J.L.: 1982, *Astrophys. J.* **263**, 791.
Ayres, T.R., Linsky, J.L., Vaiana, G.S., Golub, L., Rosner, R.: 1981, *Astrophys. J.* **250**, 293.
Ayres, T.R., Marstad, N.C., Linsky, J.L.: 1981, *Astrophys. J.* **247**, 545.
Ayres, T.R., Testerman, L.: 1881, *Astrophys. J.* **245**, 1124.
Ayres, T.R., Testerman, L., Brault, J.W.: 1986, *Astrophys. J.* **304**, 542.
Ayres, T.R., Wiedemann, G.: 1989, *Astrophys. J.* **338**, 1033.
Berger, M.A.: 1991, see Ulmschneider *et al.*
Bohn, U.H.: 1984, *Astron. Astrophys.* **136**, 338.
Cally, P.S.: 1991, see Ulmschneider *et al.*
Cram, L.E.: 1991, see Ulmschneider *et al.*
Cram, L.E., Damé, L.: 1983, *Astrophys. J.* **272**, 355.
Cuntz, M.: 1990, *Astrophys. J.* **353**, 255.
Cuntz, M., Luttermoser, M.: 1990, *Astrophys. J. Letters* **353**, L39.
Cuntz, M., Muchmore, D.: 1989, *Astron. Astrophys.* **209**, 305.
Cuntz, M., Stencel, R.E.: 1991, see Ulmschneider *et al.*
Dupree, A.K.: 1991, see Ulmschneider *et al.*
Feldman, U.: 1983, *Astrophys. J.* **275**, 367.
Golub, L.: 1991, see Ulmschneider *et al.*
Habbal, S.R.: 1991, see Ulmschneider *et al.*
Haisch, B.M., Bookbinder, J.A., Maggio, A., Vaiana, G.S., Bennett, J.O.: 1990, *Astrophys. J.*, in press.
Hammer, R., Ulmschneider, P.: 1990, in *Cool Stars, Stellar Systems, and the Sun*, ed. G. Wallerstein (San Francisco: Astronomical Soc. Pacific), p. 51.
Hollweg, J.: 1991, see Ulmschneider *et al.*
Johnson, H.R.: 1991, see Ulmschneider *et al.*

Jordan, C.: 1991, see Ulmschneider *et al.*
Jordan, C., Linsky, J.L.: 1987, in *Exploring the Universe with the IUE Satellite*, ed. Y. Kondo et al. (Dordrecht: Reidel), p. 259.
Judge, P., Stencel, R.E.: 1991, *Astrophys. J.*, in press.
Linsky, J.L.: 1991, see Ulmschneider *et al.*
Luttermoser, D.G., Johnson, H.R., Avrett, E.H., Loeser, R.: 1989, *Astrophys. J.* **345**, 543.
Maggio, A., Vaiana, G.S., Haisch, B.M., Stern, R.A., Bookbinder, J., Harnden, F.R. Jr., Rosner, R.: 1990, *Astrophys. J.* **348**, 253.
Maltby, P., Avrett, E.H., Carlsson, M., Kjeldseth-Moe, O., Kurucz, R.L., Loeser, R.: 1986, *Astrophys. J.* **306**, 284.
Muchmore, D., Nuth III, J.A., Stencel, R.E.: 1987, *Astrophys. J. Letters* **315**, L141.
Musielak, Z.: 1991, see Ulmschneider *et al.*
Narain, U., Ulmschneider, P.: 1990, *Space Science Reviews*, in press.
Pasquini, L., Brocato, E., Pallavicini, R.: 1991, see Ulmschneider *et al.*
Pasachoff, J.M.: 1991, see Ulmschneider *et al.*
Rosner, R.: 1991, see Ulmschneider *et al.*
Rodonò, M.: 1986, in *Cool Stars, Stellar Systems, and the Sun*, ed. M. Zeilik and D.M. Gibson, Springer-Verlag, Berlin, p. 475.
Rutten, R.J. 1991, see Ulmschneider *et al.*
Saar, S.H.: 1987, in *Cool Stars, Stellar Systems, and the Sun*, Lecture Notes in Physics 291, ed. J.L. Linsky, R.E. Stencel, Springer-Verlag, Berlin, p. 10.
Schrijver, C.J.: 1987a, in *Cool Stars, Stellar Systems, and the Sun*, Lecture Notes in Physics 291, ed. J.L. Linsky, R.E. Stencel, Springer-Verlag, Berlin, p. 135.
Schrijver, C.J.: 1987b, *Astron. Astrophys.* **180**, 241.
Schrijver, C.J.: 1991, see Ulmschneider *et al.*
Simon, T., Drake, S.A.: 1989, *Astrophys. J.* **346**, 303.
Sivaraman, K.R.: 1991, see Ulmschneider *et al.*
Sivaraman, K.R., Livingston, W.C.: 1982, *Solar Phys.* **82**, 227.
Stępień, K., Ulmschneider, P.: 1989, *Astron. Astrophys.* **216**, 139.
Swank, J.H., Johnson, H.M.: 1982, *Astrophys. J. Letters* **259**, L67.
Swank, J.H., White, N.E., Holt, S.S., Becker, R.H.: 1981, *Astrophys. J.* **246**, 208.
van den Oord, G.H.J., Mewe, R., Brinkman, A.C.: 1988, *Astron. Astrophys.* **205**, 181.
Ulmschneider, P., Priest, E., Rosner, R. (eds.): *Mechanisms of Chromospheric and Coronal Heating*, Springer-Verlag, Berlin, in press.
Vilhu, O.: 1987, in *Cool Stars, Stellar Systems, and the Sun*, Lecture Notes in Physics 291, ed. J.L. Linsky, R.E. Stencel, Springer-Verlag, Berlin, p. 110.
Vilhu, O., Walter, F.M.: 1987, *Astrophys. J.* **321**, 958.
Zwaan, C.: 1991, see Ulmschneider *et al.*

Ca II H High-Resolution Spectral Monitoring of Active Late-type Dwarfs

J. E. Beckman [1], R. J. García López [1], R. Rebolo [1],

and L. Crivellari [1,2]

[1]Instituto de Astrofísica de Canarias, E-38200 La Laguna, Tenerife, Spain
[2]Osservatorio Astronomico di Trieste, Via G. B. Tiepolo 11, 34131 Trieste, Italy

Abstract: We have monitored Ca II H with a spectral resolution of 8×10^4, in a sample of late-type MS stars over a four-year period. The high resolution enables us to add information on velocity fields to the usual flux monitoring. We detect changes in wavelength of different parts of the Ca II H feature, which can be interpreted as velocity fields in the lower chromosphere, with downflow and upflow of order $0.5\,\mathrm{km\,s^{-1}}$, depending on the star. Flux variations in Ca II H emission can be ascribed, via velocity tagging, to long-term change in plage cover rather than short-term modulation by (incompletely sampled) rotation cycles.

1. Observations and objectives

Over a four year period, from 1984 to 1988, we used the 1.4 m CAT plus Coudé Echelle Spectrograph of ESO, La Silla (effective spectral resolution $\lambda/\Delta\lambda \sim 8\times10^4$ at 3970 Å) to observe systematically, at least once per year, the Ca II H line in a sample of late-type dwarfs (late G and K). The objective: to see how much more we could learn from variability in the resolved profile of an emission core than from flux monitoring alone.

2. Wavelength and flux calibration

Given the high S:N ratios (at least 200) we could use 5 sharp photospheric absorption lines to establish a photospheric rest frame for each star (see Crivellari *et al.*, 1987 for further details) to a precision of ± 2 mÅ, i.e. ± 160 m $\mathrm{s^{-1}}$; the basic limit is the formation process of these lines in the photosphere. Chromospheric velocities (wavelength displacements detected in the Ca II H feature) are absolute, in sense and in amplitude, with respect to this frame.

Relative calibration of a Ca II H profile with respect to the nearby pseudo-continuum is good to a few percent, being limited by non-uniformity ("tilt") in the focal plane illumination. This is the order of the error in comparing chromospheric emission fluxes of the same star at different epochs. The absolute calibration is estimated good to $\pm$ 20 % (see Rebolo *et al.*, 1989, for more details). This uncertainty applies to intercomparisons of fluxes between different stars.

3. Velocity gradients and flux variations

Parameterizing a chromospheric emission line into a positive-going gaussian representing the emission core, and a superposed negative-going gaussian representing the self-absorption, we can specify: λ_0 the photospheric rest wavelength of Ca II H, λ_{em} the wavelength of the emission peak, and λ_3 the wavelength of the self-absorption dip. We can also specify the bisector locus of the measured emission core. We should emphasize one key result: the linear dependence of $\lambda_{em} - \lambda_3$ on the intensity ratio I_{2V}/I_{2R} of the blue and red peaks (H_{2V} and H_{2R}); a single linear relation encompassing all the data for all the stars.

Figure 1 shows how the velocity field differs from star to star, when $\lambda_{em} - \lambda_0$ and $\lambda_3 - \lambda_0$ are shown on different arbitrary height scales above the photosphere. These representative plots show that in general $|\lambda_{em} - \lambda_0|$ is greater than $|\lambda_3 - \lambda_0|$, and that in general the sense of motion is steady for a given star. We can summarize the bisector information by stating that, in general, the emission core bisectors show negligible velocity gradients.

When we plot the emission flux, measured in a band of width 0.8 Å centred on λ_{em}, against $\lambda_{em} - \lambda_0$, for all the stars, two notable observational trends emerge: (a) There is, in general, an absence of correlation between flux and wavelength shift. (b) For a given star the $\lambda_{em} - \lambda_0$ values are not symmetrically dispersed around zero.

4. Conclusions

Measuring the profiles has given us much more information than flux monitoring alone could give, although the sparse and infrequent sampling limits the validity of our inferences. We can summarize the main conclusions as follows:

(i) The flux variability observed cannot be due mainly to rotational modulation (a result of our sparse sampling), as shown by the trends (a) and (b) in the previous paragraph, and also because the $\lambda_{em} - \lambda_3$ values are an order of magnitude greater than would be expected from differential rotation with height during the passage of a plage from centre to limb.

(ii) A given star maintains a velocity field in its lower chromosphere whose sense and gradient remain rather stable over periods of years. These velocities are of order $500\,\mathrm{m\,s^{-1}}$, and the sense sampled by the Ca II H emission differs from star to star (although in the majority, as for α Cen B, upflow is sampled).

(iii) The ratio of intensities in the H_{2V} and H_{2R} emission peaks is linearly proportional to the displacement of its self-absorption trough from the overall emission peak, a relation encompassing all stars at all epochs.

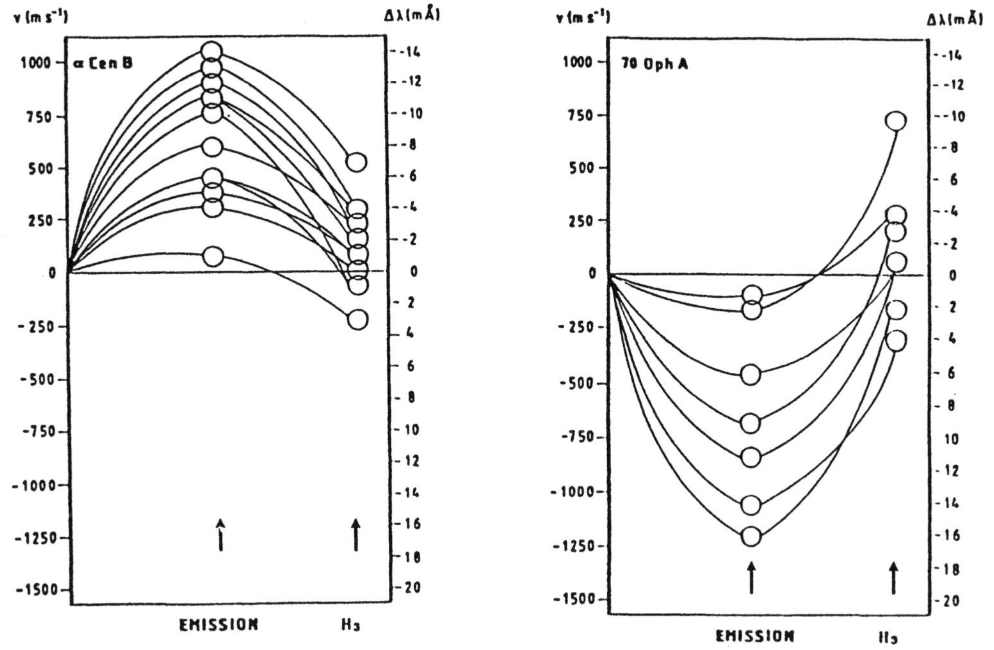

Fig. 1. Shifts $\lambda_{em} - \lambda_0$ of the emission from the photospheric rest wavelength, and $\lambda_3 - \lambda_0$ of the self-absorption from the rest wavelength, against relative height above the photosphere for α Cen B and 70 Oph A.

References

Crivellari, L., Beckman, J. E., Foing, B. H., Vladilo, G.: 1987, *Astron. Astrophys.* **174**, 127

Rebolo, R., García López, R. J., Beckman, J. E., Vladilo, G., Foing, B. H., Crivellari, L.: 1989, *Astron. Astrophys. Suppl. Ser.* **80**, 135

The Onset of Chromospheres in A-type Stars: The Altair Affair

S. Catalano [1], P. Gouttebroze [2], E. Marilli [3],

R. Freire Ferrero [2,4]

[1]Istituto di Astronomia, Università di Catania, Viale A. Doria 6, I-95125 Catania, Italy
[2]Institut d'Astrophysique Spatiale, IAS/LPSP CNRS, BP N° 10, 91371 Verrieres-le Buisson, France
[3]Osservatorio Astrofisico di Catania, Viale A. Doria 6, I-95125 Catania, Italy
[4]Observatorie Astronomique, 11, Rue de l'Université, 67000 Strasbourg, France

Abstract: Here we present preliminary results on the study of the chromosphere of Altair from $L\alpha$ high dispersion profiles. We report also the detection of $L\alpha$ chromospheric emission from α Cep (A7 V) and β TrA (F0 V) with the IUE. We show that chromosphere may exist up to $B - V = 0.22$. The possible heating mechanism, magnetic or non-magnetic, for the chromosphere of these stars is also discussed.

1. Introduction

Several studies of activity in F stars, mainly based on the detection of the UV emission lines C II, C IV and $L\alpha$ and of the optical He I 5876 Å absorption line (see Wolff *et al.*, 1986; Simon and Landsman, 1987; Walter and Schrijver, 1987), place the onset of stellar chromospheric activity near $B - V = 0.28$, which corresponds approximately to spectral type F0. However, X-ray observations of A and F stars (Schmitt *et al.*, 1985) show that coronal activity may occur at a significantly earlier spectral type. With the exception of the $L\alpha$ in Altair (A7 V–IV, $B - V = 0.22$) (Blanco *et al.*, 1980), no significant chromospheric or transition region emission has been detected in these earlier type stars. In a study of $L\alpha$ emission, using low dispersion IUE spectra, Simon and Landsman (1987) did confirm the detection of emission from Altair but did not find any emission in the spectra of 10 other late A-type stars.

Should one consider Altair a special case? Has Altair a solar type chromosphere? In an effort of answering these questions and defining more clearly the location of the high temperature boundary of chromospheric detection we have started an observing program of high dispersion $L\alpha$ spectra with IUE. Here we report some preliminary results on the chromospheric model of Altair and $L\alpha$ observations of Altair-like stars.

2. Altair's chromosphere

The absence of detectable transition region emission lines connected with the detection of X-ray and $L\alpha$ emission, poses the question of whether or not Altair has a solar type chromosphere. We have two basic tests to answer this question:
i) Analyse the $L\alpha$ profile: The $L\alpha$ profile has been obtained by different authors from high dispersion IUE spectra. However the main problem in defining the $L\alpha$ profile from that spectra, is the high geocoronal contamination. We have solved this problem making specific observations of the sky background, which allow us to make a correct cleaning of the stellar profile (Catalano *et al.*, 1990). The emission profile [1] we have obtained as average of four different spectra is quite similar to the solar $L\alpha$ profile suggesting a similar chromospheric structure.
ii) Calculate chromospheric models: We are computing $L\alpha$ profile from model atmospheres. The calculations are performed with the following assumptions: 1) NLTE; 2) partial redistribution; 3) hydrogen atom with five level plus continuum; 4) rotational broadening $v \sin i = 220$ km/sec; 5) HI interstellar absorption. The main characteristic of the models giving the best agreement with the observed profile is a temperature increase after a minimum value and a large plateau, i.e. a basically solar-type chromospheric structure.

3. Are there other Altairs?

We have already obtained high resolution, long exposure IUE spectra for two stars of spectral type very close to that of Altair: α Cep (A7 V) and β TrA (F0 – F2V). Even though the spectra are rather noisy, the $L\alpha$ emission component is clearly detected. These profiles including the averaged Altair profile are shown in Fig. 1. The over-imposed thin curve represents a best-fit gaussian profile, which has been used to make a preliminary rough estimate of the energy released in that line.

The integrated $L\alpha$ emission luminosities are given in Table 1, together with the relevant parameters of the three stars. β TrA, which is of a little later spectral type, has a $L\alpha$ luminosity larger by a factor of about two than Altair, as it would be expected from the deeper convection zone. α Cep, with parameters of spectral type and rotation speed that make it just a "twin" of Altair, has a $L\alpha$ luminosity

[1] Based on observations by the IUE, collected at the Villafranca Satellite Tracking Station of the ESA.

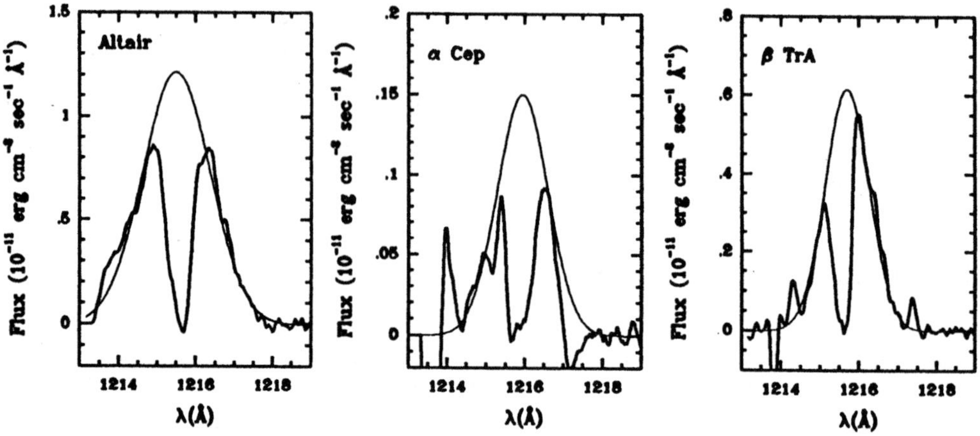

Fig. 1. Smoothed $L\alpha$ profiles of Altair, α Cep and β TrA (solid lines) characterized by a deep interstellar HI absorption. The thin lines are gaussian curves fitted to the external profiles.

in very good agreement with Altair. The $L\alpha$ luminosities of Altair and α Cep, although are close to the lowest luminosity values of later spectral type stars, do not indicate a clear fading of chromospheric emission toward hot stars. From such few data we are not yet able to establish whether there is a sharp boundary or a smooth disappearance of chromospheres.

Table 1.

star	SP	$B-V$	$v \sin i$ [km/sec]	$\log L_{L\alpha}$
Altair	A7V–IV	0.22	220	28.89
α Cep	A7V	0.22(0.24)	245	28.77
β TrA	F0-F2V	0.29	90	29.17

4. Nature of chromospheric heating

Although it is generally agreed that chromospheric and coronal emission in late type stars is due to magnetic activity, the suggestion that magnetic dynamos exist in stars as early as A7 is still controversial. Theoretical models suggest that magnetic dynamos cannot be sustained in stars earlier than F6 (Durney and Latour, 1978). As a matter of fact simple evaluation shows that the Rossby number is about 20 for stars like Altair, while theory requires a value smaller than 1 to have an efficient dynamo. On the other hand Mangeney and Praderie (1984) have shown that magnetic dynamos can be efficient if a local Rossby number is defined. The break-down of the X-ray emission-rotation correlation and the absence of rotation braking for spectral types earlier than F5, have been invoked in support of a drop of magnetic dynamos in those stars (Walter, 1983). However Schmitt *et al.* (1985) from a larger sample of X-ray data have shown that a dependence of X-ray luminosity on the Rossby number may exist, and concluded that the magnetic activity is responsible for the heating of chromospheres and corona even in late A type stars.

Magnetic and acoustic heating may coexist in late type stars (Ulmschneider 1990). However it is very difficult or impossible to separate the two contributions in the observed emission fluxes. Schrijver (1987) suggests that there exists a "basal" flux in the chromospheric emission lines, of non-magnetic origin, which is negligible in the X-ray emission. The basal flux is also negligible in late type stars. He shows that the excess chromospheric flux, above the basal flux, is well correlated with the X-ray flux, while the total emission flux is not. Let us adopt this simple model to determine whether or not the chromospheres and coronae of our Altair-like stars are magnetically heated. In Fig. 2 we have plotted X-ray luminosity against $L\alpha$ luminosity for a sample of late type stars for which both quantities are available. From the figure we see that Altair and α Cep, for which only an upper-limit on the X-ray flux is available, lie well below the regression line of late type stars. If we assume that the X-ray emission is completely magnetic origin and that the chromospheric $L\alpha$ emission results from a combination of magnetic and non magnetic heating we can estimate the former as the value one would expect for a star situated on the regression line. It turns out that only about 10% of the $L\alpha$ emission in Altair, and perhaps also in α Cep, can be accounted for by magnetic heating. The largest amount should be of non magnetic (acoustic?) origin. Calculations of acoustic flux show that for these spectral type there is plenty of energy available to heat the chromosphere (Bohn, 1984). However detailed chromospheric models and energy balance calculations are needed to clarify this problem.

5. Conclusions

We have shown from the analysis of high dispersion $L\alpha$ profiles and preliminary atmospheric models that Altair has a solar type chromosphere. In addition we have found that Altair is not an unique case, because, at least its "twin" star α Cep possesses a chromosphere, as signalled by the detection of $L\alpha$ emission. These results definitely extend the presence of chromospheres up to $B - V = 0.22$. Finally we have shown that the chromospheric $L\alpha$ emission can mostly be of non-magnetic origin.

Acknowledgements. This work has been possible thanks to the support of an international cooperation programme between the CNR (Italy) and DRCI-CNRS (France). The work was also supported by the MURST through the University of Catania, the Osservatorio

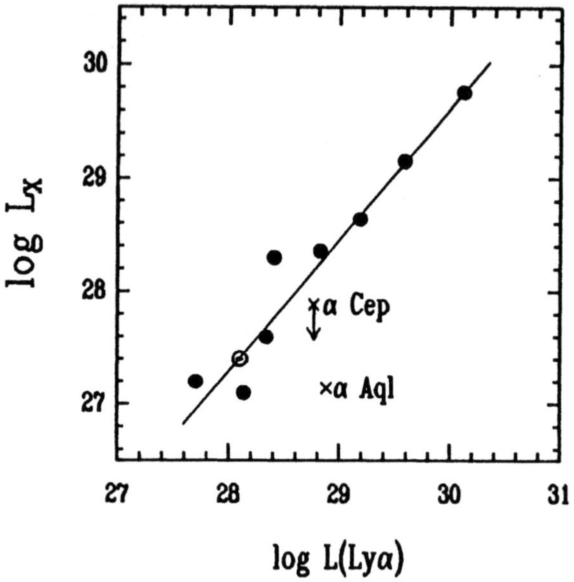

Fig. 2. Logarithms of X-ray luminosities vs. $L\alpha$ luminosities. Late-type stars (filled points) and the Sun define the best-fit line.

Astrofisico di Catania and the GNA (CNR). The extensive use of the computer facilities of the Catania ASTRONET site is also acknowledged.

References

Blanco, C., Catalano, S., Marilli E.: 1980, ESA SP-157, p. 67

Bohn, H.U.: 1984, *Astron. Astrophys.* **136**, 338

Catalano, S., Marilli, E., Freire-Ferrero, R., Gouttebroze, P.: 1990, *Astron. Astrophys.*, submitted

Durney, B.R., Latour, J.: 1978, *Geophys. Astrophys. Fluid Dyn.* **9**, 241

Mangeney, A., Praderie, F.: 1984, *Astron. Astrophys.* **130**, 143

Schmitt, J.H.M.M., Golub, L., Harnden, F.H.Jr., Maxon, C.W., Vaiana, S.G.: 1985, *Astrophys. J.* **290**, 307

Schrijver, C.J.: 1987, in *Cool Stars, Stellar Systems and the Sun*, eds. J.L.Linsky, R.E., Stencel, Springer-Verlag, Berlin, p. 241

Simon, T., Landsman, W.: 1987, in *Cool Stars, Stellar Systems and the Sun*, eds. J.L. Linsky, R.E. Stencel, Springer-Verlag, Berlin, p. 265

Walter, F.M.: 1983, *Astrophys. J.* **274**, 794

Walter, F.M., Schrijver, C.J.: 1987, in *Cool Stars, Stellar Systems and the Sun*, eds. J.L. Linsky, R.E. Stencel, Springer-Verlag, Berlin, p. 262

Ulmschneider, P.: 1990, in *Mechanism of Chromospheric and Coronal Heating*, eds. P. Ulmschneider *et al.*, Springer, Berlin (in press)

Wolff, S.C., Boesgaard, A.M., Simon, T.: 1986, *Astrophys. J.* **310**, 360

Hα and NaD Line Variability in RY Tauri

P. Petrov [1] and O. Vilhu [2]

[1]Crimean Astrophysical Observatory, 334413 Crimea, USSR
[2]Observatory and Astrophysics Laboratory, University of Helsinki, SF-00130 Helsinki, Finland

Abstract: Using high resolution spectroscopy of the T Tau star RY Tau in the regions of Hα and Na D lines, we discovered line profile variability, which is interpreted by a hypothesis of stellar 'prominences': cool gas clouds are moving inside the stellar wind, ascending and descending on a time scale of a few days. These clouds are supposed to be controlled by the magnetic field of the star. The range of radial velocities of the clouds (−100 to +100 km/s) is twice as large as the rotational velocity ($v \sin i$) of the star. The form of variability of the Hα profile suggests the existence of two components: the (variable) stellar wind and the circumstellar disc.

1. Introduction

Since the pioneering work by Joy (1945), who first noticed the similarity between the emission spectrum of T Tau stars and the chromospheric spectrum of the Sun, the idea of the close physical connection between these two phenomena has been a leading one for many investigators. During the following half of the century much more data in favour of this concept were obtained: winds, spots, UV and X-ray radiation. In all these phenomena the magnetic field plays the major role.

Besides this pattern of solar-like activity, T Tau stars exhibit another kind of spectroscopic and photometric variability, caused by gaseous and dust discs around the star (Bertout, 1988). The Sun also has a relic of the circumstellar disc (the solar system) but the solar activity is now controlled rather by internal than external forces. In T Tau stars the role of discs may be more important at certain stages of evolution.

The gas dynamics around the star can be studied by means of high resolution spectroscopy at certain wavelength regions. In this paper we discuss briefly the data of the Na D lines, published recently by one of us (Petrov, 1990). We also present new data of the Hα line variability in RY Tau. This star is a well known T Tau star with emission spectrum of moderate intensity and with underlying absorption spectrum of spectral type K1 IV-V ($v \sin i = 53 km/s$, Herbig and Bell, 1988).

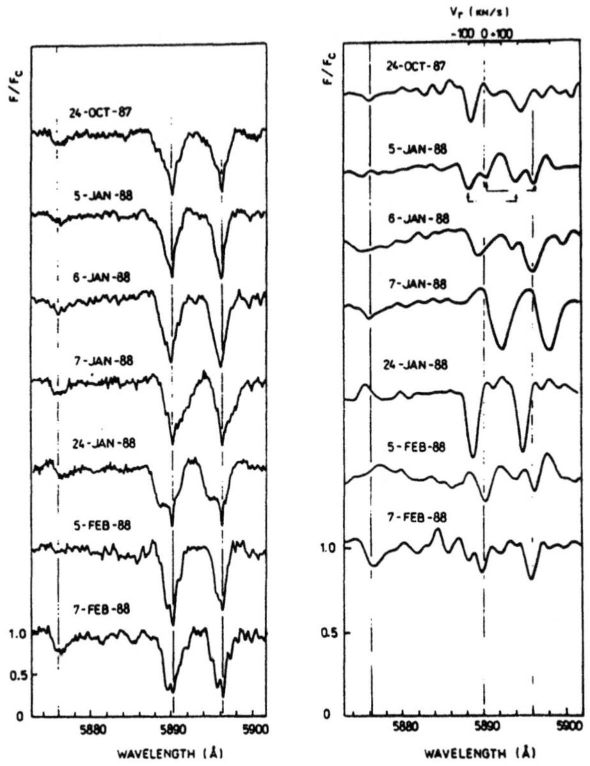

Fig. 1. a.) Spectra of RY Tau in the region of the Na D lines. b.) Differential spectra, derived by subtracting the average spectrum from the individual spectra (from Petrov, 1990).

2. Variability of the NaD lines

Spectroscopic observations of RY Tau in the regions of the Na D lines were carried out with the coudé spectrograph of the 2.6 m Shajn reflector of the Crimean Astrophysical Observatory, using the CCD camera of the Helsinki University. The spectral resolution was set by the entrance slit of 1 arc second, which corresponds to 0.37 Å on the CCD. During the winter season of 1987/88 21 spectra of RY Tau were taken. A fragment of the data set, showing the most obvious variability of the profiles, is shown in Fig. 1a. Besides the photospheric components of the Na D1 and D2 absorptions (broadened to 53 km/s by stellar rotation) and the narrow interstellar component, variable absorptions in the blue and red wings can be seen (see e.g. the red wing on 7-Jan-88 and the blue wing on 24-Jan-88). In order to extract these variable absorption components, we subtracted the average spectrum of RY Tau from each of the individual spectra. The differential (residual) spectra, smoothed to resolution of about 40 km/s, are shown in Fig. 1b.

The residual absorptions are variable both in intensity and radial velocity, moving within the range from −100 to +100 km/s, which is twice as large as the $v \sin i$ of the star. This implies that the absorption does not originate from the stellar surface but more likely from the circumstellar environment. It could be clouds or streams of relatively cool gas inside the stellar wind. If the clouds were co-rotating with the star (trapped in the magnetic field loops as proposed by Cameron and Robinson (1989) for AB Dor), their radial velocities should not exceed the $v \sin i$ –value of the star. In addition, if the clouds were orbiting the star, they must be on a very low Keplerian orbit with a period of about 8h, but the observations did not show any variability on a time scale of hours.

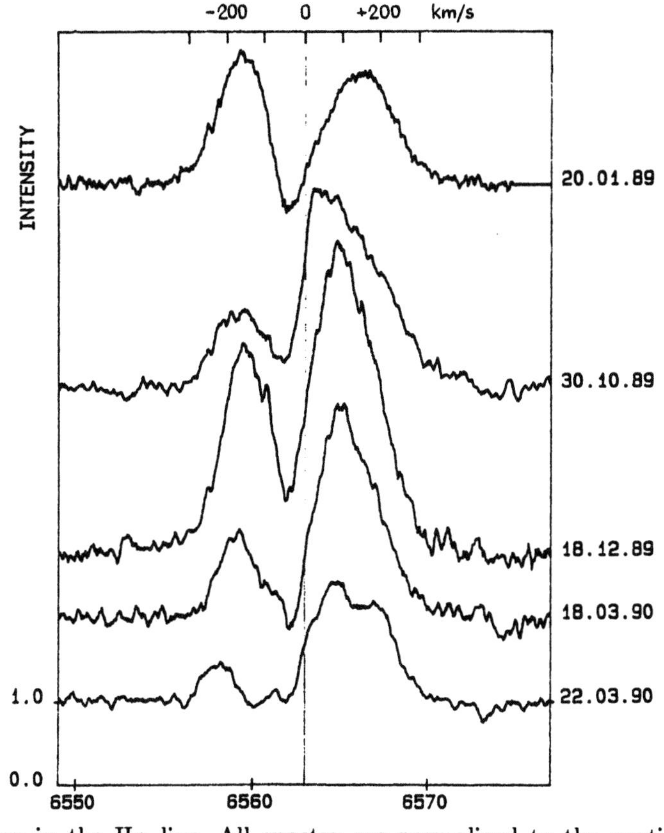

Fig. 2. Variations in the Hα line. All spectra are normalized to the continuum level. Intensity scale is shown for the lowest spectrum.

Consequently, we have to conclude, that the clouds are co-rotating with the star and have some radial velocity components. This is reminiscent of the phenomena of eruptive solar prominences and could be considered as being 'stellar prominences' on a young star. The general appearance of the residual spectra suggests that the temperature of the clouds might be in the range of 5000–7000 K, which is also close to that of the solar prominences. We did not find any of the periodicity expected from rotational modulation. Probably the life time of these 'prominences' is not much longer than the rotational period of the star (about 5 days, Zaitseva *et al.*, 1985).

3. Variability of the Hα line

In order to investigate the gas dynamics in the outer region of the circumstellar environment, we observed the Hα line between January 1989 and March 1990. The observations were a part of a more extended program of T Tau stars (in collaboration with Mark Giampapa).

Several typical Hα profiles are presented in Fig. 2, exhibiting the pattern of line variability. As in many other rapidly rotating T Tau stars, the Hα emission is broad and is split into two parts by the blue-shifted absorption. The wings of the emission extend from −300 to +300 km/s. This is a very stable parameter of the line. Another stable parameter is the absorption at −50 km/s. The range of the profile variability is shown in Fig. 3, where all the profiles are plotted together. At the bottom of Fig. 3, the dispersion of line intensities is drawn. This dispersion function reveals that the most variable part of the Hα line is located at +50 km/s. This does not fit exactly with the red emission peak. In the red wing of the emission, the dispersion drops to zero already by the half maximum of the line, i.e. the red emission wing is not variable. The blue peak is often disturbed by a broad absorption centered at −150 km/s.

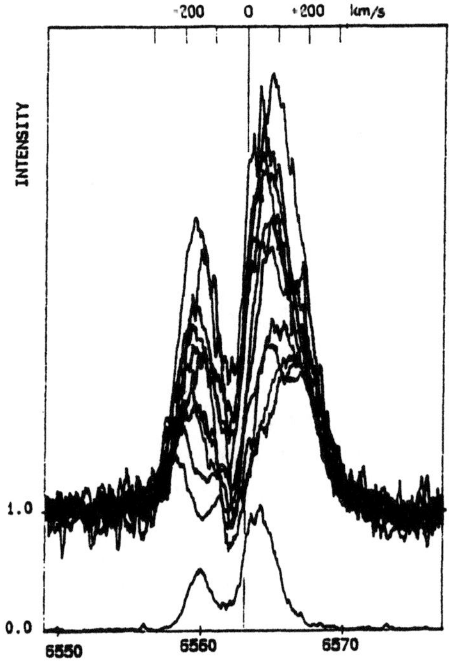

Fig. 3. Range of variability in the Hα profile. The lower curve is the dispersion of intensities.

A qualitative interpretation could be the following. The stable absorption at −50 km/s is formed at the most extended outflowing part of the wind, while the highly variable broad absorption at −150 km/s originates from the innermost part of the wind. It could be produced by the same streams or clouds ('prominences') that we observed in the Na D lines. Fig. 4 shows this effect both in the Hα and Na D lines observed during the same night.

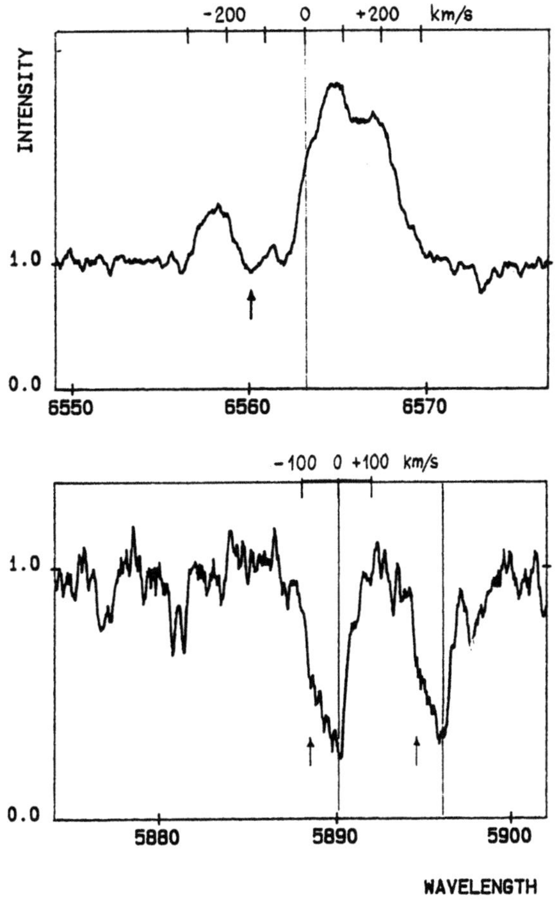

Fig. 4. Appearance of the broad blue shifted absorption in the Hα and Na D lines. Both spectra were taken on March 22, 1990.

The co-existence of the broad stable wings and highly variable emission peaks suggests, that we observe two different components: one (stable) from the disk and another (variable) from the structured wind (see Fig. 5).

4. Conclusions

Our observations of the Na D and Hα line variability in RY Tau reveal the complicated structure of the circumstellar gas environment. Closer to the star we observed 'stellar prominences', while the extended parts of the gas envelope can be described in terms of a stellar wind of variable power. The structure of the Hα line also suggests the existence of a disc around the star.

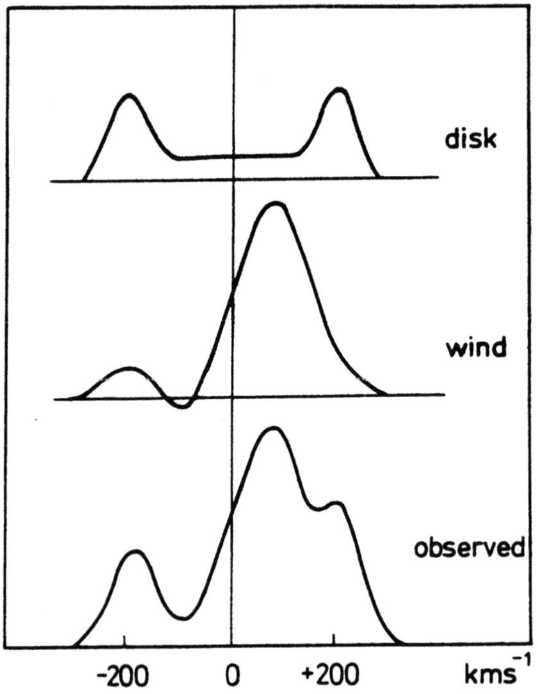

Fig. 5. A schematic explanation of the observed Hα-profile as a sum of the disk and wind.

References

Bertout, C.: 1988, in A.K. Dupree and M.T.V.T. Lago (eds.), *Formation and Evolution of Low Mass Stars*, NATO ASI Series, Kluwer Academic Publishers, Dordrecht, Holland

Cameron, A.C., Robinson, R.D.: 1989, *Monthly Notices Roy. Astron. Soc.* **236**, 57

Joy, A.H.: 1945, *Astrophys. J.* **102**, 168

Herbig, G.H., Bell, K.R.: 1988, Lick Observatory Bull. No. 1111.

Petrov, P.P.: 1990, *Astrophysics and Space Sci.* **169**, 61

Zaitseva, G.V., Kolotilov, E.A., Petrov, P.P., Tarasov, E.A., Shenavrin, V.I., Shcherbakov, A.G.: 1985, *Soviet Astron. Letters* **11**, 271

He I 5876 Å Line in the Study of Chromospheric Activity in F-type MS Stars

R. J. García López [1], R. Rebolo [1], J. E. Beckman [1],

and C. D. McKeith [2]

[1]Instituto de Astrofísica de Canarias, E-38200 La Laguna, Tenerife, Spain.
[2]Department of Pure and Applied Physics, The Queen's University of Belfast, Belfast BT7 1NN, Northern Ireland.

Abstract: We have measured the He I D_3 line, λ5876 Å, in 32 F-type young open cluster stars and in 4 field stars. Combining our results with previous observations of D_3 we strengthen the following conclusions: (a) Sub-surface convective zones appear, on the main sequence, around F0 ($B-V \sim 0.3$). (b) The stars between F0 and F5 ($0.3 \lesssim B-V \lesssim 0.42$) are chromospherically "active" irrespective of age, which implies a heating mechanism not affected by evolutionary effects. In this range the activity in D_3 and X-rays are uncorrelated. (c) For stars later than F5 the correlation among all the activity indices, chromospheric and coronal, shows a common energy source. Convective zones here should be deep enough to generate magnetic fields capable of supplying this energy, and also for braking the rotation.

1. The λ 5876 Å D_3 line of He I

The utility of He I D_3 as a chromospheric activity indicator was strongly suggested by studies such as that of Landman (1981) who showed the ratio of the line depth of D_3 quiet sun:plage to be ~ 0.09. This line is easier to observe in practice than the stronger He I IR line at 10830 Å, and although its excitation mechanism is disputed (Milkey *et al.*, 1973; Zirin, 1975), and its equivalent width usually small ($\lesssim 50$ mÅ), the advent of modern detectors and high resolution spectrographs has led to its use especially for the F spectral range, where indicators such as Ca II H and K emission have lost their sensitivity.

We observed the D_3 line in F-type stars of two open clusters, 17 in the Hyades (age $\sim 6\times10^8$ yr) and 15 in the Ursa Major Group ($\sim 3\times10^8$ yr) and in 4 field stars, all with previously known ages, metallicities and Li abundances. The observations were made with the QUBES spectrograph ($\lambda/\Delta\lambda \sim 6\times10^4$, 43 mÅ/pixel) at the

Cass. focus of the 4.2 m William Herschel telescope, La Palma. A typical S:N ratio was 200.

2. Flux in D_3 v. spectral type

We present in Fig. 1 the flux absorbed in D_3: $F(D_3)$ (computed from the equivalent width using neighbouring continuum flux from the appropriate Kurucz 1979 model) v. $(B-V)$ for all measured MS objects (classes V and IV–V) with $(B-V) < 0.8$, both field and open cluster stars, from the present work and the literature (Lambert and O'Brien, 1983; Wolff and Heasley, 1984; Danks and Lambert, 1985; Wolff *et al.*, 1985; Wolff *et al.*, 1986; Wolff and Heasley, 1987). This plot confirms the following division of the D_3–$(B-V)$ parameter space (Wolff *et al.*, 1986): (i) For $(B-V) \lesssim 0.3$ no measurable D_3 is found, agreeing with previous results for this and other chromospheric and coronal indicators (Wolff *et al.*, 1986; Schmitt *et al.*, 1985; Walter *et al.*, 1984). Models suggest (Vandenberg and Bridges, 1984) the "switch on" of convective envelopes around this $B-V$ value. (ii) All stars with $0.3 \lesssim B-V \lesssim 0.42$ (known spectroscopic binaries removed) are "D_3 active", with no apparent age dependence. In this enlarged sample selection effects appear unlikely. (iii) In the range $(B-V) \gtrsim 0.42$ cluster stars (all young) show D_3 activity, while some field stars do not. This is consistent with rotation-dependent (hence age-dependent) activity, stimulated by magnetic fields induced by rotation. Around the value $(B-V) \sim 0.4$ is where stellar envelope theory predicts a "switch on" of braking by stellar winds. D_3 appears as an excellent activity indicator for this effective temperature range.

The conclusions drawn here are strengthened when use is made of the Rossby number to parameterize magnetic activity (see García López *et al.*, 1990).

3. He I D_3 v. other chromospheric and coronal indicators

(a) C IV λ 1549 Å. There is a clear correlation between the IUE-measured emission flux in C IV λ 1549 Å (from the literature) and the absorption flux in D_3, with no obvious dependence on spectral type or on whether the star is in a cluster. (b) Ca II H and K. Similarly F'_{HK}, measuring the Ca II H+K emission core flux, correlates very well with $F(D_3)$ for those stars for which both are measurable. (c) X-ray luminosity. However a plot of the soft X-ray luminosity in units of bolometric luminosity from Einstein satellite observations, against the equivalent widht of D_3, divides into two zones: that with $(B-V) > 0.44$, with a linear correlation coefficient r of 0.89, and that with $(B-V) \lesssim 0.42$, where r is only 0.33. The difference in the behaviour of C IV λ 1549 Å and Ca II H and K, on the one hand, and X-rays on the other is because the latter is a coronal activity index (see e. g. Dobson and Radick, 1989), while the other two, and D_3, are essentially chromospheric.

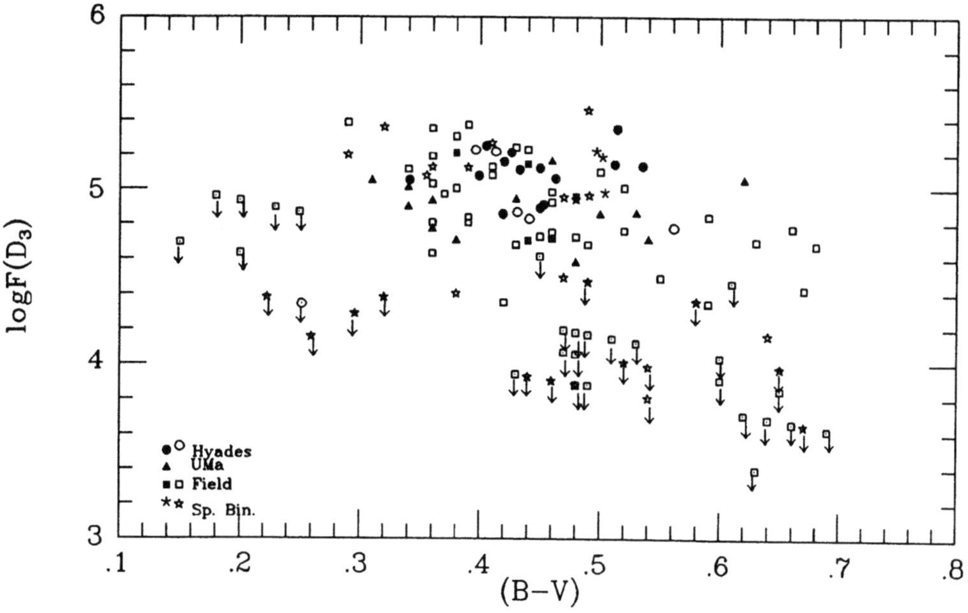

Fig. 1. Flux absorbed by the He I D_3 line (in erg cm^{-2} s^{-1}) v. $(B-V)$ for our stellar sample (filled symbols) and the literature (open symbols).

References

Danks, A. C., Lambert, D. L.: 1985, *Astron. Astrophys.* **148**, 293

Dobson, A. K., Radick, R. R.: 1989, *Astrophys. J.* **344**, 907

García López, R. J., Rebolo, R., Beckman, J. E., McKeith, C. D.: 1990, in *Heidelberg Conference on Chromospheric and Coronal Heating*, Springer-Verlag, in press

Kurucz, R. L.: 1979, *Astrophys. J. Suppl. Ser.* **40**, 1

Lambert, D. L., O'Brien, G. T.: 1983, *Astron. Astrophys.* **128**, 110

Landman, D. A.: 1981, *Astrophys. J.* **244**, 345

Milkey, R. W., Heasley, J. N., Beebe, H. A.: 1973, *Astrophys. J.* **186**, 1043

Schmitt, J. H. M. M. *et al.*: 1985, *Astrophys. J.* **290**, 307

VandenBerg, D. A., Bridges, T. J.: 1984, *Astrophys. J.* **278**, 679

Walter, F. M. *et al.*: 1984, *Astrophys. J.* **281**, 815

Wolff, S. C., and Heasley, J. N.: 1984, *Pub. A. S. P.* **96**, 231

Wolff, S. C., Heasley, J. N.: 1987, *Pub. A. S. P.* **99**, 957

Wolff, S. C., Boesgaard, A. M., Simon, T.: 1986, *Astrophys. J.* **310**, 360

Wolff, S. C., Heasley, J. N., Varsik, J.: 1985, *Pub. A. S. P.* **97**, 707

Zirin, H.: 1975, *Astrophys. J. Letters* **199**, L63

The Age Evolution of Ca II Emission in Late Type Stars

C. Trigilio [1], S. Catalano [2], E. Marilli [3], V. Reglero [4], G. Umana [1]

[1]Istituto di Radioastronomia CNR, Stazione di Noto, C.P.169, I 96017 Noto (SR), Italy
[2]Istituto di Astronomia, Università di Catania, V.le A.Doria 6, I 95125, Catania, Italy
[3]Osservatorio Astrofisico, Città Universitaria, V.le A.Doria 6, I 95125, Catania, Italy
[4]Departamento de Matematica Applicada y Astronomia, Universidad de Valencia, Spain

1. Introduction

It has been widely demonstrated that chromospheric emission is dependent on rotation rate (Hartmann and Noyes, 1987, Catalano, 1990, for recent reviews). Since the rotation decays with time, the chromospheric activity of single main sequence stars is also dependent on age. Catalano and Marilli (1983) suggested that the chromospheric emission of solar mass stars decays exponentially with the square root of the time, as recently confirmed by Barry (1988). Magnetic activity depends also on the convection parameters, i.e. on the mass. Since the evolution of surface rotation is faster for lower mass stars (Catalano *et al.*, 1988), the internal rotation distribution, and therefore the dynamo action, should change with time in a different way for different masses.

In order to get insight into the evolution of dynamo action we have studied the CaII chromospheric emission as a function of mass and age. Here we present some preliminary results of observations of late type stars in young clusters and old moving groups.

2. Observations

The data we discuss refer to:
• photographic spectra obtained at La Silla ESO Observatory with the 1.5 m telescope coudé spectrograph (resolution from 0.45 Å to 0.28 Å);
• CCD spectra obtained at Calar Alto German-Spanish Astronomical Center with the 2.2 m telescope coudé spectrograph (resolution of 0.25 Å);
• reticon spectra obtained at Calar Alto with the 1.23 m telescope Cassegrain spectrograph (resolution of 2.5 Å).

The higher resolution has been used to observe 20 stars from F8V to K5V of low chromospheric emission belonging to old kinematic groups of solar age (ζ Her, ϵ Indi, σ Puppis, Arcturus group) or slowly rotating stars of the old disk population. Lower resolution has been used for the observation of 11 Hyades stars from $B - V = 0.55$ to $B - V = 0.9$ and 8 Pleiades stars from $B - V = 0.7$ to $B - V = 1.1$. Field stars of known high activity level have been observed at low resolution to test our extracted emission fluxes. All the spectra have been calibrated in absolute units at the stellar surface using the spectrophotometric atlas of Gunn and Stryker (1983) and the Barnes and Evans relation.

In addition to our Ca II emission, we have used data from the literature (i.e. from high resolution spectra, from Mt. Wilson spectrophotometer data and from Mg II h and k data converted to Ca II H and K luminosity). The Mt. Wilson data have been corrected for a revised minimum basal flux.

3. Flux extraction

The evaluation of the net chromospheric flux above the absorption profile has been done in different ways. For high resolution spectra of low emission stars we estimated the chromospheric flux as the difference between the observed spectra and the computed synthetic spectrum obtained from Kurucz (1979) RE model, using the SYNTH procedure (Castelli, 1987) convolved with a 0.35 Å Gaussian for the cumulative effects of resolution, microturbulence and granulation.

For lower resolution spectra of active stars, the net chromospheric flux has been estimated from a direct comparison with spectra of old stars of similar spectral type convolved to reproduce the low resolution.

4. Results

The Ca II emission luminosity, for a given age, decreases smoothly with stellar mass, showing a knee at some critical mass. The knee moves from lower to higher masses as the age increases. It seems to disappear by the age of the Sun (Fig. 1).

The dependence of the Ca II luminosity from the mass can be fitted by power laws of the form

$$L_{HK} \propto M^{\gamma}$$

with different values of γ for masses lower or higher than turnover point of the knee at a given age. For stars of $M \geq 0.8 M_{\odot}$ the emission can be represented as decaying exponentially with the square root of the age and with a rate increasing as the mass decreases (Fig. 2). At masses lower than the knee turnover value the decay rate of the emission seems to be independent of the mass, as already suggested by Catalano and Marilli (1983).

Acknowledgements. This work has been supported by the CNR, the M.U.R.S. through the University of Catania, the Osservatorio Astrofisico di Catania and the GNA(CNR). The extensive use of the computer facilities of the Catania ASTRONET site is also acknowledged.

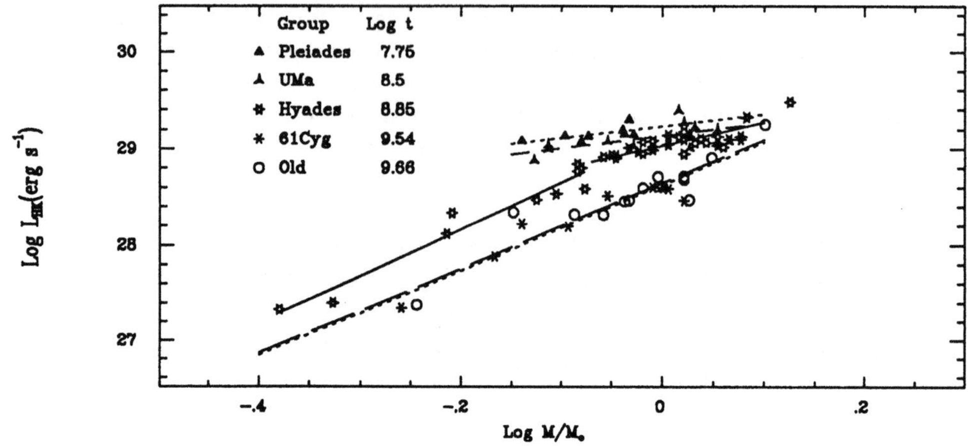

Fig. 1. H and K luminosity for MS stars of known age as a function of mass. Lines represent the fits to data for different groups of age and mass.

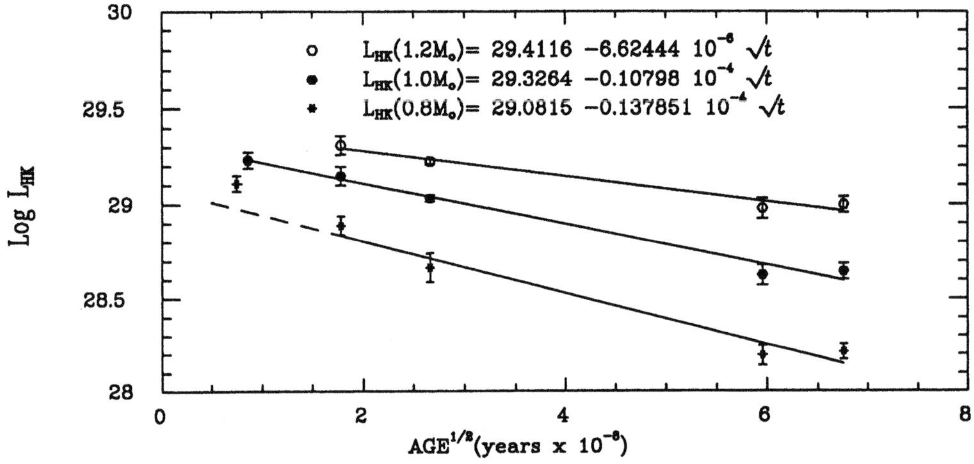

Fig. 2. Log of the H and K averaged luminosity versus square root of the age for three mass groups. A faster decay for low mass stars is apparent. Pleiades of 0.8 $M_{\odot}$ do not seem to follow the decay rate of older stars.

References

Barry, D.C., 1987, *Astrophys. J.* **334**, 436

Castelli, *Publ. OAT.*, no. 984

Catalano S. and Marilli E.: 1983, *Astron. Astrophys.* **121**, 190

Catalano S.: 1990, in *NATO ASI on Active Close Binaries*, eds. C. Ibanough and I. Yavez, Kluwer Academic Publisher

Catalano S., Marilli E., Trigilio C.: 1988 in *NATO ASI on Formation and Evolution of Low Mass Stars*, ed. A.K. Dupree, M.T.V. Lagoeds, Kluwer Academic Publishers

Gunn, J.E., Stryker, L.L.: 1983, *Astrophys. J. Suppl. Ser.* **52**, 121

Hartmann, L.W., Noyes, R.W.: 1987, *Ann. Rev. Astron. Astrophys.* **25**, 271

Kurucz R.L.: 1979, *Astrophys. J.* **40**, 1

Ca II H and K Spectroscopy of ER Vul

M.J. Fernández-Figueroa [1], J.E. Armentia [1], E. de Castro [1], M. Cornide [1], V. Reglero [2]

[1]Dept. Astrofísica, Universidad Complutense, E-28040 Madrid, Spain
[2]Dept. Matemática Aplicada y Astronomía, Universidad de Valencia, E-46100 Burjassot, Spain

1. Introduction

ER Vul (HD 200391) is a double-lined spectroscopic binary, at a distance of 45 pc, catalogued as a short-period RS CVn binary, and has been found to present eclipses. The components have spectral types G0V/G5V, masses 1.07/0.98 $M_{\odot}$, radii 1.23/1.23 $R_{\odot}$ and M_V 4.8/4.6^{m} (Strassmeier *et al.*, 1988, and references therein). The system presents emission reversals of the H and K absorption lines (Bond, 1970; Eggen, 1978), but no information on the character of these emissions has been previously reported.

2. Observations and data reduction

We have taken 4 spectra of ER Vul centered at 3950 Å using the INT telescope (2.5 m) at the Observatorio del Roque de los Muchachos (La Palma, Spain) in July 1988. The spectrograph used was the IDS with the 2400B grating and camera 2 mounting the IPCS detector. Three additional spectra were taken in July 1989 at the Centro Astronomico Hispano-Aleman of Calar Alto (Almeria, Spain) using the 2.2 m telescope equipped with the Coude B&C spectrograph and a CCD (RCA) as a detector. The spectral resolution achieved in both instrumental configurations is 0.3 Å/pixel determined by the slit aperture. The journal of observations is given in Table 1. The phase has been computed with the ephemeris from Hall and Kreiner (1980):

$$JD = 2440182.2593 + 0.69809510\,E$$

The wavelength-calibrated spectra have been extracted from the IPCS and CCD images, using standard reduction procedures. The spectra are then corrected from atmospheric extinction by means of the semiempirical method by Hayes and

Latham (1975). The flux calibration has been performed using 4 standard stars from the compilations of Oke and Gunn (1983). Night to night discrepancies found for the standard star measurements are within 5% of accuracy. After this calibration the flux units are erg cm^{-2} s^{-1} $Å^{-1}$. The spectra are shown in Fig. 1. Using photometric data and the calibration of Catalano *et al.* (1990), the continuum spectrum (above lines) can be synthesized for the components of the system, reproducing very well the expected continuum level over the absorption lines, showing that the flux calibration has been done properly. The emission fluxes for the Ca II H and K lines are obtained by reconstruction of the absorption line profile below the emission peak, as discussed by Armentia *et al.* (1989). Errors in flux measurements are about 25%, in good agreement with similar calculations from Fernández-Figueroa *et al.* (1986) using the same equipment and data reduction techniques. Final results for the emission fluxes are given in Table.

3. Spectroscopic analysis

The spectra obtained at the INT have a lower SNR than those from Calar Alto, a situation increased by the possible eclipse of the hotter star in three of those spectra (labeled a, c and d). As can be easily seen in Fig. 1, the emission reversals of the lines show that both stars are active. In the three above mentioned spectra with phases near 0.0, a gaussian deconvolution of the emission peak has been performed, whose results are presented in Table; for the other cases, the two peaks are separated. The presence of Hε in the neighbourhood of H line made a similar analysis for this line unreliable. To indentify which component is responsible for each peak we have used the orbital parameters from Strassmeier *et al.* (1988) and from new photometric determinations (Arevalo, 1990). The computed radial velocities from Doppler shifts of the peaks lead to similar results. Although this identification is not always certain, possibly due to uncertaintities in the ephemeris available, it seems that the hot star was less active then the cool component during the 1988 July observations, while the emission levels were similar in 1989 July. We have checked the Wilson-Bappu relationship (Rutten, 1984), using the K emission reversals, and the average values for the FWHM in both epochs. The cool star does not fit in any epoch (by an amount of about 4^m), and the hot star fitted in 1988 July, but not in 1989 July, due to the different activity level.

Acknowledgements: The Isaac Newton Telescope 2.5 m on the island of La Palma and the 2.2 m telescope at Calar Alto Observatory are operated, respectively, by the Royal Greenwich Observatory at the Spanish Roque de los Muchachos Observatory (Instituto de Astrofisica de Canarias), and the Max Plank Institut für Astronomie at the Centro Astronomico Hispano-Aleman of Calar Alto (Almeria, Spain). This work has been supported by the Spanish Comission Interministerial de Ciencia y Tecnologia under grant PB86-0536-C02.

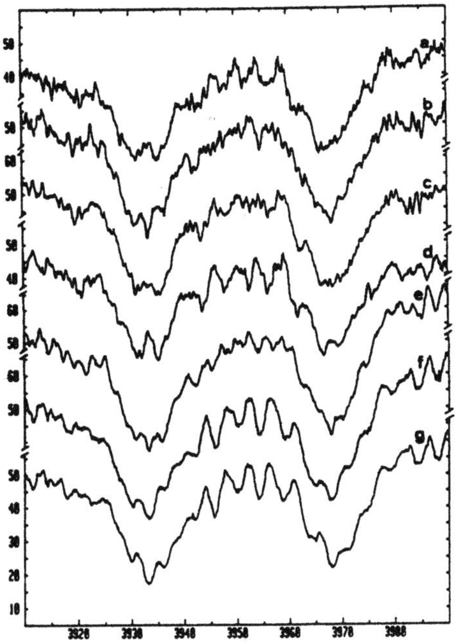

Fig. 1. Spectra of ER Vul. Fluxes in units of 10^{-13} erg cm^{-2} s^{-1}. Phases given in Table

Spec.	JD	Phase	line	λ	F_{obs}	FWHM	Notes
a	244 7370.57	0.04	K	3932.9	3.6	0.7	
	(INT)			3933.9	6.2	1.0	
			H	3968.5	3.4	-	(1)
b	244 7370.61	0.10	K	3932.2	8.2	1.0	
	(INT)			3934.6	13.6	-	(3)
			H	3967.3	1.1	-	(3)
				3969.7	5.3	-	(3)
c	244 7372.58	0.92	K	3932.9	3.7	0.7	
	(INT)			3933.9	6.1	1.0	
			H	3968.7	4.5	-	(1)
d	244 7372.64	0.00	K	3933.1	11.5	1.1	
	(INT)			3934.2	6.2	0.6	
			H	-	-	-	(2)
e	244 7721.59	0.87	K	3931.7	5.2	1.4	
	(CAHA)			3934.8	8.9	1.4	
			H	3966.3	4.3	-	(3)
				3968.8	5.6	-	(3)
f	244 7722.55	0.24	K	3931.2	9.0	1.5	
	(CAHA)			3935.3	9.7	1.6	
			H	3966.0	8.8	-	(3)
				3969.7	12.9	-	(3)
g	244 7723.58	0.71	K	3930.8	11.3	1.3	
	(CAHA)			3935.0	8.1	1.4	
			H	3965.6	8.1	-	(3)
				3969.3	7.2	-	(3)

Notes:

(1) No separation of the components is possible

(2) confuse, no measurements performed

(3) No gaussian fit possible (noisy)

Wavelength and FWHM in A

Fluxes in units of 10^{-13} erg cm^{-2}s^{-1}

INT: 2.5m Isaac Newton Telescope (Roque de los Muchachos)

CAHA: 2.2m Telescope (Calar Alto)

Table: Observations and results

References

Arevalo, M.J.: 1990, private communication

Armentia, J.E., Fernández-Figueroa, M.J., Cornide, M., De Castro, E., Fabregat, J.: 1989, NATO-ASI *Active Close Binaries*, Kusadasi, Turkey, in press.

Bond, H.E.: 1970, *Publ. Astron. Soc. Pacific* **82**, 321

Catalano, S., Trigilio, C.: 1990, private communication

Eggen, O.J.: 1978, *Inf. Bull. Var. Stars*, No 1426

Fabregat, J., Reglero, V.: 1990, *Astron. Astrophys. Suppl. Ser.* **82**, 531

Fernández-Figueroa, M.J., Montesinos, B., De Castro, E., Rego, M., Gimenéz, A., Reglego, V.: 1986, *Astron. Astrophys.* **169**, 219

Hall, D.S., Kreiner, J.M.: 1980, *Acta Astr.* **30**, 387

Hayers, D.S., Latham, D.W.: 1975, *Astrophys. J.* **197**, 593

Horne, K.: 1986, *Publ. Astron. Soc. Pacific* **98**, 609

Oke, J.B., Gunn, J.E.: 1983, *Astrophys. J.* **266**, 713

Rutten, R.: 1987, *Astron. Astrophys.* **177**, 131

The Variations of the Hα line of HD199178

Lauri Jetsu [1], Igor Savanov [2], Ilkka Tuominen [1]

[1]Observatory and Astrophysics Laboratory, University of Helsinki, Tähtitorninmäki, SF-00130 Helsinki, Finland

[2]Crimean Astrophysical Observatory, P/O Nauchny, 334413 Crimea, USSR

Abstract: Changes of emission cause variations of the Hα absorption line of HD 199178. The correlation with the simultaneous light curve may depend on the relative strength of the axisymmetric and non-axisymmetric components of the evolving spot configurations.

1. Introduction

Bopp and Stencel (1981) defined a new class of objects: FK Comae-type stars, which are apparently single, rapidly rotating, late-type giants of strong chromospheric activity. HD 199178 (G5 III–IV; $v \sin i = 80 \pm 3.3$ km s^{-1} ; $P_{\rm phot} = 3\overset{\rm d}{.}337484 \pm 0\overset{\rm d}{.}000043$) is one of the original candidates of this class (Herbig, 1958; Huenemoerder, 1986; Jetsu *et al.*, 1990a,b). A double period spot cycle of was proposed for HD 199178 by Jetsu *et al.* (1990a). The FK Comae stars show short-term changes of the Hα line profile (*e.g.* Walter and Basri, 1982; Bopp *et al.*, 1984; Huenemoerder, 1986). This paper presents some results of a search for correlations between the changes of Hα line profile and the simultaneous rotational modulation of brightness, *i.e.* spot activity. Spectroscopic observations were made at the Crimean Astrophysical Observatory during two seasons: September, 1986 and October – November, 1989. The nearly simultaneous photometry is from Jetsu *et al.* (1990a,b,c).

2. The Hα absorption and emission

The parameters derived from the Hα absorption line profile were: the equivalent width of absorption EW, the ratio of blue absorption equivalent width to red absorption equivalent width EW_B/EW_R and the centre C and peak P of a gaussian fit to the absorption line profile, which are plotted in Fig. 1. as a function of phase derived from the ephemeris of Jetsu *et al.* (1990b).

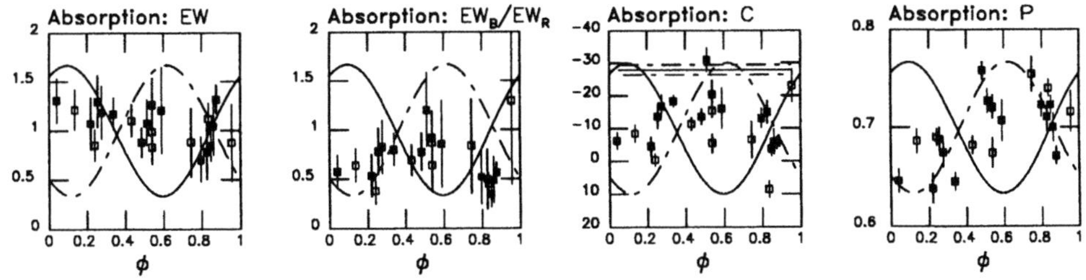

Fig. 1. The parameters EW, EW_B/EW_R, C and P: the results during 1986 are denoted by filled squares and those during 1989 by open squares. The light curves in V during both seasons are outlined in every picture on an arbitrary scale (1986: dashed curve, 1989: continuous curve). The radial velocity of HD 199178 published by Huenemoerder (1986) is shown as a straight line with error limits indicated in the second picture from the right.

An estimate of the chromospheric emission was obtained by substracting the artificially broadened spectrum of the chromospherically inactive giant ρ Cyg (G8 III) from the spectra of HD 199178 (G5 III–IV). The parameters of emission (Fig. 2.) derived during both season were: the equivalent width of emission EW_{em}, the ratio of blue emission equivalent width to the red $EW_{B,em}/EW_{R,em}$, the peak intensity of emission P_{em} and the respective of wavelength of this peak C_{em}. (Note that the amplitudes of the modulation of both C and C_{em} during 1986 are of the same order as those published by Neff *et al.* (1988) for the variation of the centre of Mg II k line).

3. Conclusions

During 1986 the parameters EW_B/EW_R, $EW_{B,em}/EW_{R,em}$, C, C_{em}, P and P_{em} do show more than marginal rotational modulation, and consequently quite a good correlation with the simultaneous light curve. During 1989 no clearly rotational modulation of emission or absorption is present, which naturally leads to a non-existent correlation with the light curve.

However, the mean brightness during 1989 is considerably lower (0.06^m) than during 1986. The difference is about 60 % of the peak to peak amplitude of varia-

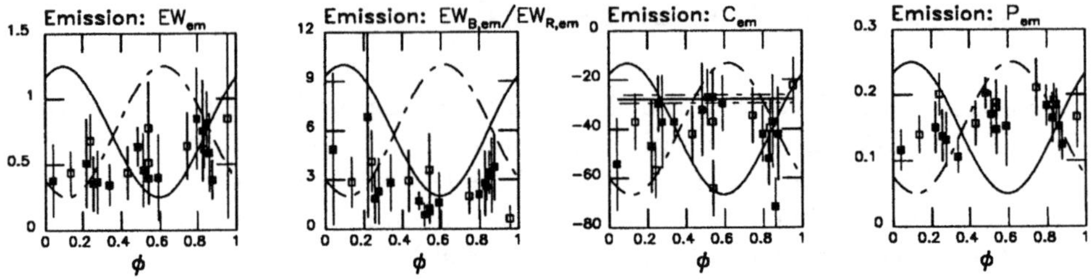

Fig. 2. The parameters EW_{em}, $EW_{B,em}/EW_{R,em}$, C_{em} and P_{em}. The notations are the same as in Fig 1.

tion of the mean brightness during the nine year cycle (Jetsu *et al.* , 1990a). The absence of detectable rotational modulation of Hα and of any correlation with the simultaneous light curve may thus be caused by the stronger axisymmetric component of spot distribution (*i.e.* spot activity is present on practically every observed longitude). Hence a detectable correlation with the light curve could only be expected when the axisymmetric part is weak compared non-axisymmetric part.

References

Bopp, B.W., Stencel, R.E.: 1981, *Astrophys. J. Letters* **247**, 131
Bopp, B.W., Goodrich, B.D., Africano, J.L., Noah, P.V., Meredith, R.J., Palmer, L.H., Quigley, R.J.: 1984, *Astrophys. J.* **285**, 202
Herbig, G. H.: 1958, *Astrophys. J.* **128**, 295
Huenemoerder, D. P.: 1986, *Astron. J.* **92**, 673
Jetsu, L., Huovelin, J., Tuominen, I., Vilhu, O., Bopp, B.W., Piirola, P.: 1990a, *Astron. Astrophys.* **236**, 423
Jetsu, L., Huovelin, J., Tuominen, I., Bopp, B.W., Efimov, Yu.S., Linnaluoto, S., Nations, H. L., Piirola, P., Shakhovskoy, N.M., Virtanen, H.: 1990b, *Astron. Astrophys. Suppl.* **85**, 813
Jetsu, L., Huovelin, J., Savanov, I., Tuominen, I.: 1990c, *Astron. Astrophys.*, submitted
Neff, J.E., Vilhu, O., Walter, F.M.: 1988, in *A decade of UV Astronomy with IUE, ESA SP-281*, **1.**, p. 291
Walter, F.M., Basri, G.S.: 1982, *Astrophys. J.* **260**, 735

Hα Emission in the Giant Components of RS CVn Systems

A. Frasca, S. Catalano

Istituto di Astronomia, Università di Catania
viale A. Doria 6, I-95125 Catania, Italy

1. Introduction

The Hα line of hydrogen is a very powerful tool for the study of the solar chromosphere (plages, flares, prominences). Although it is difficult to derive in a simple way quantitative information, like chromospheric temperature and electron density, because of the very extensive depth of its formation and the very complex mechanism of excitation, Hα has been successfully used to study the chromospheres of active stars (Herbig, 1985). Moreover this diagnostic has proved to be very useful for the study of surface inhomogeneities and transient phenomena like flares in systems of the RS CVn type and other active stars (Bopp, 1981; Bopp *et al.*, 1988; Fraquelli, 1982).

With the aim of investigating the behaviour of chromospheric activity, we have undertaken an extensive monitoring of Hα emission in a selected group of RS CVn binaries (Strassmeier *et al.*, 1988). We report here preliminary results of the observations obtained in 1988–1989.

2. Observations and reduction

All the spectra here discussed were acquired with a CCD camera at the Echelle spectrograph of the 91 cm telescope of Catania Observatory. We used the spectrograph in a low-dispersion configuration which gives a spectral resolution of about 0.9 Å/pixel. Exposure times were chosen to obtain spectra with a S/N ratio of 100–150. The spectrum extraction from the CCD images was performed following a standard procedure (subtraction of the bias level, division by a continuous spectrum given by a halogenic lamp etc.). Repeated spectra of a standard star allowed us to verify that an accuracy of about 1 % was reached in each point of the well exposed spectra, as expected for a ratio S/N $\sim$ 100.

The net chromospheric Hα emission of the binary systems was defined as the difference of the observed spectrum and a "composite" spectrum made up of the weighted sum of spectra of non active stars similar in spectral type to the primary and secondary star of each system. For each spectrum we have shifted in wavelength the two components of this composite spectrum to compensate for the radial velocity changes due to the orbital motion. We have also taken into account the rotational broadening of the reference star Hα profiles every time this was larger than the spectral resolution. This method enable us to identify the component in the binary responsible for the Hα emission.

3. Results

Time variability of the net Hα emission modulated with the orbital period was detected in nearly all of the 10 systems for which repeated observations were obtained. Equivalent width variations ranged from 0.1 Å to 1.3 Å (Table).

Table: Net Hα equivalent width variations

Star	$P_{\rm orb}$ (days)	Spectral types (Hot/Cool)	$\Delta W_{H\alpha}$(Å)
RS CVn	4.80	F5 IV / K0 IV	0.4
HR 5110	2.61	F2 IV / K2 IV	0.3
HR 5321	605.8	K0 III	0.1
HD 163621	?	F / G5 IV	0.2
Z Her	3.99	F4 V–IV / K0 IV	0.2
AR Lac	1.98	G2 IV / K0 IV	0.4
RT Lac	5.07	G9 IV / K1 IV	1.1
HK Lac	24.42	F1 IV / K0 III	2.3
SZ Psc	3.97	F8 IV / K1 IV	1.3
UX Ari	6.44	G5 V / K0 IV	0.7
HR 1099	2.84	G5 IV / K1 IV	1.1

The variation with the orbital phase of the net Hα emission equivalent width for HR 1099 is reported in Fig. 1, as example of the present results. For systems like Z Her or HR 5110 such variations do not seem to be correlated with the orbital phase. Although no significant variation (above the 2 % level) were found with time scales of a few hours, random or flare-like variations with time scales of days were detected for many systems. In particular we observed a giant Hα flare on HK Lac (a system with an orbital period of about 24 days), which showed an emission enhancement by a factor of ~ 2 and lasted about 6 days (Fig. 2). The flare peak flux at the stellar surface has been estimated to be $\sim 3 \times 10^7$ erg cm^{-2} s^{-1} and the total energy released was about 10^{38} erg. This seems to be one of the most powerful flares observed until now in systems of this kind (Catalano, 1990).

Acknowledgements: This work was supported by the MURST through the University of Catania, the Osservatorio Astrofisico di Catania and the GNA (CNR). The extensive use of the computer facilities of the Catania ASTRONET Site is also acknowledged.

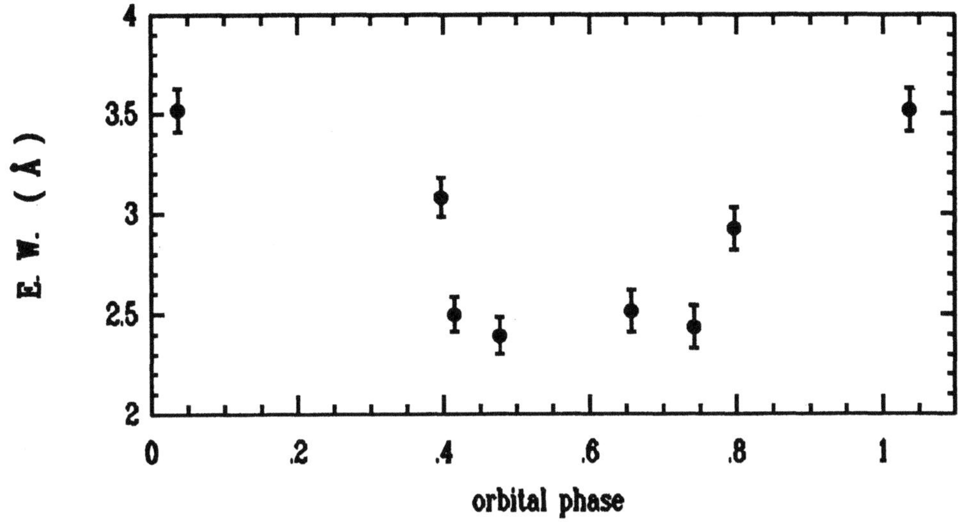

Fig. 1. Net Hα equivalent width of HR 1099 plotted against orbital phase.

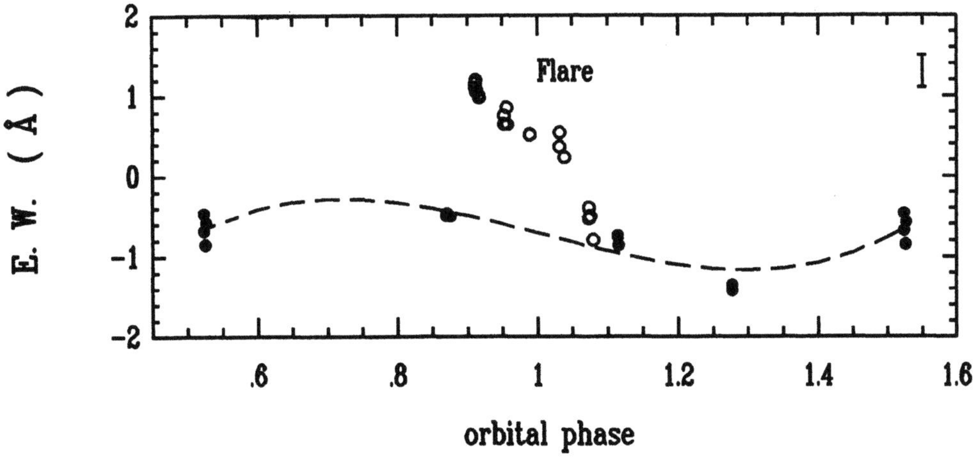

Fig. 2. Observed Hα emission of HK Lac versus orbital phase. The open circles relate to the flare-like event observed in 1989 between 25 and 29 of September.

References

Bopp, B.W.: 1981, *Astron. J.* **86**, 771

Bopp, B.W., Dempsey, R.C., Maniak, S.: 1988, *Astrophys. J. Suppl. Ser.* **68**, 803

Catalano, S.: 1990, in *Active Close Binaries*, eds. C. Ibanoglu and I. Yavuz, in press

Fraquelli, D.A.: 1982, *Astrophys. J. Lett.* **254**, L41

Herbig, G. H.: 1985, *Astrophys. J.* **289**, 269

Strassmeier, K.G., Hall, D.S., Zeilik, M., Nelson, E., Eker, Z., Fekel, F.C.: 1987, *Astron. Astrophys. Suppl. Ser.* **72**, 291

Spectroscopy of Southern Active Stars

O. Vilhu [1], B. Gustafsson [2], and F.M. Walter [3]

[1]Observatory and Astrophysics Laboratory, University of Helsinki, SF-00130 Helsinki, Finland
[2]Uppsala Astronomical Observatory, Box 515, S-75120 Uppsala, Sweden
[3]SUNY, Earth and Space Sciences, Stony Brook, NY 11794-2100, USA

Abstract: Hα and HeI D3 lines were observed in five active cool stars over their relatively short rotation periods. Three of the stars (AB Dor, Rst137B, HD82558) are in the pre main sequence phase (as deduced from their lithium abundances), while two are post main sequence FK Comae type stars (HD32918 and BD-223467) (Vilhu *et al.*, 1991). We present the results of these observations and point out the physical similarity of the Hα and He D3 transitions (Fig.1).

1. Observations

We have observed the southern active cool stars HD 32918 (K1 IIIp), HD 82558 (K2 V), BD -22°3467 (G8 III), AB Dor (HD 36705, K1 IV) and Rst 137B (M3-5), each of which lies close to the chromospheric-coronal saturation limit. The fractional Hα-luminosities $L_{H\alpha}/L_{bol}$ of the stars were close to 10^{-4}. No significant radial velocity variations were detected in any of the stars. Hence, they are most probably single stars or else the companion mass is very low.

The existence of a cool spot region of HD 32918 was traced (Piskunov *et al.*, 1990) by the variations of the FeI 6546-line profiles within the 9.55 d rotation period by a surface imaging technique. The Hα-emission of HD 32918 was variable and asymmetric, comprising possibly of chromospheric and wind (or expanding chromosphere) components.

The He I 5876 (D3) absorption was remarkably strong in HD 32918, pointing to the existence of an extended and low density chromosphere. BD-22°3467, the cool companion to the hot central star of the planetary nebula Abell 35, exhibits a photometric wave with a 0.77 day period (Jasniewich and Acker, 1988). Our observations, performed around the same time as this photometry, resolved a stronger blue-shifted Hα-emission at the minimum light (periodic R/V-variations). This behaviour was opposite to that found in HD 32918. The $v \sin i$-value (90 km/s) of BD-22°3467 is compatible with the interpretation that the 0.77 day photometric cycle is the star's rotation period.

The Hα-emission of HD 82558 was stronger than in the previous observations by Fekel *et al.* (1986) and variable, but with no obvious rotational modulation. The Hα-emission strengths of the young pre main sequence pair AB Dor - Rst137B are compatible with those of the Pleiades stars ($W_{H\alpha}$ vs. R-I, Stauffer and Hartmann, 1986), giving additional support to the small age of the system. Rst137B has the He I 5876 triplet in strong emission (EW = 750±80 mÅ), and its photospheric lines are broad ($v \sin i$ around 50 km/s). In AB Dor we detected two Hα-transients (the first in emission and the second one in absorption). The normal quiescent state of AB Dor is the filled-in Hα (no emission nor absorption).

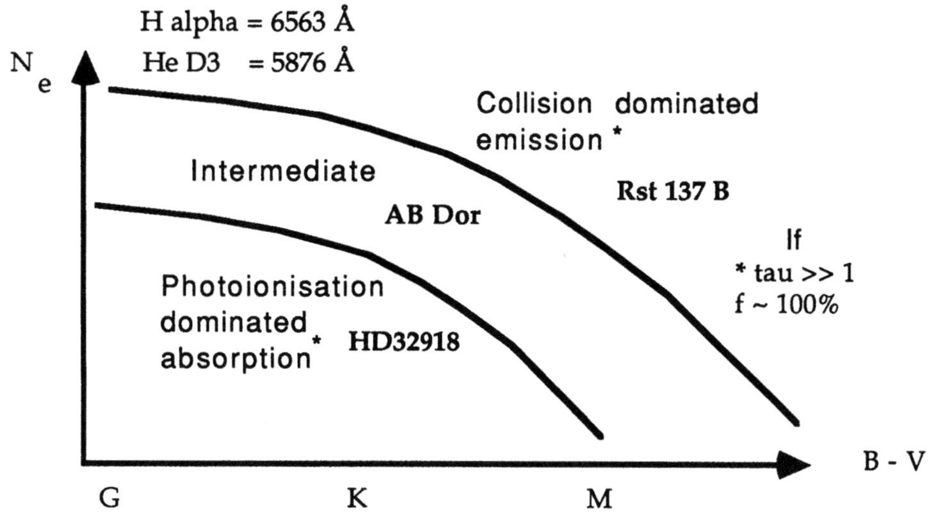

Fig. 1. A schematic representation of the Hα and He D3 lines in the spectral type - chromospheric density diagram (following Giampapa and Liebert, 1986). In a low density chromosphere the lines are in absorption, while in the high density ones they appear in emission (provided that the optical depths (τ) and surface coverage filling factors (f) are large enough). The curves, separating the domains, are somewhat different for Hα and He D3.

2. Similarities between Hα and HeI D3

In the fastest rotators of our sample, AB Dor (0.51 d) and HD 82558 (1.66 d), we detected neither HeI D3 absorption nor emission. The physics of the Hα and HeD3 transitions seem to be related. In active and low density chromospheres (with large optical thickness and filling factor) the lines appear in absorption (photo-ionization dominated plasma), while in very high density chromospheres (Rst137B) they are in emission (collisionally dominated plasma). Intermediate cases show neither absorption nor emission (filled-in level, AB Dor as a typical example, see Fig. 1).

References

Fekel F.C., Bopp B.W., Africano J.L., Goodrich B.D., Palmer L.H., Quigley L.H., Simon T.: 1986, *Astron. J.* **92**, 1150

Giampapa M.S., Liebert J.: 1986, *Astrophys. J.* **305**, 784

Jasniewich G., Acker A.: 1988, *Astron. Astrophys.* **189**, L7

Piskunov N.E., Tuominen I., Vilhu O.: 1990, *Astron. Astrophys.* **230**, 363

Stauffer J.R., Hartmann L.: 1986, *Astrophys. J. Suppl.* **61**, 531

Vilhu O., Gustafsson B., Walter F. 1991, *Astron. Astrophys.*, in press

Full Phase Coverage of BY Dra with the VLA

Jean-Pierre Caillault [1] and Stephen Drake [2]

[1]University of Georgia, Athens, GA 30602 USA

[2]SMM Data Analysis Center, Greenbelt, MD 20770 USA

Abstract: We have obtained 21 ~one-hour observations of the prototypical spotted star, BY Dra, with the VLA. These data are sufficiently well spaced in time to allow for complete rotational and orbital phase coverage; the resultant radio light curves are compared to their optical counterpart. We also plot the emission versus time without phase-folding, in order to search for longer periods of activity. Splitting the data into smaller (three minute) time bins allows for the determination of the normalized amplitude distribution of the flux; we compare this with known stellar optical and X-ray flare distributions.

1. Observations

We observed BY Dra with the VLA in April 1988. Twenty-one ~one-hour integrations were made at a wavelength of 6 cm. We show these data in Fig. 1. Although it is possible that a general decline in the level of emission occurred over the course of the month, it is likely that random large flaring was responsible for the higher fluxes detected; in fact, a large flare (peak flux $\sim$ 6.5 mJy) was detected on 6 April. We never failed to detect BY Dra (at the 3σ level), with the lowest one-hour integrated flux being ~0.30 mJy. The degree of polarization is generally small; hence, we interpret this low-level flux as quiescent emission from BY Dra.

2. Low-level flaring

Each of the 21 ~one-hour observations was broken up into three-minute "snapshots". By defining a quantity F_r as the ratio of the flux in any one snapshot to the minimum snapshot flux observed (~0.20 mJy), we have constructed a histogram $N(F_r)$ of the normalized amplitude distribution of three-minute snapshot fluxes of BY Dra; this is presented in Fig. 2. This distribution is almost flat, differing greatly from that found by Kunkel (1973) for U-band flares in dMe stars, by Drake (1971) for solar X-ray flares, by Montmerle *et al.* (1983) for the ρ Oph X-ray sources, and by Caillault and Zoonematkermani (1990) for the Orion X-ray sources (all roughly described by a power-law with exponent $\alpha \sim -1.4 \pm 0.2$). Hence, the quiescent emission does not seem to result from low-level flaring.

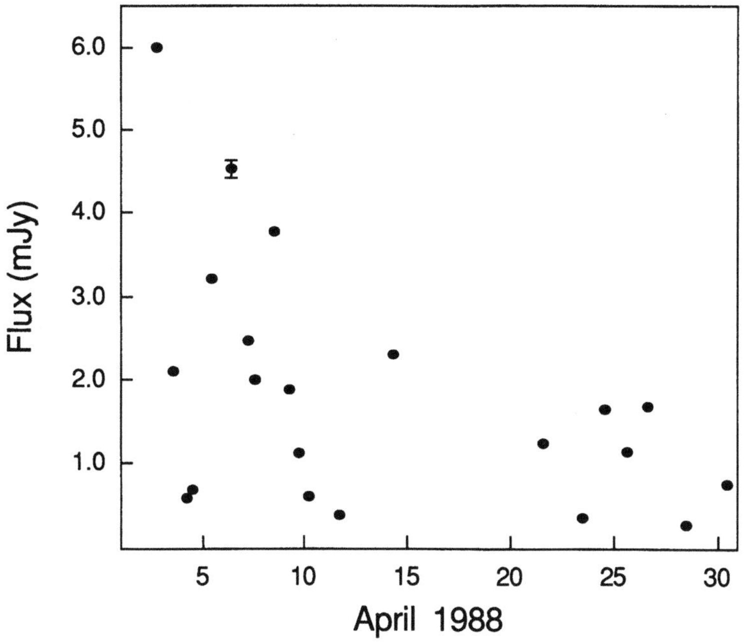

Fig. 1. The 21 ~one-hour integrations of BY Dra taken during April 1988 are plotted as filled circles. A representative error bar for these data is shown on the point corresponding to April 6 (~4.5 mJy). (This point is also the "flare" which is mentioned in the text.)

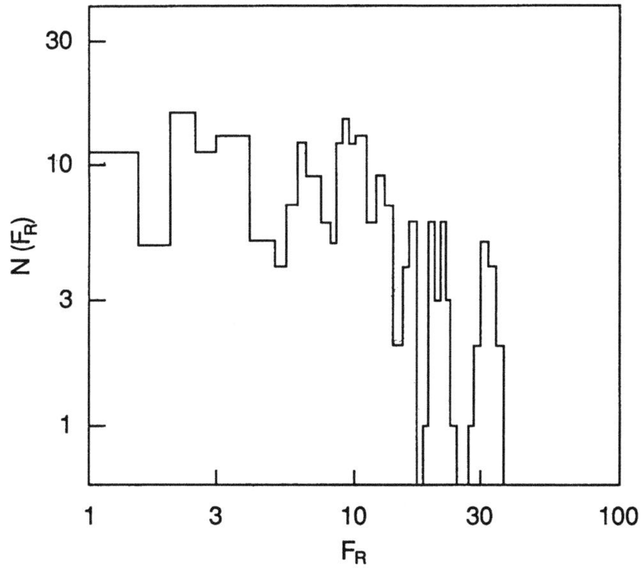

Fig. 2. A histogram $N(F_r)$ of the normalized amplitude distribution of three-minute "snapshot" fluxes of BY Dra.

3. Rotational modulation

Rotational modulation of the flux is a way of determining the size and location of the quiescent source (Gary, 1985). This has been attempted by Linsky and Gary (1983) and Gary (1986) for the eclipsing binary YY Gem and by Caillault *et al.* (1989) for the white-dwarf, K-dwarf eclipsing binary V471 Tau. The VLA data obtained in the present study are sufficiently well spaced in time to allow for complete rotational ($P = 3.83^d$) phase coverage of BY Dra; the resultant radio light curve is shown in Fig. 3 (this figure also includes data from 1986, Caillault *et al.*, 1988). An anti-correlation of the radio emission with BY Dra's optical emission (Cutispoto *et al.*, 1987) would provide direct evidence of the presumed magnetic origin of the 6 cm emission; no such pattern is seen). (We have also plotted the emission versus the orbital [$P = 5.98^d$] phase; again, no pattern is seen). The optical observations were performed in 1986, though; simultaneous, or even contemporaneous, data would have been better suited to this study. As it turns out, the inclination angle of BY Dra is $\sim 30°$ and the amplitude of optical variation is $\lesssim 0.1$ mag, requiring a very meticulous experiment be performed even without the complication provided by the random large flaring. In addition, if the radio emission results from a coronal gyrosynchrotron mechanism, then large loops might preclude any rotational modulation.

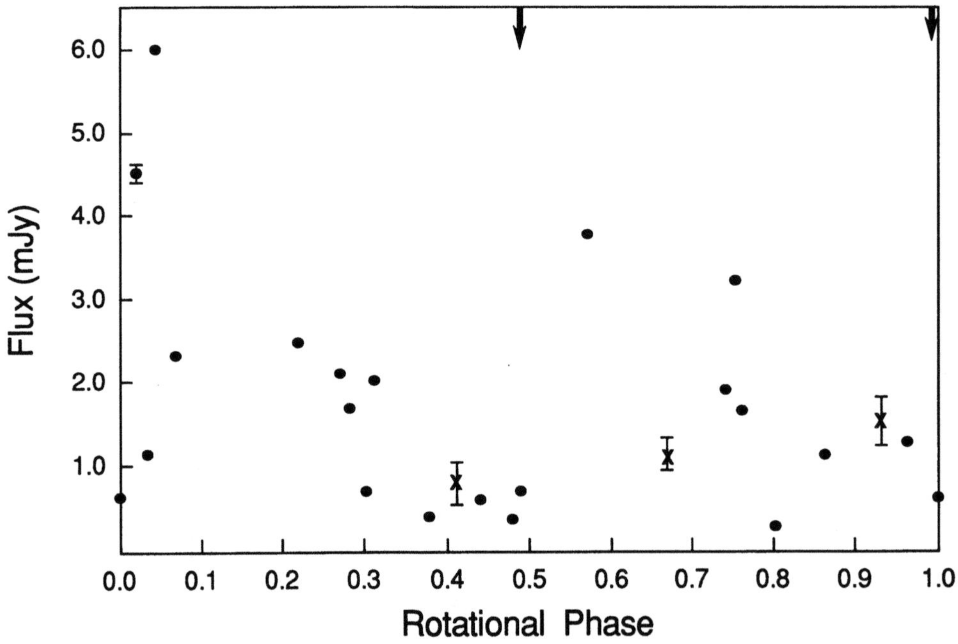

Fig. 3. The same data as in Fig. 1, but with the addition of crosses representing data obtained in February 1986. The optical light maximum and minimum represented by the two arrows at the top of the figure, are at $\phi = 0.49$ and $\phi = 0.99$, respectively.

References

Caillault, J.-P., Drake, S., Florkowski, D.: 1988, *Astron. J.* **95**, 887

Caillault, J.-P., Patterson, J.., Skillman, D.: 1989, in IAU Colloquium No 104 *Solar and Stellar Flares*, eds. B.M. Haisch and M. Rodonò, Catania p. 119

Caillault, J.-P., Zoonematkermani, S.: 1990, in *Flare Stars in Star Clusters, Associations and in the Solar Vicinity*, eds. L.V. Mirzoyan *et al.*, Kluwer p. 159

Cutispoto, G., Leto, G., Pagano, I., Santagati, G., Ventura, R.: 1987, *Inf. Bull. Var. Stars*, No 3102

Drake, J.F.: 1971, *Solar Physics* **16**, 152

Gary, D.E.: 1985, in *Radio Stars*, eds. R. Hjellming, D.M. Gibson, Reidel, Dordrect, p. 185

Gary, D.E.: 1986, in *Fourth Cambridge Workshop on Cool Stars, Stellar Systems, and the Sun*, eds. M. Zeilik and D.M. Gibson, Springer, Berlin, p. 235

Kunkel, W.E.: 1973, *Astrophys. J. Suppl.* **25**, 1

Linsky, J.L., Gary, D.E.: 1983, *Astrophys. J.* **274**, 776

Montmerle, T., Koch-Miramond, L., Falgarone, E., Grindlay, J.E.: 1983, *Astrophys. J.* **269**, 182

Coronal Magnetic Fields of Algol Binaries from Microwave Spectra

G. Umana [1], C. Trigilio [1], R. M. Hjellming [2], S. Catalano [3], M. Rodonò [3,4]

[1]Istituto di Radioastronomia CNR, stazione di Noto, C.P. 169, Noto, Italy
[2]National Radio Astronomy Observatory, Socorro, N. M., USA
[3]Istituto di Astronomia, Università di Catania, V. A. Doria 6, 95125 Catania, Italy
[4]Osservatorio Astrofisico, Università di Catania, V. A. Doria 6, 95125 Catania, Italy

1. Introduction

Algol-type binaries are basically known to undergo hydrodynamic processes related to mass exchange between components. Recent observations on radio, X-ray emission and flare-like events have raised the question of possible magnetic activity in the secondary component of these systems (Hall, 1989).

From a microwave emission survey we have shown that the radio emission from Algol systems cannot be accounted for by thermal emission from an hot corona ($T \geq 10^7 K$) and that their radio luminosities compare very well with those of the magnetically active RS CVn systems (Umana *et al.*, 1990).

2. Observations and results

Radio spectra of a small sample of Algol-type binary systems at four wavelengths (20, 6, 3.6, and 2 cm) were obtained with the VLA in two different observing runs (18 Feb. and 6 Mar. 1989). As in the case of RS CVn-type binary systems, the radio emission seems to be due to the interaction of mildly relativistic electron with the magnetic field of the active component (gyrosynchrotron emission). Calculations for an homogeneous structure, such as a corona surrounding the secondary component with uniform magnetic field strength and uniform density of the energetic particles, lead to spectra that are inconsistent with the shape of the observed spectra, for each choice of parameters (magnetic field strength, particles energy spectrum and density).

Instead, a *core-halo* structure for the radio sources with a compact *core*, with uniform magnetic field strength and density of energetic particles, together with an extended *halo*, with a radially decreasing particle density and a constant magnetic field strength, is able to reproduce the observed characteristics of the radio spectra.

In Table a summary of the parameters derived for the two regions for the observed systems is given. The average magnetic field strengths we derive are 80–100 Gauss for the core and 10–20 Gauss for the halo with typical electron densities of 5–10$\times 10^6$ cm^{-3} and 2–5$\times 10^6$ cm^{-3} for the core and the halo respectively. Characteristic sizes of the core are comparable with the size of the secondary component while those of the halo turn out to be as large as the whole system.

The good agreement between the two spectra observed at two different epochs of the binary system HR 5110 and the computed gyrosynchrotron spectra, assuming a core-halo structure for the radio source (thick line), is shown in Fig. 1, as an example of our results. The error bar represents the *rms* on the maps plus the error in the flux determination. The dashed lines represent the different contributions to the total spectrum from the halo (low frequencies) and from the core (high frequencies).

The 6 Mar 1989 spectrum of HR 5110 is consistent with an ejection of energetic particles in the lower corona (the total number of energetic particles has increased with respect to 18 Feb 1989, see Table).

Component	Magnetic Field (Gauss)	Linear Dimension (cm)	Particle Density cm^{-3}	α Index $N(r) \propto N_O(\frac{R_0}{R})^\alpha$
		Algol		
Core	60	8×10^{10}	10^6	0
Halo	20	4×10^{11}	5×10^6	2
		V 505 Sgr		
Core	40	1.8×10^{11}	2×10^6	0
Halo	10	7×10^{11}	5×10^5	2
		HR 5110 (18 Feb)		
Core	80	1.2×10^{10}	6×10^6	0
Halo	40	7×10^{11}	3×10^5	2
		HR 5110 (6 Mar.)		
Core	60	1.8×10^{10}	10^7	0
Halo	40	7×10^{11}	10^7	2
		RZ Cas (18 Feb.)		
Core	100	6.8×10^{10}	10^6	0
Halo	10	7×10^{11}	3×10^5	2
		RZ Cas (6 Mar.)		
Core	80	7×10^{10}	7×10^5	0
Halo	10	7×10^{11}	1.5×10^5	2
		TW Dra (18 Feb.)		
Core	80	4.8×10^{10}	2×10^7	0
Halo	10	3.2×10^{11}	4×10^5	2
		TW Dra (6 Mar)		
Core	80	8.0×10^{10}	1.5×10^6	0
Halo	10	3.2×10^{11}	4×10^5	2

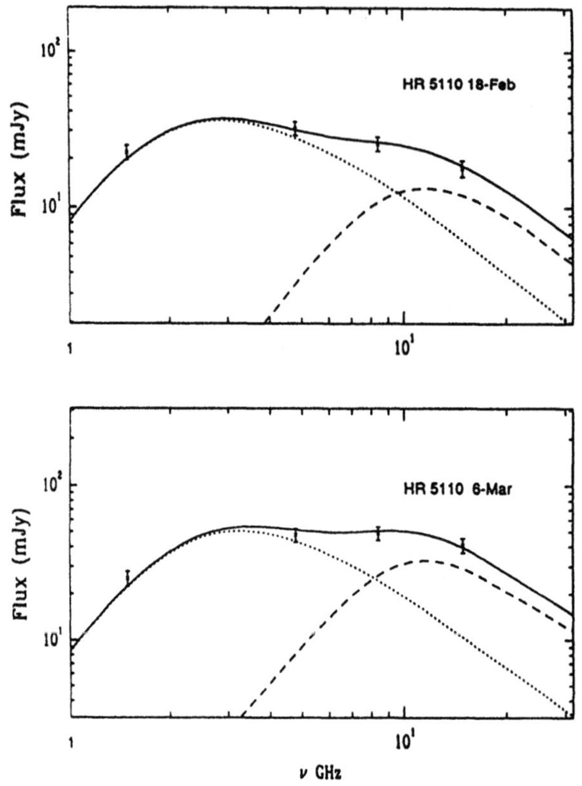

Fig. 1. Comparison between the observed radio spectra of the binary system HR 5110 and the computed spectrum obtained assuming a core-halo structure for the radio source as described in the text.

The proposed core-halo structure has not to be assumed as a rigorous model of the radio source, but it means that the probably complex topology of the coronal magnetic fields could be reduced to two basic magnetic structures: small, compact loops with high values of magnetic field strengths that we have described as the core, and larger loops, extending to the dimension of the whole systems, that we have described as the halo.

Acknowledgements. The Very Large Array (VLA) is a facility of the National Radio Astronomy Observatory operated by the Associated Universities, Inc., under contract with the National Science Foundation. This work has been supported by the CNR, the M.U.R.S. through the University of Catania, the Osservatorio Astrofisico di Catania and the GNA(CNR). The extensive use of the computer facilities of the Catania ASTRONET site is also acknowledged.

References

Hall, D. S.: 1989, *Space Sci. Rev.* **50**, 219
Umana G., Catalano S., Rodonó M.: 1990, *Astron. Astrophys.*, submitted

On the Coronal Activity of RS CVn Systems

Osman Demircan and Ethem Derman

University Observatory, Science Faculty
06100 Tandoğan, Ankara, Turkey

An analysis of the X-ray emission of a large sample of single MS stars shows that X-ray luminosity is strongly correlated with the photospheric radius (Fleming *et al.*, 1989) which implies the dependence of activity on bolometric luminosity L_{bol}, effective temperature T_{eff}, and thus stellar mass M (Micela *et al.*, 1985; Bookbinder *et al.*, 1986). The role of the emission area on the coronal activity of RS CVn systems has been noted independently by Majer *et al.* (1986), and Demircan (1986).

By using new data on some well known RS CVn systems, we produce new evidence that above idea is indeed valid. A plot of X-ray luminosity L_{x} versus $(B-V)$ for the sample of well known RS CVn systems in Fig. 1 shows that cooler, long period, more evolved systems with larger surface area have stronger X-ray emission.

Surprisingly, no clear correlation between the activity luminosity and rotation period was found in RS CVn binaries (Majer *et al.*, 1986; Demircan, 1986, 1987a, b, 1988; Basri, 1987; Morris and Mutel, 1988), unless the activity flux or the normalized activity luminosity $(L_{\text{x}}/L_{\text{bol}})$ is used as the activity indicator.

In order to understand the cause(s) of the scatter in $L_{\text{x}} - P$ diagram we divide the systems in X-ray emission into two groups. The higher level X-ray emission group is formed by the systems RZ Cnc, RT Lac, SZ Psc, RZ Eri, GK Hya, RS CVn, AR Mon, AR Lac, CF Tuc, UV Psc, V711 Tau and CQ Aur. It is interesting to see that the component stars of a system in this group differ more in physical characteristics than is the case of the other group. The larger, more evolved components in the higher level X-ray emission group have filled more than 60% of their Roche lobes, while in other group, the more evolved components never exceed 60% of their Roche lobes. The short period RS CVn systems (such as ER Vul, SV Cam and UV Psc) with MS components do not obey this rule. The Roche lobe filling percentage $RL\%$ contains information on the radius R, mass ratio q, separation a between the components and orbital period P, which is equal to the rotation period of the components in synchronized binaries. It is known that a larger $RL\%$ requires in general a smaller distance between the components and thus a shorter

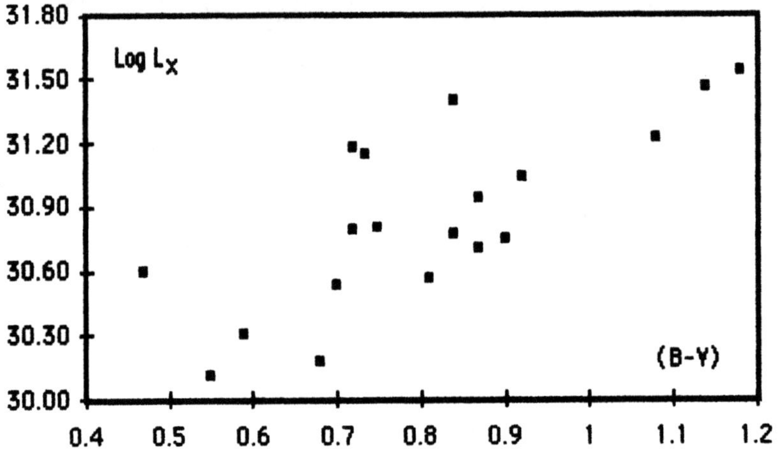

Fig. 1. A plot of X-ray luminosity versus colour index $(B-V)$ for the well known RS CVn systems. The cooler, long period, more evolved systems with larger surface area have more X-ray emission.

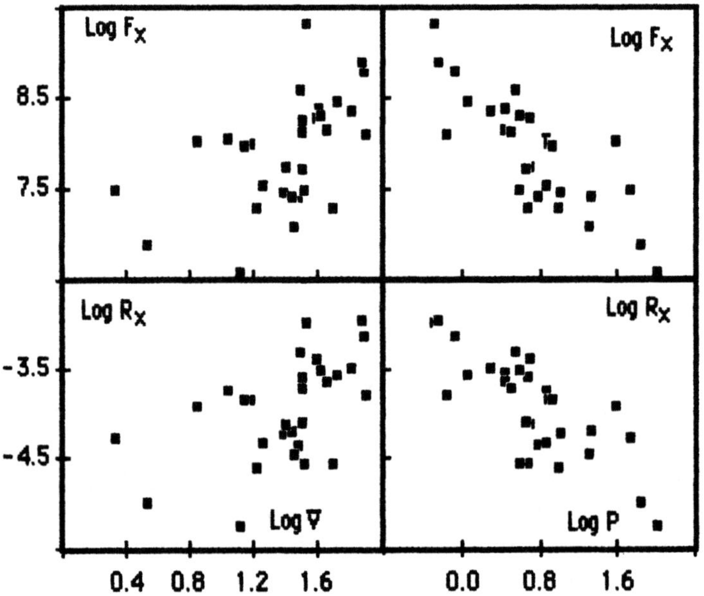

Fig. 2. The mean relative activity flux F_x, and the normalized flux R_x versus orbital period P, and mean equatorial rotation velocity $\bar{v}$ for the well known RS CVn systems. F_x and R_x tend to increase with increasing $v \propto 1/P$, as indicated by dynamo models.

period by Kepler's law. The higher level X-ray emission group have indeed smaller a, shorter P and larger equatorial rotation velocity v, on average.

We can safely conclude that the higher X-ray emission level of RS CVn type binaries is not only due to enhanced rotation by tidal interaction but is in general related to the larger $RL\%$ of one of the component stars. It should be kept in mind that the high X-ray emission level does not mean a high degree of activity. Remember also that the best measure of activity is not activity luminosity but the surface activity flux (e.g. Zwaan, 1986).

We have calculated the mean relative activity flux $F_x = L_x/(R_h^2 + R_c^2)$ and the normalized activity flux $R_x = L_x/[(L_{bol})_h + (L_{bol})_c]$ for each system of our sample. A plot of these two activity parameters, against orbital period P and mean equatorial rotation velocity $\bar{v}$, in Fig. 2 shows that an activity-rotation correlation exists for RS CVn type binaries. However a similar correlation exists between F_x (or R_x) and any system parameter such as total mass, distance a between the component stars, orbital angular momentum J, and mean density ρ of the component stars. Such correlations are clearly induced by the interrelations between mentioned parameters above. It is not clear from the present data whether one or more parameters are responsible in producing stellar activity. It seems that the activity in late-type binaries is directly related to their tidal interaction (Demircan, 1986). This conclusion is further strengthened by the findings of Glebocki and Stawikowski (1988), and Young *et al.* (1989). Evidence has been found (Demircan, 1990) that the X-ray surface flux F_x first increases with increasing $RL\%$ until a value of about 60%, and then decreases as $RL\%$ increases further.

References

Basri, G.: 1987, *Astrophys. J.* **316**, 377

Bookbinder, J., Golub, L., Rosner, R.: 1986, in *Cool Stars, Stellar Systems and the Sun*, eds. M.Zeilik, D. Gibson, p. 97

Demircan, O.: 1986, *I.B.V.S.* No 2969

Demircan, O.: 1987a, *Astrophys. Space Sci.* **136**, 201

Demircan, O.: 1987b. *Astrophys. Space Sci.* **137**, 195

Demircan, O.: 1988, in *Hot Thin Plasmas in Astrophysics*, ed. R. Pallavicini, Kluwer Acad. Publ. Dordrecht, p. 147

Demircan, O.: 1990, in *Active Close Binaries*, ed. Ibanoğlu, Kluwer, Holland.

Fleming, T.A., Gioia, I.M., Maccacoro, T.: 1989, *Astrophys. J.* **340**, 1011

Glebocki, R., Stawikowski, A.: 1988, *Astron. Astrophys.* **189**, 199

Majer, P., Schmitt, J.H., Golub, L., Harnden, F.R., Rosner, R.: 1986, *Astrophys. J.* **300**, 360

Micela, G., Sciortino, S., Serio, S., Vaiana, G.S., Bookbinder, J., Golub, L., Harnden, F.R., Rosner, R.: 1985, *Astrophys. J.* **292**, 172

Morris, D.H., Muttel, R.L.: 1988, *Astron. J.* **95**, 204

Schrijver, C.J.: 1985, *Space Sci. Rev.* **40**, 3

Young, A., Ajir, F., Thurmann, G.: 1988, *Publ. Astron. Soc. of Pacific* **101**, 1017

Zwaan, C.: 1986, in *Cool Stars, Stellar Systems and the Sun*, eds. M. Zeilik, D. Gibson, Springer-Verlag, Berlin, p. 19

Flare Rate Spectra as a Possibility of Diagnostics of Convection Zones in Stars

L.A. Pustil'nik

Special Astrophysical Observatory,
USSR Academy of Sciences, N.Arkhyz, 357147, USSR

Abstract: A theoretical interpretation is offered for the power-law dependence of the flare occurrence rate in red flare dwarfs. Flares on red dwarfs are regarded as being physically akin to solar flares, with the energy source being the magnetic field. A power law flare energy relation then emerges naturally. Model estimates of the corresponding spectral index β fit the observed values in most flare stars. Within the bounds of the interpretation offered the energy dependence of a flare event is determined by the character of the turbulence in the convection zone. The observed parameters of the dependence can be used to diagnose the character of the turbulence in the convection zones of flare stars.-

1. The physical nature of the flare rate spectrum

Early investigations of flare activity in stars addressed the problem of how the flare rate depends on the emitted flare energy (Gershberg, 1972; Lacy *et al.*, 1976). The flare rate spectrum was defined as the number of flares with energy greater than a threshold energy E_0, such that $\nu(E > E_0) \propto E_0^{\beta}$.

A theoretical analysis (Pustil'nik, 1988) has shown that the observed universality can be explained as a manifestation of the dissipation of magnetic fields in stellar atmospheres (Gershberg and Pikel'ner, 1972). The initial magnetic fields are considered to be the result of dynamo processes. The magnetic structures are expelled from the convection zone by turbulent motions and magnetic buoyancy. Physically the approach is rather transparent since the flare energy and the rate depend much on the same parameter, namely a dimension L of the magnetic structure. The flare energy is given by

$$E_H(L) = \int \frac{\Delta(H^2)}{8\pi}\, dV \propto \begin{cases} H^2 L^3 & \text{– for a field weakly decreasing in altitude, i.e. } L < z_H. \\ H^2 L^2 z_H & \text{– for a field rapidly decreasing in altitude, i.e. } L > z_H. \end{cases}$$

On the other hand the rate of occurrence of flares is determined both by the number of such structures on the star disk $N_{str} \propto R_\star^2/L^2$ and by their lifetime $\tau(L) \cong L/V(L)$. Here $R_\star$ is the dimension of of the star and $V(L)$ is a turbulent velocity in the corresponding turbulent cell of dimension L. z_H is the length-scale of magnetic field. $V(L)$ is determined by a specified turbulence spectrum for the convection zone of the star. Existing theories (see Dibay and Kaplan, 1976; Vainstein, 1983) give unambiguously a power spectrum of turbulent velocity $V(L) \propto L^{+s}$, with the spectral index s depending on the physical conditions in the convection zone (from $s = 1/2$ for Vajan convection with a strong magnetic field, to $s = 1/4$ for acoustic turbulence). The resulting dependence is

$$\nu(L > L_0) \propto \int N(L)\frac{V(L)}{L}dL \propto L^{-2+s}.$$

Using $E(L)$ we obtain

$$\nu(E > E_0) \propto \begin{cases} E_0^{-1+s/2} \propto E_0^{-(0.88-0.75)} & \text{for rapidly decreasing magnetic field} \\ E_0^{-0.66+s/3} \propto E_0^{-(0.58-0.5)} & \text{for slowly decreasing magnetic field.} \end{cases}$$

This theoretical consideration gives us not only the power-law character of the flare rate spectrum, but also indices that are similar to the observed ones. Observations of all nearby flaring red dwarfs, as well as solar flares in Hα and in microwaves, and of relatively older flare stars in Praesepe and Pleiades (see Gershberg and Shakhovskaya, 1984; Gershberg, 1988, 1989) comply with this theoretical result. However, the younger flare stars in the Orion cluster have spectra with steeper slopes than theoretically predicted.

2. Possible astrophysical consequences

The approach allows us to investigate turbulence in convection zones from the properties of the flare rate spectrum.

a) The dependence of flare rate spectrum upon activity phase. The long-term variations of the magnetic field during an activity cycle must influence the turbulence and thus also the flare rate spectrum. The ratio of the magnetic energy to the energy of the turbulent motions on different scales must be a basic parameter.

In this connection Kasinskij and Sotnikova (1990) and Gershberg (1990) have pointed out that the solar flare rate spectral index in the radiowave range changes during the 11-year solar activity cycle. for some of them. There are indications of changes in the flare rate spectrum index during several years for the classical flare star UV Ceti (Lacy *et al.*, 1976).

b) Evolutionary changes in convection zones. There are possible systematic changes in the flare rate spectrum index with the age of flare stars, as detected by Korotin and Krasnobabtsev (1985); see also Gershberg (1989). If a flare rate spectrum reflects a turbulence velocity spectrum, then the deviating Orion results indicate possible momentary realizations of anomalous turbulence spectra, suggesting that the maximum occurs in small scales during early stages of stellar evolution. This phase is extremely transient ($\leq 10^7$ years) and the spectrum changes quickly to that of Kolmogorov or Kraynovsky turbulence. In principle, the possibility of such exotic spectra is pointed out in a number of papers (Vainstein, 1983), though their realization is unlikely.

I sincerely thank Dr. R.E. Gershberg for his continuing support.

References

Dibai, E.A. and Kaplan, S.A., 1976, *Dimensionality and Similarity of Astrophysical Quantities* (in Russian), Nauka, Moscow, p. 224

Gershberg, R.E.: 1972, *Astrophys. Space Sci.* **19**, 75

Gershberg, R.E., Pikel'ner, S.B.: 1972, *Comments Astrophys. Space Sci.* **4**, 113

Gershberg, R.E., Shahovskaya, N.I.: 1984, *Astrophys. Space Sci.* **95**, 235

Gershberg, R.E.: 1988, *Mem. Soc. Astron. Ital.* **59**, 1

Gershberg, R.E.: 1989, *Mem. Soc. Astron. Ital.* **60**, 263

Gershberg, R.E.: 1990, a private communication

Kasinskij, V.V., Sotnikova, R.T.: 1990, *Issled. Geomagn. Aeron. Fiz. Solntsa*, Moscow, No 83, p. 99

Korotin, S.A., Krasnobabtsev, V.I.: 1985, *Izv. Krym. Astrophys. Obs.* **73**, 131

Lacy, C.N., Moffet, T.J., Evans, D.S.: 1976, *Astrophys. J. Suppl. Ser.* **30**, 85

Pustil'nik, L.A.: 1988, *Sov. Astron. Lett.* **14**. 5, 398

Vainstein, S.A.: 1983, *Magnetic Fields in the Space*, Nauka, Moscow

Concluding summaries

Observations Relating to Stellar Activity, Magnetism and Dynamos

C. Jordan

Department of Theoretical Physics, University of Oxford, 1 Keble Road, Oxford OX1 3NP, UK

Abstract: Dynamo theory can in principle be tested by comparing its predictions with a variety of types of solar and stellar observations. The progress apparent from papers presented at this Colloquium is summarized, and areas where further observations would be of particular value are stressed.

1. Introduction

The aim of IAU Colloquium 130 was to bring together the scientists working on stellar dynamo theory with those making observations related to stellar activity. The Colloquium was timely in that there have been recent theoretical developments, particularly in the study of non-linear effects, and a wide range of observations is now possible, from X-ray to radio wavelengths. In particular, helioseismology is providing new ways of probing convection zone properties, and the techniques for determining stellar surface magnetic fields have substantially improved over the past few years. Here we examine the extent to which the existing observations can be used to confront dynamo theory, and what further observations are required.

2. Internal conditions: the convection zone

Theoretical arguments concerning the generation of magnetic fields, and the timescale for their emergence through the stellar surface, require the region of generation to be placed within about one pressure scale height above the base of the convection zone (Parker, 1975). The properties of the surface fields then depend on the rotation rate, and the rate of differential rotation, with depth in the convection zone. Until recently there was no way of making observations of these regions, but the situation has been changed significantly by the advent of helioseismology.

The papers by Goode and Paternó, for example, addressed this area. The global oscillations can be used to sample the region at $R > 0.5R_\odot$ and the splittings according to l allow the differential rotation to be investigated. The high l splittings reflect conditions near the surface, whilst the low l splittings reflect conditions near the base of the convection zone. The available data sets (see Goode for references) show that near the equator the rotation rate decreases only slowly from the surface value until near the base of the convection zone. The data for non-equatorial latitudes are less reliable, but show tendencies that depart from the surface law. Belvedere *et al.* also use helioseismology data to deduce that the gradient of the angular velocity in the boundary layer at the base of the convection zone should change from positive to negative at a latitude of around 37°, and relate this result to observations of the latitude dependence of solar activity and to their non-linear dynamo model.

Near the base of the convection zone there seems to be a rapid transition to solid body rotation, but different data sets give conflicting evidence on the core rotation rate. However, the data support the concept of dynamo action near the base of the convection zone. On the other hand, they already exclude rotation on cylinders, predicted by some theories.

Future observations, both from space (e.g. from SOHO) and from ground-based networks, (e.g. GONG and the Birmingham network), should allow a wider variety of modes to be observed, and the accuracy of the data to be improved. Thus there is a real expectation of a secure knowledge of the solar differential rotation as a function of depth and latitude, and as a function of the solar activity cycle, in the not too distant future. These data will provide a foundation for extending dynamo theory to other stars.

3. Solar surface magnetic fields

Observations of solar magnetic fields at high spatial resolution also provide valuable boundary conditions to constrain dynamo theories, and several speakers addressed the interpretation of these data. For example, Ruzmaikin *et al.* analyzed magnetograms obtained at Kitt Peak and La Palma in terms of fractal structure within a turbulent cell, such as a supergranule. They show that such a fractal distribution can be produced in an inhomogeneous "stretch-twist-fold" dynamo model, and investigate the resulting degree of inhomogeneity in the velocity field.

Stenflo discussed synoptic observations obtained from Kitt Peak and Mount Wilson, and how they can be used to infer the rotational phase velocity. He finds that daily longitude displacements and the recurrence of patterns at the central meridian lead to completely different rotation laws. He suggests that these can be reconciled if the observed pattern of surface fields is renewed from deep layers in a time scale of less than a month, and again argues that this is evidence for dynamo action at the base of the convection zone. The quasi-rigid rotation found from the recurrence method would then represent the phase velocity of the source pattern.

Another way of studying the magnetic field and the velocity fields is to use measurements of the positions of sunspots as a function of time, a method pioneered by Tuominen (1945). Ribes *et al.* reported the results of analyzing spectroheliograms obtained at Meudon during solar cycle 21, using techniques involving digitized data (Mein and Ribes, 1990). They find that there is a zonal and time-dependent meridional circulation, which coincides with the large scale azimuthal field as measured at Stanford by Hoeksema, and that the rotation rates for each year vary from the mean differential rotation. They interpret these results in terms of magnetic convective tori and azimuthal rolls, and suggest that behaviour of the latter could explain the torsional pattern observed by Howard and LaBonte (1980) and Snodgrass (1986) from measurements of velocity fields.

Although some of the observations discussed in Sections 2 and 3 lead to results that are preliminary, or tentative, they are clearly well worth pursuing. The studies of the magnetic fields, together with the helioseismological data, are revealing the complexity of the solar dynamo, and cast doubts on the simple $\alpha\Omega$ model. Parker (1987) has reviewed the difficulties produced by recent observations, in particular the sign of the migration of the azimuthal field in models where the the angular velocity decreases inwards, and discusses ways in which the simple model may have to be modified.

4. Other solar observations

It is now generally accepted that the energy emitted from the transition region and corona originates from the dissipation of magnetic energy. Although there may be a non-magnetic component to the energy deposited in the chromosphere, the correlation between the spatial location of strong regions of emission in chromospheric lines, such as Ca II H and K, and regions of high magnetic flux, is taken to indicate heating via the magnetic field in this region too. Schrijver (1987) has shown that the "excess" flux (i.e. the component heated via the magnetic field) in the Ca II and Mg II emission lines correlates linearly with the flux in transition region lines, such as C IV, which are also stronger over the supergranulation boundaries. The overlying coronal emission does not show clearly the supergranulation pattern but is more amorphous (Reeves, 1976). A study of emission line intensities by Schrijver *et al.* (1985) showed tight correlations between transition region and coronal fluxes, in both the quiet sun and integrated active region fluxes. The coronal emission is related to the transition region emission approximately according to

$$F(\text{Corona}) \propto F(\text{Transition Region})^{1.5} \qquad (1)$$

The same correlation is found for other cool stars which have X-ray emission. (See Section 5). These flux-flux correlations appear to be the result of the energy balance within the atmosphere (Jordan *et al.*, 1987; Jordan, 1990).

Shcherbakov and Shcherbakova discussed their observations of the He I 10830 Å absorption line, which is formed in the high chromosphere. These, and previous observations (see their paper for references), show that there is a good correlation between the depth of the He I line and the Ca II K index, and also with the Fe XIV coronal index. They give details on the variation of the He I line with the solar cycle. One interesting result (Sitnik *et al.*, 1986) is that during the solar maximum the helium chromospheric network is enhanced, as well as there being additional contributions from active regions. Others, (e.g. Skumanich *et al.*, 1984) have come to a similar conclusion from the variations of the Ca II K index.

The variation of chromospheric and coronal lines with the solar cycle, and as a function of the type of spatial structure observed, contains potentially a great deal of useful information on the magnetic fields that extend beyond the photosphere. Reviews of these data can be found in the Proceedings of COSPAR Symposium 12 (Chapman, 1988). The interpretation is at present limited by our lack of detailed knowledge concerning the processes heating the chromosphere and corona and their precise dependence on the magnetic field.

The emission line fluxes can also be correlated with the underlying photospheric magnetic field, and can be used to find the energy required to be balanced by the as yet unknown energy dissipation processes. Since the chromospheric and coronal magnetic fields must be related to the dynamo generated fields, correlations between the energy requirements and the solar cycle should eventually help to understand the solar dynamo.

5. Stellar Observations

Observations of the chromospheric and coronal properties of main sequence stars which have outer convection zones can provide important additional information on dynamo action. In particular, the dependence of emission line fluxes on Ω, the stellar rotation rate, can be found for stars of similar spectral type and structure, as well as the dependence on spectral type, and hence convection zone structure, for stars of a similar Ω. Rutten (1987) has carried out studies of this type using the Ca II flux and the Ca II "excess" flux. The Ca II flux shows a systematic dependence on $B - V$ for a given range of Ω, but also shows the presence of a "minimum" flux at each $B-V$. When the Ca II excess flux is examined, it is found that, for a chosen range of $B - V$, the excess flux depends mainly on Ω. Noyes *et al.* (1984) have also found that their Ca II index depends on the inverse Rossby number, although for main sequence stars of type G0 to K9 the dependence on just Ω is very similar. Using a simple power law fit the dependence is roughly

$$\Delta F_{\mathrm{CaII}} \propto \Omega^{1.2} \qquad (2)$$

The X-ray fluxes show a stronger dependence on rotation, more like

$$F_X \propto \Omega^2 \qquad (3)$$

These different dependences on Ω suggest the presence of two separate energy dissipation processes for the chromosphere and corona. The dependence of the chromospheric and coronal magnetic fields on Ω can be found only if specific heating processes are postulated. (See, for example, Montesinos and Jordan, 1988; Jordan, 1990). Further chromospheric modelling and improved data for X-ray fluxes are required to pursue this approach in a more quantitative manner.

It is also vital to continue and improve the present work on measuring stellar magnetic fields and their dependence on the rotation rate. Saar reviewed recent progress. The sample available at present contains a mixture of observing methods and different degrees of sophistication in the modelling of the lines used. There are also uncertainties in separating the observed mean fields, $\langle Bf \rangle$, into the values of B and the filling factor f. At present the dependence of $\langle Bf \rangle$ on Ω found by different authors (Schrijver *et al.*, 1989; Saar, 1990) is between

$$\langle Bf \rangle \propto \Omega^2 \quad \text{and} \quad \Omega^{1.3} \tag{4}$$

A comparison between the observed dependence of $\langle Bf \rangle$ on Ω and that predicted by a simple dynamo theory, was presented by Montesinos *et al.* In particular, the results of adopting two different variations of the angular velocity with depth, and two different ways of estimating the filling factor, f, were explored. The different approaches affect both the absolute values of $\langle Bf \rangle$ at a given Ω, and the dependence of $\langle Bf \rangle$ on Ω.

There is also considerable interest in Ap, CP and F stars, to investigate where convection zone dynamos take over from fossil fields (e.g. Landstreet; Khokhlova). Observations of magnetic fields on pre-main-sequence stars are also important in this respect (e.g. Basri and Marcy).

Studies of the rotation and activity of giant stars have begun, but the sample of single stars for which the essential data are known is rather small. The situation should improve when further observations become available fron ROSAT and the Hubble Space Telescope. The theory of dynamos in giant stars is also less advanced. However, Dravins *et al.* have studied β Hydri (G2 IV) as an example of the post-main-sequence evolution of a solar type star and Gray presented evidence for a dynamo generated magnetic brake operating to cause a rapid loss of angular momentum between stars of type G0 III and G3 III. If stars cooler than G3 III have a unique dependence of rotation on spectral type then the inclination can be found and it can be deduced that the activity occurs towards equatorial latitudes. Changes in stellar structure, giving lower surface gravities, may then be the main cause of the absence of hot coronae for stars cooler than about K1 III.

The most direct information on stellar magnetic fields and star spots comes from observations of RS CVn and other binary systems. A number of papers were presented on methods of surface imaging (e.g. Piskunov; Hackman *et al.*) on results for particular systems. For example, Vogt and Hatzes find that in the system UX Arietis the spotted primary has an equatorial rotation rate which is synchronized to the orbital angular velocity, while the differential rotation is in the opposite sense to, and an order of magnitude smaller than, that of the Sun. In spite of the lower differential rotation the primary is very active. Zeilik finds that in several systems the active regions are clustered at two longitudes and tend to occur at high latitudes. This distribution may reflect the magnetic field interactions occurring in close binaries with synchronous rotation. The complexity of the interpretation is apparent from the work of Donati and Semel, who find that in HR 1099 and σ^2 CrB the magnetic field is correlated with bright photospheric regions rather than with dark spots. Schrijver and Zwaan (and Demircan and Derman) discussed the role of binarity in causing "overactivity" and suggest that this somehow enhances the dynamo action in the primary star. Hall has studied a large sample of other systems and finds that in many the differential rotation decreases rapidly with the Rossby number, and is not inversely proportional to it, as predicted by linear theory. This could be consistent with nonlinear theories where rapid rotation suppresses the differential rotation. These binary systems may eventually allow dynamo theory to be tested in situations very different from those in single main-sequence stars.

Finally, the interpretation of stellar UV and X-ray observations depends critically on the availability of accurate measurements of basic stellar parameters and distances; results from the Hipparcos satellite should eventually improve these data.

References

Chapman, G.A.: 1988, Editor, *Adv. in Space Res.* Vol. 8, No. 7.

Howard, R.H., LaBonte, B.J.: 1980, *Astrophys. J. Letts.* **239**, L33.

Howard, R.H., Gilman, P.A., Gilman, P.I.: 1984, *Astrophys. J.* **283**, 373.

Jordan, C.: 1990, In Proc. of Heidelberg Conf. on *Mechanisms of Coronal Heating*, June, 1990. In Press.

Jordan, C., Ayres. T.R., Brown, A., Linsky, J.L., Simon, T.: 1987, *Mon. Not. R. astr. Soc.* **225**, 903.

Mein, P., Ribes, E.: 1990, *Astron. Astrophys.* **227**, 577.

Montesinos, B., Jordan, C.: 1988, In *A Decade of UV Astronomy with IUE*, ESA SP-281, Vol. 1, 283.

Noyes, R.W., Hartmann, L.W., Baliunas, S.L., Duncan, D.K., Vaughan, A.H.: 1984, *Astrophys. J.* **279**, 763.

Parker, E.N.: 1975, *Astrophys. J.* **198**, 205.

Parker, E.N.: 1987, *Sol. Phys.* **110**, 11.

Reeves, E.M.: 1976, *Sol. Phys.* **46**, 53.

Rutten, R.G.M.: 1987, *Astron. Astrophys.* **177**, 131.

Saar, S.H.: 1990, In *Solar Photosphere: Structure, Convection and Magnetic Fields*, Proc. IAU Symp. 138, ed. J.O. Stenflo, Kluwer, Dordrecht, p. 427.

Schrijver, C.J.: 1987, *Astron. Astrophys.* **172**, 111.

Schrijver, C.J., Cote, J., Zwaan, C., Saar, S.H.: 1989, *Astrophys. J.* **337**, 964.

Schrijver, C.J., Zwaan, C., Maxson, C.W., Noyes, R.W.: 1985, *Astr. Astrophys.* **149**, 123.

Sitnik, G.F., Kozlova, L.M., Divlikeev, M.I., Porfiryeva, G.A.: 1986, *Soviet Solar Data*, No. **12**, 67.

Skumanich, A., Lean, J.L., White, O.R., Livingston, W.C.: 1984, *Astrophys. J.* **282**, 776.

Snodgrass, H.B.: 1986, *Astrophys. J. Letts.* **316**, L91.

Tuominen, J.: 1945, *The Observatory* **66**, 160.

Concluding Summary: Theoretical Aspects

L. Mestel

Astronomy Centre, University of Sussex,
Falmer, Brighton BN1 9QH, England

To begin this formidable task I find it convenient to look back at the earliest ideas on the dynamical and thermal properties of the solar convection zone and of the mutual interaction with rotation and magnetism. The simplest picture of turbulent convection has the mixing length just an enormously increased mean-free-path, limited by the kinematic consequences of compressibility to a value of the order of the pressure scale-height. The turbulent heat transport down a superadiabatic temperature gradient is so efficient that the transport equation has to be read from right to left: Biermann and Cowling showed that in a dense convective zone the relative degree of superadiabaticity required to carry a solar luminosity is only $\cong 10^{-6}$, yielding the adiabatic p-ρ relation as an excellent approximation for stellar model construction. The same mean-free-path was provisionally assumed to yield a large effective *isotropic* viscosity, so that the Reynolds stresses, acting alone on an arbitrary rotation field, would tend rapidly to convert it into one of uniform rotation.

A significant modification, now called the "Λ-effect", was introduced in a seminal paper by Biermann. The preferred direction of gravity builds an essential *an*isotropy into the turbulence, so that the Reynolds stresses no longer vanish for rigid rotation. In the original linear version, the consequent redistribution of angular momentum produces instead a zero-order radius-dependent $\Omega_0(r)$. This however is inconsistent with the hydrostatic equilibrium of a nearly adiabatic domain, which requires that Ω be more nearly a function of the axial distance ϖ ("Taylor columns"). The consequent imbalance of meridional forces drives a laminar meridional circulation, restrained by the frictional drag due to the same anisotropic turbulence. This circulation advects angular momentum, yielding as the next approximation to the solution of the azimuthal equation of motion a latitude-dependent form for Ω, which by suitable choice of parameters can be made to agree with the observed equatorial acceleration of the Sun (Kippenhahn, Köhler).

We heard from Dr Goode an up-to-date account of the overall solar differential rotation as inferred from helioseismology. In particular, the convection zone is now believed to rotate very similarly to the surface (with near solid body rotation beneath), without any obvious presence of Taylor columns. This is broadly consistent with the explicit ordering in the Biermann-Kippenhahn approach, which takes as the dominant constraint on the rotation field the azimuthal component of the equation of motion (with the anisotropic turbulence incorporated). Drs Rüdiger and Tuominen have made significant modifications, in particular by including non-linear terms which force Ω to be a function of both r and θ in advance of the effect of meridian circulation. To get good agreement with observation, however, the dominant motions affecting the angular momentum distribution must be the giant cells rather than the small-scale eddies.

Unfortunately one can study helioseismology only for the Sun. For the observed rapidly rotating late-type stars, I wonder whether the greatly increased centrifugal acceleration might not enforce a changed ordering, with the Taylor-Proudman effect dominating. A significantly differing internal rotation law would presumably modify the preferred stellar dynamo mode.

Drs Rieutord and Zahn question the validity of parametrizing the mean stress generated by the small-scale motions as an effective anisotropic viscosity (a complementary argument to the Rüdiger-Tuominen emphasis on the role of giant cells rather than small-scale eddies). However, they also state that their porous medium model does seem to give results similar to those resulting from anisotropic viscosity.

Several papers reported on new "first-principles", linear stability analysis. Kwing Chan and H.G. Mayr have introduced density stratification into the model of differential rotation as the zonal wind present in a heliostrophic eigenmode. Paul Roberts and colleagues have responded to the challenge set by Elisabeth Ribes's observations which suggest that near the solar surface the postulated solar giant cells form a "toroidal" or "doughnut" pattern rather than the predicted "cartridge-belt" or "banana-cell" pattern. They include in their zero-order state a (dynamo-generated), primarily toroidal magnetic field $\boldsymbol{B}$. The crucial parameter is the Elsasser number $\sigma B^2/2\Omega\rho$ with σ the (turbulent) conductivity. In the low-density surface regions the Lorentz force dominates over the Coriolis force, and convective cells do indeed align themselves about the toroidal $\boldsymbol{B}$. Such linear analysis gives valuable insight and is complementary to the phenomenological attempts to model developed turbulence. I recall earlier work by Busse which can be interpreted as showing how unstable modes in a rotating fluid with (isotropic) micro-viscosity leads to an angular momentum transport which can be mimicked by an anisotropic macro-viscosity.

I was particularly intrigued by Vittorio Canuto's critique of standard mixing length theory, because of its use of a one-eddy model (appropriate for systems with high rather than low micro-viscosity) and for its postulating rather than calculating the mixing length. His prescription for a more sophisticated model includes a whole spectrum of turbulent eddies, with compressibility effects represented by taking the mixing length to be equal to the local depth. He is able to predict very closely the solar effective temperature, without any free parameters, and finds a significant reduction in the age of a globular cluster as is found from the mass at the giant branch turn-off. This work may therefore have reverberations in the ongoing debate about the value of Hubble's constant.

Dynamo theory grew out of the need to evade the consequences of Cowling's theorem, which illustrates (in Elsasser's words) the topological asymmetry between poloidal fields $\boldsymbol{B}_\mathrm{p}$ and toroidal fields $\boldsymbol{B}_\mathrm{t}$. In axial symmetry, it is trivially easy to generate $\boldsymbol{B}_\mathrm{t}$ from $\boldsymbol{B}_\mathrm{p}$ via non-uniform rotation, whereas strictly axisymmetric motions merely advect both $\boldsymbol{B}_\mathrm{p}$ and $\boldsymbol{B}_\mathrm{t}$ but do not enable one to convert $\boldsymbol{B}_\mathrm{t}$ into $\boldsymbol{B}_\mathrm{p}$. Without deviation at some level from axisymmetry, the systematic shrinkage of poloidal loops into O-type neutral points, pointed out by Cowling, cannot be offset by the generation of new poloidal flux. In a seminal paper, Gene Parker showed that non-axisymmetric motions in turbulent cells, subject to Coriolis force, twist $\boldsymbol{B}_\mathrm{t}$ so as to yield new $\boldsymbol{B}_\mathrm{p}$ lines, so that in a stratified medium there is a net generation of poloidal flux. The process can be parametrized as yielding an emf along $\boldsymbol{B}$. The theory was later developed independently by Steenbeck, Krause and Rädler with the crucial new term in Ohm's law written as $\alpha\boldsymbol{B}$, whence the "α-effect" nomenclature. A whole industry has grown up studying in particular the solar dynamo, the solar cycle and its "butterfly diagram", dynamos in other solar-type stars, in early-type stars with convective cores, in planets and in galaxies and galactic clusters.

We heard many variations on this theme, showing that the richness of the mhd equations with the α-effect included is nowhere near being fully explored. Of the poster papers I liked particularly those on the "$\alpha\Lambda$-dynamos" of the Helsinki-Manchester-Toulouse-Potsdam collaboration, which use in the kinematic dynamo equation a differential rotation and meridian circulation derived from the Λ-effect. (However, I would like the authors to iron out any differences among themselves – should the Sun have Taylor columns or not?). As emphasized in several papers, estimates of the field amplitudes and of filling factors must be through effective deviation from linear kinematic theory, eg by non-linear quenching of the α-effect (David Moss, Karl-Heinz Rädler), or by flux loss by buoyancy (Durney and Robinson, Montesinos *et al.*).

So much is coming out of the "standard" dynamo equations that it is salutary to be reminded by one of the founding fathers – Fritz Krause – of the probably over-simple parametrization employed, eg in the adoption of $\alpha \propto$ local helicity. The advent of new computers may enable one to avoid such parametrization by direct solution of the basic mhd equations in low-symmetry geometry. Once solutions with growing large-scale fields have been found, then one can perform the same averaging over the solutions as is at present done to the *equations* before their solution is attempted, if only to test the reliability of the standard procedure. I am however glad that Dr Chan, with his wide experience of the opportunities afforded by computers, went out of his way to emphasize the pitfalls of numerical work. In another area – interstellar magnetogasdynamics and star formation – the effect of numerical viscosity (whether explicit or implicit) is often apparent, when eg numerical solutions of the infinitely conducting mhd equations show spurious deviations from flux freezing. One might imagine that in problems with strong phenomenological turbulent viscosity or resistivity, such consequences of numerical approximation would be ignorable. But Dr Chan noted other dangers: for example, numerical simulation of solar convection can exaggerate the approximation to adiabaticity and so lead to a spurious prediction of Taylor columns. The lesson is, as always, that the computer is no substitute for the brain, and that it must be kept as a slave, not allowed to become a master.

Paul Roberts's discussion of both the similarities and the differences between the geo- and solar dynamos was particularly enlightening. Another high spot of the meeting for me was the review by Axel Brandenburg and Ilkka Tuominen of the current state of the solar dynamo. Again recalling Fritz Krause's caveats, it is salutary to have some correction to the euphoria that periodically overtakes us. Their paper stressed less the considerable achievements than the persisting problems – "there is as yet no good solar dynamo". Of special interest is the buoyancy question which has been for so long at the centre of discussion, focusing attention on the interface between the radiative core and the convective envelope. I note that Axel and Ilkka argue that loss by buoyancy is not in fact a real problem – there is rather a tendency for downward suction, so that the case for dynamo action in a boundary layer instead of in the bulk of the convective core is weakened. This is contrary both to the earlier general opinion and to the claim by Dr Petrovay in his paper that topological pumping will strengthen rather than weaken the buoyancy. Resolution of this disagreement on a fundamental issue is urgent. Dr Petrovay infers from his analysis a much deeper effective convection zone for the Sun. Some increase over the standard depth could be convenient for other questions, such as Li-burning, but as emphasized by Dr Goode, helioseismology sets severe limitations.

Extrapolation of solar-type dynamos to stars with higher rotation rates is encouraged by the apparent correlation between chromospheric emission and the dynamo number $\propto (\Omega\tau_c)^2 \propto$ (Rossby number)^2, where τ_c is the convective turnover time at the base of the convective zone (Noyes, Vaughan *et al.*). Elsewhere Noyes, Weiss and Vaughan argue for a "solar cycle" period $\propto 1/\Omega$ in those solar-type stars which do show quasi-periodic activity. However Marcello Rodono at Tromsø in 1987, and some participants here have queried whether the observations do point clearly to Rossby number dependence rather than just correlation with Ω. I would like to see a concensus from the observers on the extent to which these relations are good guides for the theorist. Note also the warning from Dr Zeilik and others against using the solar dynamo as a universal paradigm; in particular, RS CVn activity is not well described as of solar type.

Dynamo theorists have tended to wait anxiously for information about the Ω-gradient within the solar convection zone, for the sign of $\alpha\partial\Omega/\partial r$ determines the direction of dynamo wave propagation. I note the paper by Gaetano Belvedere and colleagues which exploits the helioseismological inference of a latitude dependence in $\partial\Omega/\partial r$. Nevertheless, I continue to wonder whether interpretation of the sunspot zone migration as a pure dynamo wave phenomenon is absolutely compelling. A laminar circulation of quite modest speed would easily offset a "wrong" sense of wave propagation – in Gene Parker's words at Heidelberg (1989), one could walk at the rate required without any difficulty. This underlines the importance of having a reliable theory not only of the rotation law but of the dynamically linked meridian circulation field.

I won't disguise my (hopefully not premature) relief on hearing from Dr Goode that the 1984 hint of a rapidly rotating solar core appears now as an artefact of the method of data inversion, for such a feature if long lived sets severe if not implausible limits on the strength of any relic B-field. To date it appears that we have no reliable information about the rotation of regions with $r < 0.3R$.

The organizers rightly decided not to allocate time to the detailed dynamics of magnetic braking, as there was a whole meeting later in the summer at Noto, Sicily devoted to the Angular Momentum of Young Stars. Nevertheless there was inevitably some discussion of observations and the quotation of simple theoretical formulae. Evidence for dynamo-generated magnetic control of stellar rotation was well brought out by David Gray in his paper on G-giants. The basic theoretical result on braking by a magnetically coupled wind is that of *effective* corotation: the rate of transport of angular momentum carried *jointly* by the magnetic and material stresses is *equivalent* to what would be carried by the gas alone if it were kept in strict corotation with the star out to the surface S_A where the wind speed reaches the local Alfvén speed. The simplest form of the theory assumes just thermal driving of the wind, strong core-envelope coupling, no significant dependence of the wind temperature on rotation, a linear dynamo law relating the surface field B_s with Ω, and the whole corona's participation in the wind. The consequent braking rate $\dot{\Omega} \propto -\Omega^3$ yields the relation $\Omega \propto t^{-\frac{1}{2}}$, *prima facie* consistent with observations of line widths in the solar-type stars in the Hyades, Pleiades and α Persei clusters. For very young stars, however, this relation appears inappropriate: David Gray and more recently Stauffer and Hartmann have produced evidence for a rapid early spin-down on the main sequence, after which something like the Skumanich $t^{-\frac{1}{2}}$ law takes over. The various corrections to the simple model suggested by theory and observation – coronal dead zones, centrifugal wind driving, a possible increase of wind temperature with B_s and hence with Ω – all act to reduce the braking rate somewhat.

During these discussions reference was made to the Stauffer-Hartmann suggestion that the apparent rapid initial spin-down is just of the convective envelope, which has only $\cong 1/20$ of the moment of inertia of the whole Sun. But as noted by Ian Roxburgh, a star of the age of the Sun still has had to get rid of most of its angular momentum (recall that there is at the moment no clear evidence of even a moderate differential rotation between the solar core and its convective envelope). The picture would require the transfer time of angular momentum from the core to the envelope to be longer than the time for the initial rapid braking of the envelope, which therefore spins down until the loss of angular momentum from the surface falls to a value comparable with the supply from the core. Calculations made after the meeting by my student Li Jianke suggest that such a model may indeed be consistent with observation, provided effective core-envelope coupling is re-established soon after the early rapid braking phase, and careful account is taken of the actual distributions of rotation rates among stars of a given type in clusters of different ages.

The rapidly accumulating wealth of observational data will require for detailed interpretation the construction of the coronal magnetic, density, temperature and velocity fields in terms of the dynamo-built magnetic flux distribution over the photosphere. It is by no means obvious that a simple dipolar field is appropriate for all rotation rates up to the maximum (*cf.* the paper by Dr Katsova), or that B_s continues to scale with Ω. One can however expect all fields to have a multiple structure, with closed field-line "dead zones", of high temperature and so bright in X-rays, immersed in the cooler open-field wind regions observed as coronal holes. Recent observations of particular interest are of the rapidly rotating dwarf star AB Doradus by Cameron and Robinson, referred to in the paper by Dr Stewart and colleagues, and by Dr Vilhu during the discussions. X-ray observations reveal an extensive, very hot zone ($T > 10^7$ K) containing cool condensations picked up in Hα absorption, corotating with the star, and showing also evidence of slow *outward* motion. The results can be understood semi-quantitatively in terms of compression of gas within a "dead zone" but beyond the centrifugal-gravitational balance point, and of consequent radiative cooling. The "dead zones" are thus not fully dead, but contribute to magnetic braking through the bursting out of these corotating condensations. However, I think it will be difficult to predict a *total* angular momentum loss greater than that given by a Weber-Davis type field, without any dead zones and, so with the whole corona taking part in the wind. The efficiency of braking depends most critically on the strength of the surface field, which fixes the field strength at the Alfvénic surface. Detailed comparison with observations of rapidly rotating young stars could supply a hint on the $B_s(\Omega)$ dynamo relation at high rotation.

Most of the discussion has been of magnetic fields generated and maintained by contemporary dynamo action in convective zones. These zones are by definition spontaneously unstable to the continuous conversion of thermal energy into turbulent energy, and interaction of the turbulent cells with the rotation yields the degree of anisotropy capable of generating a large-scale dipolar field. There has been some limited discussion also of "fossil" magnetic fields, slowly decaying relics either of the galactic field threading the gas from which the stars formed, or of a field built up by dynamo action in an earlier epoch of the star's life. A weak fossil field threading a turbulent zone will at least initially be passive: the back-reaction of the field on the turbulence is unimportant, and the compression of the field into filaments and the possible expulsion of the field from the zone can be described in terms of the same "turbulent resistivity" as appears in the dynamo equations. A sub-adiabatic zone (without a large rotational shear) is stable against spontaneous convection: perturbing motions with a radial component are prevented from growing by the adiabatic stabilization term in the energy principle. But the presence in a sub-adiabatic zone of a magnetic field can lead to new instabilities: for motions that are non-axisymmetric but essentially horizontal, the adiabatic term vanishes, and fields of simple topology – eg purely poloidal or toroidal fields, with O-type neutral points – will spontaneously convert magnetic into kinetic energy. There is no nonlinear theory extant but it is at least plausible that a necessary condition for dynamical stability is that the field be of complex topology, with poloidal flux linking toroidal loop and vice versa. Fields that have been built up by standard dynamo action will automatically have acquired this topology.

We were reminded by John Landstreet of the other broad class of magnetic stars – the early type magnetic Ap stars with sub-photospheric *radiative* envelopes. They have large-scale, well-ordered fields, apparently quasi-steady in the frames rotating with the stars, and with much higher mean strengths than the fields in the late-type main sequence stars. They can clearly take up a Conference on their own; even the debate about the origin and nature of the fields – as fossils or as the result of contemporary dynamo action in the stellar convective cores – is unresolved. Dr Landstreet drew attention to problems that may be awaiting us in the boundary area between the Ap and the solar-type stars. The rapid decrease at type F in the extent of the outer convection zone with increasing T_e is consistent with the disappearance of the solar-type phenomenon, but it is not obvious how to explain the apparent absence of the Ap phenomenon among mid-F stars. We should remain aware of this contiguous area of magnetic star research, noting also that fossil fields may be significant in the late-type main sequence and pre-main sequence stars, eg (as already noted) by coupling the rotations of core and envelope. Dr Dudorov argues that fossil flux in fact plays an important role in the observable layers of the Sun and other solar-type stars.

I found particularly exciting the report by Drs Basri and Marcy of a strong field in a T Tauri star. It is a difficult but important task to try and decide whether all this flux is dynamo-built, or whether (as suggested by Roger Tayler) there is evidence from the activity/Rossby number curve for pre-main sequence stars of a superposition of dynamo-built and fossil fields. An Alfvén-wave driven wind is the essential feature of the T Tauri model of Teresa Lago, Michael Penston and others. And there remains the possibility (first suggested I think by Fred Hoyle) that the energy source in the T Tauri phenomenon is the dissipation of galactic magnetic flux trapped in the star during its formation.

There are many other papers in the published proceedings which will repay careful study. Those by Ed Spiegel on waves in the solar cycle, and Sasha Ruzmaikin and colleagues on the fractal distribution of solar fields impressed me particularly as being forward-looking. Our collective thanks to Ilkka Tuominen and colleagues for organizing an enjoyable and scientifically fruitful meeting.

List of participants

J. E. Armentia
Dept. Astrofisica
Universidad Complutense de Madrid
E-28040 Madrid
Spain

J. Arsenijević
Astronomical Observatory
Volgina 7
11050 Beograd
Yugoslavia

J. L. Ballester Mortes
Department de Fisica
Universitat de les Illes Balears
E-07071 Palma de Mallorca
Spain

G. Basri
Dept. Of Astronomy
601 Campbell Hall
Berkeley, CA 94720
USA

J. E. Beckman
Inst. de Astrofisica de Canarias
Camino de la Hornera
38200 La Laguna Tenerife
Spain

G. Belvedere
Istituto di Astronomia
Citta Universitaria
95125 Catania
Italia

E. Benevolenskaya
Pulkovo Astronomical Observatory
196140, Leningrad
USSR

A. Brandenburg
Observatory and Astrophys. Lab.
University of Helsinki
Tähtitorinmäki
SF-00130 Helsinki
Finland

J.P. Caillault
Dept. of Physics and Astronomy
Univ. of Georgia
Athens, GA 30602
USA

D. Callebaut
Phys. Dept. UIA
B-2610 Antwerpen
Belgium

V. M. Canuto
NASA Goddard Space Flight Center
Institute of Space Studies
2880 Broadway
New York, N.Y. 10025
USA

S. Catalano
Istituto di Astronomia
Citta Universitaria
Viale A. Doria
95125 Catania
Italia

K. L. Chan
Code 910.2
Goddard Space Flight Center
Greenbelt, MD 20771
USA

F. D'Antona
Osservatorio Astronomico di Roma
Monte Porzio
00136 Roma
Italia

O. Demircan
A.U.Science Faculty, Astronomy Dept.
06100 Tandogan, Ankara
Turkey

V. N. Dermendjiev
Dept. Of Astronomy and NAO
72 Lenin Blvd
1784 Sofia
Bulgaria

E. S. Dmitrienko
Crimean Astrophysical Observatory
Nauchny, Crimea, 334413
USSR

A. Z. Dolginov
A.I.Ioffe Phys.-Tech. Institute
Academy of Sciences of USSR
194021 Leningrad
USSR

J. F. Donati
DESPA, Observatoire de Meudon
5 Place Janssen
F-92195 Meudon-Cedex
France

K. J. Donner
Observatory and Astrophys. Lab.
University of Helsinki
Tähtitorinmäki
SF-00130 Helsinki
Finland

D. Dravins
Lund Observatory
Box 43
S-22100 Lund
Sweden

A. E. Dudorov
Sternberg Astronomical Institute
13, Universitetskij pr.
119899 Moscow V-234
USSR

E. Ergma
Ulikooli 18
Tartu University
202400 Tartu
Estonia, USSR

J. Fabregat
Physics Dept.,The University
Southampton S09 5NH
U.K.

M. J. Fernández-Figueroa
Dpto. Astrofisica
Facultad de fisicas
Univ. Complutense de Madrid
28040 Madrid
Spain

B. H. Foing
European Space Agency/ESTEC
P.O. Box 299
2200 AG Noordwijk
The Netherlands

P. A. Fox
Yale University
Center for Solar & Space Research
P.O. Box 6666
New Haven, CT 06511
USA

U. Frisch
Observatoire de Nice
BP 139
06003 Nice cedex
France

A. Galal
National Research Institute of
Astronomy and Geophysics
Helwan
Cairo
Egypt

K. Galsgaard
Astronomisk Observatorium
Øster Voldgade 3
1350 Copenhagen K
Denmark

J.H.G.M. van Geffen
Lab. for Space Research
Sorbonnelaan 2
3584 CA Utrecht
The Netherlands

P. Goode
Department of Physics
New Jersey Institute of Technology
Newark, New Jersey 07102
USA

D. Gray
Astronomy Dept.
University of Western Ontario
London, Ontario N6A 3K7
Canada

H. Grosser
Univ.-Sternwarte
Geismarlandstr. 11
3400 Göttingen
F.R. Germany

T. Hackman
Observatory and Astrophys. Lab.
University of Helsinki
Tähtitorinmäki
SF-00130 Helsinki
Finland

D.S. Hall
Dyer Observatory
Vanderbilt University
Nashville, TN 37235
USA

Han Jiling
Dept. of Physics
Beijing Normal University
Beijing 100875
P.R. China

A. Hatzes
Astronomy Dept.
University of Texas
Austin, TX 78712
USA

L. Hejna
Dept. of Computer Technique
Faculty of Math. and Phys.
Charles University
Ke Karlovu 3
12116 Prague 2
Czechoslovakia

A. S. Hojaev
Astronomical Institute
Astronomicheskaya 33
700052 Tashkent
USSR

P. Hoyng
Lab. for Space Research
Sorbonnelaan 2
3584 CA Utrecht
The Netherlands

J. Huovelin
Observatory and Astrophys. Lab.
University of Helsinki
Tähtitorinmäki
SF-00130 Helsinki
Finland

T. Jarzębowski
ul. Jelenia 28 m 31
54-242 Wroclaw
Poland

R. D. Jeffries
Dept. Space Research
Univ. of Birmingham
P.O. Box 363
Birmingham
U.K.

R. Jennings
D.A.M.T.P.
University of Cambridge
Silver Street
Cambridge CB3 9EW
United Kingdom

L. Jetsu
Observatory and Astrophys. Lab.
University of Helsinki
Tähtitorinmäki
SF-00130 Helsinki
Finland

C. Jordan
Dept. Theoretical Physics
Univ. of Oxford, 1 Keble Road
Oxford OX1 3NP
U.K.

P. Jorås
Inst. of Theoret. Astrophys.
University of Oslo
P.O. Box 1029, Blindern
N-0315 Oslo 3
Norway

M. Katsova
Sternberg Astronomical Institute
Moscow State University
119899 Moscow V-234
USSR

V. L. Khokhlova
Astron. Council of the Acad Sci. USSR
Pyatnitskaya 48
109017 Moscow
USSR

N. I. Kleeorin
Lenin's State Pedagogical Institute
Malaya Pirogovskaya 29
119435 Moscow
USSR

F. Krause
Sternwarte Babelsberg and
Astrophysikalisches Observatorium
Rosa-Luxemburg-Str. 17a
O-1591 Potsdam
F.R.G.

V.N. Krivodubskij
Kiev Univ. Astron. Observatory
Observatornaya Street 3
Kiev-53, 252053 Ukraine
USSR

V. Kršljanin
Astronomical Observatory
Volgina 7
11050 Beograd
Yugoslavia

M. Kürster
MPI. f. Extraterr. Physik
Karl-Schwarzschild-Str. 1
W-8046 Garching bei München
F.R.G.

J. D. Landstreet
Dept. of Astronomy
University of Western Ontario
London, Ontario N6A 3K7
Canada

A.P. Linnell
Dept. of Physics and Astronomy
Michigan State University
E. Lansing, MI 48824
USA

J.L. Linsky
JILA, University of Colorado
Campus Box 440
Boulder, CO 80309-0440
USA

W. Livingston
NSO/NOAO, P.O.Box 26732
Tucson, AZ 85726
USA

M. A. Livshits
Inst. of Terrestrial Magnetism
Ionosphere and Radio Wave Prop.
142092, Troitsk, Moscow region
USSR

V.I. Makarov
Pulkovo Astronomical Observatory
196140, Leningrad
USSR

A. Mangeney
Observatoire de Paris
Meudon - DESPA
5 Place J. Jansen
92195 Meudon Cedex
France

G. W. Marcy
San Francisco State University
1600 Holloway Avenue
San Francisco, CA 94132
USA

E. Marilli
Istituto di Astronomia
Citta Universitaria
Viale A. Doria
95125 Catania
Italia

T. Markkanen
Observatory and Astrophys. Lab.
University of Helsinki
Tähtitorinmäki
SF-00130 Helsinki
Finland

V. Martínez Pillet
Inst. de Astrofísica de Canarias
Camino de la Hornera
La Laguna Tenerife
Spain

J. M. Massaguer
Dept. Fisica Aplicada - UPC
Ap.Correus 30.002
08080 Barcelona
Spain

I. Mazzitelli
Istituto di astrofisica spaziale
C.P. 67, Via E. Fermi 21
00044 Frascati (Roma)
Italy

R. Meinel
Sternwarte Babelsberg and
Astrophysikalisches Observatorium
Rosa-Luxemburg-Str. 17a
O-1591 Potsdam
F.R.G.

L. Mestel
Astronomy Centre
Physics and Astronomy Division
University of Sussex
Falmer, Brighton BN1 9QH
England

B. Montesinos
Dept. Theoretical Physics
Univ. of Oxford, 1 Keble Road
Oxford OX1 3NP
U.K.

D. Moss
Mathematics Dept.
The University
Manchester M13 9PL
U. K.

J.E. Neff
Code 681
NASA Goddard Space Flight Center
Greenbelt, MD 207719
USA

K. Panov
Dept. of Astronomy and NAO
Bulgaricm Acad. Sci
72 Lenin Blvd
Sofia-1784
Bulgaria

J. M. Pasachoff
Institute for Advanced Study
School of Natural Sciences
Princeton, NJ 08540
USA

L. Pasquini
ESO, La Silla
19001 Santiago 19
Chile

L. Paternò
Istituto di Astronomia
Citta Universitaria
95125 Catania
Italia

J. Pelt
Tartu Astrophysical Observatory
Tõravere 1-6, Tartu district
202444 Estonia

P. P. Petrov
Crimean Astrophysical Observatory
334413 Nauchny, Crimea
USSR

K. Petrovay
Eötvös University
Dept. of Astronomy
Kun Bela Ter 2
H-1083 Budapest
Hungary

B. R. Pettersen
Inst. of Theor. Astrophysics
University of Oslo
P.O.Box 1029, Blindern
N-0315 Oslo 3
Norway

N. E. Piskunov
Astronomical Council
Pyatnitskaya 48
109017 Moscow
USSR

S. Pohjolainen
Metsähovi Radio Research Station
TKK, Otakaari 5A
02150 Espoo
Finland

M. Poutanen
Finnish Geodetic Institute
Ilmalankatu 1A
00240 Helsinki
Finland

P. Pulkkinen
Dept. of Theoretical Physics
University of Helsinki
Siltavuorenpenger 20c
SF-00170 Helsinki
Finland

L. Pustil'nik
Special Astrophysical Observatory
Nizhnij Arkhyz
Stavropol Territory 357147
USSR

K.-H. Rädler
Sternwarte Babelsberg and
Astrophysikalisches Observatorium
Rosa-Luxemburg-Str. 17a
O-1591 Potsdam
F.R.G.

M. Rego
Dpto. Astrofisica
Facultad de fisicas
Univ. Complutense de Madrid
28040 Madrid
Spain

S. R. Restaino
NSO/SAC Peak, Sunspot
NM 88349
USA

E. Ribes
DASOP, Observatoire de Paris
5 Place Janssen
92195 Meudon
France

J.B. Rice
Dept. of Physics
Brandon University
Brandon, MB. R7A 6A9
Canada

M. Rieutord
Observatoire de Toulouse
14 av. E. Belin
F-31400 Toulouse
France

P. H. Roberts
Inst. of Geophys. and Plan. Phys.
University of California LA
Los Angeles, CA 90024
USA

I. W. Roxburgh
Astronomy Unit
School of Mathematical Sciences
Queen Mary & Westfield College
Mile End Road,
London E1 4NS, England

S. Rucinski
Inst. for Space and Terr. Science
c/o Dept. of Physics
Petrie Bldg. 206
York University, 4700 Keele Street
Toronto, Ontario M3J IP3
Canada

A.A. Ruzmaikin
IZMIRAN
142092 Troitsk
Moscow Region
USSR

T. V. Ruzmaikina
Institute of Phys. of the Earth
123810 Moscow
USSR

G. Rüdiger
Sternwarte Babelsberg and
Astrophysikalisches Observatorium
Rosa-Luxemburg-Str. 17a
O-1591 Potsdam
F.R.G.

T.A. Ryabchikova
Astronomical Council
Pyatnitskaya 48
109017 Moscow
USSR

S. Saar
Center for Astrophysics, MS 58
60 Garden Street
Cambridge, MA 02138
USA

J. Sánchez Almeida
Inst. de Astrofisica de Canarias
38200, La Laguna Tenerife
Spain

I.S. Savanov
Crimean Astrophysical Observatory
p/o Nauchny
334413 Crimea
USSR

M. Shaltout
National Research Inst. of Astronomy
Helwan
Cairo
Egypt

F. Scaltriti
Osservatorio Astron. di Torino
Strada Osservatorio 20
I-10025 Pino Torinese
Italia

C.J. Schrijver
ESTEC, Mail box 299
2200 AG Noordwijk
The Netherlands

M. Semel
DASOP-Observatoire de Meudon
92190 Meudon
France

N. I. Shakhovskaya
Crimean Astrophysical Observatory
Nauchny, Crimea, 334413
USSR

A. G. Shcherbakov
Crimean Astrophysical Observatory
Nauchny, Crimea, 334413
USSR

Z. A. Shcherbakova
Crimean Astrophysical Observatory
Nauchny, Crimea, 334413
USSR

A. M. Shukurov
IZMIRAN
142092 Troitsk
Moscow Region
USSR

D. Sokoloff
Dept. of Physics
Moscow University
119899 Moscow
USSR

Qin Songnian
Yunnan Observatory
P.O.Box 110
Kunming, Yunnan province
P.R. China

E. Spiegel
Dept. of Astronomy
Pupin Hall
Columbia University
New York
NY 10027 USA

J. O. Stenflo
Institute of Astronomy
ETH-Zentrum, CH-8092 Zurich
Switzerland

K. Stępień
Warsaw Univ. Observatory
Al.Ujazdowskie 4
00-478 Warszawa
Poland

R. T. Stewart
ATNF CSIRO
PO Box 76
Epping 2121
Australia

K. G. Strassmeier
Inst. for Astronomy
Univ. of Vienna
Türkenschanzstrasse 17
A-1180 Wien
Austria

J. Suso
Dept. de Mat. Apl. y Astronomia
Universidad de Valencia
46100 Burjassot
Spain

M. Szydlowski
Astronomical Observatory
Jagiellonian University
Orla 171
30-244 Cracow, Poland

Tong Yi
Dept. of Astronomy
Beijing Normal University
Beijing 100875
P.R. China

C. Trigilio
Istituto di Astronomia
Citta Universitaria
Viale A. Doria
95125 Catania
Italia

I. Tuominen
Observatory and Astrophys. Lab.
University of Helsinki
Tähtitorinmäki
SF-00130 Helsinki
Finland

S. Urpo
Metsähovi Radio Research Station
TKK, Otakaari 5A
02150 Espoo
Finland

L. Valdettaro
CERFACS
42, Avenue Coriolis
31057 Toulouse
France

J. A. Valenti
Dept. Of Astronomy
601 Campbell Hall
Berkeley, CA 94720
USA

O. Vilhu
Observatory and Astrophys. Lab.
University of Helsinki
Tähtitorinmäki
SF-00130 Helsinki
Finland

A. Vincent
CERFACS
42, Avenue Coriolis
31057 Toulouse
France

S. S. Vogt
Lick Observatory
Univ. Of California
Santa Cruz, CA 95064
USA

B. Vršnak
Faculty of Geodesy
Hvar Observatory
Kaciceva 26
41000 Zagreb
Yugoslavia

W. H. Wehlau
Dept. of Astronomy
The Univ. of Western Ontario
London, Ontario
Canada N6A 3K7

S. M. Yousef
Astron. Dept.
Cairo University
Cairo
Egypt

M. Zeilik
Dept.Physics and Astronomy
Univ. of New Mexico
Albuquerque, NM 87131
USA

Zhang Hongqi
Beijing Astronomical Observatory
Beijing 100080
P.R. China

Chen Zhen-cheng
Beijing Astronomical Observatory
Chinese Academy Of Sciences
Beijing 100080
P.R. China

Li Zhongyuan
Dept. of Earth and Space Science
Univ. of Sci. and Tech. of China
Hefei, Anhui 230026
P.R. China

Lecture Notes in Mathematics

Vol. 1236: Stochastic Partial Differential Equations and Applications. Proceedings, 1985. Edited by G. Da Prato and L. Tubaro. V, 257 pages. 1987.

Vol. 1237: Rational Approximation and its Applications in Mathematics and Physics. Proceedings, 1985. Edited by J. Gilewicz, M. Pindor and W. Siemaszko. XII, 350 pages. 1987.

Vol. 1250: Stochastic Processes – Mathematics and Physics II. Proceedings 1985. Edited by S. Albeverio, Ph. Blanchard and L. Streit. VI, 359 pages. 1987.

Vol. 1251: Differential Geometric Methods in Mathematical Physics. Proceedings, 1985. Edited by P.L. García and A. Pérez-Rendón. VII, 300 pages. 1987.

Vol. 1255: Differential Geometry and Differential Equations. Proceedings, 1985. Edited by C. Gu, M. Berger and R.L. Bryant. XII, 243 pages. 1987.

Vol. 1256: Pseudo-Differential Operators. Proceedings, 1986. Edited by H.O. Cordes, B. Gramsch and H. Widom. X, 479 pages. 1987.

Vol. 1258: J. Weidmann, Spectral Theory of Ordinary Differential Operators. VI, 303 pages. 1987.

Vol. 1260: N.H. Pavel, Nonlinear Evolution Operators and Semigroups. VI, 285 pages. 1987.

Vol. 1263: V.L. Hansen (Ed.), Differential Geometry. Proceedings, 1985. XI, 288 pages. 1987.

Vol. 1265: W. Van Assche, Asymptotics for Orthogonal Polynomials. VI, 201 pages. 1987.

Vol. 1267: J. Lindenstrauss, V.D. Milman (Eds.), Geometrical Aspects of Functional Analysis. Seminar. VII, 212 pages. 1987.

Vol. 1269: M. Shiota, Nash Manifolds. VI, 223 pages. 1987.

Vol. 1270: C. Carasso, P.-A. Raviart, D. Serre (Eds.), Nonlinear Hyperbolic Problems. Proceedings, 1986. XV, 341 pages. 1987.

Vol. 1272: M.S. Livšic, L.L. Waksman, Commuting Nonselfadjoint Operators in Hilbert Space. III, 115 pages. 1987.

Vol. 1273: G.-M. Greuel, G. Trautmann (Eds.), Singularities, Representation of Algebras, and Vector Bundles. Proceedings, 1985. XIV, 383 pages. 1987.

Vol. 1275: C.A. Berenstein (Ed.), Complex Analysis I. Proceedings, 1985–86. XV, 331 pages. 1987.

Vol. 1276: C.A. Berenstein (Ed.), Complex Analysis II. Proceedings, 1985–86. IX, 320 pages. 1987.

Vol. 1277: C.A. Berenstein (Ed.), Complex Analysis III. Proceedings, 1985–86. X, 350 pages. 1987.

Vol. 1283: S. Mardešić, J. Segal (Eds.), Geometric Topology and Shape Theory. Proceedings, 1986. V, 261 pages. 1987.

Vol. 1285: I.W. Knowles, Y. Saitō (Eds.), Differential Equations and Mathematical Physics. Proceedings, 1986. XVI, 499 pages. 1987.

Vol. 1287: E.B. Saff (Ed.), Approximation Theory, Tampa. Proceedings, 1985–1986. V, 228 pages. 1987.

Vol. 1288: Yu. L. Rodin, Generalized Analytic Functions on Riemann Surfaces. V, 128 pages, 1987.

Vol. 1294: M. Queffélec, Substitution Dynamical Systems – Spectral Analysis. XIII, 240 pages. 1987.

Vol. 1299: S. Watanabe, Yu.V. Prokhorov (Eds.), Probability Theory and Mathematical Statistics. Proceedings, 1986. VIII, 589 pages. 1988.

Vol. 1300: G.B. Seligman, Constructions of Lie Algebras and their Modules. VI, 190 pages. 1988.

Vol. 1302: M. Cwikel, J. Peetre, Y. Sagher, H. Wallin (Eds.), Function Spaces and Applications. Proceedings, 1986. VI, 445 pages. 1988.

Vol. 1303: L. Accardi, W. von Waldenfels (Eds.), Quantum Probability and Applications III. Proceedings, 1987. VI, 373 pages. 1988.

Lecture Notes in Physics

Vol. 359: K. Kirchgässner (Ed.), Problems Involving Change of Type. Proceedings, 1988. XII, 207 pages. 1990.

Vol. 360: Y. Mellier, B. Fort, G. Soucail (Eds.), Gravitational Lensing. Proceedings 1989. XV, 315 pages. 1990.

Vol. 361: P. Gaigg, W. Kummer, M. Schweda (Eds.), Physical and Nonstandard Ganges. Proceedings, 1989. IX, 310 pages. 1990.

Vol. 362: N.E. Kassim, K.W. Weiler (Eds.), Low Frequency Astrophysics from Space. Proceedings. XII, 280 pages. 1990.

Vol. 363: V. Ruždjak, E. Tandberg-Hanssen (Eds.), Dynamics of Quiescent Prominences. Proceedings, 1989. XI, 304 pages. 1990.

Vol. 364: T.T.S. Kuo, E. Osnes, Folded-Diagram Theory of the Effective Interaction in Nuclei, Atoms and Molecules. V, 175 pages. 1990.

Vol. 365: M. Schumacher, G. Tamas (Eds.), Perspectives of Photon Interactions with Hadrons and Nuclei. Proceedings. IX, 251 pages. 1990.

Vol. 366: M.-J. Goupil, J.-P. Zahn (Eds.), Rotation and Mixing in Stellar Interiors. Proceedings, 1989. XIII, 183 pages. 1990.

Vol. 367: Y. Osaki, H. Shibahashi (Eds.), Progress of Seismology of the Sun and Stars. Proceedings, 1989. XIII, 467 pages. 1990.

Vol. 368: L. Garrido (Ed.), Statistical Mechanics of Neural Networks. Proceedings. VI, 477 pages. 1990.

Vol. 369: A. Cassatella, R. Viotti (Eds.), Physics of Classical Novae. Proceedings, 1989. XII, 462 pages. 1990.

Vol. 370: H.-D. Doebner, J.-D. Hennig (Eds.), Quantum Groups. Proceedings, 1989. X, 434 pages. 1990.

Vol. 371: K.W. Morton (Ed.), Twelfth International Conference on Numerical Methods in Fluid Dynamics. Proceedings, 1990. XIV, 562 pages. 1990.

Vol. 372: F. Dobran, Theory of Structured Multiphase Mixtures. IX, 223 pages. 1991.

Vol. 373: C.B. de Loore (Ed.), Late Stages of Stellar Evolution. Computational Methods in Astrophysical Hydrodynamics. Proceedings, 1989. VIII, 390 pages. 1991.

Vol. 374: L. Ting, R. Klein, Viscous Nortical Flows. V, 222 pages. 1991.

Vol. 375: C. Bartocci, U. Bruzzo, R. Cianci (Eds.), Differential Geometric Methods in Theoretical Physics. Proceedings, 1990. XIX, 401 pages. 1991.

Vol. 376: D. Berényi, G. Hock (Eds.), High-Energy Ion-Atom Collisions. Proceedings, 1990. IX, 364 pages. 1991.

Vol. 377: W.J. Duschl, S.J. Wagner, M. Camenzind (Eds.), Variability of Active Galaxies. Proceedings, 1990. XII, 312 pages. 1991.

Vol. 378: C. Bendjaballah, O. Hirota, S. Reynaud (Eds.), Quantum Aspects of Optical Communications. Proceedings 1990. VII, 389 pages. 1991.

Vol. 379: J.D. Hennig, W. Lücke, J. Tolar (Eds.), Differential Geometry, Group Representations, and Quantization. XI, 280 pages. 1991.

Vol. 380: I. Tuominen, D. Moss, G. Rüdiger (Eds.), The Sune and Cool Stars: activity, magnetism, dynamos. Proceedings, 1990. X, 530 pages. 1991.

GPSR Compliance

The European Union's (EU) General Product Safety Regulation (GPSR) is a set of rules that requires consumer products to be safe and our obligations to ensure this.

If you have any concerns about our products, you can contact us on ProductSafety@springernature.com

In case Publisher is established outside the EU, the EU authorized representative is:

Springer Nature Customer Service Center GmbH
Europaplatz 3
69115 Heidelberg, Germany

Batch number: 10838062

Printed by Printforce, the Netherlands